GEOGRAPHY AND TECHNOLOGY

Geography and Technology

Edited by

Stanley D. Brunn
Department of Geography,
University of Kentucky, Lexington, U.S.A.

Susan L. Cutter
Department of Geography,
University of South Carolina, Columbia, U.S.A.

and

J.W. Harrington, Jr.
Department of Geography,
University of Washington, Seattle, U.S.A.

KLUWER ACADEMIC PUBLISHERS
DORDRECHT / BOSTON / LONDON

A C.I.P. Catalogue record for this book is available from the Library of Congress

ISBN 1-4020-1857-6 (HB)
ISBN 1-4020-1871-1 (PB)

Published by Kluwer Academic Publishers,
P.O. Box 17, 3300 AA Dordrecht, The Netherlands.

Sold and distributed in North, Central and South America
by Kluwer Academic Publishers,
101 Philip Drive, Norwell, MA 02061, U.S.A.

In all other countries, sold and distributed
by Kluwer Academic Publishers,
P.O. Box 322, 3300 AH Dordrecht, The Netherlands.

The cover image, "A Digital Sunset over Europe and Asia," is a digital composite from daytime land images taken by MODIS instruments on NASA's Terra satellite and nighttime images taken by DMSP satellites. Copyright: The Living Earth, Inc. Used with Permission.

Printed on acid-free paper

TABLE OF CONTENTS

Part III - New Geographies with New Technologies

Part IV - The Environment and Technology

Part V - The Worlds Before Us

FIGURES

TABLES

FOREWORD

It is particularly appropriate that the AAG's Centennial Celebration should prompt the publication of a volume devoted to Geography and Technology. New technologies have always been important in advancing geographic understanding, but never have they been so thoroughly and rapidly transformative of the discipline as at this stage in geography's evolution.

Just as new technologies have profoundly expanded both research possibilities and the knowledge base of other disciplines, such as biology, physics or medicine, so too are the revolutionary new geographic technologies developed during the past few decades extending frontiers in geographic research, education and applications. They are also creating new and resurgent roles for geography in both society and in the university. This trend is still accelerating, as the integration of geographic technologies, such as the global positioning system and geographic information systems (GPS/GIS), is creating an explosion of new "real-time, real-world" applications and research capabilities. The resultant dynamic space/time interactive research and management environments created by interactive GPS/GIS, among other technologies, places geography squarely at the forefront of advanced multidisciplinary research and modeling programs, and has created core organization management tools (geographic management systems) which will dramatically change the way governments and businesses work in the decades ahead.

While these and other important geographic technologies, including remote sensing, location-based services, and many others addressed in this book, are forging new opportunities for geography and geographers, they also pose challenges. Inherent within all advanced technology is the potential for its abuse, as well as for creative and beneficial uses within science and society. As geographers and as developers of new geographic technologies, we have an obligation to employ our expertise to help ensure that appropriate regulatory and legal frameworks are implemented to safeguard civil liberties and locational privacy as these new technologies become ever more widespread in research and applications. We must also work to ensure that these technologies are accessible to community-based groups, and that their benefits accrue to those historically dispossessed around the world.

As this book illustrates, our new geographic technologies are also embedded in and magnified in their impacts by parallel development in technology generally, including the broad advances in computers, the Internet, wireless communications, and many other areas. It is also the case that a great deal of cutting-edge research and innovation related to the new geographic technologies has originated in geography's burgeoning private sector, highlighting the need to foster stronger linkages and coordination among private, public, and university geographic researchers and research agendas, as is common in other disciplines blessed with strong private or public sector research components.

Perhaps most importantly, there remains a need to better integrate geography's transformational new technologies with geography's traditional strengths and its characteristic diversity. Technology in geography does not pose a threat to our traditions; it offers a way to extend and revitalize these traditions. Just as the microscope and DNA sequencing have revolutionized research, education, and applications in biology, and in so doing made the work of Linnaeus and Darwin ever more important to modern science and to modern medical applications, so too will new geographic technologies such as interactive GPS/GIS extend research horizons in traditional areas across the full breadth of geography, and make our applications more central to the needs of our society and our rapidly changing world.

This volume also makes clear that geographic technologies are integral to the intellectual core of our discipline, and that an understanding of their evolution and impact is essential to understanding the history and philosophy of geography as a discipline. Our ways of thinking and doing as geographers always have been and will continue to be intertwined with advances in technologies which, while neither intrinsically good nor bad, in the best of hands help us to see beyond, to integrate the disparate, to visualize complexity, to communicate the remarkable commonplace as well as the merely extraordinary, to bridge continents and disciplines, and to create geographic understanding.

I commend the editors and the authors of *Geography and Technology,* for their foresight and their insight at this Centennial moment in the AAG's history, and for this important publication. The topics and issues addressed in this Centennial publication will be critical to geography's future and to that of our world during the AAG's second century.

—Douglas Richardson

PREFACE

Anticipating the excitement of a centennial celebration, the Council of the Association of American Geographers formed a Centennial Coordinating Committee in 1997, seven years before the event. Among the Committee's charges was to devise and produce appropriate book publications for the 2004 centennial year. One suggestion that generated much interest would address geography, technology, and society issues, the subject matter of this book. Inasmuch as technologies have always been integral in the production of geographical knowledge, it was deemed only fitting to examine those technologies that have affected geography as a discipline, and how various fields and subfields of inquiry have been impacted by these several technologies. Among the technologies the committee considered worthy of examining were cartography, the camera, aerial photography, computers, and other computer-related technologies that have been important in the production, presentation, and dissemination of geographical information.

As this book project on geography and technology interfaces evolved in committee discussions, we were interested in potential contributions coming from colleagues with interests in technological innovations, the impacts of technologies on geography and society, disciplinary inquiries into the social/ technology interfaces, high-tech as well as low-tech societies, and applications of technologies to the public and private sectors.

A description of this project appeared in several issues of the *AAG Newsletter* in April 2002 and was supplemented by many formal and informal calls for contributions. As editors, we also identified topics that we thought merited inclusion in any final product and solicited contributions to fill these gaps. This volume represents the contributions of individuals who submitted working titles and abstracts as well as those who responded to our suggestions for subjects we considered important in any tome on the geography/technology interfaces.

The twenty-five chapters are divided into five major sections. Part I, entitled "Geography and Technology Interfaces," includes chapters that discuss an overview of geography and technology by Wilbanks, how various communications technologies have affected the production of geographical knowledge by Johnston, and a history of federal funding for geographic research and geographical technologies during the past century by Shelley, Biglar, and Aspinall. The last chapter also describes the federally funded research that has appeared in major U.S. geographical journals.

Part II, "Technologies That Changed Geography," includes four chapters about specific technologies. Harvey and Chrisman examine the early disciplinary and academic roots of automated geography and geographical information systems. Their chapter is followed by Sui and Morrill's wide-ranging discussion on the impact of computers on the discipline of geography and specific fields during the past three decades. Remote sensing, and especially its use in studying physical and urban phenomena, are the focus of the Jensen and Hodgson chapter. The final chapter in this section by Zook, Dodge, Aoyama, and Townsend looks at the new digital technologies and how these are affecting communication, communities, and society.

Part III, "New Geographies with New Technologies," includes eleven chapters that look at how specific technologies have changed the content of what we study in geography. Individual chapters consider tools and technologies used in the classroom (Downs), conducting fieldwork in non-Western contexts (Porter and Grossman), the camera (Jakle), film and cinema (Dixon and Zonn), the automobile (Rubenstein), and aircraft (Leinbach and Bowen). Additional chapters examine the 24-hour news phenomenon (Rain and Brooker-Gross), democratization (Regulska), public health (Greenberg), gender (Moss and Kwan), and the U.S. military and war (Corson and Palka).

Part IV, "The Environment and Technology," includes five chapters on technology/environment themes. Technologies used to measure and study earth surface processes are discussed by Sherman and Bass, weather and climate by Winkler, and the land cover/land use dynamics by Walsh, Turner, and Evans. The final two chapters look at many of these same human/environmental/technology themes and current geographic technologies within an African context (Taylor) and their application in the study of natural hazard preparedness, risk, and human response (Tobin and Montz).

The volume concludes with Part V, "The Worlds Before Us," and two chapters on major technology/society themes. The first, by Dobson, ponders what might be some future uses and applications of GIS, not only in geography, but in related disciplines studying the earth and biosphere. The final chapter, by Curry, asks readers to reflect on what kinds of geographical technologies were important to early human societies and those that are affecting current generations. Many of the questions about observing, narrating, representing, describing, analyzing, and predicting the geographical have long intrigued the scientist, artist, and practitioner. While geographers have made many advances in the past millennia, as we proceed into the 21st century and beyond, we will continue to ask some of the same questions, but address them with some new technologies. In short, information and communications technologies (ICTs) and the processing and representation of georeferenced data of all kinds, including remote sensing and satellite imagery, utilizing GIS (geographical information systems) and GISc (geographical information science), will continue to affect what and how geographers understand and represent the world. The traditional and emerging subjects of our inquiry will continue to advance by incorporating new and unknown technologies into their presentation, analysis, description, and forecasting of earth features and data.

Acknowledgements

This centennial publication could not have seen the light of day without the support of many individuals. We first acknowledge the support of others on the Centennial Coordinating Committee (Don Janelle, Kathy Hansen, Pat Gober, Alice Rechlin-Perkins, Marilyn Silberfein, and Barney Warf). Also we acknowledge the strong encouragement provided by the present AAG Executive Director (Doug Richardson) and his predecessor (Ron Abler) during various stages of discussion and production, and AAG presidents (Pat Gober, Will Graf, Susan Cutter, Reg Golledge, Jan Monk, Duane Nellis, and Alec Murphy). AAG Council's continued support for this centennial activity is also recognized and appreciated. The Kluwer staff included Myriam Poort, Susan Jones and André Tournois.

Various colleagues and friends provided suggestions for potential contributions and also reviewed manuscripts. These include Karl Raitz, Tony De Souza, Jonathan Phillips, John Pickles, Harm de Blij, and Joni Seager. Four others also merit mention. First is Martin Kenzer who, in an e-mail in early 2003, included an attachment of a satellite image which is now the cover. Second is Debra Lackas of The Living Earth, Inc., who granted us permission to use this image, which is a digital composite of archived images taken by several ocean-faring ships and earth-orbiting satellites. Third is Rose Canon who, as in her nearly sixteen years as editorial assistant for *The Professional Geographer* and *Annals of the Association of American Geographers*, again performed admirably in copy-editing these chapters. Fourth is Donna Gilbreath who efficiently and with her professional graphics and cartographic skills prepared the camera-ready copy of the entire manuscript for the publisher and designed the cover.

We are very pleased with the original contributions of these senior and junior scholars. All met our deadlines and responded to suggestions provided by us or others to strengthen the chapter with added conceptual, theoretical, or bibliographic materials. As we reflect on assembling these chapters and what they might contribute to our future, we see two fruitful directions. The first is that we believe firmly, as demonstrated in the book, that geographers have much to contribute to questions about the social impacts of technology, whether as a disciplinary perspective or from a broader base, focused on improvements in the human condition. The second is that the geography/technology interfaces have only begun to be investigated in these twenty-five chapters. Each contribution provides a valuable assessment of the state-of-the-art of the his/her/their inquiry. There are clearly many more substantive ideas that merit study, whether related to specific technologies, methodologies and paradigms, social impacts, or their role in our disciplinary histories. Perhaps in another decade or two, the technology/geography interfaces will be as important in disciplinary and transdisciplinary training, instruction, and research as are the current environmental/geography fields and subfields. If this volume contributes to this "turn," we believe the contributors will have served a major and useful purpose. Geographers clearly are not alone in seeking to explore the technology/society linkages and networks. We know we have much to learn from others and for others to learn from us.

—Stanley D. Brunn
—Susan L. Cutter
—J.W. Harrington, jr.

September 2003

CONTRIBUTING AUTHORS

Editors, Affiliations, and Email Addresses

Stanley D. Brunn, Department of Geography, University of Kentucky, Lexington, KY 40506-0027 brunn@uky.edu

Susan L. Cutter, Department of Geography, University of South Carolina, Columbia, SC 29208 scutter@gwm.sc.edu

J. W. Harrington, Department of Geography, University of Washington, Seattle, WA 98195 jwh@u.washington.edu

Authors, Affiliations, and Email Addresses

Yuko Aoyama, Graduate School of Geography, Clark University, Worcester, MA 01610 yaoyama@clarku.edu

Richard Aspinall, Department of Geography, Montana State University, Bozeman, MT 59717 aspinall@montana.edu

Andreas C.W. Baas, Department of Geography, King's College London, Strand, London WC2R 2 LS, UK andreas.baas@kcl.ac.uk

Wendy Bigler, Department of Geography, Arizona State University, Tempe, AZ 85287-0104 Wendy.Bigler@asu.edu

John J. Bowen, Jr. Department of Geography & Urban Planning, University of Wisconsin-Oshkosh, WI 54901 bowenj@uwosh.edu

Susan R. Brooker-Gross, Department of Geography, Virginia Tech, Blacksburg, VA 24061 srb144@vt.edu

Mark W. Corson, Department of Geography, Northwest Missouri State University, Maryville, MO 64468 mcorson@mail.nwmissouri.edu

Nicholas R. Chrisman, Department of Geography, University of Washington, Seattle, WA 98195 chrisman@u.washington.edu

Michael R. Curry, Department of Geography, University of California, Los Angeles, CA 90095 curry@geog.ucla.edu

Deborah P. Dixon, Department of Geography, University of Aberystwyth, Aberystwyth, Wales, UK dxd@aber.ac.uk

Jerome E. Dobson, Department of Geography, University of Kansas, Lawrence, KS 66045 Dobson@ku.edu

Martin Dodge, Centre for Advanced Spatial Analysis, University College London Gower Street, London WC1E 6BT UK m.dodge@ucl.ac.uk

Roger M. Downs, Department of Geography, Pennsylvania State University, University Park, PA 16802 rd7@psu.edu

Tom P. Evans, Department of Geography, Indiana University, Bloomington, IN 46405 evans@indiana.edu

Michael Greenberg, Edward J. Bloustein School of Planning and Public Policy Rutgers University, New Brunswick, NJ 08903 mrg@rci.rutgers.edu

Lawrence S. Grossman, Department of Geography, Virginia Tech, Blacksburg, VA 24061 lgrossmn@vt.edu

John A. Jakle, Department of Geography, University of Illinois, Urbana, IL 61801 j-jakle@uiuc.edu

John R. Jensen, Department of Geography, University of South Carolina, Columbia, SC 29208 jrjensen@sc.edu

Francis J. Harvey, Department of Geography, University of Minnesota, Minneapolis, MN 55455 fharvey@fharvey.email.umn.edu

Michael E. Hodgson, Department of Geography, University of South Carolina, Columbia, SC 29208 hodgsonm@gwm.sc.edu

Ron Johnston, School of Geographical Sciences, University of Bristol, Bristol, UK BS8 1SS r.johnston@bristol.ac.uk

Mei-Po Kwan, Department of Geography, Ohio State University, Columbus, OH 43210 kwan.8@osu.edu

Thomas R. Leinbach, Department of Geography, University of Kentucky, Lexington, KY 40506-0027 leinbach@uky.edu

Burrell Montz, Department of Geography, SUNY Binghamton, Binghamton, NY bmontz@binghamton.edu

Richard Morrill, Department of Geography, University of Washington, Seattle, WA 98195 morrill@u.washington.edu

Pamela Moss, Faculty of Human and Social Development, University of Victoria, Victoria, B.C., Canada V8Y 2Y2 pamelam@uvic.ca

Eugene J. Palka, Department of Geography and Environmental Engineering, U.S. Military Academy, West Point, NY 10996-1695 gene.palka@usma.edu

Philip W. Porter, Department of Geography, University of Minnesota, Minneapolis, MN 55455 pwporter@tds.net

David R. Rain, George Washington University, 1957 E. St. Suite 512 NW, Washington, DC 20052 david.r.rain@census.gov

Joanna Regulska, Department of Women's and Gender Studies and Geography Rutgers University, New Brunswick, NJ 08901 regulska@rci.rutgers.edu

James M. Rubenstein, Department of Geography, Miami University, Oxford OH 45056 rubensjm@muohio.edu

Fred M. Shelley, Department of Geography, Texas State University –San Marcos, TX 77843 Fs03@swt.edu

Douglas J. Sherman, Department of Geography, Texas A & M University, College Station, TX 77843 Sherman@geog.tamu.edu

Daniel Sui, Department of Geography, Texas A and M University, College Station, TX 77843 dsui@geog.tamu.edu

D.R.F. Taylor, Department of Geography and Environmental Studies, Carleton University, Ottawa, Canada K1S 5B6 fraser_taylor@carleton.ca

Graham Tobin, Department of Geography, University of South Florida, Tampa, FL gtobin@luna.cas.usf.edu

Anthony M. Townsend, School of Urban Planning and Communications, New York University, NY 10011 anthony.townsend@nyu.edu

B. L. Turner, II, Graduate School of Geography and Marsh Institute, Clark University, Worcester, MA 01610 bturner@clarku.edu

Stephen J. Walsh, Department of Geography, University of North Carolina, Chapel Hill, NC 27599-3220 swalsh@email.unc.edu

Thomas J. Wilbanks, Oak Ridge National Laboratory, Building 4500N/Box 2008 Oak Ridge, TN 37831-6184 wilbankstj@ornl.gov

Julie A. Winkler, Department of Geography, Michigan State University, East Lansing, MI 48823 winkler@pilot.msu.edu

Leo E. Zonn, Department of Geography, University of North Carolina, Chapel Hill, NC 27599-3220 zonn@email.unc.edu

Matthew Zook, Department of Geography, University of Kentucky, Lexington, KY 40506-0027 zook@uky.edu

COLOR PLATES

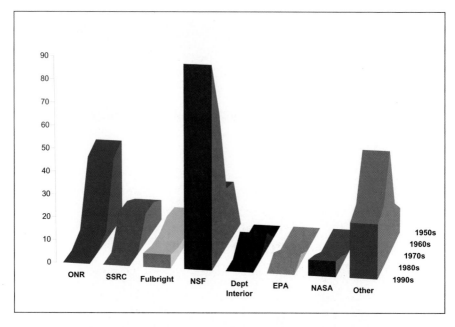

Figure 3-1. The change in federal funding for geographic research: 1950-99 by major sources and publications in Annals of the Association of American Geographers, The Professional Geographer, *and* Geographical Review.

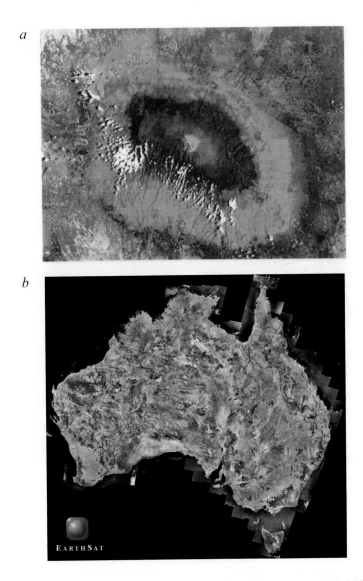

Figure 6-7. GeoCover-Ortho images: a–Mount Kilimanjaro at 28.5 Ч 28.5 m spatial resolution [RGB = Landsat TM bands 7,4,2 (mid-infrared, near-infrared, and green)]; b–Mosaic of hundreds of GeoCover-Ortho Landsat Thematic Mapper images of Australia obtained between 1987 and 1993 (courtesy of Earth Satellite Corporation and NASA).

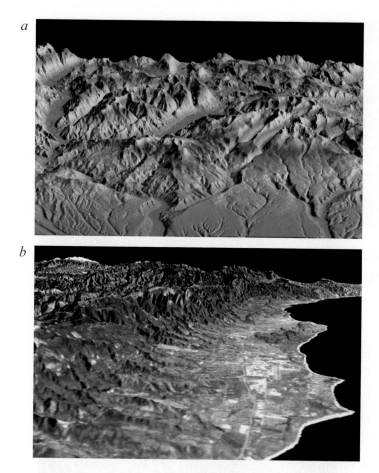

Figure 6-10. Shuttle Radar Topography Mission (SRTM)-derived digital elevation models: a–portion of San Martin de Los Andes, Argentina. Elevations range from 700 to 2,400 m (2,300 to 8,000 ft.); b–Santa Barbara, California. Mount Abel (2,526 m; 8,286 ft.) is seen in the distance. The DEM is draped with a Landsat Thematic Mapper image (RGB = bands 7,4,2) (courtesy NASA Jet Propulsion Lab and NIMA).

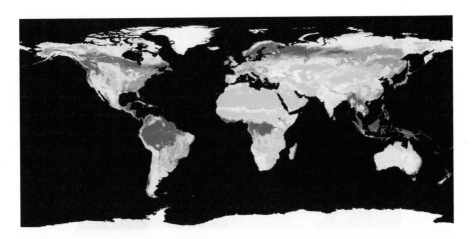

Figure 6-13. Global land cover 1 x 1 km dataset derived from MODIS data (courtesy NASA Goddard Space Flight Center and Boston University).

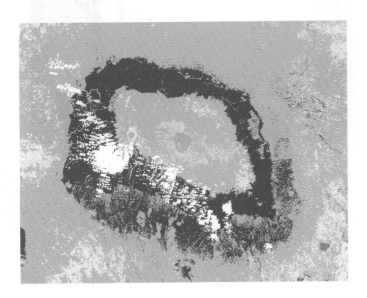

Figure 6-14. Thirteen class land cover map of Mount Kilimanjaro derived from Landsat Thematic Mapper imagery. This is part of the EarthSat GeoCover land-cover mapping project. Compare this thematic map with the original imagery found in Figure 6-7a (courtesy Earth Satellite Corporation).

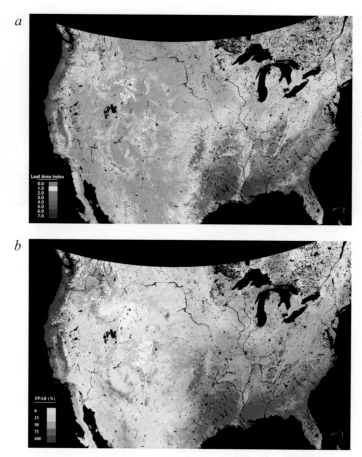

Figure 6-16. Leaf Area Index (a), and Fraction of Photosynthetically Active Radiation (b) maps of the U.S. derived from MODIS hyperspectral imagery (courtesy of NASA Earth Observatory).

Figure 6-15. NASA Enhanced Vegetation Index (EVI) composite obtained June 1-16, 2003. EVI is a surrogate for the amount of biomass present (courtesy of NASA Earth Observatory).

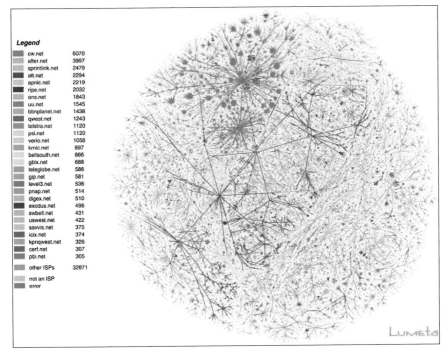

Figure 7-2. A graph visualization of the topology of network connections of the core of the Internet, December 11, 2000. (Source: Bill Cheswick, http://www.lumeta.com)

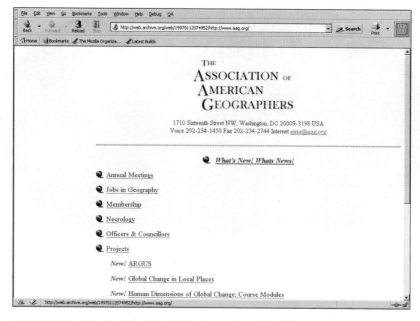

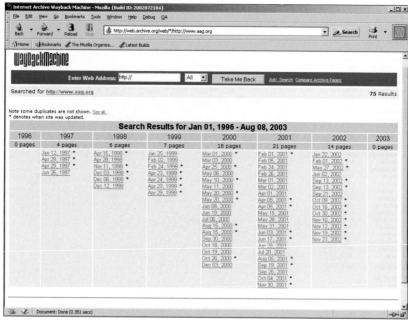

Figure 7-3. Sample of Wayback Machine webpages: top–a screenshot of the AAG homepage as it was in January 1997; bottom–a listing of all available views of the AAG web pages from different dates stored in the Wayback Machine. (Source: http://www.archive.org/)

Graphics in the *Annals of the Association of American Geographers*

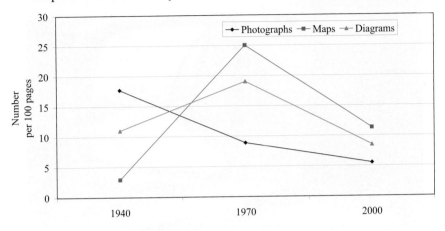

Graphics in *Economic Geography*

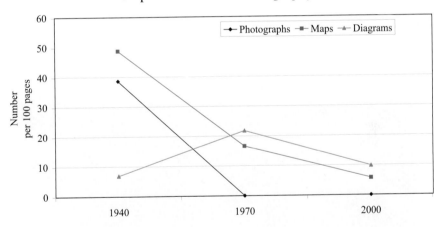

Figure 10-1. Comparison of the number of graphics included in the Annals of the Association of American Geographers and in Economic Geography, 1940-2000.

Figure 10-2. View through a rearview mirror on U.S. 30 west of Toledo, Iowa, 2002.

Figure 10-3. Windmill farm northwest of Palm Springs, California, 2002.

Figure 10-4. Log house north of Andover, Pennsylvania, 1991.

Figure 10-5. The image of former Mayor Frank Rizzo on South Ninth Street in largely Italian-American South Philadelphia, 1995.

Figure 10-6. Tour bus on "The Strip" in Las Vegas, 2002.

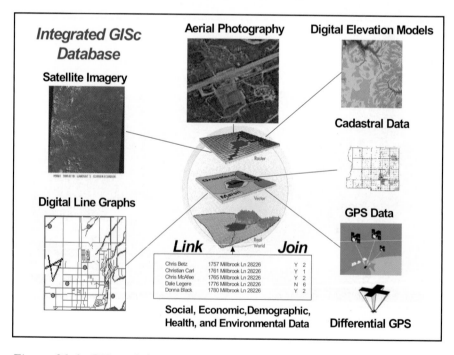

Figure 21-1. GIS and the integration of data and technologies for linking people, place, and environment.

PART I

GEOGRAPHY AND TECHNOLOGY INTERFACES

CHAPTER 1

THOMAS J. WILBANKS

GEOGRAPHY AND TECHNOLOGY

It can be argued that geography's origins as a category of knowledge had technology as a cornerstone. According to James (1972), geography arose from two fundamental desires in human nature: (1) to find out what lies over the next hill, i.e., to explore unknown places and report what we find; and (2) to know where we are in order both to get home and to return to that spot if we wish, i.e., to navigate on the face of the earth. In the first case, moving from one place to another always depended on meeting needs for water, shelter, fire, weapons for hunting and self-defense, and the transport of supplies, all of which are related to technology, however primitive. In the second case, successful navigation depended on observing the stars and recording landmarks, also related to technology, and in many cases stimulating attention to technology needs. In both cases leaps of technology have time and again changed how geography is done and what it means for society, and in both cases, maps have been a tool that is a foundation of our discipline and its concepts (Curry, Chapter 25).

In this introductory essay, by "technology" is meant the devices and techniques employed by society to sustain its existence and improve its quality of life. Technology refers most often to tangible, practical objects—hardware and software—that satisfy social definitions of usefulness.

Abundant historical evidence has shown that technology and society evolve together. In his classic study, for instance, Mumford (1963) showed how uses of the technologies responsible for the industrial revolution were enabled by centuries of prior social change, from the evolution of monastic practice to the organization of massed armies. If so, a salient question for today's world is whether ever more rapid technological change may be stretching boundaries of social acceptance (from bioengineering to the invasion of privacy) or of social existence (such as weapons of mass destruction).

Stanley D. Brunn, Susan L. Cutter, and J.W. Harrington, Jr. (Eds.), Geography and Technology,
3- 16. © 2004 Kluwer Academic Publishers. Printed in the Netherlands.

The fact is that, although professional colleagues may argue whether *science* can be socially neutral, no one responsibly argues that *technology* can be socially neutral. It is by definition the result of social choices. It is by definition useful, for good or ill. And its use is responsible for ripples of social change, large and small, including unintended consequences along with expected ones (e.g., Tenner 1997).

It is not unreasonable to ask why—if the past century has seen an unprecedented and apparently accelerating rate of technological change—so many social indicators show little improvement, even in a relatively affluent country such as the U.S. If technology is so great, why then are the lives of so many of our fellow citizens, especially in the global community, not up to a level to which any human being would aspire? For most of the readers of this book, our levels of comfort, convenience, mobility, labor productivity, and access to diversions is far higher than any of our ancestors. If our great-grandparents were to be dropped into our world, they would be dazzled. But such indicators of happiness as mental health and the stability of family relationships, along with many indicators of the incidence of crime and morbidities, tell us that in important respects, our lives are not necessarily better.

Technology is an enabler, not a solution, whether our frame of reference is society in general or the discipline of geography in particular. It can be produced by research and development systems wrapped in their own supply-push imperatives, delivered for our consumption whether we asked for it or not. It can come with a kind of overwhelming momentum that sweeps us along like flotsam in a flood, while we compete with each other to show how skillful we are in using its new tools.

This does not mean, of course, that because, in many ways new technology is irresistible, it is unfortunate, even malign. Technology is what we make of it, and for geography it offers us pathways for intellectual discovery and social relevance that are as unprecedented as the technology itself. If, two generations ago, geographers could have imagined remotely sensed and in situ electronically instrumented data sources, geographic information science and its powerful visualizations for learning and communication, GPS and its capacity to pin down locations of field observations, the Internet for information access and professional networking, and computer-based spatial and statistical analysis—not to mention a public with a growing taste for information delivered in graphic forms—our predecessors would have wished they were in our shoes. These are days of both historic opportunities and daunting risks, not only that we will miss opportunities but that we will misuse them, as we contemplate what technology means in the first part of the 21st century for both the society around us and for our discipline.

It is in this powerful context that this book considers two dimensions of relationships between geography and technology: (1) geography as it helps in understanding roles and implications of technology in society, and (2) technology as it is used by geography in relating itself to society.

1. GEOGRAPHICAL DIMENSIONS OF TECHNOLOGY IN SOCIETY

The more important of the two issues is with regard to understanding society, as contrasted with understanding geography (which, after all, is a means to social ends, not an end in itself). In this regard, geography as a particular kind of human inquiry is fundamentally concerned with at least three ramifications of technology in society:

1. Technology affects the meaning of proximity, and thus the significance of location and operational definitions of efficient spatial organization, by helping to determine the effort required to overcome distance.
2. Technology affects the character of places by shaping what happens there, what it is like to live there, and how residents think about their sense of place.
3. Technology shapes nature-society relationships by changing social demands for nature's services and by changing tools for environmental management.

These dimensions interact with a wide range of overarching social issues: poverty, democratization, globalization, technological change, environmental sustainability, and others, not to mention such unexpectedly emerging issues, too recent to be addressed in this book, as "homeland security" (see Cutter et al. 2003).

1.1. Technology and the Meaning of Location

To geographers, location is intrinsically relative: location relative to what? how far from what? affecting interactions with other locations in what ways? related to structures for spatial organization how? Any changes in technologies that reshape the meaning of proximity reshape the way we view our world, and in the hundred years since 1904, we have seen changes in technologies that dwarf any other century in the earth's history. From the perspective of a person who studies energy issues, the changes are dominated by effects of the emergence of electricity and petroleum fuels (Wilbanks 1988) because of their effects on systems for movement and communication, although another person might point to equally profound changes in such fields as medicine, materials, and aerospace sciences.

Technology affects the meaning of location most directly by reshaping the meaning of location because it changes what proximity means. Geographic research has shown that technological change can change the effect of distance on human interaction, whether direct human interaction or larger spatial structures such as central place systems. Technology also affects locational choice: changes in the inputs that are important in determining what location is best, such as materials substitution as an issue for resource-intensive industrial production. Information technology affects the diffusion of innovations, from the not-quite-random pattern of invention to the patterns of interaction that cause inventions to be transformed into innovations. And hazards associated with technological innovation also have spatial patterns, whether related to vulnerabilities or to responses in the event of emergencies.

In recent decades, technology has reshaped geographical relationships in two fundamental ways: by shrinking our globe, so that people who used to be distant from each other are now neighbors (Leinbach and Bowen, Chapter 13), and by introducing alternatives for interaction that are (or appear to be) released from constraints of distance and conventional spatial pattern, e.g., cyberspace (Sui and Morrill, Chapter 5; Zook et al., Chapter 7).

Keys to these changes have been what are, by standards of historical experience, astonishingly rapid technological developments in movement systems—transportation and communication—in turn enabled by rapid technological developments in energy systems (e.g., electricity), materials, and management/organizational control systems. Geographers have contributed to understanding the impacts of such transformations by focusing attention on changes in the optimal spacing of functions (Rubenstein, Chapter 12), innovations in the spatial structuring of functions (e.g., the Federal Express approach to package transfers: Leinbach and Bowen, Chapter 13), and new patterns of diffusion of ideas, people, forms of control, economic relationships, and waste, pollution, and disease (e.g., Greenberg, Chapter 16). At the same time, geographers have advocated awareness to social and environmental consequences of this reshaping of our world (see below), including issues related to unequal access to opportunities as they emerge (Zook et al., Chapter 7).

Looking to the future, perhaps the biggest question is how the growing use of largely aspatial information and communication technologies will change the meaning of location and how we interact and arrange our activities in space. What will be the continuing value of physical proximity, of personal contact and communication, of social interactions and groupings? How will these aspects of life be modified and how will cultures be changed by new technologies and how we use them?

1.2. Technology and the Character of Places and the People Who Live There

Technology affects the character of regions and the people who live there, because technology shapes not only regional comparative advantage but also life in any place. One particularly vivid example is how new information technology has made it possible for people with striking but mobile skills to make a living in rather remote locations (Robb and Riebsame 1997). Another example is the growing attraction of electronic recreational alternatives within the household as contrasted with more conventional recreational activities outside the household, especially for many young people.

Technology, in fact, is usually a key ingredient and sometimes the defining quality of the character of regions (e.g., Silicon Valley, the Research Triangle, the "rust belt"). Technology is a foundation of economic growth by offering opportunity or threatening obsolescence. By compressing time and space, it can transform the human life experience (D. Harvey 1989). It influences the work experience by increasing labor productivity and changing skill requirements; and the geographic pattern of access to work experiences stimulates migration, including "brain drains" from developing countries (Taylor, Chapter 22). It shapes such attributes of life as comfort, convenience, mobility, and access to many leisure time activities. It underlies many structures for exercising power and control, from the expanding reach of globalization to new opportunities for local empowerment. It contributes to cultural convergence or divergence and therefore to a sense of place and community. It can represent threats through weapons of mass destruction, hazards from materials use and waste disposal, or supplies of designer drugs (Greenberg, Chapter 16). It can even affect how we view ourselves (Moss and Kwan, Chapter 17).

One familiar example of impacts of technology and places is the transformation of land uses by the automobile, and an example of impacts on the lives of people who live in places is the role of automobiles as personal spaces. Moreover, the automobile age has pioneered mass production and vertical integration as industrial approaches: thus "Fordism" (Rubenstein, Chapter 12). Another familiar example is the role of technology in catalyzing and enabling global economic integration (Leinbach and Bowen, Chapter 13).

Less familiar but profoundly important are the potential impacts of modern information and communication technologies on democratic institutions, from new opportunities for direct engagement in political activities to new ways to access (and monitor) government services. Issues that deserve more research attention include equity in access to these opportunities, whether they are likely to make democratic decisionmaking more or less consensual and integrated, and the degree to which the impacts are likely to differ according to political and cultural context (Regulska, Chapter 15).

1.3. Technology and Nature-Society Relationships

Technology transforms nature-society relationships. Consider, for example, impacts in the past century of new transportation technologies, new information technologies, new medical technologies, and new materials such as plastics. Can we imagine looking at a city from any angle in an evening and not seeing electric lights? Can we imagine not having personal automobiles for our use? Can we imagine not having plastics for packaging? Can we imagine having our favorite fresh vegetables available only a few weeks of the year? Can we imagine not having weather information available on television or the Internet? Yet each of these developments in consumer preferences has implications for the global environment.

Decades ago, Kenneth Boulding suggested that in nature-society linkages, technology is both culprit and savior (Darling and Milton 1966:717-18). On the one hand, it stimulates demands for resources, accelerating the consumption of the earth's building blocks; on the other, it can develop substitutes and enable greater resource efficiency. On the one hand, technology can contribute to environmental degradation (e.g., mechanized strip mining); on the other, it can contribute to environmental protection (e.g., monitoring technologies). On the one hand, it can produce hazardous wastes; on the other, it can improve capacities to handle wastes safely. Greenberg (Chapter 16), for instance, points out how technology has accelerated the detection and treatment of diseases but at the same time has created new health risks and increased the potency of some health hazards. He also notes that the balance between positive and negative implications of technology use depends on the political and social systems that make choices and how they view risks.

Geography has played key roles in the emergence of sustainability science as a new cross-cutting field of integrative science (Kates et al. 2001), with contributions to the global research effort focused especially on land use and land cover change (e.g., tropical deforestation). Among the perspectives that are important have been discoveries that nature-society integration is more likely to be feasible in a place-based context in a relatively small region, the associated importance of understanding dynamics at a relatively fine-grained spatial scale, and the importance of considering relationships between processes that operate at different geographic scales (Walsh et al., Chapter 21).

2. TECHNOLOGICAL DIMENSIONS OF GEOGRAPHY IN SOCIETY

Trapped in our historical syndrome of tending to look internally rather than externally, geographers have often paid more attention to what technology means to us than what it means to society, and we do indeed have a lot to consider. Although the implications of this transformation are greater for society than for geography itself, and their implications for geography as a *subject* are greater than for geography as a *discipline* (National Academy of Sciences/National Research Council 1997), the fact is that our contributions to knowledge, learning, and society at large are being reshaped by the information and computing technology revolution in ways that this book illustrates abundantly.

Clearly, technology is revolutionizing how geographic information is collected, analyzed, and communicated to audiences from classrooms to the general public, supported by federal government funding not only for geographic research but for the origins of such powerful tools as the Internet and e-mail (Shelley et al. Chapter 3; Corson and Palka, Chapter 18). Forty years ago, geographers were being trained to make maps by hand, to use printed maps as research sources, and to analyze data using mechanical calculators and punch-card inputs to very slow computers. Our students in the first decade of the 21st century probably cannot imagine such a primitive context for doing geographic research—only a generation and a half in the past.

The challenge is to look beyond the miracles of the present, which we have incompletely assimilated, down the road twenty years or more to anticipate both opportunities and consequences and to identify priorities for the discipline of geography in assuring that we will be positioned to use the new capabilities appropriately and well.

2.1. Technology and Recording and Communicating Geographic Information

Because of the centrality of maps for geographers, the first technology-impact issue is not data collection—even though it comes first in empirical research—but impacts of technology on how we record and communicate geographic information, not only to general audiences but also among ourselves. Geographic Information Science (GISc), the next step beyond Geographic Information Systems (GIS), has rapidly become geography's most salient practical contribution to society and the discipline's most active area for student job placement. Coupled with Global Positioning Systems (GPS), computers for integrating and analyzing geographic data, and electronic communication media for moving information to others, this new technology has made geography newly technology-oriented and, in the minds of many, newly relevant in applications of geographic perspectives from international business to local planning.

The importance of technology in enabling GIS and GPS, and in turn, their impacts on geography, need not be recounted here; for a powerful illustration, see Walsh et al., Chapter 21. Waldo Tobler once observed (personal communication) that the computer changed cartography from a field focused on producing general-purpose print maps to a field emphasizing the production of tools for decentralized special-purpose maps—a much more interesting professional challenge. Computer-based advances in conceptions and realizations of spatial visualization (MacEachren et al. 1992) have become especially powerful as a catalyst for inquiry and not only a magnet for interest on the part of users of geographic information but also a way to promote awareness of the role of space in phenomena and processes of interest to many fields of research (Sui and Morrill, Chapter 5).

But technology has also affected geographic communication in other ways. Television news has promoted public interest in geographic information (Rain and Gross, Chapter 14), the Internet has made maps readily available to the public for tracking weather and investigating travel opportunities, and e-mail has enabled new kinds of information transfer and collaboration.

Within the discipline, technological change has changed how we relate to each other—how we exchange messages and materials, encouraging and accelerating many kinds of interaction—although personal contacts remain essential (Johnston, Chapter 2). As in the case of society, a question yet to be answered is about the roles of face-to-face interaction in a technological era, where literatures on time geography may be relevant as a source of hypotheses (e.g., Parkes and Thrift 1980).

Technology is also changing geography in the classroom, especially as college classrooms become equipped with desktop computers for student use. Already, PowerPoint presentations of classroom material are being made available on Internet sites for student access, changing the nature of classroom note-taking. Laboratory exercises are in many cases carried out on computers. GIS-based problems are bringing vitality into interactions between instructors and students. Downs (Chapter 8) notes that, historically, the introduction of new tools into geography classrooms (such as globes and wall maps) has often emphasized factual knowledge rather than intellectual challenges, and geography as a significant subject for learning has suffered. But the far more interactive potentials of our new technologies for teaching and learning are exciting, if we can meet the challenges of teacher training and equipment purchase and maintenance.

Underlying these developments are a number of reasons for concern as well as excitement. Access to the fruits of many of our new technologies for mapping and information exchange involves access to financial resources and technical skills, and as a result parties endowed with these assets have advantages that others can find difficult to overcome (Dobson, Chapter 24; Zook et al., Chapter 7). One example is the control of much of the GIS data and tools relevant to developing countries by external parties, which calls for a new emphasis on endogenous initiatives (Taylor, Chapter 22; National Academy of Sciences/National Research Council 2002). Another reason for concern, related to origins of GIS in defense programs, is that its social construction may not be appropriate for many of society's pressing social issues (Harvey and Chrisman, Chapter 4; Harvey 2000). Yet another concern is that new information technologies may bring threats to privacy, the integrity of databases, and the protection of intellectual property.

2.2. Technology and the Gathering of Geographic Information

Besides changing what we do with geographic information once we have it, technology has had a powerful impact on how we collect the information in the first place. Such traditional approaches as fieldwork and field-mapping continue to be relevant, with GPS systems helping to document locations, and laptop computers helping in recording observations (Porter and Grossman, Chapter 9). Earlier technological innovations such as photography and digital data from aircraft (Tobin and Montz, Chapter 23; Walsh et al., Chapter 21), the camera (Jakle, Chapter 10), and film (Dixon and Zonn, Chapter 11) are also still important. One example is the power of photography in communicating images (e.g., the work of the National Geographic Society); another is the power of film in representing current events and nonfamiliar cultures, spaces, and places.

But imagery from earth satellites, together with technologies for observing and recording phenomena on or near the earth, has immersed us in a flood of data that begs conversion into useful information. Sherman and Bass (Chapter 19) describe how electronic instrumentation has increased the capacity to investigate sediment transfer in air and water, and Winkler (Chapter 20) indicates how technology has changed atmospheric observations and climate science. Walsh et al. (Chapter 21) note how technology has improved change detection in land use and land cover. In many cases, vegetation and other easily observable land uses have been the phenomena where we have learned to apply the new GIS tools (Jensen, Chapter 6), although the growing availability of data with very high spatial resolution is increasing attention to human settlement phenomena as well.

This array of potentials, however, raises fundamental issues about selective supply and use of data, protocols for condensation of data, storage of and access to data, and most importantly how we apply our theories in making sense out of the data (Hodgson and Jensen, Chapter 6). For example, we tend to have to accept decisions of others about the spatial and temporal scale of observation, as well as their decisions about what to observe; consider the capabilities of new earth observation platforms currently being designed. Many of the agendas of those who fund and support the data systems are focused on quick practical uses rather than fundamental learning, and as a result the systems and their data may not present researchers with a coherent picture of reality (Tobin and Montz, Chapter 23; Winkler, Chapter 20). And the ready availability of data about physically observable phenomena and processes makes it easy for researchers to focus on these topics, while in any subject area many of the most important relationships involve variables that are difficult to observe via an earth satellite or an electronic instrument. There is a danger that in this sense technology may tend to shape our research agendas, distorting the balance of attention to different geographic research issues because of differences in the effort required on the part of individual investigators to collect information as a basis for analysis and discourse.

2.3. Technology and the Analysis of Geographic Information

Finally, technology has "transformed geography" (Sui and Morrill, Chapter 5) by giving us radically new capacities to analyze both quantitative and qualitative information. Sui and Morrill (Chapter 5) suggest that the quantitative revolution in geography could not have occurred without the arrival of computers, which enabled analyses of large datasets and new approaches to spatial visualization and display.

Research by geographers on land use and land cover change is perhaps the best example to date of combining new capacities for information collection with new capacities for information analysis. For instance, surveys linked with GPS, which link dynamics to location, along with data from satellites and aircraft, have been fed into spatial and statistical models which begin to represent difficult and complex aspects of coupling human and environmental systems at multiple scales (Walsh et al., Chapter 21). There are many challenges in this regard: e.g., in relating human behavior and agency to complex multistress environmental processes (Walsh et al., Chapter 21) and representing uncertainty and fuzziness (Dobson, Chapter 24).

But we face a future where, for the first time in human existence, computational capacities will not be a constraint on the questions we can ask with our quantitative methods. We can begin to imagine new approaches to dealing with nonlinearities and discontinuities, new capacities to examine qualitative notions and information, new abilities to examine "optima" considering multiple criteria. If we use these new technologies thoughtfully and, even more important, engage actively in discourses that consider new technology development and information system agendas, we can look forward to opportunities that will truly stretch all of our capabilities as a discipline.

3. CONCLUDING REFLECTIONS

Over the past several generations, geography has made increasing use of technology and has paid significant (though rather uneven) attention to its implications for society while as a discipline being ambivalent about whether we are happy about these developments. The chapters of this book frequently communicate a healthy concern about technologies and their implications, even a tendency to be a bit sensitive about the fact that technological change has benefited geography as a research discipline more than many other fields— not only in expanding our tool chest but also increasing society's awareness of what we do and how useful it is. Perhaps to some degree this reflects our "muddy boots" traditions of hands-on fieldwork. In part, it reflects our sensitivity to social-theoretic concerns about agendas, values, and consequences.

Contributions to the book suggest several cautionary notes, all of them very important:

1. Technology should not become a major determinant of the questions we ask and the messages that we develop from our research (Harvey and Chrisman, Chapter 4).
2. New technologies are tending to focus attention on the observable and the computable, when these aspects of learning are not sufficient for understanding most important issues.
3. The increasing availability to decentralized users of powerful capacities for geographic visualization (e.g., moving from one historical period to another, zooming in and out from one scale to another) can cause us all to become "bewitched by an image" rather than seeking more fundamental understanding (Curry, Chapter 25).
4. Applications of new technologies are not always socially benevolent (Dobson, Chapter 24; Curry, Chapter 25)

But we will be guilty of intellectual narrowness if we fail to appreciate the positive potentials of technological change in our admirable determination to keep it in perspective. Reservations are always essential, and skepticism is often called for. At the same time, however, we need to show professional sophistication in understanding what the technologies can do, how they work, how we can apply them in addressing the questions that we believe to be interesting and important, and how we can do our part in helping to assure that the steady stream of new technology is supportive of sustainability on this earth rather than a threat to it.

Meanwhile, we will be guilty of disciplinary parochialism and social insensitivity if we fail to recognize that the central issues about technological change in the early decades of the 21st century are not about geography but about society. Onrushing changes in information and computing technology are likely to reshape our economic and political systems, our workplaces, our households, and even how we think about ourselves and others (Moss and Kwan, Chapter 17). While they empower us with awesome access to information, they will increase concerns about freedom and control, and they are likely to affect our conceptions of community. Changes in medical and pharmaceutical technology will extend lifetimes, increasing questions about the meaning of (and support for) retirement. While bioengineering offers new potentials for food supply and biologically based pollution management, it presents risks of unintended consequences. Efforts to prevent the use of weapons of mass destruction will seek to outrace the development of new dangers. The list of potential advances and impacts is even longer than our imagination. In virtually every line of inquiry in geographic research, including our examinations of the past, we need to be especially alert to technology-related issues and their interactions with human and environmental systems of importance.

Looking down the road, say, twenty years (or fifty, or one hundred), what kind of future do we see for geography as we are swept along by further technological changes? One thing is certain. Our grandchildren will be more technically proficient than are we, as comfortable with electronic information access and technologies as we are with kitchen appliances. They will be surrounded by instrumentation and electronic controls and integrated television/electronic communication/computing terminal devices, getting much of their learning from electronically based or enhanced media; and they will consider these technologies entitlements, not commodities. They will push older generations to keep up.

In this future, geography will be far more data-rich and computationally far more sophisticated, It will use ingenious new visualizations as sources of research ideas and ways to answer research questions, with decentralized users able to choose from a vast array of options. It will mainly teach through machines, mainly publish electronically, and increasingly define peer groups and disciplinary communities through cyberspace networks. Geography will be more international, as it becomes easier to interact over distances and as colleagues in developing countries get more or less equal access to information and communication technologies. Geography will, I believe, continue to be a watchful user of technology, a discipline that expresses caution as well as enthusiasm; but the interest of new generations in the new tools will be a driving force. In this future, universities will have to rethink how they organize themselves and carry out their missions. Conventional academic disciplines may be largely supplanted by more integrative approaches to the search for knowledge. Definitions of criteria for success and status may change.

I believe that, in these ways and others, technological change over the next half-century will drive social and institutional change in all aspects of our lives. A challenge to us and the geographers we train—and a profoundly important one—is to assure that social and institutional change, in turn, shapes technological change for the social good, not only in geography itself but in the world around us. If this is to be possible, a high priority for the discipline is to increase our interactions with those parties who determine technology research and development agendas and resource allocations, so that our perspectives make a difference in the evolution of the technology portfolio rather than being limited to reacting to a portfolio that could be far from what we think is wise and appropriate.

REFERENCES

Cutter, S., Richardson, D., and Wilbanks, T. (2003). The Geographical Dimensions of Terrorism, with Cutter, S. and Richardson, D. (Eds.) New York: Routledge.

Darling, F., and Milton, J. (Eds.) (1966). Future Environments of North America. Garden City, NY: Natural History Press.

Harvey, D. (1989). The Growth of Postmodernity. Baltimore: Johns Hopkins University.

Harvey, F. (2000). The Social Construction of Geographical Information Systems, International Journal of Information Science 14: 711-13.

James, P.E. (1972). All Possible Worlds: A History of Geographical Ideas. Indianapolis: Odyssey Press.

Kates, R., et al. (2001). Sustainability Science, Science 292: 641-42.

MacEachren, A., et al. (1992). Visualization. In Abler, R., Marcus, M., and Olson, J. (Eds.) Geography's Inner Worlds, 99-137. New Brunswick: Rutgers University.

Mumford, L. (1963). Technics and Civilization. New York: Harcourt.

National Academy of Sciences/National Research Council (AS/NRC) (1997). Rediscovering Geography: New Relevance for Science and Society, Washington, DC: National Academy Press.

National Academy of Sciences/National Research Council (AS/NRC) (2002). Down to Earth: Geographical Information for Sustainable Development in Africa, Washington, DC: National Academy Press.

Parkes, D., and Thrift, N. (1980). Times, Spaces, and Places. New York: Wiley.

Robb, J.J., and Riebsame, W. (1997). Atlas of the New west: Portrait of a Changing Region. New York: Norton.

Tenner, E. (1997). Technology and the Revenge of Unintended Consequences. New York: Random House.

Wilbanks, T. (1988). Impacts of Energy Development and Use, 1888-2088. In Earth '88: Changing Geographic Perspectives, 96-114. Washington: National Geographic Society.

CHAPTER 2

RON JOHNSTON

COMMUNICATIONS TECHNOLOGY AND THE PRODUCTION OF GEOGRAPHICAL KNOWLEDGE

A step forward in science is of little value unless it is made known to others, especially other scientists, so that they can examine it critically, duplicate it, and eventually improve upon it. (Wilford 2002, 428)

Abstract   Academic knowledge is produced through interaction between those involved in its creation and circulation. Such interaction involves two interpenetrating networks: face-to-face interpersonal contact, and the circulation of manuscripts and other materials. The operation of each of these networks is facilitated by technological changes, especially in communications media. Their role is outlined here with particular reference to the recent history of geography, with the concluding section suggesting that whereas those media have speeded up the circulation of materials and widened their spatial scope, nevertheless it is probably the case that interpersonal face-to-face contacts remain crucial to many aspects of knowledge dissemination.

Keywords   communications media, academic communities, networks, history of geography

Academic disciplines comprise communities of scholars working on linked problems. It is rare for all members of a discipline to agree on every aspect of its activities—on epistemologies, ontologies, and methodologies; rather there is disagreement, sometimes over details of what questions should be posed and how they should be addressed, and occasionally about larger issues regarding the discipline's orientation, even its raison d'être. There may be a division into subcommunities, therefore, each comprising groups with shared interests and approaches to their own subject matter. Members of those groups may, for a variety of reasons, both intellectual and "political" (Johnston 2000a), seek to convert others to their way of thinking. Thus the history of an academic discipline is, to a considerable extent, a history of its debates and

Stanley D. Brunn, Susan L.Cutter, and J.W. Harrington, Jr. (Eds.), Geography and Technology,
17-36. © 2004 Kluwer Academic Publishers. Printed in the Netherlands.

controversies, of the issues tackled in attempts to better understand those components of the world on which it focuses.

Progress in a discipline can thus be assessed at two main scales. The first involves additions to understanding as new research findings are added to the body of accepted knowledge—although it is generally accepted that all knowledge is provisional, that all of our understandings are open to continued organized skepticism and potential reassessment. At this scale, progress occurs by slow accretion, by extending the range of understanding within what is frequently termed an accepted paradigm—the generally agreed procedures by which we create new research findings in the context of what we already know.

The second of the scales involves much broader issues than just making additions to the store of information and knowledge. Instead it comprises debates about what knowledge is and how that knowledge can be created. In these, the discussions and debates within a discipline are not just about new research findings and how they fit into our existing frameworks of understanding, but rather about the creation of new frameworks, new ways of looking at the world and of studying it. Such debates may be relatively lengthy: indeed they may not be resolved, in the sense that one point of view never eventually dominates the discipline. Change in the balance of views may occur through processes of cohort replacement, however, with new appointees to the discipline differing from those they replace. Few disciplines are in a state of intellectual equilibrium for long, and debate about their nature seems to be virtually continuous. This is certainly the case across many of the social science disciplines, as essays on their history indicate:[1] certainly geographers are not alone in regularly debating the nature of their discipline and occasionally generating major changes in its practices, if not its basic raison d'être.

Several features stand out from the history of geography as an academic discipline over the past 100-150 years. The first is the growing volume of work published. To a large extent this is a function of the increasing number of academic geographers, but only partly so: geographers publish more per person per decade now than they did previously. The second is the increasing range of that work, which is again at least in part a reflection of the larger number of geographers, each seeking a particular niche within the academic division of labor in order to enhance her/his career as a researcher. Third, and related to this point, there is growing specialization within the discipline, with most individuals involved in just a small component of it and very few ranging widely across the entire discipline (or even major segments of it). Finally, there is the growing rapidity of change, the increased rate with which the "conventional wisdom" is challenged and alternative blueprints for the discipline—or parts of it—promoted. (On these, and other, shifts, see Brunn and O'Lear 1999.)

For a while, it was popular to represent the process of change as involving a series of revolutions, whereby one paradigm—one accepted way of working—was overthrown and replaced by another. But despite some of the valuable insights that it provided (and still does) the paradigm model was soon abandoned. In some cases, proposed revolutions were themselves overthrown before they became the conventional wisdom, whereas in others a series of parallel conventional wisdoms coexisted and it was impossible to talk of a dominant paradigm–or even of a "conventional wisdom." (On the paradigm model, see Johnston and Sidaway 2004.)

Whatever the means of trying to appreciate changes within the discipline—and several have been on offer—there is no doubt that in the last third of the twentieth century there was much less consensus about the nature of geography than was the case in preceding decades. The nature of geographical knowledge and how it was to be produced (the practice of geography) became much more contested than previously, when there was greater attention to producing more knowledge by the same routes. Ferment within the discipline increased–as it did in many others–and competing views fought for attention, as their proponents contested for status and positions within the discipline.

How has this atmosphere of change been brought about? Is it in any way related to parallel technological changes? In this essay, I explore the impact of changing communications technology on the practice of geography and its knowledge-production activities, with particular reference to two moments during the second half of the twentieth century. That exploration is set within a tentative model of interaction networks and the role of publications in the mobilization of change.

1. NETWORKS OF INTERACTION AND THE MOBILIZATION OF CHANGE

Debates and discussions about the nature of an academic discipline involve interpersonal interaction at a variety of scales and through a number of media. Individual members of a discipline are parts of several overlapping networks, including:

1. Their colleagues and students in the department or other unit where they work, and with whom they are in regular, frequent contact;
2. Colleagues in other departments in the same institution, with whom there may also be frequent contact;

3.   Colleagues in other institutions in the same country, whom they meet
      infrequently, most often at formal events such as conferences; and
4.   Colleagues in other institutions world-wide, whom they meet even less
      frequently if at all.

If the flow of information only occurred through interpersonal conversations
in these groups a great deal of it would be localized. Contacts between workers
in different places would be infrequent, and the stimuli that it generates rare.
New findings would circulate only slowly through the relevant community;
debates about wider issues would be rare and probably extensive in time before
any resolution was reached.

The printed word ensures that academic progress is not dependent on
such interpersonal contact patterns alone. The initial creation of written records
stored in a few libraries ensured that materials were available over both time
and space—as copies of manuscripts were made and transmitted to scholars
elsewhere. The invention of printing made much wider and more rapid
circulation of ideas possible: information and arguments could be transmitted
across both space and time, providing permanent records of what people
thought and knew at any particular moment. These records became parts of
the networks through which scholars—later, academics—interacted: they
enabled interaction-at-distance (both spatial and temporal), without either
individual movement or direct face-to-face meetings. Printed media provide
major channels whereby individuals can contact others, whom they may not
know, but who may be influenced by their work when accessing it via
impersonal means. For many centuries, however, library users were restricted
to the contents of the collections held at those which they were able to access.
Few libraries—such as the five "copyright libraries" in the United Kingdom
to which copies of all materials published there must be donated—had
complete collections, and most had only limited coverage. Only in the second
half of the twentieth century did it become possible to borrow books held in
other libraries through interlibrary loan systems: until then, unless they traveled
to consult materials not immediately available to them, scholars were very
much constrained by the contents of locally available collections. Today, as
discussed below, this constraint is even weaker as materials—especially those
published in journals—are made available on the web. Some libraries no
longer subscribe to hard copies of journals that can be accessed electronically,
thereby saving themselves the costs of housing and maintaining a collection
of printed materials.

In evaluating the history of a discipline, the relative importance of personal
and impersonal contacts may vary, but can be crucial. Many undergraduate
students, for example, are introduced to a field of study through a combination
of the two: they attend lectures and other classes given by teachers who know

the relevant material and literature and present it in structured and stimulating ways; the students also read textbooks (in many cases recommended to them by those lecturers: Johnston 2000b) which do the same thing, in more formal and comprehensive ways in many cases. Through this combination, they learn what it is we know in a given field (i.e., what is accepted as "truth"), what issues remain contested (i.e., are unresolved), and what procedures provide the accepted ways of producing "new knowledge," to resolve the uncertainties. Courses and textbooks cover both knowledge and methods of producing knowledge–and (should) encourage skepticism over both.

Having been inducted into a field of study through undergraduate courses, some students then take the next step to prepare themselves for work that will itself produce new knowledge—by attending graduate school. Here, again, they will be exposed to a range of media—courses from "experts" in their chosen fields, the published results of research by many other "experts" (many in the form of journal articles written specifically for those already well-grounded in work in that area, and subject to peer review), and the debates (published and otherwise) around those writings. Graduate students contribute to such debates—initially through seminars and formal and informal discussions with other members of the graduate community—and take up positions regarding issues relating to that area of knowledge production. Increasingly, the organized skepticism that characterizes all scholarly endeavor becomes a part of the practitioner's cast of mind.

In these early stages of becoming an academic, the individual is involved in a combination of networks of influence. The first is localized in a particular place—the graduate school—where there is intensive interaction among a small number of people, among whom one or two may be perceived as the "leaders" because of their seniority, experience, enthusiasm and/or stimulating presence. The second is nonlocalized, comprising the international (global in some cases) networks of impersonal interaction sustained through the circulation of printed media. The two are interdependent, with individuals recommending reading to each other.

The next career stages involve widening the face-to-face networks, as individual scholars join groups that meet, with varying regularity and frequency, at conferences and other arenas: regional and national networks are grafted on to the localized ones.[2] New interpersonal contacts are made, and the social components of the academic networks both broaden and deepen. Alongside this, the individual begins to contribute to the "nonhuman" components of the network–by writing papers and books that enter circulation. These bring her/him into impersonal contact with a wider range of individuals, who are thus loosely connected to the emerging network—and who over time might become more firmly linked through a personal contact.

As academics interact through these networks, they learn about new research findings and projects, about questions raised regarding previously reported findings and the methods involved in their production, and about debates regarding the nature of the research in their discipline and even wider within academia. Much of their reading and other interactions is undertaken to reinforce their own attitudes and agenda: they focus on the work of others researching in the same or similar fields. But, given the open-minded skepticism that characterizes academia, they may encounter and consider alternate—even opposing—views, some of which may directly challenge their own work, either in detail (the validity of a certain piece of research and its findings) or more generally (the value of a particular approach to knowledge production). Debate is stimulated through such interactions, and support for competing positions is mobilized, through a combination of face-to-face meetings and reading. What we have yet to fully appreciate is the relative role of these two sets of interactions in mobilizing change and, subsequent to that, the degree to which technological changes in the circulation of information through academic networks have influenced the nature and rate of such mobilization.

2. PUBLICATION IN ACADEMIC GEOGRAPHY

No education without publication (Anon).

Publications are central to the interactive processes involved in the production and circulation of knowledge. Over time, the ease and speed of circulating publications have altered drastically, with clear implications for the process of knowledge production and, even more so, the obsolescence of particular pieces of knowledge. How does this affect the entire process? Two particular episodes in geography's recent history illustrate this.

2.1. The Mimeograph, the Xerox Machine, and the "Quantitative and Theoretical Revolutions"

Until the late 1950s, there was a relatively long time-lag between the production of a piece of knowledge and its wide circulation. An individual completed a piece of research, discussed it with colleagues in her/his department-university, wrote it up, and presented it to wider audiences–in seminars at different universities, academic conferences, etc. Those who heard the oral presentations became aware of the results prior to their formal circulation, therefore, and may, through formal and informal discussions, have influenced the shape of later, pre-publication versions. One of these was then submitted and, after the usual peer review process (which may take several months), accepted for publication–perhaps in revised form. It was then set and appeared in an issue of the journal—almost certainly at least a year after the first seminar and/or conference presentations. It was then circulated through libraries and personal subscriptions (widely if placed in a journal with an extensive circulation), and occupied a permanent place on the network.[3]

Such processes still operate, but they are no longer the predominant means of knowledge circulation. The relatively leisured pace at which new knowledge was circulated and made widely available to others of necessity meant that change was not rapid. New "facts" were slow to replace the "old"; the rebuttal and counterrebuttal of interpretations took time, and fundamental debates about the nature of a discipline could continue for several years. This was illustrated in the 1950s in the debates over "exceptionalism" and "positivism" as bases for geographical scholarship. The first comments on Schaefer's (1953) critique of the "Hartshornian orthodoxy" appeared in the *Annals* some nine months after the original (Hartshorne 1954), for example, with longer responses in 1955 and 1958, and a book-length reconsideration in 1959 (Hartshorne 1955, 1958, 1959). Of course, debates of that type are frequently not resolved for several years—if ever—but the delays in the appearance of contributions ensured long periods when the formal procedure was in a form of limbo, even if there was much informal discussion in localized networks during the intervening period.

From the late 1950s on, however, printing and reprographic technologies became available which allowed much more rapid circulation of new results and ideas, thereby not only speeding up, but also supplementing the ways of propagating ideas and conducting debates. These allowed ideas and research results to be disseminated without the twin delays of peer review and the formalities of publication. Materials could be added to the networks at short notice, and people beyond the reach of frequent face-to-face conversations could learn much more rapidly and readily about ideas almost as soon as they entered circulation. This "revolution" was initially facilitated by the method of mimeographing, in which multiple copies of manuscripts typed on "skins," rather than conventional paper, could be produced and circulated immediately to interested parties. Nowhere was this change more apparent than in the Department of Geography at the University of Washington, where free access to the mimeograph machine allowed papers to be circulated (Barnes 2003). The speed at which new ideas could be shared, at least informally with those on mailing lists, was thus reduced to the speed of the postal service.[4]

Initially, these mimeographed publications were produced and circulated privately, to friends and colleagues, but soon others, including people personally unknown to the authors, "discovered" them and asked for individual copies and to be put on mailing lists for future such "publications." They were soon formalized. However, Morrill (1984, 61-62), for example, writing of the academic year 1957-58 at Washington, claims that:

> Perhaps the most significant innovation of the year was the establishment of the Discussion Paper series, an outlet for seminar papers, preliminary thesis findings, etc. Their circulation to other outposts of the "new geography," especially Iowa and Northwestern at first, was of great immediate (feedback, encouragement) and long-term benefit (spreading the word!).

This was especially important to the work of the group then focused on Seattle, who were actively challenging the current disciplinary orthodoxy (Johnston and Sidaway 2004) and, as Berry (1993) and others describe it, having considerable difficulties getting their work accepted for publication in established journals. Indeed, such was the perceived hostility to their work, and that of colleagues with similar views, that within a decade they had launched a new journal—*Geographical Analysis*—which would focus on such styles of work. Publication in peer-reviewed journals, which were subscribed to by libraries and so obtained permanent positions in the communication networks, remained the major route to professional recognition and status, however. If the established journals would not accept certain types of work, then alternatives had to be launched which would. Then—by their professionalism—those journals could win the desired credibility with promotions and other committees.

There were many debates over whether circulating one's ideas through informal media involved sufficient rigor in the refereeing and related processes of peer review to ensure high standards and therefore assist in academics' attempts to obtain tenure and promotion within their universities. This was to some extent allayed by the later decision to register some discussion paper series with ISSN numbers. But it was never fully resolved. Eventually, the issue faded away as new technology and media—notably the Internet—offered alternative methods of circulating material, reigniting the debates in these new contexts. But with these, new concerns developed, such as the "acceptability" of publications in on-line journals for appointment and promotion evaluations.

Discussion Paper series, such as that launched at Washington, benefited from further technological advancements in the reproduction of the printed page—notably the Xerox machine and its competitors, which removed the limits on the number of copies that could be made. Through their institutionalization, Discussion Paper series also became part of the formal disciplinary literature—albeit only partially accepted. They could be used for the rapid circulation of ideas in preliminary form, but not for the more formal procedures of establishing professional status and credibility—and some journal editors refused to allow citations to such literature (on the grounds of its inaccessibility to a wider audience, a form of opposition that was challenged when libraries began subscribing to the various series).

Instead of new ideas being initially discussed with a relatively small number of individuals at a localized seminar, and then later perhaps at a conference session with a larger audience drawn from wider afield, the seminar and conference audiences could, in effect, be substantially enlarged (potentially, world-wide) and rapidly accessed, albeit without the added advantage of those contacted being able to participate in the seminar/conference discussions. Mobilizing people to new ways of producing knowledge was speeded up enormously, even though the knowledge itself was only "accepted" more widely when it was placed in the formal publication media (albeit grudgingly, as memoirs of contacts with editors reveal: Berry 1993; Wheeler 2001).

These processes are illustrated by a Discussion Paper series established in 1963 by a group of like-minded geographers working at several neighboring universities. The Michigan Inter-University Community of Mathematical Geographers (formed by geographers at the University of Michigan, Michigan State University, and Wayne State University) produced a series of papers—including some by visitors to their seminar series—which, through their rapid circulation not only in North America but also the UK, Australia, and New Zealand, enhanced awareness of contemporary developments much more rapidly than had ever been the case before. One was the first appearance (in 1966) of Peter Gould's seminal work on mental maps.

Such developments in "home-made" publication outlets moved further at the end of the 1960s with the creation of "home-made journals." The most important of these by far was *Antipode*, whose origins lay in the "radical revolution" of the late 1960s and thereafter, which challenged not only the status quo in Anglo-American geography that predated the "theoretical and quantitative revolutions" of the late 1950s and early 1960s, but those revolutions too. Based in the Graduate School of Geography at Clark University, *Antipode* was produced "in-house" until 1986 when it became "respectable," being taken over by a commercial publisher and gaining the credibility associated with standard, peer-reviewed journals (Editorial 1986).[5]

2.2. The Internet, the World Wide Web, and Disciplinary Fragmentation

The second technological development in the circulation of information considered here is the creation and rapid expansion of the Internet and the World Wide Web.[6] By the late 1980s, most universities had links to these. Through attachments to e-mail messages, academics could send manuscripts (draft and otherwise) to anybody in any country who was similarly linked.

New findings and ideas could be circulated instantaneously, and those linked to such networks need experience no delays in discovering what people elsewhere were doing, thinking, and reporting.[7] As this technology developed, it was not only written "manuscripts" that could be transmitted over the Internet but original data, diagrams, photographs, video clips, music, and other media too. Given the importance of visual communication in geography, this was especially valuable to the discipline.[8]

The developments continue. The greater speed and storage space of computers, the rapidity with which networks could be searched for material, and the ability of broadband connections to transmit large amounts of information quickly led to the massive expansion of Internet use for "publishing" and accessing material at a distance, with virtual libraries of material in all forms of media being accessible from personal computers on individuals' desks, and then from portable laptop computers. A massive range of individuals, groups, and institutions have established their own websites to make information—such as meeting announcements and discussion papers— freely available (though some have restricted access), and the ability to link directly from one website to another via hyperlinks enables rapid searches for related materials. Learned societies such as the Association of American Geographers and the Royal Geographical Society also make wide use of their websites to promote members' interests. Academics can also put their teaching materials on web sites to enable students—even those at other institutions— to access them at any time. In addition, collections of such materials can be assembled into larger bodies of material. At Ken Foote's "virtual geography department"[9] it is even possible to go on "virtual field trips."[10] The "invention" of e-science in the early twenty-first century is taking this even further, with analysts being able to access and work on datasets housed on distant computers without having to transfer materials to their "home sites" (Foster 2003).

Again, although some of the early development was informal—with individuals sending copies of their latest manuscripts to known colleagues (perhaps in even-less-completed form than was the case with their discussion papers a decade or so earlier)—it was soon formalized as the World Wide Web became available for academic communication. This made it possible to place manuscripts on websites, both personal and institutional, from which they could be downloaded without any contact with the author. It became known that some authors placed all of their ongoing manuscripts on such sites (as well as providing links to the journals where completed pieces had been published), and periodic visits could explore their latest work. Institutions too, such as learned societies, created sites where members could put their latest offerings, including the papers they were giving at its upcoming conferences. And powerful "search engines"—such as Google, Yahoo, and

Altavista—have enabled researchers to discover such materials very readily indeed.

Alongside these largely reactive means of increasing the circulation of ideas were the more active ones, where individuals could join discussions of contemporary issues relating to their research interests. Many such electronic mailing lists were created. Brunn et al. (1997) identified 6,374 that were known to the LISTSERV organization in September 1995—of which 4,601 were in the U.S. The memberships of seven of the lists oriented to geographers ranged from 182 to 1,359, with subscribers from between 9 and 50 countries. Detailed study of one such list, GEOGRAPH, found that it was dominated by university faculty and graduate students who mainly used it to ask and respond to requests for information. Many recent developments in the discipline—such as those linked to social theory and critical geography—have been facilitated by such "informal" publication media, and it may be that their activities may not have "taken off," certainly not in the ways that they have, without that ease of communication without the more formal (and often more conservative) methods offered by learned societies and other institutions.[11]

The Internet has also made access to formal publications much easier, and has encouraged their growth, for example, the creation of on-line journals, which have yet to become widespread and popular among geographers but are more common in some other disciplines, with learned societies creating them in some instances. The number of new journals in geography and related fields has expanded rapidly over the last two decades, in part reflecting the expansion in the number of practicing geographers and in part the pressures on them to publish regularly and frequently in recognized, peer-reviewed outlets. (In the last decade, for example, journals such as *Social Geography*, *Ethics, Place and Environment*, *Space and Polity*, and *Journal of River Basin Management* have been launched, among many others.)[12] The balance of work undertaken by the commercial firms publishing almost all of these has changed substantially. The academic decisions (on which papers to accept for publication) have remained with the editors, academics who work for the commercial concerns. But publishers no longer have to undertake a massive amount of effort to prepare the copy provided by the authors and editors into formats ready for printing. A decade ago, many asked authors to provide camera-ready copy in formats that could be prepared for photographic reproduction. The costs of the mechanical processes of journal (and book) production were reduced substantially, allowing publishers to take on journals with relatively small sales, but which nevertheless were soon profitable. Now most journals ask for the finished papers (if not the initial submissions) to be submitted in electronic formats, which they edit lightly (into their "house styles") before creating the journal issues ready for printing. And their print runs are decreasing. Instead of selling journal subscriptions and circulating "hard copies," they are now selling access to journals on password-protected websites. Many libraries no longer obtain the hard copies of such journals, but subscribe to the electronic versions (often "packages" of journals produced by large-scale publishers rather than individual items),[13] and their "subscribers" (faculty and students) can download copies of relevant articles—meeting the printing costs themselves. Large numbers of libraries do, however, subscribe to specialist citation services which enable searches for related literature on a subject, as well as the widely used ISI databases, which collect citation data from some 8,712 journals worldwide.[14]

Web-mounted journals do not necessarily increase the rapidity with which academic material is circulated—though there are benefits, given the time that some postal services take. But they do make access much easier.[15] Individuals who want to read an article do not have to go to a library to see it—and perhaps photocopy it.[16] They can obtain their own copy without going anywhere near the library. And publishers make this easier for them by circulating tables of contents for the latest issues through e-mail channels. At one time, researchers had to search for papers that were relevant to their interests. Now they need only log on to either their e-mail server or the web to be told what has recently been published—or is about to be published. Eventually, this may lead to them working in paperless offices, simply calling up what they want on screen when they want it, instead of searching their shelves—though such is the attachment to books and journals that this is unlikely to happen too soon!

It is now very easy for academics to circulate their draft papers and other materials and, assuming that they are of high quality, to find specialized formal outlets in which to place them—thereby gaining the charisma and status from publication in "recognized, peer-reviewed journals" that promotion and appointment panels are looking for, as are outside evaluators (such as the notorious UK Research Assessment Exercise Panels). This has occurred during a period when geography is both flowering and yet fragmenting. A great deal of research activity is taking place in specialist subdisciplinary communities— many of them with stronger inter- than intra-disciplinary connections. Increasingly, those communities are served by their own journals, many of which have rapidly achieved the wider standing required for bestowing status on those who publish in them. It remains the case that the longer-established, discipline-wide journals, notably those published by learned societies, continue to get the highest "impact factors" in citation analyses,[17] although editors of such journals regularly complain that the materials they receive do not reflect the breadth of research interests and practices within the contemporary discipline. Many (most?) academic geographers, it seems, are content to publish in the specialist journals that serve their own communities, and to "converse" with like-minded scholars in the relevant web-based discussion groups and specialist study group conferences (plus their sessions, which rarely attract many "outsiders," at the large, learned society conferences). They are less concerned, or so it seems, to reach wider audiences through publication strategies aimed at less specialized audiences. Mobilizing support widely within their discipline for particular forms of knowledge production apparently has a lower priority than it did 30-40 years ago, and instead there is an anarchic situation of many smaller communities each relatively secure in its own intellectual cocoon. (On the publication practices of UK geographers see Johnston 2003b.) The disciplinary framework remains—important in an institutional sense for the organization of teaching in universities and for sustaining the standing of the loosely related communities of geographers— but research practices are now both fragmented and, increasingly, interdisciplinary. The map of research is not homologous with the map of research activity, with the intellectual and political projects (like those of other disciplines) becoming increasingly difficult to sustain (Johnston 2004).

3. ON FACE-TO-FACE AND DISTANT INTERACTION

By focusing on the role of "printed" materials in the communication networks, the argument so far in this essay may appear to be that technological changes in the circulation of material have made interpersonal face-to-face contacts increasingly redundant. Knowledge advances through reading and writing materials that are accessed via the net. Indeed, there are now cases of researchers in different places—even different countries—collaborating in research and publication without ever meeting.[18] But this seems to be countered by other evidence regarding, in particular, the role of conferences in the transmission of knowledge, both the major, inclusive conferences—of which the annual meetings of the Association of American Geographers are by far the largest in the English-speaking world—and the many specialist, smaller conferences now being held, both those run by (formal and informal) interest groups within the discipline and those which are convened ad hoc to address particular themes. There is an increasing number of such conferences and many are attracting increased attendances—in part at least because the relative cost of travel has been declining in recent years.

The continued importance of conferences suggests that hearing about new work remains important, alongside reading it. Why might this be so? One major suggestion is that for many academics, attendance at a conference brings to their attention ideas and material that otherwise might evade it–if for no other reason than that there is an information overload (both on and off the web). So much is available for reading that we need filters to help us decide which is likely to be most beneficial. Those filters may be other individuals that we meet at conferences. Through informal conversations there we learn what they currently consider important and worth spending time on. And the filters probably include the formal presentations attended. These could have more impact on those who decide to attend (and they may not, if the title/abstract is unappealing) than lists of titles in journal Tables of Contents, or lists of new papers posted on the Internet.

Conferences are only one format for face-to-face contacts, where we hear rather than read about new knowledge and arguments regarding its production. The ease and relative cheapness of movement means that most geography departments run regular seminar series to which outsiders are invited to contribute—not only outsiders from nearby institutions but also visitors from longer distances who may be flown-in for a few weeks or are on leave in another country and prepared to travel around discussing their current work and ideas. These, too, appear to operate as "floating subcommunities." Many departments either have separate sets of seminars for physical and human geography, for example, or have very different audiences for seminars depending on the subject matter—and the subdivision may extend within both physical and human geography. Further, there are many examples of clusters within departments meeting together informally as "reading groups" to discuss recent publications. The fragmentation of publication within the discipline is paralleled by the fragmentation of meetings.

Some years ago, Allan Pred (1979, 1984) presented time geography as a framework with which to analyze the importance of where he had been, and with whom, on the progress of his academic career. Contacts with individuals at Chicago and Berkeley and in Sweden were crucial to him. As he put it with regard to one part of his learning experience (Pred 1984, 88: his emphasis):

> it was necessary for me in each instance to bring my path into convergence— *usually at specific times and places*—with those of specific students, professors, journals, desks etc.

The implication is that in many cases reading about others' work is not enough: to realize its full import and potential, some reinforcement is necessary, and that may well be best achieved through interpersonal, face-to-face contact— whether at conferences or, more likely in Pred's argument, through visits to other institutions where conversations can be stretched, punctuated, and, perhaps with a shifting membership, in informal settings.

This argument is illustrated by a further example, linked to Pred's own. Torsten Hägerstrand's original work on the diffusion of innovations was first published in 1953, but had little initial impact outside Sweden (Duncan 1974).

In part this was because it was seen as irrelevant to the way in which contemporary geography was practiced in the U.S. and, especially, the UK. But then Hägerstrand visited the Department of Geography at the University of Washington in 1959, stimulating a great deal of interest and excitement among the graduate students (Getis 1993)–including Morrill (1984), for whom Hägerstrand's visit was the most influential feature of 1959 (it had an "electrifying impact;" p. 62) and led to him spending a year at Lund, out of which his first major research publications emerged.[19] Hägerstrand became a much-respected and -honored researcher internationally over the subsequent decades. Would that have happened if his work had remained in print but never discussed by him personally with leading geographers elsewhere?

4. NETWORKS AND DISCIPLINARY CHANGE: NOT YET A CONCLUSION

Our understanding of the history of geography—even its most recent history—is far from complete,[20] and as a consequence we are unsure what strategies work best in mobilizing either interest in the work of any individual (including ourselves!) or concerns about wider issues regarding the discipline's nature, directions, and agenda. We are aware that our discipline operates, intellectually at least, as a series of subdisciplinary, quasi-independent, communities, with relatively little cross-community interest, even though these tend to come together for political reasons of disciplinary (and so career) defense and promotion (Johnston 1996, 2002a). It comprises a massive network of people and publications, subdivided into a number of (partially overlapping in some cases) subnetworks, some much weaker than others, some much longer-lived than others. How that overarching network and its subnetworks are sustained is only poorly appreciated.

There is much else that we do not understand. One aspect is that of "selective reading and citing." There are many papers published of considerable relevance to an area of work which nevertheless rarely get cited (and, we assume, are rarely read), whereas others—on the same subject matter, perhaps appearing later—are clearly much more influential. Does that reflect the status of the authors, or the degree to which they are connected to the subnetworks—perhaps through their personal contact fields, and those of their students (Johnston 2002b)? How do we decide whether something is worth reading, given the time pressures that ensure we can only access a portion of what is available and potentially relevant to our work? Undoubtedly our selection criteria vary, but what seems to be the case—albeit as yet without any formal testing of the idea—is that direct interpersonal contacts (including face-to-face) remain important in the maintenance of the academic networks (or "invisible colleges").[21] Information and debates may be circulating rapidly and widely, but in selecting what to read from that plethora of riches, many seem to concentrate on pieces by people they know and whose work they respect. If time is limited, it seems, we stick with the "tried and trusted," just as we do in so many other choice situations.

Furthermore, the contexts for network creation, sustenance, and break-up are themselves changing, as I have argued here with regard to communication (especially printing and circulation) technologies. The importance of those changing contexts is certainly poorly understood, and this essay has but touched on some of the important issues involved. Can it be said, for example, that the reprographic innovations of the 1950s influenced the course of the "quantitative and theoretical revolutions"? In the longer term, probably not. Something very similar would have happened, somewhere, somewhen, because there were wider social influences pressing disciplines towards such reorientations of their work, and there were individuals "willing" to be convinced of the agenda's validity. But did it facilitate it happening when it did and, even more so, the role of the faculty and graduate students at the University of Washington in that process? Undoubtedly so. The revolution would probably have happened anyway—but its timing and form, including its spatial form, reflected the enterprise of those at Seattle, followed by others elsewhere, to mobilize support for a new form of knowledge production via a new form of knowledge circulation.

Similarly, have the Internet, the World Wide Web, virtual libraries, and the reduced costs of journal publication ensured the growing fragmentation of the discipline? Almost certainly the answer is no. Those technological developments were neither necessary nor sufficient for the changes to occur; there were already tendencies towards fragmentation in the 1980s. But, again, the technological developments have undoubtedly facilitated them. The directions in which the discipline—indeed academia in general—was moving were eased by changes in communications media.

Understanding ourselves, where we have come from and how that has happened, is important to appreciating our identities in all walks of life and certainly so with regard to the disciplinary tribal territories with which we are associated. Identity formation is a consequence of interpersonal interaction—both direct (or face-to-face) and indirect (through media such as publications). Mobilization towards changing identities relies crucially on such interactions. This essay has explored in an introductory way the role of changed communication technologies in those interactions, and the consequent changes in the ways that geography is practiced as an academic discipline. Only further exploration will evaluate the underlying arguments in any detail, identifying the role of different types of encounter, formal and informal, planned and serendipitous, in the history of an academic enterprise called geography.

ACKNOWLEDGMENTS

I am grateful to Rita Johnston and Stan Brunn for many valuable suggestions regarding the arguments developed here.

NOTES

1. See the essays on individual disciplines in the recently published International Encyclopedia of the Social and Behavioral Sciences (Smelser and Balte 2001).
2. Some members of a graduate community may already be members of incipient networks, containing teachers and fellow-students they have interacted with at previous institutions, such as those where they obtained their undergraduate degrees.
3. The slowness of this procedure led a group of academic geographers in the UK to form a new journal–Geographical Studies–in the 1950s to circumvent the time-lags: they were concerned not only at the slowness with which ideas gained wide circulation but also at the impact this could have on their career prospects (Johnston 2003a).
4. I was working in Australia and New Zealand in the late 1960s and early 1970s, and my mobilization into the "quantitative revolution" was much advanced by such samizdat publications–even though sea-mail took two months or more!
5. The same happened with another journal, launched a little later–Journal of Geography in Higher Education–which sought to stimulate interest in pedagogy rather than research in academic communications.
6. For a brief resume of the history of the Internet, see Brunn et al (1999).
7. Of course, as Taylor (1990) argued with regard to GIS, there were major international inequalities in this access, depending on the availability of the relevant links.
8. There was an intermediate stage, still widely used though not for the transmission of major volumes of information. The FAX machine allowed hard copies of materials to be transmitted through telephone systems, at relatively low cost, so that scholars could exchange ideas in written form. Although still widely used for transmitting relatively short documents, FAX machines are increasingly being superseded by e-mail.
9. See http://www.colorado.edu/geography/foote/foote.html
10. See http://www.geog.le.ac.uk/cti/virt.html
11. The critical geographers' forum, for example, was stimulated by the expense of "formal" international conferences and reactions to policies of the Royal Geographical Society: see their website at http://econgeog.misc.hit-u.ac.jp/icgg/.
12. Most of these publish solely in English–and many journals in other languages have English abstracts for their papers. This has led to claims that Anglophone geographers are dominating the discipline globally–and thereby limiting its diversity; such a trend may well be assisted by the web and the predominance of English therein (see Garcia-Ramon 2003).
13. The large-scale publishers of geography journals include Blackwell, Kluwer, Elsevier, Taylor and Francis and John Wiley with others–such as Hodder, Routledge and Sage–responsible for some of the more salient journals for the discipline.
14. See http://www.isinet.com/isi/
15. And also, some would claim, plagiarism!

16.   Almost none now write to authors asking for reprints, though journals still provide them for the authors to circulate–and also to use when making their cases for promotions etc.

17.   This includes journals such as the Transactions of the Institute of British Geographers and the Annals of the Association of American Geographers, but also commercially produced journals such as Political Geography and the four series of Environment and Planning.

18.   I have written more than 150 papers with a colleague who works more than 200 miles from my university: we meet several times annually, but correspond electronically almost daily. We have also worked and published with somebody–initially in the U.S.; now in Germany–whom neither of us has met (Gschwend et al. 2003) and I have an active research programme with colleagues in Australia and New Zealand whom I meet, at most, annually.

19.   Pred also did his PhD work in Sweden, as a Chicago graduate student, but at Göteborg. Although he became aware of Hägerstrand's work then, it wasn't until he returned to Sweden in 1966 "that my path first became intertwined directly with the paths of Swedish geographers who were the generators and proponents of specific 'revolutionary' innovations" (Pred 1984, 97).

20.   Of course some–e.g. Barnett (1995) and Thrift (2002)–downplay the efforts involved in learning about that history!

21.   On the potential impact of the Internet on such "colleges," see Brunn et al. (1999).

REFERENCES

Barnes, T.J. (2003). The Place of Locational Analysis: A Selective and Interpretive History, Progress in Human Geography 27: 69-96.

Barnett, C. (1995). Awakening the Dead: Who Needs the History of Geography? Transactions, Institute of British Geographers NS20: 417-9.

Berry, B.J.L. (1993). Geography's Quantitative Revolution: Initial Conditions, 1954-1960. A Personal Memoir, Urban Geography 14: 434-41.

Brunn, S.D., Husso, K., Kokkonen, P., and Pyythiå, M. (1997). The GEOGRAPH Electronic Mailing List: The Emergence of a New Scholarly Community, Fennia 175: 97-123.

Brunn, S.D., Husso, K., and Pyyhtiä, M. (1999a). Scholarly Communication in Cyberspace: Profiles of Subscribers Using the GEOGRAPH Electronic Mailing List, Fennia 177: 171-84.

Brunn, S.D., Husso, K., and Pyyhtiä, M. (1999b). Writing and Communication in Cyberspace. In Buttimer, A., Brunn, S.D., and Wardenga, U. (Eds.) Text and Image: Social Construction of Regional Knowledges, 290-302. Leipzig: Institut für Länderkunde.

Brunn, S.D. and OLear, S.R. (1999). Research and Communication in the Invisible College of the Human Dimensions of Global Change, Global Environmental Change 9: 285-301.

Duncan, S.S. (1974). The Isolation of Scientific Discovery: Indifference and Resistance to a New Idea, Science Studies 4: 109-34.

Editorial (1986). Editorial, Antipode 18: 1-4.

Foster, I. (2003). The Grid: Computing Without Bounds, Scientific American (April): 80-85.

Garcia-Ramon, M.D. (2003). Globalization and International Geography: The Questions of Languages and Scholarly Traditions, Progress in Human Geography 27: 1-5.

Getis, A. (1993). Scholarship, Leadership and Quantitative Methods, Urban Geography 14: 517-25.

Gschwend, T., Johnston, R.J., and Pattie, C.J. (2003). Split-Ticket Patterns in Multi-Member Proportional Election Systems: Estimates and Analyses of Their Spatial Variations at the German Federal Election, 1998, British Journal of Political Science 33: 109-28.

Hartshorne, R. (1954). Comment on Exceptionalism in Geography, Annals of the Association of American Geographers 44: 108-09.

Hartshorne, R. (1955). Exceptionalism in Geography Re-Examined, Annals of the Association of American Geographers 45: 205-44.

Hartshorne, R. (1958). The Concept of Geography as aScience of Space from Kant and Humboldt to Hettner, Annals of the Association of American Geographers 48: 97-108.

Hartshorne, R. (1959). Perspective on the Nature of Geography. Chicago: Rand McNally.

Johnston, R.J. (1996). Academic Tribes, Disciplinary Containers, and the Realpolitik of Opening up the Social Sciences, Environment and Planning A 28: 1943-48.

Johnston, R.J. (2000a). On Disciplinary History and Textbooks: Or Where Has Spatial Analysis Gone, Australian Geographical Studies 38: 125-37.

Johnston, R.J. (2000b). Authors, Editors and Authority in the Postmodern Academy, Antipode 32: 271-91.

Johnston, R.J. (2002a). Reflections on Nigel Thrift's Optimism: Political Strategies to Implement His Vision, Geoforum 33: 421-25.

Johnston, R.J. (2002b). Robert E Dickinson and the Growth of Urban Geography: An Evaluation, Urban Geography 22: 702-36.

Johnston, R.J. (2003a). The Institutionalization of Geography as an Academic Discipline. In Johnston, R.J. and Williams, M. (Eds.) A Century of British Geography. Oxford: Oxford University Press and the British Academy.

Johnston, R.J. (2003b). Geography: A Different Sort of Discipline? Transactions of the Institute of British Geographers NS28: 133-41.

Johnston, R.J. (2004). Coming Apart at the Seams. In Castree, N., Rogers, A., and Sherman, D. (Eds.) Questioning Geography: Essays on a Contested Discipline. Oxford: Blackwell.

Johnston, R.J. and Sidaway, J.D. (2004). Geography and Geographers: Anglo-American Human Geography Since 1945. 6th ed. London: Arnold.

Morrill, R.L. (1984). Recollections of the Quantitative Revolutions Early Years: The University of Washington 1955-65. In Billinge, M., Gregory, D., and Martin, R. (Eds.) Recollections of a Revolution: Geography as Spatial Science, 57-72. London: Macmillan.

Pred, A. (1979). The Academic Past Through a Time-Geographic Looking Glass, Annals of the Association of American Geographers 69: 175-80.

Pred, A. (1984). From Here and Now to There and Then: Some Notes of Diffusions, Defusions and Disillusions. In Billinge, M., Gregory, D., and Martin, R. (Eds.) Recollections of a Revolution: Geography as Spatial Science, 86-103. London: Macmillan.

Schaefer, F.K. (1953). Exceptionalism in Geography: A Methodological Examination, Annals of the Association of American Geographers 43: 226-49.

Smelser, N.J. and Baltes, P.B. (Eds.). (2001). International Encyclopedia of the Social and Behavioral Sciences. Oxford: Elsevier.

Taylor, P.J. (1990). Editorial Comment: GKS, Political Geography Quarterly 9: 211-12.

Thrift, N.J. (2002). The Future of Geography? Geoforum 33: 291-98.

Wheeler, J.O. (2001). Assessing the Role of Spatial Analysis in Urban Geography in the 1960s, Urban Geography 22: 549-58.

Wilford, J.N. (2002). The Mapmakers: The Story of the Great Pioneers in Cartography–From Antiquity to the Space Age. London: Pimlico.

CHAPTER 3

FRED M. SHELLEY
WENDY BIGLER
RICHARD ASPINALL

FEDERAL FUNDING, GEOGRAPHIC RESEARCH, AND GEOGRAPHIC TECHNOLOGIES: 1904-2004

Abstract   The history of geographical research in the U.S. has been closely intertwined with the federal government throughout the past hundred years. Since World War II, the federal government has provided substantial support for geographic research. Programs such as the Office of Naval Research and the National Science Foundation have provided major investment into geographers' research activities. In this paper, we trace the history of relationships between the federal government and academic geography over the past century, and we analyze patterns of federal support for research papers published in the *Annals*, *The Professional Geographer*, and *Geographical Review*.

Keywords   Federal government, research, Office of Naval Research, National Science Foundation, history of geography.

1. INTRODUCTION

The purpose of this chapter is to trace the historical linkages between federal funding and the development of geographic technologies over the past hundred years. The first section discusses relationships between changes in the world economy and international political system, changes in the U.S. federal government and its funding patterns and priorities, the development of geography as a discipline, and the development of geographic technology. The remaining sections illustrate how these changes have been reflected in AAG publications, other major publication outlets, and the research activities of prominent AAG members.

In order to effectively examine how federal funding has impacted geographic research and technology over the past century, it is essential to examine the changing role of the federal government in American life and in geography, since the early twentieth century over five key time periods: the early twentieth century (prior to 1933), the New Deal and World War II (1933-

Stanley D. Brunn, Susan L. Cutter, and J.W. Harrington, Jr. (Eds.), Geography and Technology,
37-62. © 2004 Kluwer Academic Publishers. Printed in the Netherlands.

45), the early Cold War period (1945-70), the later Cold War (1970-89), and the post-Cold War era (1989 to the present). Each of these periods is associated with a changing conceptualization of the nature of federal government and its relationship to American life and society. These changing relationships affected priorities in federal support for geographic research, which in turn influenced linkages between federal support and the development of geographic technology.

2. CHANGING TIMES

2.1. The Early Twentieth Century

The Association of American Geographers (AAG) was founded in 1904. During the first three decades of the AAG's existence, the U.S. was undergoing transition from a rural, agrarian frontier society to a modern industrial country. In 1893, historian Frederick Jackson Turner declared that the frontier was closed (Turner 1920). Americans began to recognize that North America's land resources were limited. The poor, the dispossessed, and new immigrants from Europe could no longer count on the availability of free or cheap land to make a fresh start. Instead, Americans turned increasingly to cities in order to advance economically. The late nineteenth and early twentieth centuries represented the start of a century-long movement from rural to urban places and lifestyles. In 1900, nearly three-quarters of Americans lived on farms. By 1930, this proportion had dropped below half. The large-scale movement of Americans from rural to urban places has continued ever since, with only one percent of Americans living on farms today.

Urbanization and modernization in American life coincided with the rapid development of those technologies which, as they developed, would play important roles in geographic research in the years ahead. The founding of the AAG occurred only a few months after the Wright Brothers' historic first flight at Kitty Hawk, North Carolina. Aviation and space technology would revolutionize geography over the course of the twentieth century, not only through opportunities to obtain information via aerial photography and remote sensing, but also by facilitating travel by geographers to research sites and opportunities throughout the world (Richardson 1984). Rapid development also occurred in the automobile industry, radio, electronics, and telecommunications, foreshadowing the discipline's later emphasis on these and more sophisticated technologies in research and data collection over the course of the twentieth century.

During the early twentieth century, as more and more people moved to cities, political and economic leaders began to promote a more urban, scientific view of public policy. The first three decades of the twentieth century are often referred to as the Progressive Era (Hofstadter 1955). The Progressives emphasized the need for rational, scientific approaches to public policy issues. They regarded existing approaches to policymaking as dominated by rustic, uneducated rural folk, greedy corporations, and corrupt urban political machines. They regarded these approaches to policy formation as outmoded, archaic, irrational, and unsuited to the needs of an increasingly urbanized society. They strongly supported the efforts of President Theodore Roosevelt and others to "bust" the trusts and encourage government regulation of corporate activity, and they valued scientific expertise as a more rational and effective means of developing public policy.

The Progressives regarded government as the most appropriate and effective means of ensuring rationality and efficiency in the marketplace and public life. The Progressives were linked closely to the leading colleges and universities of their era. President Woodrow Wilson, who held a Ph.D. in political science and was a graduate school classmate of Turner, exemplified the Progressive Era. Scholars and intellectuals were encouraged to contribute their talents to improving society through rational planning and scientific management.

The formative years of the AAG, like many other scholarly associations, took place within the Progressive Era. Members of the AAG and other such organizations played important roles in the progressive movement. Prominent geographers including William Morris Davis and Isaiah Bowman were known not only for their achievements as scholars but also for their contributions to public policy. For example, the U.S. Shipping Board undertook a nationwide survey of commodity imports in order to plan for shipments of war material, and geographers such as Charles Colby and Vernor Finch played important roles in these studies (Martin and James 1993, 389). Under Roosevelt and Wilson, the U.S. maintained an active foreign policy, and geographers placed their expertise at the service of government leaders. For example, Bowman played a key role in the re-creation of Europe's boundaries after World War I (Martin 1980; Martin and James 1993, 391-92). Bowman and other geographers also lent their expertise to the settlement of other border disputes in the 1920s. In addition, geographers were involved in land classification studies, which were used to promote land conservation and the wise use of natural resources (Colby 1941).

Although the Progressives argued strongly for increasing the role of government in public policy, the role of the federal government in public policy remained small, relative to contemporary standards, throughout the early twentieth century. After World War I, however, a majority of Americans repudiated the Progressives' ideals. The 1920s were a decade of prosperity, but were associated with a small laissez-faire federal government. Not until the 1930s, in response to the Great Depression, would the federal government again play an active role in policy formation.

2.2. The New Deal and World War II

In October 1929, the U.S. stock market collapsed and the American and world economies entered a decade of severe economic depression. The optimism of the 1920s vanished, and technological progress slowed. In 1933, Franklin D. Roosevelt was inaugurated as President. Roosevelt believed that enhanced federal activity would be needed in order to reinvigorate the economy, and he created numerous new federal programs. In response to the Depression, Roosevelt and the heavily Democratic Congress established the New Deal, a series of federal programs that resulted in sustained growth in the federal government, and this growth in the public sector increased greatly after the U.S. entered World War II in December 1941.

Both the New Deal and the later war effort generated programs that involved geographers directly. During the Depression, college and university enrollments had plummeted. Many states had slashed their budgets and frozen hiring, and hundreds of would-be academic geographers found themselves jobless or underemployed. Various New Deal programs such as the Works Progress Administration and the National Youth Administration helped geographers maintain their careers. One of the first large-scale federal programs involving geographic technology was in agriculture. The agricultural sector was hit especially hard by the Great Depression of the 1930s, with net farm incomes dropping by 60 percent between 1929 and 1932 (Monmonier 2002). In response, the incoming administration of Franklin Roosevelt adopted several strategies intended to increase farm incomes by reducing supplies of crops. A voluntary acreage-reduction program was established, giving farmers incentives to reduce the number of acres planted and encouraging farmers to take marginal lands out of production.

The success of these efforts depended not only on voluntary cooperation by farmers, but also on accurate measurements of the acreage planted. Beginning in 1934, the U.S. Department of Agriculture (USDA) used aerial photography to survey the amount of acreage planted (Monmonier 2002). By the outbreak of World War II in 1941, more than 90 percent of the U.S. had been photographed by air (Moyer 1949). Forty-six separate federal agencies were involved in land classification studies, which were coordinated by the Land Committee of the U.S. Natural Resources Planning Board (Colby 1941). After the war began, the expertise developed in this effort to photograph agricultural lands was applied to collecting military intelligence. The USDA contributions included laboratories for photo processing (Monmonier 2002).

Photogrammetry would continue as an important intelligence-gathering information during World War II and throughout the Cold War. For example, color infrared technology was initiated in 1943, in order to detect camouflage. The U.S. Geological Survey (USDA) was also active in promoting the development of photogrammetry and remote sensing during this period. For example, in 1933, USDA developed a method to survey and map the area covered by the Tennessee Valley Authority (Southard 1984).

After Pearl Harbor, the war effort brought an entire generation of young geographers into the federal government (Stone 1979; Martin and James 1993). Hundreds of geographers served on active duty in the armed forces. Many others, especially those with expertise in technology and in foreign areas, worked in military or civilian capacities for the federal government. The Office of Strategic Services (OSS), which would later evolve into the Central Intelligence Agency, was empowered "to collect and analyze such strategic information as was required by the Joint Chiefs of Staff for military operational planning" (Stone 1979, 90). The OSS hired more than 200 professional geographers. Others worked in the Office of Naval Intelligence, the Board of Economic Warfare, the State Department, and the U.S. Census Bureau, which provide statistical information and analysis for wartime and postwar planning purposes (Stone 1979). The needs of the war effort required geographers and others to work in collaboration and under deadline pressure: "a team approach was used in imaginative organizations improvised for each assignment" (Stone 1979, 91). The war effort had the long-run effect of encouraging more applied geography and collaboration in research within the geographic community as well as between geographers and persons in other disciplines.

The influx of geographers into federal service had the long-run effect of restructuring the AAG. Prior to World War II, membership in the AAG was by invitation only. During the war, however, young geographers in federal service in Washington and elsewhere chafed at exclusion from the Association and started a new organization, the Society of Professional Geographers. The Society merged with the AAG in 1948, and the newly merged AAG dropped its invitation-only membership policy and adopted its current policy of allowing anyone with an interest in American professional geography to join (Martin and James 1993).

Relative to the older, more established membership of the AAG, the Society of Professional Geographers was dominated by younger geographers who were frequently veterans of the war effort, either through military or civilian service. After the war, the ranks of professional geographers and AAG members were augmented greatly by military veterans who took advantage of the opportunity to attend college through the GI Bill of Rights and subsequently qualified for academic and professional positions in geography.

Not surprisingly, the younger generation was much more oriented to new research technologies than was the older generation; indeed, many in the younger generation had worked directly on the development and implementation of new technologies for the Armed Services or the Federal Government prior to entering academic life. The war effort and its aftermath saw a rapid influx of technological development in mapping, remote sensing, aerial photography, and statistics. It was during this period also that computers began to be developed, which within two decades would become a basic research tool for the discipline.

2.3. The Early Cold War Period

The decades following the end of World War II were years of profound change for American society and the discipline of geography. After World War I, the U.S. withdrew into isolationism and did not play a major role in global affairs until the onset of World War II two decades later. The situation after the Great Depression and World War II , however, was very different. The war left the U.S. and the Soviet Union as the world's leading powers. The U.S. adopted an activist approach to foreign policy (Trubowitz 1998). Within a few years, as Sir Winston Churchill put it, an Iron Curtain had descended across the continent of Europe. The Cold War, pitting the interests of the capitalist and democratic U.S. and its allies against those of the communist Soviet Union and its satellites, was in full swing.

After transition to peacetime following World War I, federal spending had shrunk dramatically, but the federal budget increased in size after the end of World War II. In keeping with the U.S.'s activist role in global affairs, many post-World War II federal expenditures were associated with military and defense-related activities. Even before the war ended, leaders in government, the armed services, and private industry advocated a large-scale role for the federal government in research. Mazuzan (1994, 3) noted that

> There had been numerous, if modest, government-science interactions throughout the history of the Republic, but the Second World War vastly intensified that environment. Not only was government support of scientific endeavors sharply escalated, but the relationships among government agencies, universities, private foundations, and industry were altered in ways that disallowed a return to prewar times. The war greatly strengthened, for example, the link between the nation's universities and the government.

These changed linkages had considerable impact on geography as a discipline. While large numbers of professional geographers left government service after the war to return to academic life, some stayed on in federal service. As Hart (1979, 109) pointed out,

> Right through the 1950s the geographic communities of Washington and Academe were linked by close ties of professional respect and personal friendship that had been forged by working side by side during the war.

Of the numerous federal programs established or expanded after World War II, many involved geography, directly or indirectly. Research programs involving the development of atomic energy, aviation, space, electronics, computers, and other technologies provided funding for many projects, including those supporting the work of geographers. Foreign policy activism brought about renewed demand for foreign area specialists, and the Department of Defense and other agencies funded numerous geographic research projects and doctoral dissertations outside North America. Much of this attention went into the development of new cartographic and photogrammetric techniques. As Cloud (2002, 261) states,

> Most of the fundamental technologies of contemporary American cartography were devised in the last half of the twentieth century and shaped by the exigencies and opportunities of the Cold War.

These included the development of global positioning systems, remote sensing, and intelligence mapping.

The Cold War also directly affected the process of education. The period after World War II was one of dramatic expansion of college and university enrollments. The GI Bill of Rights provided government support for higher education for millions of ex-military personnel, creating a rapid increase in faculty positions. At the precollege level, educators raised concerns that American public education was archaic, outmoded, and insufficiently organized to meet the needs of the postwar world. In particular, it was argued that American public schools were failing to teach science and mathematics adequately, and that continued failure to train future scientists and engineers would hold back the U.S. in its effort to maintain global hegemony (Conant 1959). Accordingly, the National Defense Education Act provided large levels of federal support for education and educational projects, including many by professional geographers.

Immediately after the war ended, many economists and politicians predicted that the economic boom of the war years would end and that the U.S. would slide back into its prewar depression. These fears proved unfounded. The U.S. and global economies experienced a long, steady economic boom encompassing much of the 1950s and 1960s. To a considerable extent, this boom was associated with the application of new technologies originally developed for military purposes to civilian uses. Television and computers were two of the major technological innovations of the postwar period coming into common use. Many of the technological advances associated with the war and the Cold War became useful in the production of civilian goods and services and in nonmilitary programs. For example, satellite technology, which was originally designed to track movement of foreign military personnel, became useful in weather forecasting, planning, analysis of land use change, and many other nonmilitary activities, many of which directly involved geographers and geographic research.

The discipline itself—its ranks augmented greatly as the post-World War II generation completed their educations and moved into academic and nonacademic positions—also underwent considerable transformation. During the 1950s and 1960s, the functionalist, areal-differentiation approach to geography associated with Richard Hartshorne (1939) gave way to the quantitative revolution. Geographers began to adopt a more nomothetic, scientific approach in their research. For many, the goal of geographic research was to develop models and theories applicable to a wide variety of situations and places. This quantitative and scientific approach to geographic research was enhanced greatly by new technologies that enabled them to use complex statistical methods and mathematical models, to obtain data from satellite images and other remote sensing technologies, and to manipulate large datasets. Government support played an important role in the development and diffusion of these technologies.

2.4. The Later Cold War

By the late 1960s, the quantitative revolution had been completed, and many geographers regarded themselves as practitioners of scientific methods. As the 1960s gave way to the 1970s and 1980s, however, a variety of social trends as well as trends in the discipline caused many to rethink the dominant quantitative paradigm.

Many of these changes were driven by fundamental changes within U.S. society. The decade of the 1960s was marked by the civil rights movement, opposition to an unpopular war in Vietnam, and an awareness of large-scale environmental degradation. As Americans became more and more aware of racial discrimination, foreign policy concerns, and environmental problems, many came to question the Cold War paradigm that had dominated American life since the end of World War II. By 1970, the U.S. had moved from an extrovert to an introvert phase in its foreign policy (Trubowitz 1998), and large-scale military expenditures and activities were scrutinized carefully and questioned closely. No longer did most Americans blindly accept the dominance of what President Dwight D. Eisenhower had called the "military-industrial complex." This skepticism deepened in the 1970s, as the booming economy of the 1950s and 1960s gave way to economic recession and the energy crisis of that decade unfolded.

These social changes had profound effects on the discipline of geography and on the relationships between professional geography and the federal government. Prior to the mid-1960s, research funds for foreign area studies and quantitative analyses were plentiful. In many departments, doctoral students in geography had been expected or required to develop expertise in a region of the world outside of the U.S. and to do foreign-area fieldwork in order to complete dissertations and other research projects. Many of these funds were provided by the Department of Defense, the Armed Services, and related agencies. By the 1970s, funds for these programs had been cut considerably, in response to reduced foreign policy activism and economic recession. The consequent lack of available funds induced many doctoral students to do more theoretical research or to undertake local rather than foreign field work to complete their doctoral degree requirements. At the same time, the federal government adjusted its priorities. Military spending declined, while domestic spending increased. For example, the Environmental Protection Agency (EPA) was founded in 1970. This agency became an important source of funding and information for geographic research and technology, and as documented below, the EPA provided substantial support for geographers.

Geographers who were concerned about the state of the world began to question more openly the impacts of their research on society. Some began to express concern that geographic research and its applications could reinforce racism, inequality, and environmental degradation throughout the world. Such concerns were raised with particular respect to the use of technology. For example, some argued that remotely sensed data could be used by unscrupulous dictators and rapacious corporations to reinforce the suppression of persons in less developed countries, or to promote development projects that when completed would accelerate declines in environmental quality (Blaut 1979; Peet 1977). As these concerns became more widespread, many began to question the neoclassical, positivist paradigm that had dominated geography in the early Cold War years, and many embraced alternative intellectual approaches to geographic thought, including Marxism, structuralism, and world-systems theory. These approaches emphasized the historical understanding of observed inequalities in society.

Despite these concerns, during the 1970s and 1980s, geographic technology continued to develop. Many technologies that are standard in geographic research and communications today, including the Internet, basic geographic information systems, and microcomputers, were developed during the later Cold War period. Interestingly, the space program proved to be a boon to geographic research even after its heyday in the 1960s. During the 1960s, the federal government began investigating the use of satellites to observe the earth's surface routinely. For several reasons, Congress was slow to allocate funds for such a program. These reasons included concern about Soviet reaction to surveillance and doubt about the long-run economic benefits of such a program above and beyond military uses. Nor was there a clear institutional structure to manage and distribute data. The National Aeronautics and Space Administration (NASA), the National Oceanic and Atmospheric Administration (NOAA), the Department of the Interior, and other agencies all had jurisdiction over various aspects of the earth surface observation program (Williamson 1997).

Once these problems were overcome, the first Earth Resources Technology Satellite (ERTS-1) was launched and in the same year, construction of the EROS Data Center near Sioux Falls, South Dakota—a facility that today employs dozens of geographers—began. The ERTS project was soon renamed Landsat. Today, up to thirty years worth of Landsat data are available for many locations, providing valuable information to researchers interested in temporal change in land use, land cover, and other applications. Additional Landsat satellites were launched later in the 1970s. A variety of projects involving the use of these Landsat data were funded by the federal government, which also made Landsat data available to researchers. In the 1980s, the Reagan administration undertook efforts to privatize the system. Of course, Landsat data have been used in many geographic studies ever since.

Another important federal effort in the 1980s was the establishment of the National Center for Geographic Information and Analysis (NCGIA). NCGIA was established with funding from the National Science Foundation (NSF) in 1988 (Goodchild 1992). The establishment of NCGIA encouraged many geographers to become more heavily involved in GIS, which had previously been regarded as relatively esoteric and inaccessible. NCGIA was charged with promoting research in spatial analysis and spatial statistics, spatial relationships and database structures, artificial intelligence and expert systems, visualization of spatial data, and social, economic, and institutional issues associated with GIS technology. Many frequently cited and important books and papers were published by AAG members and other geographers under the auspices of NCGIA (i.e., Goodchild and Gopal 1989). The establishment of NCGIA induced several prominent geography departments to focus their attention on geographic information science. Geography departments including the University of California at Santa Barbara, Ohio State University, the State University of New York at Buffalo, the University of South Carolina, and Syracuse University placed increasing emphasis on geographic information science in order to secure the Center, which was eventually located jointly at Santa Barbara, Buffalo, and the University of Maine. This emphasis on geographic information science would eventually diffuse across the discipline, with GIS courses developed and taught to wider audiences (Frank et al. 1991).

2.5. After the Cold War

The Cold War came to an abrupt end in the late 1980s and early 1990s, in association with the collapse of Communism in the Soviet Union and eastern Europe. The end of the Cold War resulted in a new global political-economic order (Taylor 1996; Cohen 1999). For the previous forty years, international geopolitics had been conceptualized in terms of conflict between East and West. The collapse of Communism ended this conflict, and a new world order associated with globalization began (O'Tuathail and Shelley 2003). The General Agreement on Tariffs and Trade and the World Trade Organization (its more comprehensive successor), the North American Free Trade Agreement, the expanded European Union, and other international organizations achieved expanded influence in an increasingly globalized world.

Recognizing these changes in the global economy, many geographers began to examine the dimensions, impacts, and effects of globalization. Many engaged in studies of migration and population change, international trade, capital flows, and information exchange. Others examined the effects of globalization on local places, economies, and physical environments. These efforts were enhanced by the large-scale declassification of millions of satellite images and aerial photographs by the U.S. government. In 1995, all of the imagery that had been obtained through three Cold War satellite reconnaissance programs were formally declassified and made available to researchers, and plans were made to declassify additional images from other programs (McDonald 1996).

These and other research efforts by geographers were informed by new and exciting developments in technology. The Internet revolutionized geographic research, both by facilitating instantaneous long-distance communication and also by encouraging an exponential increase in the availability of geographic data. Dramatic improvements in GIS technology took place, allowing geographers to obtain, manipulate, interpret, and map data with increasing ease. Interest in foreign areas, which had declined during the 1970s and 1980s, began to increase once again, especially after the terrorist attacks on the World Trade Center and the Pentagon in September 2001.

3. CHANGES IN GEOGRAPHIC TECHNOLOGY AS REFLECTED IN GEOGRAPHIC PUBLICATION

How have the changes in geographic technology and the changing relationships between geographic technology, the global economy, and federal government priorities affected the professional priorities and achievements of AAG members? To a considerable extent, the prestige of any academic organization and its members is reflected in the quality, quantity, and influence of published research. Especially since the 1960s, consistent publication has been required for tenure and promotion at most academic institutions. Because the research of many geographers has been supported by federal funds, we expected that the relative importance of various federal agencies and programs supporting geographic research would change over the years in a manner consistent with the time line described above. To what extent have various federal agencies and programs increased or decreased their support of geographic research over the past century, and especially since World War II?

In order to address this question, we undertook systematic examination of published research in geographic journals that was supported by various federal programs (Table 3.1). We first examined the AAG's two major journals, the *Annals* and *The Professional Geographer*, along with the *Geographical Review*, which has been regarded as a premier, "mainstream" geographic journal for many years.

3.1. Federal Support for Research In Geography Journals in the 1950s and 1960s

We began our analysis by examining all articles published in the *Annals of the Association of American Geographers*, *The Professional Geographer*, and the *Geographical Review* in the fifty-year period between 1950 and 1999. This 50-year time horizon was chosen so that each of the five decades could be compared readily. For each of these journals, we recorded author acknowledgement of federal government support, along with the identity of the agency or program responsible for that support. In doing so, we grouped the acknowledgments into decades. Thus the first two periods (the 1950s and 1960s) are part of the Early Cold War period; the next two (the 1970s and 1980s) are part of the Later Cold War; and the final period (the 1990s) reflects post-Cold War geographical research. It should be noted that *The Professional Geographer* at first served a function more analogous to that of the *AAG Newsletter* today. *The Professional Geographer* did not begin to publish research papers until the mid-1960s, and the first research papers acknowledging federal support and published in *The Professional Geographer* appeared in 1967. Thus the discussion below reflects content analysis of the *Annals* and the *Geographical Review* from 1950 to 1966, and all three journals between 1967 and 1999.

During the 1950s, the dominant federal program associated with geographic research was the Office of Naval Research (ONR). The ONR was established by act of Congress in 1946 to provide federal support for research of relevance to the U.S. Navy: its mission is to plan, foster and encourage scientific research in recognition of its paramount importance as related to the maintenance of future naval power, forced entry capability, and the preservation of national security.

In a speech at the University of Illinois in 1947, Captain Robert Conrad, Director of Planning for the ONR, talked about computers and their potential application:

[World War II] brought great advances in the art of computation, and the future is now bright for the mathematicians and those who must use mathematics in their studies. A machine called the ENIAC, built at the University of Pennsylvania for the Army Ordnance Department, can compute the instantaneous position of a projectile and determine its point of fall in a shorter time than the projectile is actually in the air. This machine occupies a room 30 feet by 50 feet and weighs 30 tons. It can multiply two ten-digit numbers in 1/360th of a second. Both the Army and Navy are supporting the research and development of still faster and more flexible machines. One big job for a computing machine is in weather forecasting. You all are familiar with the weather maps which form the basis for the daily forecasts. We know something of the mathematics of meteorology, but it takes six months to do a complete mathematical job of predicting tomorrow's weather from today's observations. A computing machine might do it in a few minutes. Whether the weatherman would be more reliable if he had such a machine is open to debate, but at least he could pass the buck to a machine whenever it rains on your picnics. (Conrad 1947)

Table 3-1. Federal funding for geographic research: 1950-99 by major sources and publications in Annals of the Association of American Geographers, The Professional Geographer, *and* Geographical Review.

	1950s	1960s	1970s	1980s	1990s
NSF	1	24	36	64	89
SSRC	8	13	13	2	3
Fulbright	6	7	4	4	6
ONR	35	32	4	0	0
Army	2	6	0	0	2
USAF	0	4	3	0	0
NASA	0	1	4	4	7
USGS	1	2	3	1	1
State Dept	2	0	4	1	1
EPA	0	0	1	4	3
Other	2	13	20	21	14

Source: Compiled by Jeanette Gara from issues of the journals, 1950-1999.

During the Cold War, many of the projects under the auspices of this mission involved geography and geographers. In the 1950s, 61 articles acknowledging federal support appeared in the *Annals* and *Geographical Review*. Of these 61 articles, 35, or nearly 60 percent, were sponsored by the ONR. Many ONR-sponsored projects involved research conducted on the oceans and in ports, for example studies of the ports of Bordeaux (Weigand 1955), Dar es Salaam (Hance 1958) and Genoa (Rodgers 1958). Procedures to study and map land form surfaces were also sponsored by ONR (Hammond 1958; 1964). Indeed, an address commemorating the tenth anniversary of the establishment of ONR mentions geography and geographers specifically:

> The Navy needs to know about environmental conditions in all parts of the world to prepare for possible land, sea and air operations; it needs to apply geography to military problems. ONR's geographic research on little-known areas, particularly coastal areas, has been used by both the Navy and the Marine Corps. New information obtained through geographic research provides a source of reliable, basic knowledge, ready and available as the Navy's needs arise...As a result of ONR's ceaseless efforts, geographers are thinking increasingly in terms of research that will benefit the Navy and are prepared to apply their skills and knowledge to the solution of naval problems. (Ten Years of Naval... 1956)

Not all studies published in the *Annals* and *Geographical Review* fit these criteria. Some appeared to have nothing to do with the presumed priorities of the Navy or foreign policy: examples included studies of farming in the Great Plains (Kollmorgen and Jenks 1958) and transportation across the Sahara Desert (Thomas 1952). In explaining this apparent anomaly, Pruitt (1979) pointed out that the ONR in the 1950s did not restrict its funding to topics of specific interest to the Navy:

> There was no restriction on topics, no favored fields, and no requirement for research to be applicable to naval problems: the only demand was that the work have promise that it would make a significant contribution to the world's store of geographic knowledge. (Pruitt 1979, 104)

Pruitt also pointed out that ONR was one of the first large-scale federal programs that provided funding to support geographic field work and writing; previously,

> a person was expected to pay the costs of the studies he was conducting, and to finance his own travel, photography, drafting, and attendance at professional meetings. (Pruitt 1979, 103)

Thus ONR was a pioneer in moving geographers toward large-scale, problem-oriented research projects. Most of these studies were idiographic studies of individual places, in keeping with the prevailing Hartshornian view of geography at that time. Many of the articles supported by ONR, however, dealt with quantitative methods, more general theory, and theory as potentially applicable to research technology. For example, ONR supported research on map classification techniques (Kuchler 1954; Weaver 1956) and cartographic symbolization (Williams 1958); such research on mapping techniques presaged the later development of GIS and indeed can be considered early efforts at the development of modern-day geographic information science.

The 26 articles not supported by the ONR were funded by a variety of other agencies. Nine were supported by the Social Science Research Council (SSRC) and six by the newly established Fulbright program, which encouraged scholarly exchange between the U.S. and other countries. All of these papers dealt with specific places or themes within foreign cities or countries, many of which were places of considerable strategic importance at the time. These included studies in China (Trewartha 1952), Japan (Eyre 1955), and Russia (Jackson 1959). The U.S. Army, the Department of State, and the USDA were among the other supporters of research in the *Annals* and *Geographical Review* during the 1950s.

The federally-supported papers of the 1950s were heavily oriented to economic and urban geography and to studies of specific places. Of the 61 articles, nine (16 percent) were devoted to technical subjects. Most, including the studies of map classification and cartographic symbolization, were devoted to cartography. Of the 52 nontechniques articles, only seven (12 percent) were in physical geography. Thirty-eight of the remaining 45 papers dealt with conditions or analyses of specific places, while only seven were devoted to more general theory. Clearly, federal funding in the 1950s was reflected in published contributions to the "world's store of geographic knowledge."

During the 1960s, the three journals published a total of 94 papers supported by federal government funding. The ONR, with 32 acknowledgments, was again the most common source but its proportion dropped from more than half in the previous decade to more than one-third Pruitt (1979, 107) pointed out that

> By the late 1960s the ONR program had begun to lean heavily toward coastal research and physical-geomorphological subjects, in addition to the remote sensing and geographic data management studies.

This change in philosophy is reflected in the fact that many ONR-sponsored projects reported in geography journals involved coastal research (i.e., Psuty 1965; Dolan 1966; Snead 1967). The ONR also continued to support more theoretical research that advanced quantitative methodologies (i.e., Byrne 1964; Dacey 1966).

NSF, which had had only a minimal impact on funded geographic research in the 1950s, began to play a significant impact in the 1960s and has been the dominant program supporting research in the *Annals, The Professional*

Geographer, and *Geographical Review* ever since. Members of Congress advocated formal federal support for scientific research as early as 1942, and advocacy for federal support intensified after the war. The National Science Foundation was established by act of Congress in 1950 as an independent agency of the federal government. Its mission is "to promote the progress of science; to advance the national health, prosperity, and welfare; and to secure the national defense." By 2003, the Foundation had an annual budget of over $4 billion and disbursed more than 10,000 new awards annually.

Interestingly, one of the major debates associated with the establishment of NSF was whether geography should play a role in the disbursement of research grants. In other words, should the amount of funding for scientific research be determined on a geographic basis? Vannevar Bush, who had been the scientific advisor to President Roosevelt during World War II, argued against quotas for the amount of funds distributed in different regions. Bush argued that funds should support "the best basic research in the colleges, universities, and research institutes, both in medicine and the natural sciences" regardless of where these institutions were located (Bush 1945). A related debate involved the extent to which the government would supervise and oversee the distribution of funds and the activities of the Foundation. In fact, an earlier proposal, supported by Bush, to establish NSF was vetoed in 1947 by President Harry S Truman on the grounds that it did not include any public accountability or scrutiny over the agency's activities and funding priorities. NSF was established three years later, after this issue of public control was resolved (Mazuzan 1994). The director is now appointed by the President subject to confirmation by the Senate.

Another major debate within NSF, and within the scholarly community more generally, during the 1950s was the extent to which support for scientific research should be limited to the natural and physical sciences. Although many "hard" scientists opposed including social science in NSF, a program in "sociophysical" sciences was established in 1955. This program included "mathematical social science, human geography, economic engineering, statistical design, and the history, philosophy, and sociology of science" (Mazuzan 1994, 10). In 1958, an office of social science was formally established, and social science research conducted under the auspices of NSF was required to meet "rigorous standards of objectivity, verifiability, and generality." This office eventually became the Social and Behavioral Sciences Division of NSF, and the Geography and Regional Science Program, which is a major source of federal funding for geographic research, is housed within this division.

By no means has NSF support for geographical research been limited to the Geography and Regional Science Division. Many other divisions and programs have also supported geographical research, including programs in Climate Dynamics, Cultural Anthropology, Decision, Risk and Management Science, Ecology, Ecosystems Studies, Law and Social Science, and Natural and Technological Hazards Mitigation. Since the 1960s, NSF has also sponsored numerous doctoral dissertation grants. Many of these grants supported research projects that led to the successful completion of doctoral dissertations, and the results of many of these projects eventually found their way into the geographic literature.

The fact that geography was not fully established within NSF until the 1960s is illustrated by the rapid increase in NSF acknowledgment after 1960. In the 1950s, NSF was acknowledged only once in the *Annals*, *The Professional Geographer*, and *Geographical Review*. This acknowledgment was not for research support but rather for NSF's support of a meeting of the International Geographical Union (Hitchcock 1957). By the 1960s, however, twenty-four (26 percent) of the 94 papers acknowledging federal support and published in the three journals were NSF-supported. Many NSF-supported papers were foreign area studies (i.e., Sopher 1963; Vandermeer 1968), whereas NSF also supported research in physical geography (Nelson 1965; Winslow 1968) and in techniques (Tobler 1962; Dacey 1964). The remaining 38 papers were supported by agencies other than ONR and NSF. Thirteen were supported by SSRC and seven by Fulbright grants. As was the case in the previous decade, most of these papers were foreign area studies (i.e., Winsberg 1964; Dickinson 1969).

The subdisciplinary breakdown of federally funded projects varied somewhat from the previous decade. Eleven (12 percent) of the 94 papers were specifically devoted to techniques. Many were oriented to cartography and quantitative methods, but one study, supported by the National Research Council, investigated the potential of computers in geographic research (Kao 1963) and another, supported by NASA, examined remote sensing (McCoy 1969). The percentage of papers devoted to physical geography increased to 26 percent (24 papers). The percentage of papers devoted to specific places declined substantially, with a much larger number of papers devoted to theory in both human (i.e., Wolpert 1965; Janelle 1969) and physical geography (McIntire and Walker 1964).

3.2. Federal Support for Geographic Research in the 1970s, 1980s, and 1990s

In the 1970s, the three journals published a total of 100 papers acknowledging federal support. Given the fact that *The Professional Geographer* was now publishing a significant number of research articles, it is clear that the percentage of papers published in the three journals that were supported by federal funding had declined since the previous decade—a fact perhaps not surprising given the country's sluggish economy and transition to introversion in foreign policy during the 1970s. Twenty-six (26 percent) were supported by NSF. As in the 1960s, NSF-supported projects resulting in papers in the *Annals*, *The Professional Geographer*, and *Geographical Review* covered the entire spectrum of geographic research, including foreign area studies (Winberry 1974; Leinbach 1975) and physical geography (Kalnicky 1974; Helgren and Butzer 1977). NSF support for articles associated with advances in geographic theory, methods, and techniques became increasingly prevalent, including contributions to diffusion theory (Spector and Brown 1976) and population (Vining and Luow 1978). As was the case previously, however, the three journals continued to emphasize contributions to systematic and regional geography as opposed to articles specifically devoted to advancing the development and use of specific technologies.

The SSRC was next with 13 articles. The ONR and Fulbright program, however, dropped to only four each. The others represented a wide variety of agencies and programs. The U.S. Department of the Interior became an increasingly important source of funds during the 1970s, supporting seven published papers. Given the mission of the Interior Department, it is not surprising that most dealt with domestic topics (Ter Jung and Louie 1973, Vreeken 1975). In general, the decade of the 1970s reflects a shift from foreign to domestic interests and agencies. Seven of the 100 papers in the 1970s were specifically technical, but of course, many others involved application of remote sensing, mapping, and quantitative methods to geographical research problems. Physical geography accounted for 26 of the 93 nontechniques papers (27 percent).

In the 1980s, a total of 103 articles in the *Annals*, *The Professional Geographer*, and *Geographical Review* acknowledged federal support. By far the largest number of acknowledgments were to NSF, which supported 64 (62 percent) of all papers acknowledging federal support, including 73 percent of those in the *Annals*. The topics analyzed by NSF-supported researchers included a very wide range of issues. Those dealing with policy-related issues, including analysis of relationships between the U.S. and other countries, began to increase in number. Examples included studies of foreign direct investment in the U.S. (McConnell 1980), regional inequalities (Clark 1980), and refugee resettlement (Desbarats 1985). Physical geography was also well represented, including studies of sedimentation (Knox 1987), permafrost (Nelson 1986), coastal zone dynamics (Sherman 1988), and dendroclimatology (Stahle and Hehr 1984). Significantly, an increased percentage of NSF-sponsored research papers were devoted specifically to analysis of technical issues and development. For example, NSF sponsored research leading to Goodchild and Mark's theories involving the use of fractals (Goodchild and Mark 1987), methods of representing geographic space (Peuquet 1988), and images of maps (Steinke and Lloyd 1985). The research reported in these and other NSF-supported papers would eventually contribute significantly to the development of geographic information science in the 1990s and the early twenty-first century.

The remaining 39 articles were sponsored by some twenty different federal agencies, with no agency responsible for more than four. Those with four citations each included the EPA, NASA, and the Fulbright program. The Office of Naval Research, which had supported so much geographic research between 1950 and the early 1970s, now supported no papers at all. Only six of the 103 papers were specifically technique-oriented. Interestingly, only a minority (34 percent) were devoted to analysis of a specific place or region, continuing a pattern of decline from the 1950s in which nearly all research reported was oriented to specific places.

The pattern of domination by NSF continued into the 1990s. The total number of federally supported articles rose to 126. Interestingly, federal support increased in the *Annals* from 52 articles in the 1980s to 75 in the 1990s, whereas acknowledged federal support for research in *Geographical Review* declined from 26 articles to 14. NSF remained the largest supporter of geographic research, with a total of 89 acknowledgments or 70 percent of the total. These included numerous studies in human geography (i.e., Murphy 1992; Pandit 1997; Dear and Flusty 1998; O'Loughlin 1998) and in physical systems and processes (Pope et al. 1995; Walters and Winkler 1999). NSF support of papers related specifically to geographic methodologies, technologies, and techniques continued to increase as well (i.e., MacEachren 1992; Golledge et al. 1995; Blaut 1999). The remaining 37 papers (30 percent) were supported by many different federal agencies and programs. The most frequently acknowledged federal sources were NASA (7 acknowledgments) and the Fulbright program (6).

A summary of the trends over the second half of the twentieth century is in order (Figure 3.1; see Table 3.1). Over the period 1950-1999, the number of articles in the major disciplinary journals increased steadily, more than doubling between the 1950s and the 1990s. Interestingly, most of this increase was accounted for by the *Annals*. Whether acknowledgment of federal support was a factor in editors accepting submissions is an interesting but unanswerable question. The amount of federal support for papers in *The Professional Geographer* also increased steadily—not surprisingly, given the fact that *The Professional Geographer* evolved from a newsletter to a research journal over the course of the study period. On the other hand, federal support became less

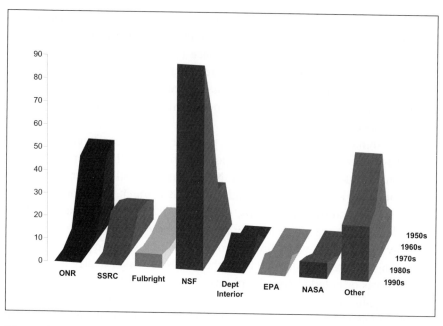

Figure 3-1. The change in federal funding for geographic research: 1950-99 by major sources and publications in Annals of the Association of American Geographers, The Professional Geographer, *and* Geographical Review.

critical to publication in the *Geographical Review*, perhaps because less funding was available for foreign area research and for research on cultural topics.

The sources of support also changed dramatically. The Office of Naval Research was dominant in the 1950s and 1960s; the National Science Foundation in the 1980s and 1990s with the 1970s as a transitional period. Small but consistent numbers of papers were supported by agencies such as NASA, the EPA, the Fulbright program, and the Social Science Research Council.

Although ONR, NSF, NASA, EPA, the Fulbright program, and the Social Science Research Council were the dominant federal agencies supporting research by geographers in the last half of the twentieth century, other agencies also contributed. Two deserving of mention are the National Institutes of Health (NIH) and the National Endowment for the Humanities (NEH). NIH supported research in various aspects of demography, medical geography, and epidemiology. Such research tended to be published outside of the AAG's journals and *Geographical Review*, although research by Meade (1976, 1977) was supported by NIH. Funds from NEH tend to go primarily to supporting the publication of books, atlases, and larger projects that are outside the scope of journals, and thus the survey of the *Annals*, *The Professional Geographer*, and *Geographical Review* underrepresents the importance of NEH's contributions to geographical research. Among the important contributions of NEH to geographical scholarship in recent years are atlases of congressional representation by Martis (1985, 1989) and the History of Cartography Project, a multivolume effort also sponsored by NSF (Harley and Woodward 1987; Woodward and Lewis 1998). Geographers employed directly by federal agencies, and those working for federally sponsored research organizations have contributed substantially to the geographic literature in many ways as well. Examples include the Office of the Geographer, U.S. Department of State (Wood 1996, 1999) and Oak Ridge National Laboratory (i.e., Wilbanks and Kates 1999).

4. CONCLUSION

Over the past hundred years, the history of American geography has been intertwined with the history of the U.S. and its changing role in the global economy. The relationship between professional geographers and the federal government has changed in accordance with landmark events in U.S. history.

The AAG was founded during the Progressive Era. For the first time in U.S. history, the federal government began to tap the expertise of scholars and scientists to address important social, economic, and environmental problems. Many prominent early members and leaders of the AAG lent their expertise to the U.S. government during this period. Travel, field work, and other research by professional geographers prior to World War II was generally supported privately.

Not until after World War II did the federal government support geographic research on a sustained basis. During the war, hundreds of professional geographers served in the armed forces and hundreds of others provided expertise to the war effort as civilians. Close ties were forged between geographers in federal service and those in academic life. At the same time, the government began to fund geographic research on an ongoing basis. Agencies such as the Office of Naval Research and later the NSF began to support dozens of research projects annually, enabling professional geographers to "contribute to the world's storehouse of geographic knowledge." Through these and other agencies including NASA, NOAA, EPA, the Department of the Interior, the National Endowment for the Humanities, and many others, geographers wrote proposals which, once funded, provided them with support for research, fieldwork, travel, and publication. The nature of research funded by federal programs has evolved in accordance with the evolution of geography as a discipline, and with the changing priorities of the federal government in a rapidly changing society.

Without a doubt, the federal government will continue to play an important role in the development of geography and geographic research in the years ahead. The importance of geography and geographic approaches to important social and environmental problems has become increasingly, if belatedly, apparent to policymakers and government officials. As Murphy (2003, 3) noted, geographers can and must continue to show

> how advanced geographical inquiry can shed critical light on the growing gap between rich and poor, the potential consequences of climatic change, the impacts of humans on the environment, the nature and implications of ethnic conflict, and much, much more.

He goes on to write that

> The more that geography becomes part of the public debate over where our society has come from and where it is going, the more geography will be strengthened, as will society at large. (p. 5)

After the tragedy of September 11, 2001, the AAG produced an NSF-funded study of the geographical dimensions of terrorism, focusing upon how a geographic perspective can shed light on the causes of terrorism, the conceptualization of possible terrorist threats as hazards, and the role of geographic data and geospatial analysis in predicting the effects of terrorist activities (Cutter et al. 2003).

The AAG and its membership must recognize the central role of geography in understanding the problems of a complex and interdependent world. As it does so, and as government agencies and elected officials become increasingly aware of the importance of geographic thinking, the federal government will have an increasingly important and symbiotic role in the development of geography in the years ahead.

ACKNOWLEDGEMENT

The research assistance of Jeanette Gara and Melissa Gray and helpful comments by Stan Brunn, J. W. Harrington, and Don Janelle on earlier drafts of this paper are gratefully acknowledged.

REFERENCES

Blaut, J. (1979). The Dissenting Tradition, Annals of the Association of American Geographers, 69: 157-64.
Blaut, J. (1999). Maps and Spaces, The Professional Geographer 51: 510-16.

Bush, V. (1945). Science: The Endless Frontier. Washington, DC: GPO.

Byrne, J.V. (1964). An Erosional Classification for the Northern Oregon Coast, Annals of the Association of American Geographers 54: 329-35.

Clark, G.L. (1980). Capitalism and Regional Inequality, Annals of the Association of American Geographers 70: 226-37.

Cloud, J. (2002). American Cartographic Transformations During the Cold War, Cartography and Geographic Information Science 29: 261-82.

Cohen, S.B. (1999). Geopolitics in the New World Era: A New Perspective on an Old Discipline. In Demko, G.J. and Wood, W.B. (Eds.) Reordering the World: Geopolitical Perspectives on the 21st Century, 40-68. Boulder, CO: Westview.

Colby, C.C. (1941). Land Classification in the United States. Washington, DC: National Resources Planning Board.

Conant, J.B. (1959). The American High School Today. New York: McGraw-Hill.

Conrad, R. (1947). The Navy Looks Forward with Research. Address on "Navy Day" to the University of Illinois at Urbana, October 27. http://www.onr.navy.mil/onr/hismis.htm

Cutter, S.C., Richardson, D.B.R., and Wilbanks, T.J.(Eds.) (2003). The Geographic Dimensions of Terrorism. New York: Routledge.

Dacey, M.F. (1964). Modified Poisson Probability Law for Point Pattern: More Regular Than Random, Annals of the Association of American Geographers 54.

Dacey, M.F. (1966). A Probability Model for Central Place Location, Annals of the Association of American Geographers 56: 559-65.

Dear, M. and Flusty, J. (1998). Postmodern Urbanism, Annals of the Association of American Geographers 88: 50-73.

Desbarats, J. (1985). Indochinese Resettlement in the United States, Annals of the Association of American Geographers 75: 522-38.

Dickinson, J.C. (1969). The Eucalypt in the Sierra of Southern Peru, Annals of the Association of American Geographers 59: 294-307.

Dolan, R. (1966). Beach Changes on the Outer Banks of North Carolina, Annals of the Association of American Geographers 56: 699-711.

Eyre, J. (1955). Water Controls in a Japanese Irrigation System, Geographical Review 45: 197-216.

Frank, A.U., Egenhofer, J.J., and Kuhn, W. (1991). A Perspective on GIS Technology in the Nineties, Photogrammetric Engineering and Remote Sensing 57: 1431-36.

Golledge, R. (1992). Route Learning and Relational Distances, Annals of the Association of American Geographers 82: 223-45.

Goodchild, M. (1992). The National Center for Geographic Information and Analysis, Photogrammetric Engineering and Remote Sensing 58: 1141-43.

Goodchild, M. and Gopal, S. (1989). Accuracy of Spatial Databases. London: Taylor and Francis.

Goodchild, M. and Mark, D. (1987). Fractals, Annals of the Association of American Geographers 77: 265-78.

Hammond, E.H. (1958). Procedures in the Descriptive Analysis of Terrain, Final Report. Madison, WI. ONR Project NR 387-015.

Hammond, E.H. (1964). Analysis of Properties in Land Form Geography: An Application to Broad-Scale Land Form Mapping, Annals of the Association of American Geographers 54: 11-19 and Map Supplement.

Hance, W.A. and Van Donegan, I.S. (1958). Dar es Salaam: The Port and Its Tributary Area, Annals of the Association of American Geographers 48: 45-68.

Harley, J.B. and Woodward, D. (1987). Cartography in Prehistoric, Ancient and Medieval Europe and the Mediterranean. Chicago: University of Chicago Press.

Hart, J.F. (1979). The 1950s, Annals of the Association of American Geographers 69: 109-14.

Hartshorne, R. (1939). The Nature of Geography. Lancaster, PA: Association of American Geographers.

Helgren, D and Butzer, K. (1977). Paleosols of the Southern Cape Coast, South Africa: Implications for Laterite Definition, Geographical Review 67: 430-45.

Hitchcock, C. (1957). The Eighteenth International Geographical Conference, Rio de Janeiro, 1957, Geographical Review 47: 118-23.

Hofstadter, R. (1955). The Age of Reform: From Bryan to FDR. New York: Vintage Books.

Jackson, W.A.D. (1959). The Russian Non-Chernozem Wheat Base, Annals of the Association of American Geographers 49: 97-109.

Janelle, D. (1969). Spatial Reorganization: A Model and Concept, Annals of the Association of American Geographers 59: 348-64.

Kalnicky, R.A. (1974). Climatic Change Since 1950, Annals of the Association of American Geographers 64: 110-12.

Kao, R. (1963). The Use of Computers in the Processing and Analysis of Geographic Information, Geographical Review 53: 530-47.

Knox, J. (1987). Valley Floor Sedimentation, Annals of the Association of American Geographers 77: 224-44.

Kollmorgen, W. and Jenks, G. (1958). Suitcase Farming in Sully County, South Dakota, Annals of the Association of American Geographers 48: 27-40.

Kuchler, A.W. (1954). Vegetation Mapping in Europe, Geographical Review 44: 91-97.

Leinbach, T.R. (1975). Transportation and the Development of Malaya, Annals of the, Association of American Geographers 65: 270-82.

MacEachren, A. (1992). Conceptual Knowledge Acquisition from Maps, Annals of the Association of American Geographers 81: 245-75.

Martin, G.J. (1980). The Life and Thought of Isaiah Bowman. Hamden, CT: Archon Books.

Martin, G.J. and James, P.E. (1993). All Possible Worlds: A History of Geographical Ideas. 3rd ed. New York: Wiley.

Martis, K.C. (1985). The Historical Atlas of United States Congressional Districts, 1789-1983. New York: Free Press.

Martis, K.C. (1989). The Historical Atlas of Political Parties in the United States Congress, 1789-1989. New York: Macmillan.

Mazuzan, G.T. (1994). The National Science Foundation: A Brief History, National Science Foundation, Report NSF 88-16.

McConnell, J.E. (1980). Foreign Direct Investment in the United States, Annals of the, Association of American Geographers 70: 259-70.

McCoy, R.M. (1969). Drainage Network Analysis with K-Band Radar Imagery, Geographical Review 59: 493-512.

McDonald, R.A. (1996). Opening the Cold War Sky to the Public: Declassifying Satellite Reconnaissance Imager, Photogrammetric Engineering and Remote Sensing 62385-91.

McIntire, W. and H.J. Walker (1964). Tropical Cyclones and Coastal Morphology, Annals of the Association of American Geographers 54: 582-96.

Meade, M. (1976). Land Development and Human Health, Annals of the Association of American Geographers 66: 428-39.

Meade, M. (1977). Medical Geography as Human Ecology, Geographical Review 66: 379-93.

Monmonier, M. (2002). Aerial Photography at the Agricultural Adjustment Administration: Acreage Controls, Conservation Benefits, and Overhead Surveillance in the 1930s, Photogrammetric Engineering and Remote Sensing 68: 1257-61.

Moyer, R.H. (1949). Use of Aerial Photographs in Connection with Farm Programs Administered by the Production and Marketing Administration, USDA, Photogrammetric Engineering and Remote Sensing 15: 536-40.

Murphy, A.B. (1992). Western Investment in East Central Europe, The Professional Geographer 44: 249-59.

Murphy, A.B. (2003). Geography in an Uncertain World, AAG Newsletter 38: 3-5.

Nelson, F. (1986). Permafrost Distribution, Annals of the Association of American Geographers 76: 550-69.

Nelson, G.J. (1965). Some Effects of Glaciation in the Susquehanna River Valley, Annals of the Association of American Geographers 55: 404-48.

O'Loughlin, J. (1998). The Diffusion of Democracy, Annals of the Association of American Geographers 88: 545-75.

O Tuathail, G. and Shelley, F.M. (2003). Political Geography and the New World Order. In Gaile, G. and Wilmott, C. (Eds.) Geography in America 2000. Washington, DC: Association of American Geographers, forthcoming.

Pandit, K. (1997). Cohort and Period Effects in U.S. Migration, Annals of the, Association of American Geographers 87: 439-51.

Peet, R. (Ed.) (1977). Radical Geography: Alternative Viewpoints on Contemporary Social Issues. Chicago: Maaroufa.

Peuquet, D. (1988). Representation of Geographical Space, Annals of the Association of American Geographers 78: 375-94.

Pope, G., Dorn, R., and Dixon, J. (1995). A Conceptual Model of Weathering, Annals of the Association of American Geographers 85: 38-64.

Pruitt, E.L. (1979). The Office of Naval Research and Geography, Annals of the, Association of American Geographers 69: 103-08

Psuty, N. (1965). Beach Ridge Development in Tabasco, Mexico, Annals of the, Association of American Geographers 55: 112-24.

Richardson, S.L. (1984). Pioneers and Problems of Early American Photogrammetry, Photogrammetric Engineering and Remote Sensing 50: 433-50.

Rodgers, A. (1958). The Port of Genoa: External and Internal Relations, Annals of the Association of American Geographers 58: 319-51.

Sherman, D. (1988). Longshore Current Models, Geographical Review 78: 158-68.

Snead, R. (1967). Recent Morphological Changes along the Coast of West Pakistan, Annals of the Association of American Geographers 57: 550-65.

Sopher, D. (1963). Population Dislocation in the Chittagong Hills, Geographical Review 53: 337-62.

Southhard, J.B. (1984). Historical Highlights of Photogrammetry in the U.S., 1904-1984, Photogrammetric Engineering and Remote Sensing 50: 1275-76.

Spector, A. and Brown (1976). Acquaintance Circles and Communication: An Exploration of Hypotheses Relating to Innovation Diffusion, The Professional Geographer 28: 267-76.

Stahle, D. W. and Hehr, J. (1984). Dendroclimatic Relationships, Annals of the Association of American Geographers 74: 561-73.

Steinke, T. and Lloyd, R. (1983). Images of Maps: A Rotation Experiment, The Professional Geographer 35: 455-61.

Stone, K.H. (1979). Geography's Wartime Service, Annals of the Association of American Geographers 69: 89-96,

Taylor, P.J. (1996). The Way the Modern World Works: World Hegemony to World Impasse. New York: Wiley.

Ten Years of Naval Research: The Decennial of ONR (1956). Naval Research Review.

Ter Jung, W. and Louie, S. (1973). Solar Radiation and Urban Heat Islands, Annals of the Association of American Geographers 63: 181-207.

Thomas, B.E. (1952). Modern Trans-Sahara Routes, Geographical Review 42: 267-82.

Tobler, W. (1962). Geographic Area and Map Projections, Geographical Review 52: 59-78.

Trewartha, G.T. (1952). Chinese Cities: Origins and Functions, Annals of the Association of American Geographers 42: 69-93.

Trubowitz, P. (1998). Defining the National Interest: Conflict and Change in American Foreign Policy. Chicago: University of Chicago Press.

Turner, F.J. (1920). The Frontier in American History. New York: H. Holt.

Vandermeer, C. (1968). Changing Water Control in a Taiwanese Rice Field Irrigation System, Annals of the Association of American Geographers 58: 720-47.

Vining, D. and Luow, S. (1978). A Cautionary Note on the Use of Allometric Functions to Estimate Urban Populations, The Professional Geographer 30: 365-70.

Vreeken, W.J. (1975). Quaternary Evolution in Tama County, Iowa, Annals of the Association of American Geographers 65: 283-96.

Walters, C.K and Winkler, J.A. (1999). Cloud to Ground Lightning, The Professional Geographer 51: 349-67.

Weaver, J.C. (1956). The County as a Spatial Average in Agricultural Geography, Geographical Review 46: 536-65.

Weigand, G.G. (1955). Bordeaux: An Example of Changing Port Functions, Geographical Review 45: 217-43.

Wilbanks, T.J. and Kates, R. (1999). Global Change in Local Places, Climatic Change 43: 601-28.

Williams, R.L. (1958). Map Symbols Appearing Equal Intervals for Printed Screens, Annals of the Association of American Geographers 48: 132-39.

Williamson, R.A. (1997). The Landsat Legacy: Remote Sensing Policy and the Development of Remote Sensing, Photogrammetric Engineering and Remote Sensing 63: 877-85.

Winberry, J.J. (1974). The Log House in Mexico, Annals of the Association of American Geographers 64: 54-69.

Winsberg, M.D. (1964). Jewish Agricultural Colonization in Argentina, Geographical Review 54: 487-501.

Winslow, J.H. (1966). Raised Submarine Canyons: An Exploratory Hypothesis, Annals of the Association of American Geographers 56: 634-72.

Wolpert, J. (1965). The Decision Process in Spatial Context, Annals of the Association of American Geographers 55: 537-58.

Wood, W.B. (1996). From Humanitarian Relief to Humanitarian Intervention: Victims, Intervenors, and Pillars, Political Geography 15: 671-95

Wood, W.B. (1999). A Lesson for Diplomats, Fletcher Forum for World Affairs 23: 5-20

Woodward, D. and Lewis, G.M. (1998). Cartography in the Traditional African, American, Arctic, Australian, and Pacific Societies. Chicago: University of Chicago Press.

PART II

TECHNOLOGIES THAT CHANGED GEOGRAPHY

CHAPTER 4

FRANCIS J. HARVEY
NICHOLAS R. CHRISMAN

THE IMBRICATION OF GEOGRAPHY AND TECHNOLOGY: THE SOCIAL CONSTRUCTION OF GEOGRAPHIC INFORMATION SYSTEMS

Abstract   This chapter examines the disciplinary interface between geography and technology from an ecological perspective. Our analysis of a few cases shows that the relationships between geography and technology are never clear-cut, but always intertwined like tree roots in a forest. The roots (or rhizomes) of each tree support an individual above ground, while the tangled roots just visible or invisible below the surface are a messy tangle that also are part of biochemical processes that sustain each tree, and the forest in which other trees grow. Rhizomes from different trees and plants interweave in symbiotic or parasitic relationship that are integral to the ecosystem. The relationships between geography and technology also can be understood in this ecological metaphor. What has made the growth of GIS possible? We argue that a highly fertile interface between geography and technology supports the profuse growth of GIS. Drawing on science and technology studies, we argue that the geography-technology relationship is no "chicken and egg" problem, but is evidenced in myriad, theoretically infinite relationships and interactions occurring worldwide. Each GIS implementation follows many roots, and we need to excavate them in order to understand the historical development of the interface between geography and technology. This approach calls for attention to specific sites and configurations of technology. The development of geography is intricately interwoven with the development of technology, but under no circumstances does technology determine a path for geography. In the fillamous rhizomatic network, many geographies and technologies connect and lead to subsequent developments.

Keywords   GIS, topology, social construction, disciplinary and intellectual history, GIS development

1. CHICKENS, EGGS, AND EGG SALAD

The topic of technology is vast and rich, and geographers have not been and are not among the most prominent explorers in the studies of the history and philosophy of technology. This volume offers an opportunity to establish

Stanley D. Brunn, Susan L. Cutter, and J.W. Harrington, Jr. (Eds.), Geography and Technology,
65-80. © 2004 Kluwer Academic Publishers. Printed in the Netherlands.

some connections that deserve our attention. While other authors discuss the development of specific technologies by or for geography, our contribution is more historiographic in nature and seeks to understand how geography and technology are intertwined. This chapter is motivated by our reading of the literature in an interdisciplinary field called Science and Technology Studies (STS). But we do not follow a technological determinist "importer scenario" (Chrisman 1987a), where advancement in geography consists of importing a tool or technique from some other discipline with the hope that it will solve a geographic problem after transplantation. We are also not providing an in-depth review of the STS literature, nor covering its relevance to all branches of geography. Our goal is far more modest. Whereas other geographers have begun to connect their work to the theoretical and empirical work in STS (Barnes 1998; Demeritt 1998; Harvey 2000), we seek to demonstrate the complex interactions between geography and technologies.

We emphasize geographic information systems (GIS) technologies, in spite of difficulties posed with using the term (Chrisman 1999). Through their role in geographic analysis and many economic applications, these technologies have special relevance to current debates about technology within the discipline of geography. Geographers of various stripes have expressed their views about technology in general through critical comments about GIS (see Harley 1989, 1990; Smith 1992; Lake 1993; Curry 1995; Pickles 1995; Sheppard 1995). While more recent work about "GIS and Society" has become more nuanced, the earlier polemical phase still influences the discipline's intellectual agenda in the discipline. In particular, we believe that geographers have all too readily adopted a rhetoric of "technological determinism." This position assumes that technology has its own internal logic that, once engaged, moves inexorably towards results and becomes a singular force without much ability to be influenced by political, economic, or social goals (Sejerstad 1997). One example of this position is the idea that democratic applications can only be served by creating some distinct "GIS II" (GIS2 or "GIS too"). This concept originated during the 1996 NCGIA Initiative 19 meeting (Harris and Weiner 1996), and has been championed by Sieber (2000) and others. They contend that current GIS is tainted at its origin and cannot be corrected. The recent developments supporting this approach share much in common with technocratic approaches to other issues which emphasize the use of technology to achieve instrumental measures of economic or social improvement (PolicyLink 2002; Wood 1999). Inherent problems in these positions have been discussed by several authors (Dunn et al. 1997; Mugerauer 2000; Stonich 1998).

Our disagreements with much of the current historical work relating GIS and technology can best be resolved empirically. Ecologically oriented empirical studies using distinct sites and developments provide useful insights into the relationships between geography and technology. Below we discuss three examples. First, we contend that origins are elusive and of little assistance in understanding how technology develops. Second, we demonstrate that technological decisions are not irreversible. Third, we contend that various forces utilize and shape technologies to their goals.

The title of this section deals with the folk aphorism about origins; "which came first, the chicken or the egg?" The linear logic of origins and heroic inventors demands that some single event be declared the original invention. Yet, inexorably, that invention in turn has its own origins. It would seem impossible or unimportant to draw a firm line at any one point to call something more original than any event in the chain. Our argument does not stop with this linear recursion. We find it more important to demonstrate that the linear causative logic conceals many lateral rhizomatic connections: that is, events interconnect in complex interactions. The whole idea of an origin vanishes in some complex culinary melange of chickens, eggs, spices, too many cooks, and no clear logic of what comes first. An examination of the rhizomes of geography and technology's interface can expand our understanding of the historical developments that give GIS its multiple forms and meanings today.

2. THEORETICAL FRAMEWORK

The theoretical basis for resolving this dilemma about technology and geography is essential to understanding the overall theme of this volume. The field of STS has been confronting the imbrications of society and technoscience for decades. Works such as Mackenzie's study of missile accuracy (MacKenzie 1990) provide clear evidence that the concepts of technological determinism are subject to political influence. While certain U.S. missile programs show a clear increase in accuracy over time, this is the result of substantial effort by actors who desired that particular result, not something inherent in the technology of gyroscopes that ensured missile guidance. During this same period, the progression of Soviet missiles involved quite distinct gyroscope models, while the pressure towards greater accuracy was not found in China or France, due to different military strategies and different financial resources available. The most telling argument is that the civilian navigational sector (using the same gyroscope technology) did not evolve toward greater accuracy, but toward lower cost and higher reliability. Mackenzie's conclusion is that one cannot study the political pressures without understanding the technology and conversely, one cannot study the technology without understanding the politics of the organizations involved.

A symmetrist approach, which considers technology and humans as equal contributors is articulated by a number of STS scholars including Bruno Latour (1987, 1990, 1993, 1996; Latour and Woolgar 1986), Michel Callon (1981), Karin Knorr-Cetina (1983), and many others (Bijker et al. 1987). It also surfaces in a paradigmatic conflict with a school of "sociology of scientific knowledge" (Barnes 1974; Collins 1975; Bloor 1976) also known as the "strong program" because it asserts that social (and political/economic) affairs underlie the controversies of science. [Note: The symmetrists argue that the strong program was a natural reaction to the even earlier approach that allows social study of scientists, but not of science.] A key point in the symmetrist approach is that attempts to maintain a division between "social" explanations and "technical" explanations can become blurred and pointless. Symmetrists offer a variety of approaches to study what Latour refers to as technoscience (Latour 1987, 1999). The rhizome concept is related to actor-network theory as it shares an interest in analyzing relationships and interactions from multiple perspectives (Law and Hassard 1999).

Before presenting our case studies, we briefly depict the rhizome theory's genesis and key theoretical contributions for our work (Deleuze and Guattari 1976). Introduced by Deleuze in 1976 in *Proust et les signes* (English title: *Proust and Signs: The Complete Text*) (Deleuze 2000), he takes on the Freudian and Lacanian psychoanalysis focus on single root causes and conceptualizes human desire as an evergrowing tangle of desire, comparable to the underground root system of a rhizome. Instead of a single tap root, rhizomes spread forming a chaotic network that connects any point with every other point. For Deleuze, a psychologist by training, desire is multidimensional and deep-rooted. Rhizomes became a central feature of the 1980 book *Milles Plateaux* (written with Felix Guattari and translated as *A Thousand Plateaus: Capitalism and Schizophrenia* (Deleuze and Guattari 1987). To some extent, the texts of these works are in themselves rhizomes; the bibliographies are also a bit convoluted.

These authors' later works have greater relevance for our constructionist approach. In *A Thousand Plateaus*, Deleuze and Guattari extend Deleuze's previous use of the concept through a series of geographic metaphors oriented to underscoring the process and dynamic characteristics of rhizomes. From their writings, we believe the concept of assemblage is especially relevant to social inquiries into technologies. Before engaging the assemblage concept, it is useful to spell out several additional characteristics of their rhizomatic thinking.

The first is connection, which refers to semiotic chains that connect diverse modes of coding and that can relate different regimes of signs and phenomena that themselves have a different status. Connections, in other words, need not be material; they can be symbolic and relate symbols and artifacts emanating from distinct domains. Connections agglomerate diverse acts and resources. The second characteristic is heterogeneity. Rhizomes are not homogeneous, but are exceedingly diverse and fluid, a "throng" in their words. Rhizomes also lack a unit but are characterized by multiplicity–the third characteristic for Deleuze and Guattari. Multiplicity is fundamental to understanding rhizomes as magnitudes and dimensions, never objects nor subjects, but the outside definition of features that reflect outside determinations of a specific rhizome's characteristic. These determinations can reflect ruptures in the rhizome that segment, organize, or stratify the rhizome itself. Deleuze and Guattari (1987) offer the example of a book, not as a material object, but a rhizome with the world. Geographer Michael Curry (1996) examines similar connections of the written "work," though without using the term rhizome. While Curry's argument contains some idealist streaks, the rhizomatic interpretation insists on a mutual reconstitution of meaning enacted at each reading.

These characteristics of rhizomes lead to assemblages, the key entities in coalescing interactions of a rhizome.

> An assemblage, in its multiplicity, necessarily acts on semiotic flows, material flows, and social flows simultaneously...an assemblage establishes connections between certain multiplicities...The book is an assemblage with the outside. (Deleuze and Guattari 1987, 22-23)

A quotidian example of an assemblage is a book that becomes part of the interactions of readers, author, cited work, institutions and disciplines. GIS technologies are also thought of as assemblages that connect users, data providers, software engineers, institutions, and cultural values. Recalling Deleuze and Guattari's rhizome concepts, our analysis below of examples of GIS development emphasizes the flows and interactions between the diverse communities of actors.

3. ORIGINS ARE ELUSIVE

In our first empirical vignette, we address one of the most common critical arguments about GIS regarding its origins. In its essentials, this argument almost comes across as a form of original sin. That is, GIS technology is judged to be contaminated at its origin and thus fundamentally flawed. Neil Smith (1992) discusses its military origins, while Pickles (1995) and Sheppard (1995) connect GIS origins to modernist philosophies. These different arguments may seem to be inherently contradictory in a linear reading of GIS development, but in the rhizomatic approach, these different stories about origins may all be "correct" at some level. The key issue relates to the idea of a unitary causative origin.

GIS technologists in general have adopted their own version of a unitary origin narrative with heroic figures in a mythic past who created "GIS" in some unitary vision. In historical treatments (Coppock and Rhind 1991; Foresman 1997), the most common presentations, Roger Tomlinson (a Canadian geographer) and Ian McHarg (a landscape planner in the U.S.) take prime billing as originators. Foresman's edited book (1997) adopts a "frontier" metaphor with "pioneers" who precede others in some ordered occupation. Origin stories feature claims about the "first GIS" (Tomlinson 1998) with the rest of the logic imputing some special status to the originator. Roger Tomlinson certainly accomplished major creative feats in constructing the Canada Geographic Information System (CGIS) for the Canadian government during the period leading up to his first published paper in 1968 (Tomlinson 1968). These accomplishments we do not seek to minimize. Yet CGIS makes a less compelling case than the Harvard Laboratory for Computer Graphics and Spatial Analysis, which at the same time had an emphasis on developing networks of research collaborators and dissemination through sponsored conferences.

Tomlinson's retrospective work (1998) contends that GIS had its origins in the government sector, aided by the consulting commercial sector. If we were interested in which project first called itself a "geographic information system," Tomlinson is correct. Yet, if we deal with the origins of the particular collections of software that currently operate our main GIS ventures, the issue of origins becomes more complex. The software written by IBM for CGIS in Ottawa is not on record as having been installed anywhere else. There were only two models of the CGIS scanner built—the original one and the one built on the IBM service site in Ottawa to keep the other one functional. At the base of its technology, CGIS created the Morton index for quadtree data structures to manage coordinates with an incremental code to represent lines in a fixed resolution framework. No current system has accepted such a limitation, opting instead for more storage-hungry vector representations. These are only a few of the ways in which CGIS technology did not influence later developments. For being first, CGIS leaves little direct legacy.

On both the raster and vector sides of GIS, there are many linkages of current software to origins at the Harvard Laboratory for Computer Graphics and Spatial Analysis. The "Spatial Analyst" extension to ESRI's ArcView is one of many implementations of Tomlin's map algebra, which has a clear heritage to the MAP package. Tomlin (1990) first wrote these functions onto Sinton's IMGRID as a Harvard graduate assistant to Carl Steinitz's National Science Foundation-funded grant to study urban growth. Also, ERDAS, founded by two other graduate students during the same era, called its first software package IMGRID, before it evolved into IMAGINE. These connections appear to link Harvard in the mid-1970s with the commercial world, yet there are clear traces of an earlier period. In the world of applications, it is common to refer to each layer as a "cost-grid" in map overlays, even for classical McHargian overlay analysis for site suitability (see La Placa 1997, for example). This usage of cost surfaces relates to Tomlin's development of incremental minimum cost approximations (Tomlin 1990). This element of GIS technology was not in Tomlinson's 1960s vision of a natural resources inventory. Nor does it come from McHarg's model of transparent overlays. Rather, it can be traced to Harvard through William Warntz, second director of the Lab, who worked on the theory of surfaces, with particular attention to geodesics of minimum cost (Warntz 1957, 1965, 1966). Warntz was trained, along with many other geographers of his generation, as a navigator for bombers in World War II. For these substratospheric bombers, "pressure pattern flying" required sophisticated map analysis skills from which Warntz formulated his highly theoretical view of maps as surfaces. In the 1960s, this surface work remained theoretical, not implemented on a computer, but as it became a part of graduate instruction, it inspired later programmers to incorporate the surface mapping skills into packages once the computers were large enough to perform analyses of sufficient size. Of course, Harvard is not an obligatory link in this chain, as other geographers at this time had the same vision, but few had the same opportunity to influence the current software in such a direct manner. Warntz was influenced by Stewart's "social physics" (Warntz 1957, 1965) as much as his service in the military.

It is useful in the context of rhizomatic thinking to review the innocent little term "cost-grid" using the three characteristics of rhizomes. La Placa (1997) selected this term instead of the more mundane (and more appropriate) "overlay" perhaps because it makes it sound more technical or sophisticated. Its main feature is that as a technical term, it invokes connection. Cost-grid invokes the high theory of Stewart, as brought into geography by Warntz. It also invokes the practical map-reading skills of an analogue era when leather-clad navigators flew bombers without any digital assistance, just paper maps. Is it tinged by the military? Perhaps it is, but not in the sense that Smith (1992) and others insist that all the choices come from the military origin. Finally, it is heterogeneous in the extreme. Also the cost-grid is not some distinct "object" judged solely by its internal characteristics. Rather, it is a component in a multiplicity. It only becomes a cost-grid through some process of interpretation, in a community of practice. Thus a cost-grid is an assemblage, a composite of symbol, metaphors, tangible representation, disciplinary expectations, and software artifact.

On the vector side, as well, current GIS software also traces specific features to other teams at the Harvard Lab. The ODYSSEY software (Dougenik 1980; Chrisman et al. 1992) demonstrated the ability to overlay complex geometries with a "fuzzy tolerance" in order to avoid slivers and other geometric problems that plagued earlier implementations (Goodchild 1978). These traits were carried to ESRI by Scott Morehouse (a member of the ODYSSEY team) and implemented by programmers trained at SUNY-Buffalo. The product was called Arc/INFO. Other vendors had to match this capability in order to gain entry into the field (for example, Intergraph's TIGRIS (Topologically Integrated Geographic Information System), Herring 1987; Burrough 1986; and Cowen 1988 for the importance of topology). There are traces of the linkage embodied in software, but the assemblage also includes people, training, ideas, software, and corporations. ODYSSEY was the product of the Harvard Lab during the 1970s, and it was done in full awareness of prior work in Canada and the U.S. It probably owes more of its topological bent to the U.S. Census Bureau than to CGIS. The Census Use Study that originated the DIME files also had close connections with Harvard as the topological technologies were being developed, but at the time, these contacts were mostly related to printing maps on line printers. There were also formal and informal connections between the ODYSSEY teams at Cambridge and Don Cooke at Geographic Data Technology in New Haven, Connecticut as the Census Use Study worked far outside the normal bureaucratic channels (Cooke 1998). The Harvard software of the 1960s did not adopt the topological data structure proposed by the Census Bureau, despite the connections. In writing about the flaws in nontopological structures, the earlier software provided sufficient examples to demonstrate the inadequacies (Peucker and Chrisman 1975). The Harvard ODYSSEY team only learned about the possibility that CGIS had some topological approach from the Australian Cook who had built his own system following a stay in Ottawa (Cook 1967). These histories mean that the connections were there, but not always in the order that they appear to occur.

In summary, the narrative of the Harvard Lab begins to resemble a twisted maze of string, with curves and twists, some of which are out of sight. This account does not have linear origins, but rather a series of interactions inside a community with common interests. The metaphor proposed by Deleuze—a rhizome—is applicable here. That is, there is no single root to the Harvard Lab, even though Howard Fisher, founder of the Lab, first heard about computer mapping at a workshop run by Edgar Horwood from the University of Washington in 1961 (Chrisman 1997). Does this fact mean all the Lab's early ideas came from the University of Washington? Rather unlikely. Like rhizomes in a patch of prairie, the world of technology has multiple connections, abrupt turns, seeming dead ends, and seemingly independent flowers appearing on the top from the maze of connections. The individuals involved in each instance of GIS can demonstrate multiple connections and the diversity of rhizomes many times. The following example shows how the technology itself also is part of the rhizomatic flows.

4. IRREVERSIBILITY: THE STORY OF TOPOLOGY

If technology is independent of its social, institutional, and political context, then it should follow that a technical decision, once made, should be irreversible. That is, the reasoning applied in the first place should continue to apply. In addition, as time goes on, a technical decision should accumulate strength and become harder to undo. For these reasons, it is important to demonstrate events when a technical decision is reversed and to consider the reasons why. This section deals with the trait of topological data structures in GIS. Topology in this context means a data model that has explicit differentiation between the points and lines that constitute an areal object in the data structure. These points and lines are organized so that each line is stored only once, though it might be used as the boundary of the polygons on either side. Technically, it is more complex than a data structure (including those used in earlier software such as SYMAP from the Harvard Lab) where each polygon was simply a list of straight line segments, and adjacent boundaries are duplicated without any overt recognition.

Topology is a fairly recent subfield of mathematics considering how basic it is. It crossed over from the scientific to the technical when it was adopted as a rationale for cartographic data structures.While Dueker's (1972) doctoral thesis considered alternative forms of topological representation, it was the U.S. Census Bureau that brought the topological approach into prominence by proposing to implement a nationwide coverage between 1967 and the 1970 Census (Cooke 1998). The theoretical argument behind the adoption of topology was clearly articulated, but not openly published until much later (Corbett 1979). In the meantime, others also proposed some topological models. It was adopted at the U.S. Geological Survey for the nationwide coverage of the land use/land cover project Geographic Information Retrieval and Analysis System (GIRAS), and it became a key part of the polygon overlay program POLYVRT, part of the ODYSSEY system at the Harvard Lab (Peucker and Chrisman 1975). But the issue of origins is not our major concern here. The point is that this approach gained allies, moving from a relatively peripheral status to a clear requirement for a full-service GIS by the mid-1980s.

In an influential paper published in 1975, Peucker and Chrisman (1975) reviewed a series of alternative data structures using a progressive spectrum from the least complex to the nuances of the full topological model. They argued that increased sophistication provided additional power, and that databases and software should adopt the topological model. This contention was presented in a number of venues, forming the rallying cry for "intelligent" data structures. The division between mere drafting systems and analytical GIS was drawn largely along this line of data structures (Cowen 1988; Dangermond 1986).

The implementation of topology took time. The prototype Harvard ODYSSEY system described above was operational in 1980 (Dougenik 1980), relying on a topological coverage model. This prototype had strong links to the first generation of Arc/INFO from ESRI, which was first delivered in 1983. The success of this software might seem inevitable when viewed retrospectively, but at that time, commercial software was far from dominant. Most government agencies either wrote their own software (as with GIRAS) or imported software from the academic sector. At this point, Harvard had more than 500 recorded users of SYMAP, many more than the clients of ESRI. Tomlinson and Boyle (1981) decided that no available commercial package was adequate for Saskatchewan natural resource planning. But two years later, the freshly completed Arc/INFO was judged adequate by the same team from Tomlinson Associates for the State of Washington Department of Natural Resources. Why the difference? There were many advances in the new Arc/INFO, and many bugs were fixed. Also at the time, the GIS community came to accept that topology was a key difference (Burrough 1986). In order to compete with ESRI, Intergraph had to create not one, but two topologically oriented software packages, TIGRIS (Herring 1987) and MGE. Intergraph, a company founded around hardware for graphic workstations, had hired David Sinton (designer of the IMGRID package at Harvard, mentioned above) to lead its GIS developments. Sinton hired mathematical programmers with degrees in algebraic topology such as John Herring. TIGRIS did not survive, but MGE (a design not very different from Arc/INFO) survived to develop into the company's product line. The main point here is not the specific history of each software package, but how the story cannot be told in a linear manner; the assemblages of people, training, companies, hardware, and software form heterogeneous connections.

The process above might appear to be a technological imperative. One explanation is that topology is simply a requirement, and once it is discovered others will have to adopt it in order to compete. Topology did afford a technically superior means to perform certain kinds of consistency checks in the quality control required for digitizing (Chrisman 1987b). The question is whether the results of this checking needed to be stored for all other uses. When using the computers of the 1970s and 1980s as a guide, the choice was clear. But there were alternatives. By 1985 Butler demonstrated some very fast geometric processing using a proprietary approach that was overtly organized on a nontopological data structure. Butler licensed his software to a handful of users, but his corporate structure could not compete with the larger firms. By the 1990s, Intergraph and ESRI both tried to acquire Butler's technology. ESRI moved more nimbly, perhaps due to its private ownership, so Butler's technology—now called Spatial Database Engine (SDE)—is a core part of ESRI's product line. The corporation, which was built on the topological approach, acquired Butler's software based on the older, dumber data model, and trumpeted SDE as the wave of the future. How could this happen?

Acquiring SDE and proclaiming it as an advance was not just a marketing move. SDE is faster because it did not have to store the more complex data structure. It also provided faster ways to calculate all the relationships of topology that are needed. In fact, SDE provides more kinds of topological relationships than the old so-called topological coverage approach. The lesson is that the choice between storing and computing depends on the computers available. In the earlier era, storage made more sense. Today computing is much cheaper. These are external factors, not eternal verities of geometry. In searching for a rationale, it is all too easy to look for a simple root, a single explanation for what happened, when, and why. Topology turns out to require a complex web of connections, that is, a rhizome.

5. DEFINING GIS

Attempts to determine origins are matched by efforts to determine disciplinary positions and relationships to other fields. At the end of the 1980s and the beginning of the 1990s, ample evidence existed that illustrates some of the more long-lasting disciplinary intellectual struggles for GIS. Several technical issues were, in fact, just as, if not more important for the development of GIS as a discipline. In the first of these, the distinction between computer-assisted design (CAD) and GIS was technically meaningful. But as the previous section shows, the distinction was less important in hindsight than the political importance to distinguish GIS, the new upstart for computer-aided planning, architecture, and related fields, from the well-rooted CAD industry. A number of articles appeared more or less simultaneously titled "CAD vs. GIS" (Dangermond 1986; Cowen 1988). By the late 1980s, topology had become the identifying assemblage for anyone with a stake in GIS to distinguish CAD from GIS (Dueker 1987). Connection of anyone new to GIS was facilitated by the rhizomatic connections to other GIS users who could also indicate their use of topology. The rapid growth of GIS, measurable in sales and jobs on the surface, was accompanied by developments under the surface in groups that supported the network of interactions between various GIS adopters. Topology was essential to a GIS for political and technical reasons, but other disciplinary rhizomes were also linked to GIS's network. Of particular note is the connection to cartography, which now has clearly oriented itself to questions of a representation (MacEachren 1995). In 1990 this issue was far from clear. Cartography was much stronger traditionally than GIS in academic geography. GIS was again an upstart with many relationships to cartography, but was developing a new assemblage and set of rhizomatic relationships (Foresman 1997).

While many articles were written at the time arguing for or against cartography's role in GIS, other debates took place (Clarke 1990). Clarke created an assemblage that linked GIS and cartography and demonstrated possibilities for fruitful collaboration between GIS and cartography. This view was supported by citing a minority tendency within cartography, viz., Tobler's analytical cartography (Tobler 1976), and setting it up as an alternative to communications-school cartography. Other authors saw far less of a connection, focusing instead on tensions and changes to dominant paradigms (Goodchild 1988; Fisher 1998).

The GIS-cartography issue also influenced attempts to establish the analytical roots of GIS. While clear and strong roots in analytical cartography are important components of GIS's rhizomatic network (Tobler 1979), attempts to define GIS as an analytical field were going on between topology and cartography. Computational approaches played a strong role, but the many attempts to create a discipline centered on analytical uses of GIS (Openshaw 1991, 1998) became quite contentious for those who sought a more amicable middle-ground between preexisting disciplines. Centrist positions regarding the intellectual core of GIS were strong enough to resist attempts to center GIS on computational analysis and strengthen the rhizomatic network with other areas of geography.

GIS came out of these intellectual debates significantly strengthened by the development of institutions such as the National Center for Geographic Information Analysis (NCGIA) and the University Consortium for Geographic Information Science (UCGIS) in the U.S., AGI in Britain, and others) that stabilized the rhizomatic networks and promoted the development of assemblages that enhanced the roles of network participants. All the while, continued debates to ensure GIS's role and stature among and with other disciplines continued. Although the semantics of GIS definitions leave much room for discussion and debate (Chrisman 1999), a substantial core of GIS became a recognized area of science (Goodchild 1992) and stabilized many disciplinary relationships. By constraining the entry points to participate in the field and enhancing its position in the scientific community, GIScience sought to fit "into a broader set of usages" (Douglas 1989, 54) and stabilize interactions between rhizomes.

6. CONCLUSIONS

The developments of geographic information technologies underscore the rhizomatic character of the relationships between geography and technology. These developments have been influenced by complex relationships between techniques, theories, politics, and economics. In the rhizomatic approach, causality can indeed account for all these relationships without constraining the analysis to a limited set of influences and the assumption of a unitary form for geographic information technologies.

Geography's technological artifacts are indeed inseparable from human activities, not only in a constructive sense, but intrinsically interwoven with our ideologies and politics. GIS technologies are locally constituted assemblages that connect multiple activities. When more activities use and engage the technologies, the rhizomatic network thickens and assemblages grow. They also become resilient to change, are transplanted, and lose parts of the network to competition, change, or new assemblages. It is no longer a question of chicken "or" egg, but questions surface about what relationships and interactions impacted the development of a particular instance of technology. What works at one time and place may not work again in the same way elsewhere. Seemingly inevitable directions can (and do) reverse themselves (again and again). Excavations of rhizomes can help geography better understand the technological roots and flowers of the field.

REFERENCES

Barnes, B. (1974). Scientific Knowledge and Sociological Theory. London: Routledge and Kegan Paul.

Barnes, T.J. (1998). A History of Regression: Actors, Networks, Machines, and Numbers, Environment and Planning A 30: 203-24.

Bijker, W.E., Hughes, T.P., and Finch, T.J. (Eds.) (1987). The Social Construction of Technological Systems: New Directions in the Sociology and History of Technology. Cambridge, MA: MIT Press.

Bloor, D. (1976). Knowledge and Social Imagery. London: Routledge and Kegan Paul.

Burrough, Peter A. (1986). Principles of Geographical Information Systems for Land Resource Assessment. Oxford: Clarendon Press.

Callon, M. (1981). Pour une Sociologie des Controverses Techniques, Fundamenta Scientiae 2: 381-99.

Chrisman, N.R. (1987a). Challenges for Research in Geographic Information Systems. Proceedings, International Geographic Information Systems Symposium 1: 101-12.

Chrisman, N.R. (1987b). Efficient Digitizing Through the Combination of Appropriate Hardware and Software for Error Detection and Editing. International Journal of Geographical Information Systems 1: 265-77.

Chrisman, N.R. (1997). Academic Origins of GIS. In The History of Geographical Information Sysetms. Foresman, T. (Ed.), 32-43. London: Taylor and Francis.

Chrisman, N.R. (1999). What Does GIS Mean? Transactions in GIS 4: 175-86.

Chrisman, N.R., Dougenik, J.A., and White, D. (1992). Lessons for the Design of Polygon Overlay Processing from the ODYSSEY WHIRLPOOL Algorithm. Proceedings 5th International Symposium on Spatial Data Handling 2: 401-10.

Clarke, K.C. (1990). Analytical and Computer Cartography. Englewood Cliffs, NJ: Prentice Hall.

Collins, H.M. (1975). The Seven Sexes: A Study of the Sociology of a Phenomenon, or the Replication of Experiments in Physics, Sociology 9: 205-24.

Cook, B.G. (1967). A Computer Representation of Plane Region Boundaries, Australian Computer Journal 1: 44-50.

Cooke, D.F. (1998). Topology and TIGER: The Census Bureaus Contribution. In The History of Geographic Information Systems. Foresman, T.W. (Ed.), 47-58. Upper Saddle River, NJ: Prentice Hall.

Coppock, J.T. and Rhind, D.W. (1991). The History of GIS. In Maguire, D.J., Goodchild, M.F., and Rhind, D.W. (Eds.) Geographical Information Systems, vol. 1, 21-43. London: Longman.

Corbett, J. (1979). Topological Principles in Cartography, Research Papers 48. U.S. Census Bureau.

Cowen, D.J. (1988). GIS versus CAD versus DBMS: What Are the Differences? Photogrammetric Engineering and Remote Sensing 54: 1551-55.

Curry, M.R. (1995). GIS and the Inevitability of Ethical Inconsistency. In Ground Truth: Social Implications of Geographic Information Systems, Pickles, J. (Ed.), 68-87. New York: Guilford.

Curry, M.R. (1996). The Work in the World: Geographical Practice and the Written Word. Minneapolis: University of Minnesota Press.

Dangermond, J. (1986). CAD versus GIS. Computer Graphics World 9: 73-74.

Deleuze, G. (2000). Proust and Signs: The Complete Text. Minneapolis: University of Minnesota Press.

Deleuze, G. and Guattari, F. (1976). Rhizome Introduction. Paris: Editions de Minuit.

Deleuze, G. and Guattari, F. (1987). A Thousand Plateaus. Capitalism and Schizophrenia. Minneapolis: University of Minnesota Press.

Demeritt, D. (1998). Science, Social Constructivism and Nature. In Remaking Reality: Nature at the Millenium, Braun, B. and Castree, N. (Eds.), 173-93. London: Routledge.

Dougenik, J.A. (1980). WHIRLPOOL: A Geometric Processor for Polygon Coverage Data. Proceedings, AUTO-CARTO IV 304-11.

Douglas, D. (1989). Letter to the Editor. GIS World, 2: 53-53.

Dueker, K. (1972). A Framework for Encoding Geographic Data. Geographical Analysis 4: 98-105.

Dueker, K.J. (1987). Geographic Information Systems and Computer-Aided Mapping, Journal of the American Planning Association 53: 383-90.

Dunn, C.E., Atkins, P.J., and Townsend, J.G. (1997). GIS for Development: A Contradiction in Terms? Area 29: 151-59.

Fisher, P. (1998). Is GIS Hidebound by the Legacy of Cartography? The Cartographic Journal 35: 5-9.

Foresman, T. (Ed.) (1997). The History of Geographic Information Systems. London: Taylor and Francis.

Goodchild, M.F. (1978). Statistical Aspects of the Polygon Overlay Problem. In Harvard Papers on Geographical Information Systems 6, Dutton, G. (Ed.) Reading, MA: Addison Wesley.

Goodchild, M.F. (1988). Stepping Over the Line: Technological Constraints and the New Cartography, The American Cartographer 15: 311-20.

Goodchild, M.F. (1992). Geographical Information Science, International Journal of Geographic Systems 6: 35-42.

Harris, T. and Weiner, D. (Eds.) (1996). Scientific Report for the Initiative 19 Specialist Meeting, March 2-5, 1996, South Haven, MN. Santa Barbara: National Center for Geographic Information and Analysis.

Harley, J.B. (1989). Deconstructing the Map, Cartographica 26: 1-20.

Harley, J.B. (1990). Cartography, Ethics and Social Theory, Cartographica 27: 1-23.

Harvey, F. (2000). The Social Construction of Geographical Information Systems, International Journal of Geographical Information Science 14: 711-13.

Herring, John (1987). TIGRIS: Topologically Integrated Geographic Information System, Proceedings, AUTO-CARTO 8 282-91.

Knorr-Cetina, K. (1983). Towards a Constructivist Interpretation of Science. In Science Obsesrved: Perspectives on the Social Study of Science, Knorr-Cetina, K. and Mulkay, M. (Eds.), 115-40. Beverly Hills: Sage.

Lake, R.W. (1993). Planning and Applied Geography: Positivisim, Ethics, and Geographic Information Systems, Progress in Human Geography 17: 404-13.

La Placa, J. (1997). Nature Park Site Analysis in Fairfax County, Virginia. Using the ArcView Spatial Analyst Extension. ESRI Users Conference. http://gis.esri.com/library/userconf/proc97/proc97/to550/pap517/p517.htm.

Latour, B. (1987). Science in Action. Cambridge, MA: Harvard University Press.

Latour, B. (1990). Drawing Things Together. In Representation in Scientific Practice, Lynch, M. and Woolgar, S. (Eds.), 19-68. Cambridge MA: MIT Press.

Latour, B. (1993). We Were Never Modern. Cambridge, MA: Harvard University Press.

Latour, B. (1996). Aramis or the Love of Technology. Cambridge, MA: Harvard University Press.

Latour, B. (1999). Pandoras Hope. Cambridge, MA: Harvard University Press.

Latour, B. and Woolgar, S. (1986). Laboratory Life: The Construction of Scientific Facts. Princeton, NJ: Princeton University Press. 2d ed.

Law, J. and Hassard, J. (Eds.) (1999). Actor Network Theory and After. Oxford: Blackwell Publishers/The Sociological Review.

MacEachren, A.M. (1995). How Maps Work. Representation, Visualization, Design. New York: Guilford Press.

MacKenzie, D.A. (1990). Inventing Accuracy: An Historical Sociology of Nuclear Missile Guidance. Cambridge, MA: MIT Press.

Mugerauer, R. (2000). Qualitative GIS: To Mediate, Not Dominate. In Janelle, D.G. and Hodge, D.C. (Eds.) Information, Place, and Cyberspace. Issues in Accessibility, 317-38. Berlin: Springer Verlag.

Openshaw, S. (1991). A View on the Crisis in Geography, or Using GIS to Put Humpty-Dumpty Back Together Again, Environment and Planning A 23: 621-28.

Openshaw, S. (1998). Towards a More Computationally Minded Scientific Human Geography, Environment and Planning A 30: 317-32.

Peucker, T.K. and Chrisman, N.R. (1975). Cartographic Data Structures. The American Cartographer 2: 55-69.

Pickles, J. (Ed.) (1995). Ground Truth: Social Implications of Geographic Information Systems. New York: Guilford.

PolicyLink (2002). Community Mapping. Using Geographic Data for Neighborhood Revitalization (Report). Oakland, CA: PolicyLink.

Sejerstad, F. (1997). Beyond Technical Determinism. Paper presented at the Society for the Social Studies of Science Annual Meeting (4S), Tucson.

Sheppard, E. (1995). GIS and Society: Towards a Research Agenda, Cartography and Geographic Information Systems 22: 5-16.

Sieber, R.E. (2000). Conforming (to) the Opposition: The Social Construction of Geographical Information Systems in Social Movements, International Journal of Geographic Information Science 14: 775-93.

Smith, N. (1992). History and Philosophy of Geography: Real Wars, Theory Wars, Progress in Human Geography 16: 257-71.

Stonich, S. (1998). Information Technologies, Advocacy, and Development: Resistance and Backlash toIndustrial Shrimp Farming, Cartography and Geographic Information Systems 25: 113-22.

Tobler, W. (1976). Analytical Cartography, The American Cartographer 3: 21-31.

Tobler, W. (1979). A Transformational View of Cartography, The American Cartographer 6: 101-06.

Tomlin, C.D. (1990). Geographic Information Systems and Cartographic Modeling. Englewood Cliffs, NJ: Prentice Hall.

Tomlinson, R.F. (1968). A Geographic Information System for Regional Planning. In Land Evaluation Stewart, G.A. (Ed.), 200-10. Melbourne: Macmillan.

Tomlinson, R.F. and Boyle, A.R. (1981). The State of Development of Systems for Handling Natural Resources Inventory Data, Cartographica 18: 65-95.

Tomlinson, R.F. (1998). The Canada Geographic Information System. In The History of Geographic Information Systems: Perspectives from the Pioneers, Foresman, T.W. (Ed.), 21-32. Upper Saddle River, NJ: Prentice-Hall.

Warntz, W. (1957). Transportation, Social Physics and the Law of Refraction, The Professional Geographer 9: 2-7.

Warntz, W. (1965). A Note on Surfaces and Paths: Applications to Geographical Problems. Discussion Papers X. Ann Arbor: Michigan Inter-University Community of Mathematical Geographers.

Warntz, W. (1966). The Topology of Socio-Economic Terrain and Spatial Flows. Papers, Regional Science Association 17: 47-61.

Wood, W.B. (1999). Geo-Analysis for the Next Century: New Data and Tools for Sustainable Development. In Demko, G.J. and Wood, W.B. (Eds.) Reordering the World. Geopolitical Perspectives on the 21st Century, 199-205. Boulder, CO: Westview Press.

CHAPTER 5

DANIEL SUI
RICHARD MORRILL

COMPUTERS AND GEOGRAPHY: FROM AUTOMATED GEOGRAPHY TO DIGITAL EARTH

Abstract   The computer has drastically transformed both the world of geography as an academic discipline and the geography of the world in which we live. This chapter traces the evolution of computers from being a tool for geographers to collect, analyze, map, and visualize data since the mid- to late-1950s to increasingly becoming an integral part of the world geographers study by the end of the 20th century. Computers have enriched the discipline of geography with the development of automated geography, GIScience, and the virtual geography department. The increasing etherealization of geography, as evidenced by the emerging digital individuals, virtual cities, and digital earth, has raised many fundamental scientific, socioeconomic, and ethical questions that need further investigation. To better understand the world, geographers must try to rely on state-of-the-art computers on the one hand, and at the same time, recognize the fundamental limits of computation and build dialogues with a variety of different scholarly traditions.

Keywords   computers, automated geography, GIS, digital earth, geographies of the information society

1. INTRODUCTION

Computers are crucial in the history of geography and technology. Computers and the computer-led information revolution have changed both scientific practices and many facets of human society in fundamental ways, as evidenced by the rapid growth of the use of computers in all the sciences (Siegfried 2000), the increasing emphasis of mapping and visualization as a discovery tool (Hall 1993; Chen 2002), and the maturing information society led by the digital economy (U.S. Department of Commerce 2000; Malecki 2002a). In most developed countries, especially in North America and Western Europe, computers penetrate almost every aspect of our lives. It is under this broad scientific and societal milieu that geographers entered the 21st century.

Stanley D. Brunn, Susan L. Cutter, and J.W. Harrington, Jr. (Eds.), Geography and Technology,
81-108. © 2004 Kluwer Academic Publishers. Printed in the Netherlands.

For the semicentennial celebration of the Association of American Geographers, Arthur Robinson (1954) contributed the only chapter dealing with geographic techniques in *American Geography: Inventory and Prospect,* edited by Preston James and Clarence Jones. In the chapter, Robinson provided a thorough review of the importance of maps for understanding geography and geographical problem solving. Although primitive computers existed in the early 1950s, nobody had yet recognized the potential of computers in cartography, much less anticipated the emergence of geographic information systems (GIS) and geographic information science (GIScience) in the second half of the 20th century.

No other technological innovation in human history has affected the practice of geography in such a profound way as the computer. It has drastically transformed both geography as an academic discipline and the geography of the world. Our ancient, ill-defined, eclectic, and traditionally conservative discipline has been radically changed by the past fifty years of the computer age, and especially the last twenty years. Computers have fundamentally changed not only how geographers study the world, but also what we study, as spaces and places, natural and human landscapes, are constantly being refigured by computers.

Our overarching concept for the broad topic of "computers and geography" is best illustrated in M.C. Escher's painting *Drawing Hands*

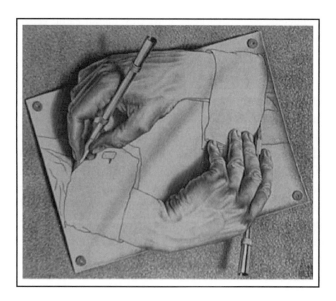

Figure 5-1. Illustration on the relationship between computers and geography (Drawing Hands by M.C. Escher).

(Figure 1). Geography is a child of its technological time as the tools we make will eventually make us. From being a tool used by geographers of the 1950s to collect, analyze, map, and visualize data, computers and their applications have evolved to become an increasingly integral part of what we study by the end of the 20th century (Sui 1997). The twin themes of this chapter reveal that there is a self-referentiality at a grand scale. What are its deeper implications? We hope this chapter will stimulate discussions among geographers on the broader issues of computers' increasing infiltration in geography and society.

In this chapter we trace the history of the dual roles played by the computer in our discipline and society. We first review how computers have revolutionized the practice of geographic research and education, followed by a discussion on the broader social, economic, and environmental impacts the computer revolution has brought to society. Next we discuss the impacts of computers on geography and society. The chapter concludes by discussing the need for geographers to improve our craft as we harness the powers of computers in our quest for truth and justice and at the same time avoid unintended consequences of technological innovations in the future (Tenner 1996; Sui and Goodchild 2003).

2. COMPUTERS AND THE WORLD OF GEOGRAPHY: FROM AUTOMATED GEOGRAPHY TO THE EMERGING VIRTUAL GEOGRAPHY DEPARTMENT

Although computers had been envisioned and experimented with for centuries (Goldstine 1993), the first serious electronic computer was not built until 1945 (ENIAC) and improved upon in 1952 (UNIVAC). But computers were not widely distributed until the 1960s with the invention of the IBM 360 series in 1964 (Freed and Ishida 1995; Williams 1997). Both military and academic computing in the 1950s and the 1960s were devoted to mathematical/ statistical estimations of relationships. Even though the early computers were unbelievably slow by today's standards, they were thousands of times faster at processing large datasets, at statistical estimation, and at simulating physical and social processes than what had been possible before. Programming languages arose in the mid-1950s, and early statistical packages, such as BMDP, appeared by 1961.

2.1. Computing Comes to Geography

According to Haggett (1969), mechanical calculators were first available to geographers around 1950, but it seems that the use of real electronic computers in geographic research did not begin until around the mid-1950s.

Although it is hard to trace who, when, and where computers were first used in geographic research, it is probably safe to claim that Bill Garrison and his students at the University of Washington in the U.S. (Morrill 1984) and Torsten Hägerstrand and his associates at the University of Lund in Sweden were among some of the earliest pioneers using computers to conduct substantive geographic research around the mid-1950s. The U.S. National Science Foundation, even before there was a Geography and Regional Science program, sponsored the seminal Summer Institutes in Quantitative Methods (1959-1962), as they did in other social sciences. The foundation later supported equipment grants.

In regard to publications, Garrison (1956) was one of the first geographers to use computers to study the impacts of rural transportation infrastructure on the reduction of rural poverty, and later on studies of highway development and geographic change (Garrison 1959). Tobler (1959) was among one of the earliest users of computers to automate certain aspects of the cartographic process. Barry (1960) reviewed the use of punched cards in geographic research. Moser and Scott (1961) pioneered in using computers to study the socioeconomic differences among British towns. Morrill (1963) explored the use of computers in migration modeling. Computers were used by Coppock (1964) to compile an agricultural atlas of England and Wales. Hägerstrand (1967) also used computers to conduct stochastic modeling using Monte Carlo simulation in diffusion studies, which convinced him that computers would become an indispensable tool for geographers in the decades to come.

There are several comprehensive reviews on the earlier use of computers in geographic studies (Kao 1963; Hägerstrand 1967; Coppock 1967; Tarrant 1968; Haggett 1969; Gould 1970; Rhind 1989). Pitts (1967) responded to numerous attacks on computing and "quantification." Marble (1967) published the first compilation of actual computer programs used by geographers. Liao and Scheidegger (1968) developed a computer-based hydrological simulation model of drainage networks. Following Tobler's (1965) lead, cartographers continued the exploration of using computers in mapmaking (Monmonier 1965; Balchin and Coleman 1967; Gaits 1969; Scripter 1969; Rosing 1969). One of the earliest primitive mapping programs SYMAP (SYnagraphic MAPping) appeared in 1966, and by the late 1960s, it was already used in substantive geographic research (Megee 1968). All these computing activities paved the way for the future development of GIS (Foresman 1998).

For about three decades from the mid-1950s to the mid-1980s, geographers used primarily mainframe computers for data processing. The key hardware revolution came with the invention of the silicon chip (INTEL) in 1971, which ushered in a continuing and competitive race for faster processing, from several hertz in 1970 to several gigahertz by 2002 (Williams

1997). The first practical home computers appeared by 1973 (Apple's Mac in 1975), even as academic computers became ever smaller, larger, faster, and more efficient. These early computing opportunities were embraced by the younger, more quantitative geographers from the start, and spread to the wider discipline after the availability of personal computers and more useful software, that is, mainly after 1975. Before the mid-1970s, computing was confined to a handful of pioneers and the students they trained (Dawson and Unwin 1984). The key revolution for the wider diffusion of computing was the development of software for the data spreadsheet (1974) and for word processing (1975). While the spreadsheet (data table) has been of utmost importance to business, economics, and government, as well as the academy (efficient data entry and analysis), word processing has been the more astounding revolution through more efficient writing and printing for both humanities and the sciences.

3. IMPACTS OF THE COMPUTER ON GEOGRAPHY

The increasing use of computers has allowed geographers faster processing of larger sets of data. The increasing speed and capacity of computers allowed and encouraged the use of datasets that were simply unmanageable and tedious only twenty years ago. Especially in physical geography (e.g., huge satellite imagery datasets, such as Landsat 1972) but also in human geography (very large census files, like the 5 percent Public Use MicroSample (PUMS), or time-hogging analyses like clustering), far more comprehensive datasets can be used and far more complex models can be run (Kirby et al. 1975; Mather 1976; MacDougal 1976). Although coding of qualitative data (recorded interviews, answers to open-ended survey questions) remains slow, direct computer answering of surveys has vastly eased data entry, data analysis, and textual search. By 1998 a task could be completed in 12 hours on a 512-processor Cray T3E parallel supercomputer that in 1980 would have taken approximately eight years of nonstop computing on a top-end workstation (Openshaw and Adrahart 2000).

The increasing use of computers has also contributed to the creation of large amounts of data with better spatial and temporal resolutions. The digital processing of satellite imagery began as early as 1965 in the military, but has exploded in the last decade with the declassification of high-quality, frequently updated images of most parts of the earth's surface. These images permit the analysis of change and processes, both at short-term (weather, traffic) and long-term (physical and human land use), and also allow recognition of complex and previously hidden patterns revealing geological and historic processes. By the late 1990s, satellite images with 1-meter resolution were commercially available for civilian applications. Corporations in insurance, credit, and marketing started collecting massive consumer data at the individual level. Also access to vast amounts of spatial and nonspatial data has become unbelievably easy and convenient. Nowadays, even personal office computers (especially if networked) can process very large datasets. Only 25 years ago the coauthor had to take boxes of punch cards to the computer, usually over in the College of Engineering, wait patiently for the bulky printed output, and only then discover that a stupid input error had ruined the whole operation. It is perhaps conservative to suggest that the efficiency of going from data entry to visualizing results has increased at least a millionfold in the last 50 years.

With faster processing of more data comes the capacity to estimate ever more complex and nuanced relationships and structures in data and to design and test far more comprehensive models of behavior. It is now possible to execute astoundingly intricate and useful simulations of the evolution of human landscapes (both past reconstructions and future scenario development) (Batty and Longley 1994; Wilson and Burrough 1999). The most recent developments in geocomputation (Longley et al. 1998; Openshaw and Abrahart 2000), especially in data mining and knowledge discovery techniques (Miller and Han 2001), have pushed the computing capabilities for spatial data to a new height.

Computers have made writing, editing, and publishing much faster and easier. Word processing perhaps affects the largest number of academics, including geographers, entrapping even those who otherwise eschew technology or quantification. It became so efficient that despite our irritation with the programs, and the real risk of long-winded, sloppy writing, the easy editing and rearrangement (and we include here presentation preparations, such as *PowerPoint*) vastly eased the writing of articles, reports, speeches, theses and dissertations, and the preparation of class materials. In addition, the productivity of nonteaching staff (administrators, secretaries, advisors) has been immensely increased, as they are freed from manuscript preparation and much classroom material preparation—all aided by a number of specialized computer programs and printers.

Early users of computers were dominantly male in a tight network focusing on quantitative methods and mathematical models (Barnes 1998). But now women are becoming the leaders in qualitative methods, and the computing gap has narrowed greatly over the years. Also computers are increasingly important in bridging qualitative and quantitative approaches (Philip 1998).

3.1. GIS

With the maturing of GIS technology and the further convergence among remote sensing, global positioning systems (GPS), and computer cartography, visualizing spatial data and mapmaking have been made much easier. Very likely the GIS revolution is the single most visible tangible consequence of the computer revolution for geographic practice and perhaps for its visibility (Dobson 1993). This theme is covered in separate chapters in this volume, but must be briefly assessed here. Only thirty years ago, map creation was a slow, tedious endeavor, which often entailed copying base maps of dubious positional accuracy and essentially hand-executed, with ink and zipatone. It may be conservative to say that it is a million times faster and probably far more spatially accurate to make the same maps today.

The components of the cartography revolution included, first, the creation of digital point, line, and area shape files (lat/long streams), beginning with the 1970 DIME files by the U.S. Bureau of the Census. These files, now covering much of the world, contain a vast amount of geography: topography, roads, structures, addresses, covering thousands of kinds of territories. They include, for example the complete content of all the USGS topographic maps, the entire street system with address matching of many countries, and millions of U.S. city blocks. Second, mapping software programs display much of this geography at any scale, with any projections, three- or two-dimensionally, with extended options for line, pattern, color, and text. Third, there is now the capacity to overlay this geography with an immense amount of independent information about the places, lines, and areas. Fourth, there is the capacity to print the maps cheaply in color or transparent paper if desired, to imbed the images in presentations and manuscripts, and to e-mail them around the world.

If as geographers we believe that the map in itself communicates patterns and reveals processes and hypothesized relationships, the research potential of the ease of generating subtly different maps is an invaluable asset. And in combination with locational analysis programs like shortest path, pattern recognition, optimum location, or political redistricting algorithms, maps are becoming a powerful tool for government and business. All of these accomplishments have made it possible to communicate geographic information in a digital age to a much broader audience with greater ease and speed (Goodchild 2000).

Computers have extended geographers' eyes (via remote sensing and air photos), their hands (via computer-assisted cartography), their mouth (via the World Wide Web and telephones), and their minds (via data processing and artificial intelligence). These extensions have greatly facilitated the acquisition, visualization, processing, and communication of unprecedented amounts of geographic data. As a great dividend of such extensions, computers, with their increasing integration with GIS and GPS, have been increasingly successful in assisting the learning and navigation experiences of persons with disabilities, like mobility assistance for the blind and visually handicapped

(Golledge 1993). In addition, rapid innovations have fostered the development of specialized instruments and software tools that can greatly facilitate geographic research. For example, GPS, laptops, portable digital assistants (PDA), handheld computers, digital cameras, portable scanners, and electronic data loggers all raise the efficiency of field work as well as library retrieval. Project Batutta is developing a new generation of field devices that potentially can greatly facilitate geographers' field work (http://dg.statlab.iastate.edu/dg).

3.2. Trends in Computer Technology

Currently there is substantial convergence between different portable technologies, including cell phones, personal digital assistants (PDAs), laptops, palm computers, and portable GPS receivers. The seamless integration of the GPS and cellular technologies has facilitated a new generation of mobile electronic devices capable of measuring their position on the Earth's surface and of using that data to modify the information they collect and present. The Wireless Communication and Public Safety Act (U.W.) of 1999 permits operators of cellular networks to release the geographic locations of users in certain emergency situations. Also, a range of electronic services is now being developed and offered to assist users in finding nearby businesses and other facilities. A location-based service (LBS) is an emerging and rapidly maturing information service that exploits the ability of technology to know where it is, and to modify the information it presents accordingly. LBS technology is inherently distributed, mobile, and potentially ubiquitous. Its services can augment the information provided directly to observers through the normal human senses by allowing them to access information in databases that represents what cannot be sensed, either because it is beyond the reach of the senses, as was true in the past, or might be true in the future. Its services can also allow data to be analyzed as they are collected, in a progressive construction of knowledge. According to Goodchild (2002), LBS technology represents only the beginning of a series of technological innovations that can potentially impact society in numerous ways, ranging from surveillance and the invasion of personal privacy, to technologically induced changes in human spatial behavior, the role of location in social networks, and the spatial structuring of retail and other services. LBS has the potential to provide novel sources of data to social science, including detailed information about daily activities and their locations. In addition, LBS technology has the potential to allow researchers to access databases and conduct sophisticated analyses of data, while located in the field, and immediately following acquisition of these data. As such, it may eventually revolutionize social science fieldwork as well as have profound social and ethical implications (Shiode et al. 2002).

3.3. The Computer Revolution and Automated Geography

With the introduction of the PC and the increasing availability of spatial data, Dobson (1983) suggested the future of geography cannot be separated from computing. Echoing earlier optimism by Hägerstrand (1967) and Haggett (1969) on the future applications of powerful computers, Dobson (1983) proposed the development of Automated Geography in 1983—the year when the "Man of the Year" in *Time Magazine* was a computer! According to Dobson (1983), geographers should rally behind the banner of automated geography to compute our way into the future. Despite the mixed responses from geographers (Marble and Peuquet 1983), the diffusion of computers proliferated in geography throughout the 1980s, largely through the growing popularity of GIS, culminating in the late 1980s with the establishment of the National Center for Geographic Information and Analysis (NCGIA) funded by the U.S. National Science Foundation (NSF). NCGIA played an important role in moving GIS from essentially a technical discipline to a respected geographic information science (GIScience) (Goodchild 1992). Project Varenius funded by NSF, further advanced the development of GIScience throughout the 1990s. In Europe a call for the establishment of computational geography by Openshaw (1994) and the recent push for GeoComputation (Openshaw and Abrahart 2000) paralleled many of the recent developments in the U.S. At the end of the 1990s, geography and computers are integrated in a more intellectual sense (Armstrong 2000).

In retrospect, it is obvious that the so-called quantitative revolution that took place (Burton 1963; Barnes 2001) in geography during the 1950s and 1960s was only possible with the invention and subsequent adoption of computers in geographic research. It was the consensus among the "space cadets" that if geography were to become respected, it had to be concerned with theory, and that the search for and testing of theory required better methods than previously used by geographers. Most, if not all, applications of quantitative methods in major geographic journals between 1956-1986 were basically computer-based (Slocum 1990). Computers have greatly facilitated geographers' quest for theories and concepts. During the first three decades, computers were used primarily as tools for data storage, analysis, and visualization/mapping. By using computers, geographers were able to investigate a whole new realm of intellectual territories that had been closed by the sheer intractability of multivariate problems and the huge quantities of information. New problem areas, whose geographic pertinence was never suspected before, have been opened by statistical insight that is now feasible precisely because it is computable.

Computers did not simply automate the manual processes that geographers have been working on for years; they stimulated new lines of inquiry, accelerated the quest for laws and theories, and addressed serious spatial issues in statistics, including the modifiable areal unit problem (Openshaw and Taylor 1979), spatial autocorrelation (Cliff 1970), and development of new spatial statistical techniques (Getis and Ord 1992; Fotheringham et al. 2000). Computers also stimulated the growth of new branches of geography. The most visible of all were the births of quantitative geography (Barnes 2003), computer-assisted cartography, analytical cartography, remote sensing, and GIS. AAG Specialty Groups were formed to further such specific areas as Mathematical Models and Quantitative Methods (now Spatial Analysis and Modeling), GIS, Microcomputers, and the World Wide Web.

The proliferation of GIS and the increasing use of georeferenced data have made the broader social science community recognize the key role that space plays in human society. The establishment of the Center for Spatially Integrated Social Sciences (CSISS) (www.csiss.org) at the University of California, Santa Barbara in 1999 aimed to promote research that advances our understanding of spatial patterns and processes. Cartographic visualization, GIS, pattern recognition, spatially sensitive statistical analysis, and place-based search methodologies are the tools of an emerging spatially integrated social science (SISS). SISS tries to integrate knowledge across disciplines and paradigms across the social and behavioral sciences. According to Goodchild et al. (2000), SISS views space as integrating social processes and sees social science problems as processes in place. CSISS uses GIS to integrate data by location and also uses spatial analysis to integrate multidiscipline views. Geographers, especially those interested in GIS and spatial analysis, played a very prominent role in expanding GIS to other branches of social sciences through the activities of CSISS. In reviewing the application history of computers in geography, it becomes apparent that computers not only helped making geography a more respected discipline among social scientists in the 1960s and 1970s, but also that geographers, as they enter the 21st century, are beginning to play the leadership role among social scientists in addressing important social issues that have a spatial dimension.

3.4. The Internet and the Virtual Geography Department

To the general public, the most revolutionary aspect of the computer age, along the widespread availability and use of personal computers, is the rise of the Internet and the World Wide Web. Although the germs of the Internet began with the Ethernet (1973), Arpanet (1975), Bitnet (1981), the NSF-funded academic (.edu), then government (.gov) e-mail communication, followed by file transfer (FTP) in 1975, Transmission Control Project (TCP) (1982) capability, and the World Wide Web (WWW) in 1992, the real potential of the web was not realized until 1995 when the net browsers were developed (Hafner and Lyon 1998). Rarely in human history has an innovation proved so popular so quickly.

Even before the Internet, the capacity to generate and print out manuscripts, carry out analyses, and generate computer maps increased scholarly communication among academics, and between academics and governmental and business clients and data sources. But e-mail and the capacity to attach files (documents, spreadsheets, maps, photos, and other graphics) was a quantum leap to collaboration, as members of a dispersed team of researchers, or between authors and editors and reviewers, between persons with related interests, and increasingly, those across disciplines interested in topics ranging from global climate change to immigration policy communicate via the Internet.

So quickly have we become dependent on the web for information, and for data files, maps, and photos to download for our research and teaching, that we can rarely remember that this has been possible for only seven or eight years. Among the most surprising aspects, thus far, is that so much of the searching, and even of the files for downloading, are ostensibly "free," even if the paper copies (as of pdf files) may be quite expensive. Amazingly large amounts of data are available from federal and state agencies, scores of nongovernmental organizations, and even from corporations (e.g., map shape files from ESRI or from the U.S. Bureau of the Census). Hundreds of maps, created by scores of agencies and organizations can be downloaded and printed for free. The quest for data is increasingly aided by ever better search engines, which also serve to generate job prospects, to find potential clients, and to identify persons with similar interests.

Geographers and geography departments across the world, especially those in the U.S., have been quick to tap into the vast potential of this new telecommunication medium, culminating in the ambitious project to build a Virtual Geography Department (VGD). With the growth in routine use of the Internet (such as e-mail and web surfing) among geographers, and with more and more courses, journals, books, and data available on-line, a new geography department (which is not yet listed in the AAG *Guide to Programs in Geography*) emerged around 1997—the so-called virtual geography department (VGD). The VGD was first proposed by Foote in his NSF-funded project to develop and link web-based geography materials to create a complete geography curriculum on the web (Foote 1999). The goal of the VGD is to produce and disseminate geographic knowledge and geography course materials. Broadly speaking, the VGD should include, but (because of the rapidity of technological innovations) not be limited to: (1) the virtual community of geographers formed via e-mail, discussion lists, bulletin boards, Internet relay chat rooms, etc.; (2) the growing number of geography-related websites on the world-wide web (the most prominent of which include the VGD and the GeographyWeb at the University of Colorado, NCGIA's on-line GIScience and Remote Sensing curriculum, the Alexandria Project (virtual library) (http://www.Alexandria.uscb.edu), and the GDN project in the U.K.); (3) the increasing number of on-line journals for knowledge production and dissemination; (4) desktop conferencing and on-line degree programs; and (5) multimedia products on CD, such as ARGUS, Human Dimensions of Global Change (HDGC), Geographic Inquiry into Global Issues, and GeographyCal in Britain (Olson 1997). As these and other programs and products continue to develop, the VGD's status and popularity among students, geographers, and the general public is likely to grow rapidly.

Although these different components of the VGD share a common characteristic (viz., use of the computer as the medium for storage, display, and communication of information) each component of the VGD utilizes a different, electronically-mediated, communication-network topology that dramatically affects the way people interact within it (Adams 1998). Since more and more web-enabled products are pouring into the market, more and more browsers incorporate all of what used to be individual technical components, such as e-mail, multimedia, video conferencing, etc.—everything is being drawn into the Web. This process is similar to the transformation brought about by writing and the printing press, viz., the emerging electronic medium is more than a technical advance—it is a source of deeper conceptual and cognitive changes among the individual users and in society as a whole. Many literary scholars have observed that the new electronic medium is creating a new form of reading and writing and thus may precipitate a sea-change in the meaning of literacy (Lanham 1993; Landow 1992). To understand the impacts of the VGD on geography, we must move beyond the current technically oriented discussions. This virtual geography department is not simply an expansion of the geographer's "invisible college," but is part of the emerging cyberspace created by the vast computer networks. Computers have not only transformed the world of geography, but also the geography of the world.

4. COMPUTERS AND THE GEOGRAPHY OF THE WORLD: FROM DIGITAL INDIVIDUALS TO DIGITAL EARTH

Computers are not simply tools, but are actually becoming part of the world that we are trying to study using the very same tools. This brave new world, different from the physical, tangible world geographers have studied for thousands of years, is virtual, digital, and ephemeral. In this virtual world, everything geographers study has become bits of information, which can be transmitted across the globe instantaneously (Brunn 1998). Each of us is not only an information processor, but also information processed. In other words, we have literally become digital individuals: our identity is more and more equated to digital information such as Social Security and credit card numbers, multiple ID and PIN numbers. Some digital individuals (cyberpunks more appropriately?) are increasingly hanging out in virtual communities, information cities, or digital places that do not have geographic propinquity (Horan 2000). The so-called E-topia (Mitchell 1999) is the city of bits, and electropolis has replaced Utopia as a new form of urban life, most of which basically lives on the screen (Turkle 1995). Indeed, even the entire earth is now becoming digital as embodied in the concept and initiative of a digital earth (www.digitalearth.gov). Such a relentless pursuit of technological progress will continue as long as humanity exists. To some technologists, seeking salvation via technology is a deeply embedded religious impulse (Noble 1999).

This brand new digital virtual world has provided geographers with a fascinating new subject of study (Kotkin 2001; Kellerman 2000, 2002a, b). Geographers have tried to visualize this basically invisible world of the cyberspace using various mapping techniques (Dodge and Kitchin 2000; Zook 2000a; Brunn and Dodge 2001); Goodchild (2001) explored the utility of classic location theory in analyzing distributed computing facilities and e-commerce. Malecki (2002a, b) presented new evidence on how economic geography as defined by the Internet's infrastructure contributes to the urban competitiveness among cities. Environmental impacts of the emerging digital economy were explored by Sui and Rejeski (2002) and Wilsdon (2001). Scholars have also tackled issues of accessibility in the age of the Internet (Janelle and Hodge 1999; Moss and Townsend 2000), the growing digital divide (Schön et al. 1999; Norris 2001; Warf 2001), ethical/legal consequences of increasing surveillance (Curry 1998; Castells 2001; Dobson and Fisher 2003), the overall impacts of the digital revolution on the landscape (Kotkin 2001) and the spatial distribution of economic activities (Zook 2000b, 2001, 2002a, b; Leanmer and Storper 2001; Leinbach and Brunn 2001).

From these empirical studies and theoretical exegesis, we learned that computers have had and will continue to have enormous impact on society, on human behavior and interaction, on human settlement and the character of places, on the meaning of distance and of regions, and on the degree of spatial variation and inequality across the landscape (Kolko 2000, 2002; Wilson and Corey 2000; Leamer and Storper 2001). As a consequence, the content of geography has changed, for example, via concepts such as globalization (or glocalization), space-time compression, and space-time distanciation (Giddens 1984; Brunn and Leinbach 1991).

4.1. The Computer as Part of Modernization

Overall, the computer is another product and facilitator of modernization, a technological complex that alters social structures, rearranges power and wealth, and indirectly, transforms the landscape and the interactions that geography seeks to understand. Computers, like the automobile and the telephone, were initially tools for the rich and powerful, but the logic of capital and their sheer utility led to a fairly rapid spread to the middle and even lower classes. Computers were available first to the elite, but have diffused rapidly across the classes and the globe, not yet to the extent of television, but potentially even more extensively (in part because of their fusion with television). The digital divide from local to global levels has enlarged the gulf between information-rich and information-poor.

As a tool of modernization, the computer acts to increase productivity and the efficiency of production, communication and trade. It also contributes to weaken the frictional power of distance yet more, and thus to encourage easier and more global communication, travel, trade and economic organization, and ultimately, information and culture. Indeed, the worldwide networked computers have created what Bill Gates (1999) called friction-free capitalism.

4.2. The Computer and Globalization

Although not the original purpose of computers, the rapid, cheap worldwide transfer of messages, documents, data, and images has brought about yet another shrinking of distance and abetting of globalization in all its economic and cultural forms. And even though access to computers for personal, corporate, or group gain reflects extant power and income differences, the computer has become sufficiently universal to have a vast democratizing effect, enabling communication and mobilization for almost any group or cause. On the other hand, the computer also extends the spatial range of control, fostering worldwide corporate mergers that enabled economic and political power to partially supercede national boundaries. In addition, alternate channels for advertising, buying, and selling are transforming the traditional retail industry as we know it. While catalogs have been around for a century, e-commerce (and more generally persuasive communication) has a vast yet partly unknown potential for selling, shopping, and delivery of goods and services, and for recruitment of persons (for jobs and for causes). The web is immensely valuable for accessing information about rare and obscure goods, services, persons, and places, even from the most remote places, as exemplified in the success of E-Bay.

Of course the global economy is hardly new. Hugill (1993) showed that world trade can be traced back to the early European trading companies in the 15th century. But the information control capacity enabled by computers permits the realization of an ever greater specialization. In this sense, concentration of high-level control functions occurs in the very highest-order world centers, while the dispersion of production and distribution functions relates to wherever offers the most cost-effective resource, labor, tax, and political advantages. Based upon what we know so far, the pronouncement on "the death of distance," as made by Cairncross (1997) in the information age, is premature (Sui 1999). The earlier speculations on an anytime-anywhere-anything paradigm turn out to be overly simplistic. A more realistic depiction of new economic activities seems to involve simultaneous concentration and dispersion at various scales (Graham 1998). The net effect of these trends is to spread economic development, incorporating everyone and everyplace into a world market, a global structure of capital, of production, of trade and transport, across countries and across sectors; but at the same time, these trends seek to maintain and increase inequality, as well as cause the erosion of local and regional (and often national) autonomy (Graham and Marvin 2001).

Despite the democratizing and homogenizing potential, the information age so far has resulted in a larger concentration of power and wealth and an increase in inequality. Information is power. The 1990s, the era of the Internet, was a decade of rising inequality, perhaps *within* as much as *between* countries, as increasing returns to education strengthened the professional and managerial, especially in the information and technology sectors. Although in theory "footloose" (and there have been many examples of entrepreneurs in rural places or smaller cities (e.g., Gateway Corporation in Sioux Falls, SD) on the strength of easy computer connection to the world, ironically the net geographic effect has been the ever greater concentration of these sectors, these professionals, and their wealth and influence, in a rather small number of global cities and their amenity playgrounds (Wilson 2001; Townsend 2001a, b). It is in these places where inequality has worsened the most. This inequality could be an effect of the vanguard of innovators, to be followed by routinization and dispersal, and even some restraints on the most excessive forms of financial and economic behavior; but there is no evidence thus far.

4.3. The Computer and Cultural Diversity

The explosion of information at the global level has also contributed to spread unequal cultural forms (DiMaggio et al. 2001). The computer, especially in the form of e-mail and file transfers, is the most recent and perhaps most effective technological agent for cultural diffusion, reinforcing the global reach of ideas, fads, values, information about products and technology that were already spread by television, movies, newspapers, magazines, and people themselves. And despite the effects of fearful conservative governments and of cultural-protection institutions, the computer (as China has found) is difficult to control. Still, while the influence and information come primarily outward from the wealthier and more technologically and culturally tolerant western centers of power, a counterflow of information and values comes from the peripheries, leading to surprising alliances among interest groups as well as efforts to preserve highly valued differences, such as languages at risk of extinction or local cultural resources.

The tendency of advanced technologies to homogenize culture also triggers antiglobalization and antiwestern reactions in an attempt to rebel against hegemonic control. As it has been true for centuries, more aggressive, technically superior ideas and products threaten customary behavior, traditions, and institutions, tending first toward a degree of homogenization of demanded goods and services, and economic forms of production and exchange, then followed by similar political, social, and artistic practices. But at the same time, those aggressive centers selectively adopt diverse products and ideas from less technically advanced areas. Again, over the centuries these invasions engender antimodern, antiwestern reactions, the defense of traditional modes, and cultural and economic autonomy. The conflict occurs deeply within and between countries, and in military (Afghanistan) as well as political (across classes), academic, and other forums. The on-going war on terrorism is in a sense a manifestation of Samuel Huntington's (1996) thesis on the *Clash of Civilization* thesis or of Robert Kaplan's (2000) anticipated coming age of anarchy. Local conflicts in the Middle East, Chechnya, the Korean Peninsula, and across the Taiwan Strait, plus various parts of Africa, and the recent financial crisis in Latin America all signal what Robert Kaplan (2000) called "the coming anarchy" in the global information age.

4.4. Digital Earth

Perhaps the most ambitious concept to exemplify the potential of computers both as tools and as reality is the concept of "digital earth." In a speech delivered at the opening of the California Science Center in Los Angeles in January 1998, then U.S. Vice President Al Gore first proposed the concept of digital earth. According to Gore, the digital earth can be envisioned as "a multi-resolution, three-dimensional representation of the planet, into which we can embed vast quantities of georeferenced data" (www.opengis.org/info/ pubaffairs/ALGORE.htm). Following Gore's call, NASA developed a plan to operationalize Gore's vision of a digital earth, an immersive environment through which a variety of users could explore the planet, its environment, and its human societies. Digital Earth could be available at museums or libraries, and a more modest version might be available through standard WWW browsers running on a simple personal computer.

Digital Earth is inherently an intriguing concept for geographers. First, it has some of the properties of a *moonshot*, or a vision that can motivate a wide range of interdisciplinary research and development activities. Digital Earth challenges our state of knowledge about the planet, not only in terms of raw data, but also in terms of data access and the ability to communicate data through visualization. Moreover, it challenges our understanding of process in the invitation to model, simulate, and predict, since the concept should not be limited to static portrayal. Second, Digital Earth is interesting because of its implications for the organization of information. The prevailing metaphor of user interface design is the office or desktop, with its filing cabinets and clipboards. Many prototype digital libraries employ the library metaphor, with its stacks and card catalogs. But Digital Earth suggests a much more powerful and compelling metaphor for the organization of geographic information, by portraying its existence on a rendering of the surface of the Earth. This idea can already be seen in limited form in many current products and services, including Microsoft's Encarta Atlas. Third, at a much deeper level, Digital Earth may be regarded as a materialization of what *Teilhard de Chardin* called the noosphere, a sphere dominated by flows of digital information. Together with geographers' focus on the atmosphere, hydrosphere, and biosphere, the study of the geography of noosphere could be a new challenge for geographers in the 21st century. Indeed, with the emergence of digital individuals, digital places, and digital earth (Curry 1998), Arnold Toynbee's (1972) earlier vision of the etherialization of history has converged with the etherialization of geography.

5. TERRAE INCOGNITAE AND THE LIMITS OF COMPUTATION: WHITHER COMPUTATIONAL AND DIGITAL GEOGRAPHY?

From the above discussions, we can see that the computer has transcended its role as tools to become part of the reality geographers have been studying. More fundamentally, information has become the driving metaphor behind much of the cutting-edge scientific research. Reality, in some increasingly meaningful sense, is information. We must be keenly aware of the danger of mistaking the map for the territory, a common mistake in the cult of information (Roszak 1994). Throughout human history, the dominant machines during a certain age often determined the dominant metaphor used to conceptualize our world (Haken et al. 1993). For example, following the growth of cosmological theory in the 17th century, Newton conceived the universe as a cosmic clock; the steam engine inspired the science of thermodynamics in the 19th century; and the computer has, in the information age, developed into a powerful metaphor for understanding the universe. Indeed, Siegfried (2000) contends that the computer has become such a powerful symbol for the universe that scientists are in danger of mistaking the metaphor for nature itself. The computer has become as all-encompassing as clocks and steam engines in earlier days (Bolter 1984).

The outer boundaries of the computing metaphor are obviously demarcated by the research on the limits of computation, which have clearly revealed that certain aspects of reality are essentially noncomputable, or at least not computable according to the concept of computation defined by current computer technology as we know it. Research on computing theory has revealed perfectly precise problems that can never be solved computationally. These problems are intrinsically unsolvable even if wholly unreasonable amounts of time and space are made available for their solution. These undecidable problems such as the NP-complete problem or the halting problem seem to indicate that what we can capture using the current computer technology is only a few aspects of reality.

Certain aspects of the universe, including the human mind as opponents of strong AI (Artificial Intelligence) argued (Dreyfus and Dreyfus 1986; Dreyfus 1992), are essentially noncomputable. Gödel's incomplete theorem further implies that the limit of computation is not only physical but also logical. These fundamental research findings on the limits of computing demand that we be vigilant about the tools we use, and not become prisoners of our own making. What lies beyond the computing metaphor and the limits of computation is a vast terra incognita, waiting to be cultivated and explored through new levels of human creative imagination. Failure to recognize the fundamental limits of computation will make all of us like drunkards—searching for lost keys underneath the street light, not because that is where we have lost them, but simply because there is light there. We seem to be up against a paradox: the more we can compute, the clearer we see the limits of computation. Indeed, as the island of our knowledge expands, our shore of ignorance also stretches.

What do these dual trends along the scientific frontier have to do with geography? Evidently, though the connections are underappreciated, research frontiers in computational geography and GIScience echo the dual trends discussed above. First, the development of computational geography is driven by innovations in computer technology. The growing interests in geocomputation reflect our continuing effort to understand nature from a computational perspective. Second, the conceptual foundations of computation geography (or automated geography) have been questioned and challenged, especially by social theorists. The gist of the criticism is this: certain aspects of reality transcend computation and thus do not lend themselves to algorithmic processes. While acknowledging the utility of computers in data storage, spatial analysis, mapping, and problem solving, critics argue that reality is too complex to be adequately represented using our current technological apparatus. In other words, social theorists argue that there exist vast terrae incognitae that need to be cultivated using the mental apparatus of human imagination as well as technical tools. Although such a critique may be yet another attempt by those who fear that science may undermine their perceived truth, the dialogue between the computable and noncomputable camp has enriched our understanding of the role of computer tools and their limitations.

Recent technological innovations have made the entire world more digital, thus more computable, but the ultimate computability of geography has been challenged. Geographers have also engaged in a more critical discourse on the impacts of computers on the discipline and society (Mercer 1985; Pickles 1995). In sharp contrast to an earlier technocratic view on computers, critical discourse on the role of computers has enabled geographers to view technology in a more sophisticated, holistic manner. Many questions raised by geographers regarding the impacts of computers transcend the technical dimensions and focus, instead, on ontological, epistemological, and ethical issues. Computer-based analyses reveal as much about the systems we are conducting our research *in* as the reality they are supposed to be *about*. In other words, computer systems can shape our understanding of social or physical reality so that effects are due, not to *the phenomena measured*, but to *the systems measuring it*. In particular, Veregin (1995) warned of the perverse consequences of the reverse adoption of computers in geography: instead of using computers to solve problems, we might have consciously or subconsciously modified our institutions and concepts to meet the technical demands of our computers. Indeed, the tools we make will eventually make us.

A social and technological revolution challenges traditional ways and differentially impacts institutions and people. While few would decry the demise of carbon paper, the broader impacts of computers may not be totally benign. In geography, the traditional mapmaker could and did sneer at the crude early computer maps, but the productivity gap was too huge; the critique of overly simple computer maps did lead to far more sophisticated and even artistically sensitive programs. Yet the tendency toward monopoly and homogenization again threatens creative diversity and uniqueness. More broadly, the head-start and near-monopoly power of American/European computer technology sectors tends to lead to an arrogant, taken-for-granted control of scientific information and communication, and in turn to a homogenization of academic practice and thought, a technological and cultural hegemony that threatens intellectual and academic diversity.

Wondrous as they are, and much as they have raised our ostensible productivity, computers are not very helpful for the basics of creating knowledge—the conception of a problem and the choice of appropriate methods to investigate that problem. Significance and meaningfulness require human perception and subjective judgment. This may indicate the computer-led technological innovations cannot and perhaps never will make profound geographic thoughts obsolete. On the contrary, new technologies can make geographic concepts and theories more potent and relevant. Some of the recent works linking GIS and computational geography with conventional as well as social geographical theories have turned out to be quite fruitful. Berry's (1964) geographic matrix has been used to develop a pedagogic framework to link GIS with geography's intellectual core (Sui 1995). Hägerstrand's space-time prism was linked to a GIS framework to analyze the spatial-temporal patterns of residents (Kwan and Lee 2003) and inspired the development of a people-based (as opposed to place-based) GIS (Miller 2002). Francis Harvey (1997) showed that Hartshorne's holism is conceptually consistent with the overlay-based methodology in the context of GIS. Peuquet (2002) explored how recent developments in cognitive models can be used to enrich the spatial-temporal representations in GIS. Kwan's (2002) recent work on integrating GIS with feminist theories also demonstrated the possibility that GIS-based empirical work can reach a great theoretical depth. Earlier works on integrating naïve geography and indigenous, feminist, and public participation theory into conventional GIS modeling processes essentially represent new ways of thinking beyond the limits of computation. At the interface of these computing and noncomputing paradigms, we can expect exciting ground-breaking development, even espouse the potential to bridge the two cultures that have plagued the sciences and humanities for so long (Gilbert 1995).

While the Internet has made readily available truly vast set of data resources, historic as well as current, the GIGO (garbage in, garbage out) principle still applies. It is highly doubtful that the quality and accuracy of data have improved (at least in human geography), so neither has the quality of analyses, however sophisticated, if based on suspect data. This point raises a number of ethical issues (Kling 1996). From the lowliest student to the prize-winning professor, academics have long succumbed to the temptation of tampering with results, of ignoring contradictory evidence, of plagiarism, including of one's own work. The volume of such cheating and dishonesty has surely increased, as the computer and Internet have increased the volume of scholarly output. It is plausible that the share of unethical behavior has increased, because it is so technically easy. Student purchase of term papers on the Internet is the most publicized, along with the inventiveness of some journalists, but the problems are probably far more subtle and pervasive (Hester and Ford 2001; Ermann and Shauf 2003).

Maps have long been a propagandistic device, whether used by "good guys" or "bad guys" for military, sales, land use, transportation, environmental, or other causes. GIS has raised the capacity for manipulation to ever higher levels. Arguably the major geographic arena of computer application and productivity gain, GIS is also the showcase of horrendous and misleading maps, easily generated by quasi-monopolistic programs, and featuring scant attention to quality of representation, such as the necessities of projection, scale, class intervals, coloring, and treatment of absolute and relative values. Perhaps the use for which GIS can be least proud is its ability to gerrymander political districts to new depths of manipulation and discrimination (Monmonier 2001). Furthermore, reflecting American and British leadership in the computer revolution, the fact that English is the lingua franca of the Internet contributes to the perception of American hegemony in the information and control sectors of the global economy.

6. COMPUTERS AND GEOGRAPHERS' CRAFT: SUMMARY AND CONCLUSION

Undoubtedly the development of information technology in general and computer technology in particular have created the most far-reaching impacts for geography as a discipline and the world we live in. As tools, computers (in combination with other technologies) have extended geographers' eyes, hands, and brains. As an integral part of reality, computers provide geographers new objects of study or new reality to contemplate. Either consciously or unconsciously, we may have plugged ourselves in with an intellectual umbilical cord and become part of the machines (Gould 1985).

In many interesting ways, the computer has also returned geographers to the age of exploration—not exploration of the physical world, but a virtual, digital world connected by computers. The computer is empowering and the basis of stupendous gains in productivity, the spread of knowledge, the enabling of long distance communication and collaboration, and changing the practice of the academy–probably far more in the future than it already has. The impacts of computers on society and on the landscape are also in their infancy; and while history ever reminds us of the intense competitive drive toward concentration of power and wealth, and while those with such power are in the best position to maintain and increase it, overall the computer may be even more subversive than the printed word, and an inevitable and unstoppable agent for democratization. Perhaps most important of all, computers have contributed to a larger audience both in the academy and society, that spatial is special.

If the ultimate goal of geography is to better understand how nature works and how we humans can better organize our activities on the surface of the earth, then we must continue to push the development of geographers' craft. We must try to make everything computable on the one hand, and at the same time, recognize the fundamental limits of computation and build dialogues with a variety of different scholarly traditions. Or to put it more succinctly, we need to continue to pixelize the social and, at the same time, socialize the pixels (Geoghegan et al. 1998).

It may be coincidence but we find it significant that John K. Wright delivered his AAG presidential address entitled "*Terrae Incognitae*: The Place of the Imagination in Geography" in 1946–the same year the first computer was built at the University of Pennsylvania. The more we read Wright, the more we think Wright is speaking to us as eloquently today as he was to his audience in 1946. Wright (1947) warned the oncoming generation of geographers against becoming too thickly encrusted in the prosaic. Our current state of knowledge brings out sharply what Wright had called the contrast between the shadows of ignorance and the light of knowledge. Wright further observed that "indeed, the more brightly the light of our personal knowledge shines upon a region or a problem, the more attracted we are by the obscurities within it or concerning its entire extent" (p. 4). Terrae incognitae of various forms and degrees stand forth clearly to arouse our curiosity. Perhaps even more profoundly, Wright concludes that "the most fascinating terrae incognitae of all are those that lie within the minds and hearts of men."

Maybe geography should be more than a computational science in search of new algorithms. It should also be a humanistic science in search of meanings for those computations and a speculation about what lies beyond the limits of computation. More than ever before, we need inner reflections on the outer world to balance the unprecedented superficiality of its greedy consumption of information. Echoing J.K. Wright's AAG presidential address, Yi-Fu Tuan (1999) observed that: "progress seems to consign us to a perpetual state of anxiety by offering us a giant menu for every conceivable need and desire; and yet to choose wisely, all we have to work with is a computer–the one in our head–that is now some 50,000 years old."

ACKNOWLEDGMENTS

We are grateful for Joanne C. Morrill's extensive editorial help on an earlier draft of this chapter. Thanks are also due to Zengwang Xu and Wei Tu for their research assistance.

REFERENCES

Adams, P. (1998). Network Topologies and Virtual Places, Annals of the Association of American Geographers 88: 88-106.

Armstrong, M. (2000). Geography and Computational Science, Annals of the Association of American Geographers 90: 146-56.

Balchin, W. and Coleman, A. (1967). Cartography and Computers, The Cartographer 4: 120-27.

Barnes, T.J. (1998). A History of Regression: Actors, Networks, Machines and Numbers, Environment and Planning A 30: 203–23.

Barnes, T.J. (2001). Lives Lived and Lives Told: Biographies of Geographys' Quantitative Revolution, Environment and Planning D: Society and Space 19: 409–29.

Barnes, T.J. (2003). The Place of Locational Analysis: A Selective and Interpretive History, Progress in Human Geography 27: 69–95.

Barry, R.G. (1960). The Punched Card and Its Application in Geographic Research, Erdkunde 15: 140-42.

Batty, M. and Longley, P. (1994). Fractal Cities: A Geometry of Form and Function. London: Academic.

Berry, B.J.L. (1964). Approaches to Regional Analysis: A synthesis, Annals of the Association of American Geographers 54: 2-11.

.Bolter, J.D. (1984). Turings Man: Western Culture in the Computer Age. Chapel Hill: University of North Carolina Press.

Brunn, S.D. and Leinbach, T.R. (1991). Collapsing Space and Time: Geographic Aspects of Communication and Information. London: HarperCollinsAcademic.

Brunn, S.D. and Dodge, M. (2001). Mapping the Worlds of the World Wide Web: (Re)Structuring Global Commerce Through Hyperlinks, American Behavioral Scientist 44: 1716-39.

Brunn, S.D. (1998). The Internet as the New World of and for Geography: Space, Structures, Volumes, Humility and Civility, GeoJournal 45: 5-15.

Burton, I. (1963). The Quantitative Revolution and Theoretical Geography, Canadian Geographer 7: 151-62.

Cairncross, F. (1997). The Death of Distance: How the Communications Revolution is Changing Our Lives. Boston: Harvard Business School Press.

Castells, M. (2001). The Internet Galaxy: Reflections on the Internet, Business, and Society. New York: Oxford University Press.

Chen, C. (2002). Mapping Scientific Frontiers: The Quest for Knowledge Visualization. London: Springer.

Cliff, A. (1970). Computing the Spatial Correspondence Between Geographic Patterns, Transactions of the IBG 50: 143-54.

Coppock, J.T. (1964). Agricultural Atlas of England and Wales. London: Faber.

Coppock, J.T. (1967). Electronic Data Processing in Geographic Research, The Professional Geographer 14: 1-3.

Curry, M.R. (1998). Digital Places: Living with Geographic Information Technologies. London and New York: Routledge.

Dawson, J.A. and Unwin, D.J. (1984). The Integration of Microcomputers into British Geography, Area 16: 323-29.

DiMaggio, P., Hargittai, E., Neuman, W.R., and Robinson, J.P. (2001). Social Implications of the Internet, Annual Review of Sociology 27: 307–36.

Dobson, J.E. and Fisher, P.F. (2003). Geo-Slavery, IEEE Transactions on Computer (in press).

Dobson, J.E. (1983). Automated Geography, The Professional Geographer 35: 135-143.

Dobson, J.E. (1993). The Geographic Revolution: A Retrospective on the Age of Automated Geography, The Professional Geographer 45: 431-39.

Dodge, M. and Kitchin, R. (2000). Mapping Cyberspace. London: Routledge.

Dreyfus, H.L. and Dreyfus, S.E. (1986). Mind Over Machine: The Power of Human Iintuition and Expertise in the Era of the Computer. New York: The Free Press.

Dreyfus, H.L. (1992). What Computers Still Cant Do? Cambridge, MA: The MIT Press.

Ermann, M.D. and Shauf, M.S. (2003). Computers, Ethics, and Society. 3rd ed. New York: Oxford University Press.

Foote, K. (1999). Building Intradisciplinary Collaborations in the World Wide Web: Strategies and Barriers, Journal of Geography 98: 192-211.

Foresman, T.W. (1998). The History of Geographic Information Systems: Perspectives from the Pioneers. Upper Saddle River, NJ: Prentice Hall.

Fortheringham, A.S., Brunsdon, C., and Charlton, M. (2000). Quantitative Geography: Perspective on Spatial Data Analysis. London: Sage.

Freed, L. and Ishida, S. (1995). The History of Computers. Ziff Davis Press.

Gaits, G. (1969). Thematic Mapping by Computer, The Cartographic Journal 6: 50-68.

Garrison, W.L. (1956). The Benefits of Rural Roads to Rural Property. Seattle, WA: Washington State Council for Highway Research.

Garrison, W.L. (1959). Studies of Highway Development and Geographic Change. Seattle, WA: University of Washington Press.

Gates, B. (1999). Business @ the Speed of Thought: Using a Digital Nervous System. New York: Warner Books.

Geoghegan, J., Prichard Jr., L., Ogneva-Himmelberger, Y., Chowdhury, R.R., Sanderson, R., and Turner II, B.L. (1998). Socializing the Pixel and Pixelizing the Social in Land-Use and Land-Cover Change. In Liverman, D., Moran, E.F., Rindfuss, R.R., and Sterm, P.C. (Eds.) People and Pixels: Linking Remote Sensing and Social Science, 69-81. Washington, DC: National Academy Press.

Getis, A. and Ord, J.K. (1992). The Analysis of Spatial Association by Use of Distance Statistics, Geographical Analysis 24: 189-206.

Giddens, A. (1984). The Constitution of Society. Cambridge: Polity Press.

Gilbert, D. (1995). Between Two Cultures: Geography, Computing, and the Humanities, Ecumene 2: 1-13.

Goldstine, H. (1993). The Computer: From Pascal to von Neumann. Princeton, NJ: Princeton University Press.

Golledge, R.G. (1993). Geography and the Disabled: A Survey with Special Reference to Vision Impaired and Blind Populations, Transactions of the Institute of British Geographers 18: 63-85.

Goodchild, M.F. (1992). Geographical Information Science, International Journal of Geographical Information Systems 6: 31-45.

Goodchild, M.F. (2000). Communicating Geographic Information in a Digital Age, Annals of the Association of American Geographers 90: 344-55.

Goodchild, M.F., Anselin, L., Appelbaum, R.P., and Harthorn, B.H. (2000). Toward Spatially Integrated Social Science, International Regional Science Review 23: 139–59.

Goodchild, M.F. (2001). Towards a Location Theory of Distributed Computing and e-Commerce. In Leinbach, T.R. and Brunn, S.D. (Eds.) Worlds of E-Commerce: Economic, Geographical and Social Dimensions, 67–86. New York: Wiley.

Goodchild, M.F. (2002). Final Report on Specialist Meeting on Location-Based Services. http://www.csiss.org/events/meetings/location-based/goodchild_lbs.htm

Gould, P. (1970). Computers and Spatial Analysis: Extensions of Geographic Research, Geoforum 1: 53-69.

Gould, P. (1985). The Geographer at Work. Boston: Routledge and K. Paul.

Graham, S. (1998). The End of Geography or the Explosion of Place? Conceptualizing Space, Place and Information Technology, Progress in Human Geography 22: 165-85.

Graham, S. and Marvin, S. (2001). Splintering Urbanism: Networked Infrastructures, Technological Mobilities and the Urban Condition. London: Routledge.

Hafner, K. and Lyon, M. (1998). Where Wizards Stay Up Late: The Origins of the Internet. New York: Touchstone Books.

Hagerstrand, T. (1967). The cComputer and the Geographer, Transactions of the Institute of British Geographers 42: 1-20.

Haggett, P. (1969). On Geographical Research in a Computer Environment, Geographical Journal 135: 497-505.

Haken, H., Karlqvist, A., and Svedin, U. (1993). The Machine as Metaphor and Tool. Berlin Springer-Verlag.

Hall, S.S. (1993). Mapping the Next Millennium: How Computer-Driven Cartography is Revolutionizing the Face of Science. New York: Vintage Books.

Harvey, F. (1997). From Geographic Holism to Geographic Information System, The Professional Geographer 49: 77-85.

Hester, D.M. and Ford, P.J. (2001). Computers and Ethics in the Cyberage. Upper Saddle River, NJ: Prentice Hall.

Horan, T. (2000). Digital Places: Building our Cities of Bits. Washington, DC: Urban Land Institute.

Hugill, P.J. (1993). World Trade Since 1431. Baltimore, MD: Johns Hopkins University Press.

Huntington, S.P. (1996). The Clash of Civilizations and the Remaking of World Order. New York: Simon and Schuster.

Janelle, D. and Hodge, D. (1999). Information, Place, and Cyberspace: Issues in Accessibility. Berlin: Springer-Verlag.

Kao, R.C. (1963). The Use of Computers in the Processing and Analysis and Geographic Information, Geographical Review 53: 530-47.

Kaplan, R. (2000). The Coming Anarchy: Shattering the Dreams of the Post-Cold War World. New York: Random House.

Kellerman, A. (2000). Where Does It Happen? The Location of the Production and Consumption of Web Information, Journal of Urban Technology 7: 45-61.

Kellerman, A. (2002a). The Internet on Earth: A Geography of Information. New York: John Wiley.

Kellerman, A. (2002b). New York and Los Angeles: Global Leaders of Information Production, Journal of Urban Technology 9: 21-35.

Kirby, M.J., Burt, T.P., Naden, P.S., and Butcher, D.P. (1975). Computer Simulation in Physical Geography. New York: John Wiley and Sons.

Kling, R. (1996). Computerization and Controversy. 2nd ed. San Diego, CA: Academic Press.

Kolko, J. (2000). The Death of Cities? The Death of Distance? Evidence from the Geography of Commercial Internet Usage. In Vogelsang, I. and Compaine, B.M (Eds.) The Internet Upheaval: Raising Questions, Seeking Answers in Communications Policy, 73-97. Cambridge, MA: MIT Press.

Kolko, J. (2002). Silicon Mountains, Silicon Molehills: Geographic Concentration and Convergence of Internet Industries in the U.S., Information Economics and Policy 14: 211-32.

Kotkin, J. (2001). The New Geography: How the Digital Revolution is Reshaping the American Landscape. New York: Random House.

Kwan, M.P. (2002). Feminist Visualization: Re-nvisioning GIS as a Method in Feminist Geographic Research, Annals of the Association of American Geographers 92: 645-61.

Kwan, M.P. and Lee, J.Y. (2003). Geovisualization of Human Activity Patterns Using 3D GIS: A Time-Geographic Approach. In Goodchild, M.F. and Janelle, D.G. (Eds.) Spatially Integrated Social Science: Examples in Best Practice (forthcoming).

Landow, G.P. (1992). Hypertext: The Convergence of Contemporary Critical Theory and Technology. Baltimore: Johns Hopkins University Press.

Lanham, R. (1993). The Electronic Word: Democracy, Technology, and the Arts. Chicago: University of Chicago Press.

Leamer, E.E. and Storper, M. (2001). The Economic Geography of the Internet age, Journal of International Business Studies 32: 641-65

Leinbach, T.R. and Brunn, S.D. (2001). Worlds of E-Commerce: Economic, Geographical and Social Dimensions. New York: John Wiley and Sons.

Liao, L.H. and Scheidegger, A.E. (1968). A Computer Model for Some Branching-Type Phenomena in Hydrology, Bulletin of the International Association of Scientific Hydrology 13: 5-13.

Longley, P., Brooks, S.M., McDonnell, R., and Macmillan, B. (1998). Geocomputation: A Primer. Chichester, John Wiley and Sons.

MacDougal, E.B. (1976). Computer Programming for Spatial Problems. London: Edward Arnold.

Malecki, E.J. (2002a). Hard and Soft Networks for Urban Competitiveness, Urban Studies 39: 929-45.

Malecki, E.J. (2002b). The Economic Geography of the Internets Infrastructure, Economic Geography 78: 399-424.

Marble, D. (1967). Some Computer Programs for Geographic Research, Northwestern University Special Publication # 1.

Marble, D.F. and Peuquet, D.J. (1983). The Computer and Geography: Some Methodological Comments, The Professional Geographer 35: 343-44.

Mather, P.M. (1976). Computational Methods for Multivariate Analysis in Physical Geography. London: John Wiley and Sons.

Megee, M. (1968). Statistical Prediction of Mortgage Risk Using SYMAP, Land Economics 44: 461-69.

Mercer, D. (1985). Unmasking Technocratic Geography. In Billinge, M., Gregory, D. and Martin, R. (Eds.) Recollections of a Revolution: Geography as Spatial Science, 98-149. New York: St. Martins Press.

Miller, H.J. (2002). How About People in Geographic Information Science? In Unwin, D. (Ed.) Re-presenting Geographic Information Systems. New York: Guilford (forthcoming).

Miller, H.J. and Han, J.W. (2001). Geographic Data Mining and Knowledge Discovery. London: Taylor and Francis.

Mitchell, W.J. (1999). E-topia: Urban Life, Jim-But Not as We Know It. Cambridge, MA: MIT Press.

Monmonier, M. (1965). The Production of Shaded Maps on the Digital Computer, The Professional Geographer 12: 13-15.

Monmonier, M. (2001). Bushmanders and Bullwinkles: How Politicians Manipulate Electronic Maps and Census Data to Win Elections. Chicago: University of Chicago Press.

Morrill, R. (1963). The Development of Models of Migration and the Role of Electronic Processing Machines. In Sutter, J. (Ed.) Human Displacements, Entretiens de Monaco en Sciences Humaines, 213-30. Paris: Hatchett, Editions Sciences Humaines.

Morrill, R. (1984). Recollections of the Quantitative Revolutions Early Years: The University of Washington 1955-65. In Billinge, M., Gregory, D., and Martin, R. (Eds.) Recollections of a Revolution: Geography as spatial science, 57-72. New York: St. Martins Press.

Moser, C.A. and Scott, W. (1961). British Towns: A Statistical Study of Their Social and Economic Differences. London: Oliver and Boyd.

Moss, M.L. and Townsend, A. (2000). The Internet Backbone and the American Metropolis, Information Society 16: 35-47.

Noble, D.F. (1999). The Religion of Technology: The Divinity of Man and the Spirit of Invention. New York: Penguin Books.

Norris, P. (2001). Digital Divide: Civic Engagement, Information Poverty, and the Internet Worldwide. Cambridge, MA: Harvard University, John F. Kennedy School of Government.

Olson, J.M. (1997). Multimedia in Geography: Good, Bad, Ugly, or Cool?, Annals of the Association of American Geographers 87: 571-78.

Openshaw, S. and Taylor, P.J. (1979). A Million or So Correlation Coefficients: Three Experiments on the Modifiable Areal Unit Problem. In Wrigley, N. (Ed.) Statistical Applications in the Spatial Sciences: , 127-44. London: Pion.

Openshaw, S. (1994). Computational Human Geography: Towards a Research Agenda, Environment and Planning A 26: 499-505.

Openshaw, S. and Abrahart, R.J. (2000). GeoComputation. London and New York: Taylor and Francis.

Peuquet, D.J. (2002). Representations of Space and Time. New York: Guilford Press.

Philip, L.J. (1998). Combining qQuantitative and Qualitative Approaches to Social Research in Human Geography—An Impossible Mixture?, Environment and Planning A 30: 261-76.

Pickles, J. (1995). Ground Truth: Social Implications of Geographic Information Systems. New York: Guilford Press.

Pitts, F. (1967). Chorology Revisited, The Professional Geographer 14: 8-11.

Rhind, D. (1989). Computing, Academic Geography and the World Outside. In Macmillan, W. (Ed.) Remodeling Geography, 177-90. Cambridge, MA.: Blackwell.

Robinson, A. (1954). Geographical Cartography. In James, P.E. and Jones, C.F. (Eds.) American Geography: Inventory and Prospect, 553-77. Syracuse, NY: Syracuse University Press.

Rosing, K. (1969). Computer Graphics, Area 11: 27.

Roszak, T. (1994). The Cult of Information. Berkeley: University of California Press.

Schön, D.A., Sanyal, B., and Mitchell, W.J. (1999). High Technology and Low-Income Communities: Prospects for the Positive Use of Advanced Information Technology. Cambridge, MA: MIT Press.

Scripter, M. (1969). Choropleth Maps on Small Digital Computers, Proceedings of the AAG 1: 133-36

Shiode, N., Li, C., Batty, M., Longley, P., and Maguire, D. (2002). The Impact and Penetration of Location-Based Services. http://www.casa.ucl.ac.uk/working_papers/Paper%2050.pdf

Siegfried, T. (2000). The Bit and The Pendulum: From Quantum Computing to M Theory—The New Physics of Information. New York: John Wiley and Sons.

Slocum, T.A. (1990). The Use of Quantitative Methods in Major Geographical Journals, 1956-1986, The Professional Geographer 42: 84-94.

Sui, D.Z. (1995). A New Pedagogic Framework to Link GIS to Geography's Intellectual Core, Journal of Geography 94: 578-91.

Sui, D.Z. (1997). Reconstructing Urban Reality: From GIS to Electropolis, Urban Geography 18: 74-89.

Sui, D.Z. (1999). The E-merging Geography of the Information Society: From Accessibility to Adaptability? In Janelle, D. and Hodge, D. (Ed.) Information, Place, and Cyberspace: Issues in Accessibility, 107-29. Berlin: Springer-Verlag.

Sui, D.Z. and Rejeski, D.J. (2002). Environmental Impacts of the Emerging Digital Economy: The E-For-Environment E-Commerce?, Environmental Management 29: 155-63.

Sui, D.Z. and Goodchild, M.F. (2003). A Tetradic Analysis of GIS and Society Using McLuhan's Law of Media, Canadian Geographers (in press).

Tarrant, J.R. (1968). Computers in Geography, IBG Newsletter 6: 11-25.

Tenner, E. (1996). Why Things Bite Back: Technology and the Revenge of Unintended Consequences. New York: Knopf.

Tobler, W.R. (1959). Automation and Cartography, Geographical Review 49: 526-34.

Tobler, W. (1965). Automatic Production of Thematic Maps, The Cartographic Journal 2: 32-8.

Townsend, A.M. (2001a). The Internet and the Rise of the New Network Cities, 1969-1999, Environment and Planning B: Planning and Design 28: 39-58.

Townsend, A.M. (2001b). Network Cities and the Global Structure of the Internet, American Behavioral Scientist 44: 1697-1716.

Toynbee, A. (1972). A Study of History (updated edition). New York: Weathervane Books.

Tuan, Y.F. (1999). Progress and Anxiety. Keynote Speech delivered to the 1999 Southwest Division of the AAG (SWAAG), College Station, TX.

Turkle, S. (1995). Life on the Screen: Identity in the Age of the Internet. New York: Simon and Schuster.

U.S. Department of Commerce (2000). The Digital Economy 2000. Washington, DC. http://www.ecommerce.gov

Veregin, H. (1995). Computer Innovation and Adoption in Geography: A Critique of Conventional Technological Models. In Pickles, J. (Ed.) Ground Truth: Social Implications of Geographic Information Systems, 88-112. New York: Guilford Press.

Warf, B. (2001). Segueways into Cyberspace: Multiple Geographies of the Digital Divide, Environment and Planning B 28: 3-19.

Williams, M.R. (1997). A History of Computing Technology. 2nd ed. Los Alamitos, CA: IEEE Computer Society Press.

Wilsdon, J.P. (2001). Digital Futures: Living in a Dot-Com World. London ; Sterling, VA: Earthscan Publications.

Wilson, J.P. and Burrough, P.A. (1999). Dynamic Modeling, Geostatistics, and Fuzzy Classification: New Sneakers for a New Geography? Annals of the Association of American Geographers 89: 736-46.

Wilson, M.I. (2001). Location, Location, Location: The Geography of the Dot Com Problem, Environment and Planning B: Planning and Design 28: 59-71.

Wilson, M.I. and Corey, K.C. (2000). Information Tectonics. New York: John Wiley.

Wright, J.K. (1947). Terrae Incognitae: The Place of the Imagination in Geography, Annals of the Association of American Geographers 37: 1-15.

Zook, M.A. (2000a). Internet Metrics: Using Host and Domain Counts to Map the Internet, Telecommunications Policy 24: 613-20.

Zook, M.A. (2000b). The Web of Production: The Economic Geography of Commercial Internet Content Production in the United States, Environment and Planning A 32: 411-26.

Zook, M.A. (2001). Old Hierarchies or New Networks of Centrality? The Global Geography of the Internet Content Market, American Behavioral Scientist 44: 1679-96.

Zook, M.A. (2002a). Grounded Capital: Venture Financing and the Geography of the Internet Industry, 1994-2000, Journal of Economic Geography 2: 151-77.

Zook, M.A. (2002b). Hubs, Nodes, and Bypassed Places: A Typology of E-commerce Regions in the United States, Tijdschrift voor Economische en Sociale Geografie 93: 509-52.

CHAPTER 6

JOHN R. JENSEN
MICHAEL E. HODGSON

REMOTE SENSING OF SELECTED BIOPHYSICAL VARIABLES AND URBAN/SUBURBAN PHENOMENA

Abstract   Remote sensing science may be used to provide spatially distributed information for a significant number of models and applications. Information is extracted from remotely sensed data using the *remote sensing process,* which includes stating the problem, data collection (in situ and remote sensing), data-to-information conversion, and information presentation. The process requires a thorough understanding of the spatial, spectral, temporal, radiometric, and angular characteristics of the remotely sensed data. Advances in remote sensing of national spatial data infrastructure (NSDI) framework foundation variables (e.g., geodetic control, orthoimagery, digital elevation models) and selected framework thematic variables (e.g., land use/cover data, vegetation type and condition) are examined. The chapter concludes with a summary of advances in information extraction techniques and the integration of GIS and remote sensing.

Keywords   Remote sensing, biophysical, urban/suburban land use/cover, remote sensing process, sensor technology, GIS

1. INTRODUCTION

Scientists and natural resource managers of wetland, forests, grassland, rangeland, and others recognize that spatially distributed biophysical information (e.g., vegetation biomass, temperature, land cover) are essential for ecological modeling and planning (Johannsen et al. 2003). It is very difficult to obtain such information using in situ measurement. Therefore, public agencies and scientists have expended significant resources developing methods to obtain the required information using remote sensing science (Goetz 2002; Nemani et al. 2003). The most common rationale for interfacing remote sensing and ecosystem models is to use remotely sensed data to generate model initialization products (Kerr and Ostrovsky 2003). These input data correspond to forcing functions or state variables in ecological modeling. Earth-observing sensors, however, measure radiation reflected, scattered, or

Stanley D. Brunn, Susan L. Cutter, and J.W. Harrington, Jr. (Eds.), Geography and Technology,
109-154. © 2004 Kluwer Academic Publishers. Printed in the Netherlands.

emitted by the Earth surface. Therefore, radiation flux *data* must be transformed into *information* of use to ecological process modelers using some form of empirically or theoretically derived transfer function (i.e., a model). This approach is frequently applied to measure incident and reflected photosynthetically active and shortwave radiation, cloud cover, temperature or precipitation, and vegetation-related information like leaf-area index, or land cover (Plummer 2000). In addition, persons responsible for conducting research and/or planning in the urban/suburban environment require detailed spatially distributed information about existing land use, urban infrastructure (e.g., transportation system, utilities, hydrology, or multipurpose cadastre), population density, and other factors (Jensen and Cowen 1999).

This chapter provides an overview of some of the advances in remote sensing of biophysical and urban/suburban characteristics in the landscape. It begins by reviewing the fundamental characteristics of remotely sensed data. Next, selected advances in biophysical remote sensing during the last decade are presented by highlighting the improved array of sensor systems and standardized products coming forth. Due to space limitations, only national spatial data infrastructures (NSDI) framework foundation variables (e.g., geodetic control, orthoimagery, digital elevation models) and selected framework thematic variables (e.g., land use/cover data, vegetation type, and condition) are examined. A summary of selected advances in remote sensing information extraction is presented along with a discussion of the integration of geographic information systems (GIS) and remote sensing science.

2. REMOTE SENSING PROCESS

Remote sensing is defined by the American Society for Photogrammetry & Remote Sensing as:

> the measurement or acquisition of information of some property of an object or phenomenon, by a recording device that is not in physical or intimate contact with the object or phenomenon under study. (Colwell 1997)

This definition, however, does not describe the scientific goals of remote sensing, which are to derive information about the biological, chemical, and physical characteristics of objects above, on, or occasionally beneath the Earth's surface (Barnsley 1999). Furthermore, remote sensing is used to obtain information about phenomena at a tremendous variety of scales. Astronomers rely heavily on remote sensing science to explore the chemical, biological, and physical characteristics of extraterrestrial bodies in outer space. Remote sensing science is used by multidisciplinary scientists including geographers, geologists, agronomists, ecologists, oceanographers, atmospheric physicists, and urban planners to study the Earth and its environments. Microbiologists and others rely on remote sensing science to obtain information on extremely small objects of interest using electron microscopy (inner space).

The typical *remote sensing process* that is followed when conducting most terrestrial remote sensing projects is shown in Figure 6-1 (Jensen 2000):

- the hypothesis to be tested is defined using a specific type of logic (Curran 1987),
- in situ data necessary to calibrate the remote sensor data and/or judge its geometric and thematic accuracy are collected,
- remote sensor data are collected using a variety of remote sensing instruments, ideally, at the same time as the in situ data,
- in situ and remotely sensed data are processed using (a) analog image processing, (b) digital image processing, (c) modeling, (d) *n*-dimensional visualization, and
- information of value are summarized and presented in a format (graphics, tables, or graphs) that effectively communicates the desired principles.

Hopefully, the information is utilized in conjunction with other spatially distributed data in a GIS or decision support system (DSS) to make wise natural and cultural resource management decisions (Jensen et al. 2002a).

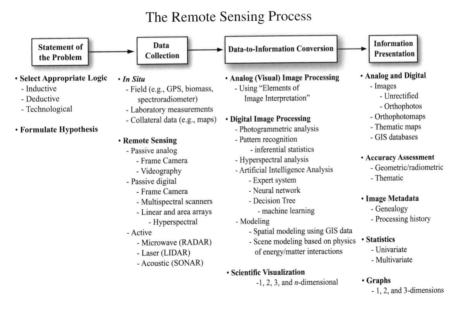

Figure 6-1. Most terrestrial remote sensing investigations utilize the remote sensing process.

3. REMOTE SENSING DATA COLLECTION

Remote sensing data collection is performed by placing a remote sensor onboard a suborbital aircraft or satellite platform. The sensor records electromagnetic radiation that is reflected, emitted, or scattered from the Earth's surface. *Passive* remote sensing systems record electromagnetic energy that is reflected or emitted from the Earth (e.g., cameras record reflected energy and a thermal infrared detector records emitted energy). *Active* sensor systems such as RADAR, LIDAR, or SONAR bathe the terrain in electromagnetic energy that was created by the remote sensing system and then record the back-scattered energy. The electromagnetic radiation R_e recorded within the instantaneous-field-of-view (IFOV) of a remote sensing system (e.g., a single silver halide crystal in an aerial photograph, a picture element in a digital image, or a unique return in a LIDAR collection) is a function of:

$$R_e = f\left(\lambda_\Omega, s_{x,y,z}, t, \theta, p\right) \qquad\qquad (6\text{-}1)$$

where,

λ = wavelength (spectral response measured in various bands or at specific frequencies) (Note: wavelength (λ) and frequency (v) may be used interchangeably based on their relationship with the speed of light (c) where $c = \lambda \times v$)

$s_{x,y,z}$ = location (x,y,z location of the silver halide crystal or picture element),

θ = angle [set of angles that describe the geometric relationship between the radiation source (e.g., the sun), the terrain target of interest (e.g., a wheat field), and the remote sensing system],

t = time (temporal information–when the information was acquired),

p = polarization of back-scattered energy, and

Ω = radiometric precision of the spectral information

There are a remarkable variety of remote sensing instruments (Jensen 2000). Characteristics of the major remote sensing systems capable of recording electromagnetic energy in the optical (ultraviolet, blue, green, red, and near-infrared), middle-infrared, and thermal infrared portion of the electromagnetic spectrum are shown in Figure 6-2.

It is instructive to briefly review characteristics of the parameters associated with Equation 6-1 and how they influence the nature of the remote sensing data collected.

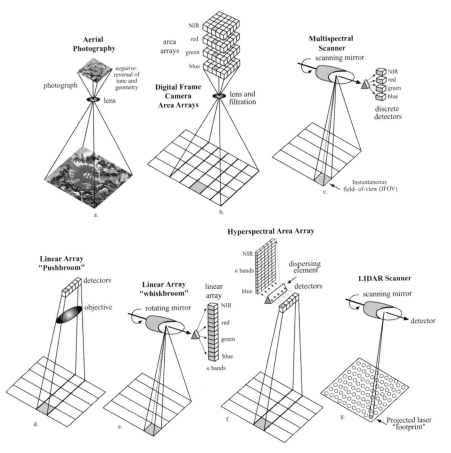

Figure 6-2. Various types of passive remote sensing systems that operate in the optical (ultraviolet, blue, green, red, near-infrared), mid-infrared, and thermal infrared portions of the electromagnetic spectrum.

3.1. Spectral Information and Resolution

The majority of remote sensing investigations in the optical portion of the electromagnetic spectrum are based on developing a deterministic relationship (i.e., a model) between the amount of electromagnetic energy reflected or emitted in specific bands and the chemical, biological, and physical characteristics of the phenomena under investigation (e.g., a wheat field canopy). *Spectral resolution* is the number and dimension (size) of specific wavelength intervals (referred to as bands or channels) in the electromagnetic spectrum to which a remote sensing instrument is sensitive. Remote sensing systems may be configured to collect data in just a single band of the electromagnetic spectrum (e.g., Space Imaging's IKONOS panchromatic band is 450–900 nm). *Multispectral* sensors record energy in multiple bands of the electromagnetic spectrum (e.g., the ADAR 5500 digital frame camera records energy in four multispectral bands: 450–515 nm; 525–605 nm; 640–690 nm; 750–900 nm (Figure 6-3). A *hyperspectral* remote sensing instrument acquires data in hundreds of spectral bands. For example, the Airborne Visible and Infrared Imaging Spectrometer (AVIRIS) has 224 bands in the region from 400–2500 nm spaced just 10 nm apart (Figure 6-4). *Ultraspectral* remote sensing involves data collection in many hundreds of bands. Certain bands of the electromagnetic spectrum are optimum for obtaining information on biophysical parameters such as biomass, chlorophyll concentration, and temperature. Careful selection of the spectral bands improves the probability that a surface feature will be uniquely detected, identified, and the appropriate information extracted.

3.2. Spatial Information and Resolution

Each silver halide crystal in an aerial photograph and each picture element in a digital remote sensor image is representative of a specific *x,y* ground location. Once rectified to a standard map projection, the spatial information associated with each pixel is of value because it allows the remote sensing

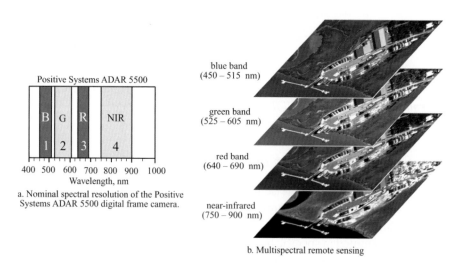

b. Multispectral remote sensing

Figure 6-3. An example of relatively coarse multispectral remote sensing (i.e., just four relatively broad bands) using the ADAR 5500 digital frame camera of an area in the Ace River Basin, South Carolina. The data were obtained at a spatial resolution of 1 × 1 ft.

derived information to be used along with other data in spatially distributed modeling efforts (e.g., in a GIS). *Spatial resolution* is the measure of the smallest angular or linear separation between two objects that can be resolved by the remote sensing system. The spatial resolution of aerial photography may be measured by (1) placing carefully calibrated, parallel black-and-white lines on tarps that are placed in the field, (2) obtaining aerial photography of the study area, and (3) analyzing the photography and computing the number of resolvable *line pairs per millimeter* in the photography. For electronic remote sensing systems, the nominal spatial resolution is the dimension in meters (or feet) of the ground-projected instantaneous-field-of-view (IFOV) (Jensen 2000). For example, the IKONOS panchromatic band has a nominal spatial resolution of 1 × 1 m and the Landsat Thematic Mapper (TM) 5 has a nominal spatial resolution of 30 × 30 m for six of its bands. A LIDAR "footprint" may only be 30 × 30 cm. Simulated examples of different spatial resolution remote

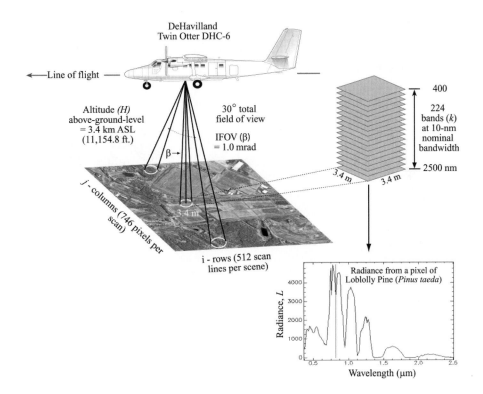

Figure 6-4. Characteristics of NASA's Advanced Visible Infrared Imaging Spectrometer (AVIRIS) flown onboard a DeHavilland Twin Otter DHC-6.

sensor data are shown in Figure 6-5. Note that it is unrealistic to expect to extract detailed residential building infrastructure information (e.g., building width, height, perimeter) from 80 × 80 m Landsat MSS data. We should also remember that because we have spatial information about the location of each pixel (x,y) in the matrix it is also possible to examine the spatial relationship between a pixel and its neighbors. Therefore, the amount of spectral autocorrelation and other spatial geostatistical measurements may be determined based on the spatial information inherent in the imagery (Walsh et al. 1998).

Some sensor systems, such as LIDAR, do not completely "image" the surface. Rather, the surface is sampled from laser pulses at some nominal time interval. The projected laser pulse on the ground may be very small (e.g., 15 cm in diameter) with samples approximately every one to five meters on the ground. Spatial resolution would appropriately describe the ground-projected laser pulse but sampling density (i.e., number of points per unit area) describes the frequency of ground observations.

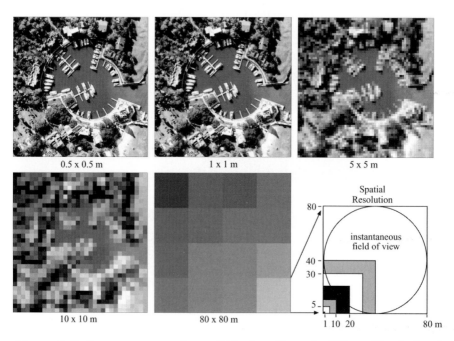

Figure 6-5. Remote sensor data of Harbor Town in Hilton Head, South Carolina collected at a nominal spatial resolution of 0.3 × 0.3 m (approximately 1 × 1 ft.) using a digital camera (courtesy of Litton Emerge, Inc.). Simulation of images of Harbor Town at resolutions from 0.5 × 0.5 m to 80 × 80 m are shown.

3.3. Temporal Information and Resolution

Remotely sensed images capture a record of the landscape at an instant in time. Such records can be used to document relatively current conditions or trends if historical imagery is available. Ideally, the remote sensing system is capable of obtaining data in the time framework we desire. *Temporal resolution* refers to how often remotely sensed data are acquired over a particular geographic area. For example, the temporal resolution of the polar-orbiting Landsat Thematic Mapper has been 16-18 days. If a remote sensing system can be pointed off-nadir (i.e., it doesn't have to look straight down) then it is possible to obtain much higher temporal resolution (e.g., the pointable SPOT, IKONOS, and Quickbird may obtain imagery every few days depending upon the latitude of the area of interest). Some satellites such as the Geostationary Operational Environmental Satellites (GOES) are located in a geostationary orbit a certain distance above a particular point on the ground. Such remote sensing systems have very high temporal resolution (e.g., they obtain imagery every one-half hour) that facilitates the tracking of frontal systems and hurricanes.

Another aspect of temporal information is how many observations are recorded from a single pulse of energy, such as in LIDAR. Most LIDAR sensors emit one pulse and record multiple responses from this pulse. Measuring the differences between multiple responses allows for the determination of object heights and structure. Also, the length of time required to emit an energy signal by an active sensor (e.g., RADAR or LIDAR) is referred to as the pulse length. Short pulse lengths allow very precise distance measurement (i.e., range).

3.4. Angular Information

Most remote sensing investigations are based on the examination of spatial, spectral, temporal, and polarization characteristics of the data. In addition, however, there are very specific angular characteristics associated with each image pixel that are important but rarely used. Every remote sensing image collected is influenced by (a) the location of the illumination source (e.g., the sun for a passive system or the sensor itself in the case of RADAR, LIDAR, and SONAR), (b) the orientation of the terrain facet (pixel) or terrain cover (e.g., vegetation) under investigation, and (c) the location of the remote sensing system. Thus, there is always an angle of incidence associated with the incoming energy that illuminates the terrain and an angle of exitance from the terrain to the sensor system. This *bidirectional* nature of remote sensing data collection is known to influence the spectral and polarization characteristics of R_e recorded by the remote sensing system. For a closed canopy, R_e is maximized when the sun is behind the sensor as all leaves are illuminated and no shadow is visible and R_e is minimized when the sun is in front of the sensor as a result of self-shadowing (Curran et al., 1998). Research continues on how to incorporate the bidirectional reflectance distribution function (BRDF) information into the digital image processing system to improve our understanding of what is recorded in the remotely sensed imagery (Jensen and Schill 2000). Utilizing BRDF information normally requires that the study area be remotely sensed from two different vantage points.

3.5. Polarization Information

Sunlight is polarized weakly. When sunlight strikes a non-metal object such as grass or forest, however, it is depolarized and the incident energy is scattered. Generally, the more smooth the surface, the greater the polarization. It is possible to utilize polarizing filters on passive systems to highlight polarized light at various angles. It is also possible to selectively send and receive polarized energy in active systems such as RADAR (e.g., horizontal send, vertical receive; vertical send, horizontal receive). The polarization characteristics of the electromagnetic energy recorded represent an important variable that can be used to discriminate the nature of objects on the Earth. Multiple-polarized RADAR imagery is an especially useful application of polarized energy.

3.6. Radiometric Precision and Resolution

Some remote sensing instruments record reflected, emitted, or back-scattered electromagnetic radiation more accurately than others. Radiometric resolution is the sensitivity of a remote sensing detector to differences in signal strength as it records the radiant flux reflected, emitted, or back-scattered from the terrain. For example, early Landsat MSS sensors had a radiometric resolution of just 6-bits (values from 0 to 63). Landsat 4 and 5 Thematic Mappers recorded data in 8-bits (values from 0 to 255). IKONOS and Quickbird sensors record information in 11-bits (values from 0 to 2,047). It is best to think of radiometric resolution as if it were a ruler. Would you rather use a ruler with just 256 subdivisions or one with 2,048 very fine subdivisions?

4. CONVERSION OF REMOTELY SENSED DATA INTO INFORMATION: THE USE OF MODELS

Electromagnetic radiation R_e recorded by a remote sensing system is dependent on the aforementioned characteristics: λ = wavelength; x, y = location; θ = angle; t = time, p = polarization, and Ω = spectral radiometric precision. But these variables are not a list of typical parameters that a geographer or ecologist would want to work with. Where is the land use, land cover, biomass, or percentage of impervious cover information that we typically see extracted from the remote sensor data? The answer lies in the fact that only a few variables (e.g., temperature, atmospheric water vapor, etc.) of interest to geographers and other environmental and social scientists can be measured *directly* using remote sensing. To obtain the information we desire, the remote sensing data R_e must function as an *indirect* surrogate for the real-world variable of interest. Therefore, we must typically *model* the electromagnetic radiation R_e recorded by a remote sensing system to transform it into useful information.

In order to do this, the scientist must understand the nature of the remote sensing-derived signal (R_e) and how it relates to the tremendous variety of variables present in the real-world. The modeling process often involves collecting in situ data and correlating these data with remote sensing measurements obtained over the same geographic location. For example, we might collect 30 in situ wheat field quadrat biomass measurements (g/m^2) on the ground and then correlate these in situ measurements with remote sensing-derived Vegetation Index (VI) values obtained over the same 30 field locations. If the simple linear regression relationship is highly correlated and significant, it may be possible to utilize the linear model (e.g., $y = ax + b$; or biomass = a [remote sensing-derived VI value] + b) to transform the R_e value of every picture element in the remotely sensed scene into a biomass estimate, creating a biomass map. Thus, the remote sensing data is transformed into biophysical information. The quality of the *model* dictates whether the remote sensing-derived information is accurate enough to use (Cohen et al. 2003). The electromagnetic energy R_e recorded by a remote sensing system contains error (noise), and our models are often relatively simplistic. Therefore, it is not surprising that most of the time the information derived from remotely sensed data is an estimate of the variable of interest (e.g., biomass) rather than a direct measurement.

5. BIOPHYSICAL AND CATEGORICAL VARIABLES AND SPATIAL DATA INFRASTRUCTURE REQUIREMENTS

Using the appropriate models, it is possible to extract a wealth of biophysical (e.g., biomass, water vapor, temperature, precipitation) and categorical (e.g., land use, land cover) spatially distributed information from remotely sensed data. Table 6-1 lists several important biophysical variables and categorical information and selected remote sensing systems that can be of value in estimating them. It is provided to inform the reader about the

Table 6-1. Biophysical and categorical variables and selected representative remote sensing systems capable of providing such information.

Biophysical Variables	Representative Remote Sensing Systems Capable of Providing Information
x,y,z **Geodetic control**	Global Positioning Systems (GPS)
x,y **Location from orthocorrected imagery**	GPS, analog/digital stereoscopic aerial photography, IKONOS, Quickbird, OrbView-3, SPOT, Landsat (TM, ETM+), Russian KVR-2000, Indian IRS-1CD, ERS-1,2 microwave, RADARSAT
z **Elevation**	
Digital elevation model (DEM)	GPS, stereo aerial photography, SPOT, RADARSAT, IKONOS, Quickbird, OrbView3, LIDAR, Shuttle Radar Topography Mission (SRTM), Interferometric Synthetic Aperture Radar (IFSAR)
Digital Bathymetric model (DBM)	SONAR, LIDAR, stereoscopic aerial photography
Land Use	
Commercial, residential, etc. Cadastral (property) Tax mapping Transportation Utilities	Stereoscopic aerial photography, high resolution satellite imagery (<1 x 1 m: IKONOS, Quickbird, OrbView-3) , SPOT (2.5 m), LIDAR, high spatial resolution hyperspectral data (AVIRIS, HyMap, CASI)
Land Cover	
Agriculture, forest, urban, etc.	Aerial photography, hyperspectral systems (AVIRIS, HyMap, CASI), Landsat (MSS, TM, ETM+), SPOT, ASTER, MODIS, AVHRR, RADARSAT, IKONOS, Quickbird, OrbView-3, LIDAR, IFSAR, SeaWiFS
Vegetation	
Temperature	ASTER, MODIS, GOES, AVHRR, SeaWiFS, suborbital thermal infrared sensors
Pigments (e.g. chlorophyll *a*)	Aerial photography, ETM+, ASTER, MODIS, IKONOS, Quickbird, OrbView-3, SeaWiFS
Canopy structure	Large-scale stereoscopic aerial photography, LIDAR, RADARSAT
Canopy height Biomass derived from vegetation indices Leaf-area-index (LAI) Absorbed photo-synthetically active radiation (APAR)	Aerial photography, Landsat TM & ETM+, ASTER, LIDAR, MODIS, MISR, IKONOS, Quickbird, OrbView-3, hyperspectral (AVIRIS, HyMap, CASI)
Water	
Temperature	Landsat (TM&ETM+), ASTER, MODIS, GOES, AVHRR, SeaWiFS, sub-orbital thermal infrared sensors

Surface hydrology	Photography, Landsat (TM, ETM+) SPOT, IKONOS, Quickbird, OrbView-3, ASTER, MODIS, SeaWiFS, hyperspectral systems (AVIRIS, HyMap, CASI), AVHRR, GOES, LIDAR, MISR
Color	
Suspended minerals	
Chlorophyll/gelbstof	TOPEX/POSEIDON
Dissolved organic matter	
Depth	SONAR, LIDAR, stereoscopic aerial photography
Soils/Rocks	
Temperature	Landsat (TM&EMT+), AVHRR, GOES, ASTER, MODIS, sub-orbital thermal sensors
Soil moisture	Passive microwave (SSM/1), RADARSAT, ASTER, MISR
Mineral composition	ASTER, MODIS, hyperspectral systems (AVIRIS, HyMap, CASI)
Taxonomy	high resolution aerial photography, hyperspectral systems (AVIRIS, HyMap, CASI)
Hydrothermal alteration	Landsat (TM, ETM+), MODIS, ASTER, hyperspectral systems (AVIRIS, HyMap, CASI)
Atmospheric	
Temperature (surface, cloud)	GOES, AVHRR, ASTER, MODIS, sub-orbital thermal infrared
Aerosols (e.g., optical depth)	MISR, GOES, AVHRR, MODIS, CERES, MOPITT
Clouds (e.g., fraction, optical thickness)	GOES, AVHRR, MODIS, MISR, CERES, MOPITT, UARS
Precipitation	Tropical Rainfall Measurement Mission (TRMM), GOES, AVHRR, Passive microwave (SSM/1)
Water vapor (precipitable water)	GOES, MODIS
Snow, ice	GOES, AVHRR, MODIS
Ozone	MODIS
Wind	ERS-2, Quickscat

increasing availability of remote sensing data that can be utilized to obtain spatially distributed information. The list is not exhaustive. As the remote sensing process can be used to measure position, height, and the biophysical variable at a location, the use of multiple dates of remote sensor data can be used to measure changes in these characteristics.

Note the first three variables in Table 6-1 and Figure 6-6: geodetic control, digital elevation and bathymetry, and orthoimagery. These were selected because they represent the framework foundation variables required by most national and global Spatial Data Infrastructures (SDI). An SDI is an umbrella of policies, standards, and procedures under which organizations and technologies interact to foster more efficient use, management, and production of geographic data (FGDC 2002). The following quotation emphasizes the importance of SDIs in developing countries (EIS-Africa 2002):

> Building infrastructure for geographic information use is becoming as important to African countries as the building of roads, telecommunications networks, and the provision of other basic services...The rationale for investing in information infrastructure is analogous to that for physical infrastructure: the provision of many other services is contingent upon their existence. The cost-effective development of a spatial data infrastructure requires the coordinated harnessing of resources and expertise residing in various government agencies, the private sector, universities, non-governmental organizations, and regional and international bodies.

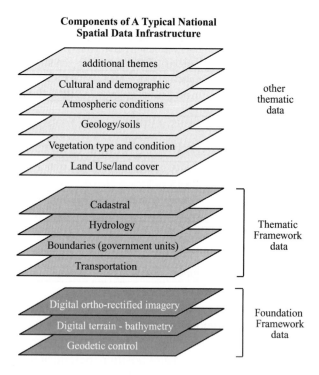

Components of A Typical National Spatial Data Infrastructure

additional themes
Cultural and demographic
Atmospheric conditions — other thematic data
Geology/soils
Vegetation type and condition
Land Use/land cover

Cadastral
Hydrology — Thematic Framework data
Boundaries (government units)
Transportation

Digital ortho-rectified imagery
Digital terrain - bathymetry — Foundation Framework data
Geodetic control

Figure 6-6. A spatial data infrastructure (SDI) consists of framework foundation data such as geodetic control, digital elevation and bathymetry, and rectified orthoimagery. SDIs also contain essential framework thematic data layers, including hydrology, political and other boundaries, transportation resources, and cadastral information. Other thematic information such as land use/land cover, vegetation, soils/geology, atmospheric conditions, demographic characteristics, and land cover may also be included in the infrastructure. (Source: Adapted from FGDC 2002.)

Also note the importance of framework foundation thematic data. These variables are judged most important by a majority of scientists and other users of spatial information on a global basis. If remote sensing-derived information is truly of value, it should be able to contribute to national and global SDIs (Jensen et al. 2002a). The following sections provide brief summaries of advancements in the collection of several of these important variables. Due to space limitations, the chapter does not summarize advancements in remote sensing of water, soils/rocks, and the atmosphere.

5.1. X,Y,Z Geodetic Control Derived from Global Positioning Systems

Geodetic control is required for projects that incorporate spatially distributed data because it is used to locate objects and features in their true geographic (x,y,z) position. Geodetic control obtained using in situ surveying techniques is very accurate but can be expensive and time-consuming. Fortunately, there is now a constellation of 24 global positioning satellites—the Navigation Satellite Timing And Ranging (NAVSTAR) Global Positioning System (GPS). This orbital satellite system is operated by the U.S. Department of Defense. It is based on radiowave-positioning and time-transfer *ranging* technology much like RADAR (RAdio Detecting And Ranging based on microwave energy) and LIDAR (LIght Detection And Ranging based on laser energy). The system was designed so that at least four satellites would be "visible" at all times on unobstructed terrain (Rizos 2002). Characteristics of the NAVSTSR GPS are summarized in Table 6-2.

The ability to obtain geographic coordinates accurate to within centimeters using GPS technology is a tremendous achievement. It makes geographic knowledge available to anyone irrespective of country of origin, race, or other factors at no cost (see Table 6-2) (Jensen et al. 2002a). It can be used to (a) obtain precision geodetic control if the proper instruments are used, and (b) to identify geographic location by lay persons and scientists throughout the world.

5.2. X,Y Geographic Location from Orthoimagery

Identifying the geographic x,y location or boundary of a road, river, or stand of trees is no trivial matter. Of course, we may use a GPS in the field but sometimes we are constrained by inhospitable terrain and the sheer magnitude of distances that must be traversed. Sometimes data collection is hampered by government restrictions. An alternative is to obtain orthorectified, remotely sensed data that can be used to extract accurate x,y information. The orthoimagery may also be used as the foundation upon which all other spatially distributed data are registered.

Table 6-2. Characteristics of the NAVSTAR global positioning system.

- Very precise time measurement is possible.
- Geographic positioning accuracy from meters down to centimeters.
- Geographic information is provided in 3 dimensions (latitude, longitude, and elevation).
- Signals available to people anywhere (air, land, or sea) without discrimination.
- Signals available free to anyone with a GPS receiver. Receiver costs continue to decline.
- Available 24 hours a day, 7 days a week.
- All-weather system not affected by clouds (intense rain or thick vegetation canopy can reduce its effectiveness).
- Geographic coordinates are tied to a single global geodetic datum.
- No interstation visibility is required for precise positioning. This means that it is not necessary for a surveyor to use a theodolite to view a distant stadia rod.
- Geographic position can be determined rapidly (in seconds to minutes).

(Source: Adapted from Rizos 2002.)

5.2.1. Orthoimagery Derived from High Spatial Resolution Imagery

Orthoimagery may be prepared from stereoscopic aerial photography or other types of remotely sensed data using photogrammetric techniques. Orthoimagery contains the rich thematic detail found in the imagery plus the rigorous geometric characteristics of a planimetric line map. Geometric distortions caused by relief-displacement and other anomalies (e.g., platform movement or vibration) are removed from the dataset and the orthoimage is adjusted (rectified) to a standard datum and map projection (Jensen 1995). Developed countries have typically produced their orthoimagery from stereoscopic aerial photography with spatial resolutions of <1 × 1 m. Orthoimagery can also be derived from high spatial resolution satellite imagery (e.g., Space Imaging's 1 × 1 m panchromatic band). Even though the imagery and processing are expensive, orthoimagery will continue to be derived using these sources because many local agencies are mandated to obtain high spatial resolution orthoimagery for cadastral and tax mapping applications.

5.2.2. Landsat GeoCover-Ortho Global Database

Many earth resource investigations cover vast geographic areas. Until recently, there was no geographically extensive moderate resolution orthoimage database available. Fortunately, we now have the Global GeoCover-Ortho database with a nominal spatial resolution of 28.5×28.5 m covering the majority of Earth's landmass with a positional accuracy of <50 m (root mean square error). For example, Figure 6-7a depicts a single frame of orthoimagery of Mount Kilimanjaro in Kenya and Tanzania. Figure 6-7b displays hundreds of mosaicked orthoimages covering Australia. Most of the remotely sensed data were acquired from 1987 to 1993 by the Landsat Thematic Mapper. The database was produced by Earth Satellite Corporation as part of NASA's Scientific Data Buy initiated in 1997. Subsequent Landsat TM images (and other types of remotely sensed data) can be overlaid on the GeoCover-Ortho imagery for change detection purposes. A GeoCover-Ortho dataset based on circa 2000 imagery is under development.

5.3. X,Y,Z Digital Elevation (Topography) and Bathymetry from Remotely Sensed Data

Elevation data are used in many environmental and engineering applications. For example, it is fundamental to many hydraulic and hydrologic studies, viewshed analysis, and transportation planning, and is essential for the creation of orthoimagery in undulating terrain. Digital elevation information may be obtained using the following remote sensing-based approaches:

- in situ surveying based on GPS technology,
- metric stereoscopic aerial photography and soft-copy (digital) photogrammetric techniques,
- LIght detection and Ranging (LIDAR) data, and
- Interferometric Synthetic Aperture radar (IFSAR) data.

The elevation data are typically summarized as (a) contours on topographic maps, (b) a raster (matrix) of equally spaced elevation values in a digital elevation model (DEM), or (c) as a triangular irregular network (TIN) of elevations derived from individual spot elevations.

5.3.1. GPS Surveying-Derived Digital Elevation Models

Individual spot elevations can be obtained on the surface of the earth using survey-grade GPS instruments and techniques. Terrain models are then constructed as a TIN or as a DEM from spatial interpolation (e.g., distance weighted, kriging). Unfortunately, each point is very expensive to collect and problematic to collect under many canopy cover types. It is also often difficult to densify an area with spot elevation measurements when the terrain is rugged or inhospitable.

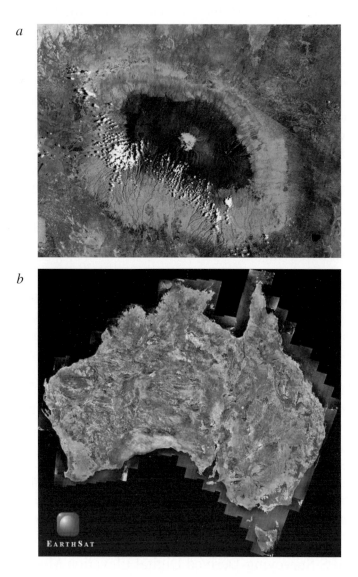

Figure 6-7. GeoCover-Ortho images: a–Mount Kilimanjaro at 28.5 × 28.5 m spatial resolution [RGB = Landsat TM bands 7,4,2 (mid-infrared, near-infrared, and green)]; b–Mosaic of hundreds of GeoCover-Ortho Landsat Thematic Mapper images of Australia obtained between 1987 and 1993 (courtesy of Earth Satellite Corporation and NASA).

5.3.2. Photogrammetrically-Derived Digital Elevation Models

It has been possible to extract digital elevation models from stereoscopic aerial photography for more than 50 years. The technology is mature and accurate but expensive. Figure 6-8 depicts a DEM of Rosslyn, Virginia derived using soft-copy (digital) photogrammetry which incorporates the elevation of the terrain as well as the urban infrastructure. Such databases are ideal for urban housing, utility routing, transportation planning, and viewshed studies.

5.3.3. LIDAR-Derived Digital Elevation Models

Many cities, counties, and states are having private engineering firms collect LIDAR data to obtain accurate DEMs. LIDAR sensors measure active laser pulse travel time from the transmitter onboard the aircraft to the target and back to the receiver (Jensen 2000; Cowen et al. 2000). As the aircraft moves forward, a scanning mirror directs the laser pulses back and forth across-track. This results in a series of data points arranged across the flightline. Multiple flightlines can be combined to cover the desired area. Datapoint density is dependent on the number of pulses transmitted per unit time, the scan angle of the instrument, the elevation of the aircraft above-ground-level, and the forward speed of the aircraft. The greater the scan angle off-nadir, the more vegetation that must be penetrated to receive a pulse from the ground assuming a uniform canopy (Jensen 2000).

Figure 6-8. Digital elevation model of Rosslyn, Virginia derived using metric aerial photography and soft-copy photogrammetry software. The aerial photography is draped over the digital elevation model.

LIDAR data avoids the problems of aerial triangulation and orthorectification because *each* LIDAR measurement is individually georeferenced. Information about the scanning mirror and aircraft attitude allow precise determination of where the LIDAR instrument was pointed at the time of an individual laser pulse. Exact aircraft position is determined from onboard GPS and inertial navigation equipment. The combination of all these factors allows three-dimensional georeferenced coordinates to be determined for each laser pulse. Although some of the LIDAR laser energy is backscattered by vegetation above the ground surface, only a portion of the energy needs to reach the ground to produce a surface measurement. Small canopy breaks may be direct pathways for the small footprint laser beams. All LIDAR sensor systems are currently on airborne (i.e., helicopter or fixed-wing aircraft) platforms. There are no operational satellite LIDAR systems.

To a certain degree, LIDAR can penetrate vegetation canopy and map the surface below. Penetrating through the canopy using aerial photography requires human image interpretation. Interferometric SAR does not penetrate most heavy cover types. A comparison of DEMs produced from automated stereocorrelation, contour-to-grid conversion of contour lines, IFSAR- and LIDAR-derived DEMs under various land covers found the tree canopy to have a profound effect on the IFSAR and stereocorrelation approaches (Hodgson et al. 2003a). When LIDAR imagery is obtained during leaf-off periods in the late winter, the effects of the tree canopy on the extraction of accurate elevation values is minimized (Hodgson and Bresnahan 2004). Most applications of LIDAR data include both automated and human interpretation methods for labeling laser returns as ground, vegetation, or other surfaces (Raber et al. 2002). The analysis of LIDAR data is now used extensively for mapping floodplains (North Carolina Flood Plain Mapping Program 2002), geomorphic and river studies (Cobby et al. 2001; Adams and Chandler 2002; Bowen and Waltermire 2002), and transportation studies (Cowen et al. 2000). For example, Figure 6-9 depicts how LIDAR-derived elevation data were used in the creation of a cumulative cost surface that was used to identify the least expensive route between a new tire plant and an existing rail line (Cowen et al. 2000).

5.3.4. IFSAR-Derived Digital Elevation Models

Surface elevation may be extracted from RADAR imagery taken from slightly different locations using Interferometric Synthetic Aperture Radar (IFSAR) techniques (SRTM 2003). RADAR interferometry yields accurate topographic data both day and night and through clouds unless the vegetation canopy is dense. When dense canopy is encountered, the IFSAR-derived elevation values may represent the top of the canopy rather than the terrain

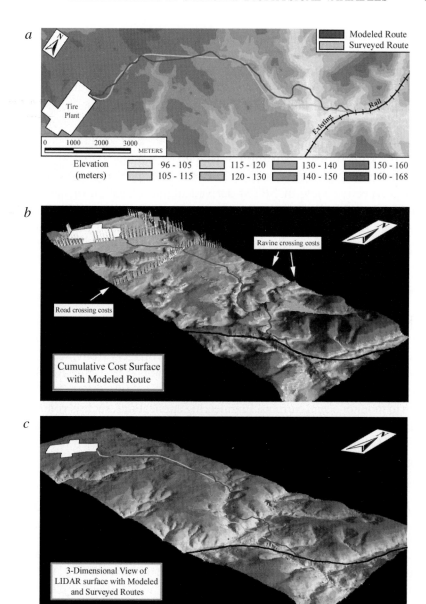

Figure 6-9. LIDAR-derived digital elevation models: a–planimetric map of LIDAR-derived DEM with a new tire plant and routes (surveyed and modeled) to an existing rail line; b–perspective view of the cumulative cost-surface created to identify the least-expensive route; c– perspective view of the route selected using traditional techniques and the proposed route identified using the LIDAR-derived DEM (Cowen et al. 2000; reproduced with permission from the American Society for Photogrammetry and Remote Sensing).

surface. The most important IFSAR-derived digital elevation model produced thus far is the *Shuttle Radar Topography Mission* (SRTM) dataset. The mission began on February 11, 2000 and lasted 11 days. During this brief period, over 80 percent of Earth's landmass was imaged at 30 × 30 m spatial resolution. The SRTM mission achieved <16-m *absolute vertical* accuracy, <10-m *relative vertical* accuracy, and <20-m *absolute horizontal* accuracy. It is currently the most accurate digital elevation model of the Earth. SRTM digital elevation data of the United States is being released at 30 × 30 m spatial resolution. Most of the rest of the world is being released at 90 × 90 m.

A perspective view of SRTM-derived digital elevation of the Andes Mountains in Argentina is shown in Figure 6-10a. An SRTM-derived DEM of the Santa Barbara region is shown in Figure 6-10b draped with a Landsat Thematic Mapper image. SRTM-derived DEMs are now a very important input to many ecological models, especially in developing countries.

5.3.5. Monitoring Change in *X, Y* or *Z*

By analyzing multiple dates of remote sensor data changes in the planimetric position of objects, elevation or vegetation height may be determined. For example, correlation of multidate imagery from the AVHRR sensor were used to map ice sheet movements (Ninnis et al. 1986). Breaker et al. (1994) used feature tracking techniques from multiple thermal AVHRR images to monitor ocean currents. Engineering applications, such as volumetric estimates, coal-pile monitoring, and volcanically induced landscape changes are routine applications. Repeat airborne laser profiling is used to monitor changes in the Greenland ice sheet. Mapping changes in horizontal position or elevation provides information on the *rates* of change.

5.4. Land Use Information Derived from Remotely Sensed Data

Land use refers to what people are doing on the surface of the earth (e.g., agriculture, forestry, commerce, recreation). *Land cover* refers to the actual type of biophysical material present on the surface of the earth (e.g., wetland, forest, bare soil, rocks, asphalt shingles, concrete, water). Land use and land cover data are central to such issues as deforestation, sustainable development, and protecting the quality and supply of water resources. In light of the human impacts on the landscape, there is a need to establish baseline datasets against which changes in land cover and land use can be assessed (Belward et al. 1999). In addition, land-cover data are needed as input to models to predict future Earth system modification. Land use and land cover data can be obtained using in situ measurements or remote-sensing technology. It is much more practical, however, to extract the information from remotely sensed data and place it in a GIS for analysis and distribution (Plummer 2000). Generally speaking, land use information is much more difficult to extract from remotely sensed imagery than land cover information, and usually requires substantial human interpretation.

The level of detail of land use and land cover information required for a particular application is normally described as being in one of n levels (Anderson et al. 1976). Lower levels of information (designated by larger

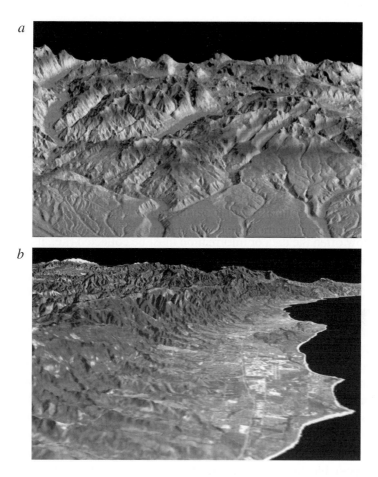

Figure 6-10. Shuttle Radar Topography Mission (SRTM)-derived digital elevation models: a–portion of San Martin de Los Andes, Argentina. Elevations range from 700 to 2,400 m (2,300 to 8,000 ft.); b–Santa Barbara, California. Mount Abel (2,526 m; 8,286 ft.) is seen in the distance. The DEM is draped with a Landsat Thematic Mapper image (RGB = bands 7,4,2) (courtesy NASA Jet Propulsion Lab and NIMA).

Roman numerals) are more specific and typically require finer spatial resolution data and/or greater spectral resolution data. For example, Level I nominal-scale forest land-cover information might be derived from the Moderate Resolution Imaging Spectroradiometer (MODIS) 500 × 500 m imagery. Level II coniferous and deciduous forest land cover information might be derived from Landsat 7 ETM+ imagery. Level III forest species information might be extracted from IKONOS 4 × 4 multispectral data. Level IV subspecies (e.g., *pinus taeda*) might be derived from 4 × 4 m hyperspectral data.

5.4.1. Urban/Suburban Land Use and Infrastructure Information Derived from High Spatial and High Spectral Resolution Imagery

Both developed and developing countries are requiring ever more detailed information about land use (e.g., commercial, residential, industrial) (Liverman et al. 1998). In addition they need spatial information about urban infrastructure for cadastral (including parcel boundaries and building perimeter, area, and height), transportation, hydrologic, and utility applications. Jensen and Cowen (1999) reviewed many urban and suburban data collection requirements stipulated by city planners and engineers used to manage cities, counties, and states. How often the data or information must be collected (i.e., its temporal resolution) and the range of remote sensing spatial resolutions (e.g., 1 × 1 m ground resolved distance) necessary to extract the desired information are summarized in Table 6-3 and as ellipses in Figure 6-11 (Jensen 2000; Jensen et al. 2002a). The temporal and spatial resolution of selected existing and proposed remote sensing systems that can be used to provide the desired information are shown as shaded rectangles in Figure 6-11.

Apparent in Table 6-3 and Figure 6-11 is that the vast majority of urban/ suburban land use applications require very high spatial resolution remote sensor data, often ≤1 × 1 m. Traditionally, large-scale (>1:10,000) vertical, analog stereoscopic aerial photography obtained by photogrammetric engineering firms has been analyzed to obtain useful information (land use, impervious surface, cadastral, tax, road centerlines, etc.). Only recently has it become possible to extract much of this information from high spatial resolution stereoscopic satellite imagery (e.g., IKONOS, Quickbird, OrbView-3). High spatial resolution hyperspectral data is useful for some urban applications such as road surface material research. Information from all these sources is extracted using a combination of visual image interpretation, soft-copy photogrammetric techniques, rule-based classification, or relatively sophisticated object-oriented image segmentation methods discussed later in this chapter (Hodgson et al. 2003b).

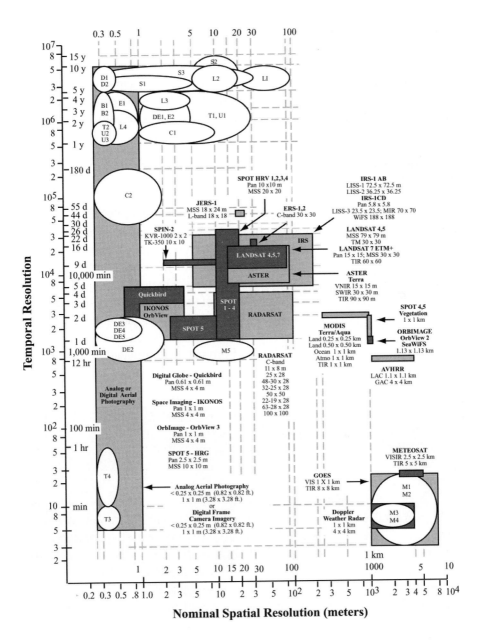

Figure 6-11. Relationship between the spatial and temporal resolution of urban and suburban attributes and the spatial and temporal resolution of various aerial and suborbital remote-sensing systems. The clear polygons represent the spatial and temporal requirements for selected urban attributes listed in Table 6-3 (Source: updated from Cowen and Jensen 1999; Jensen 2000).

For example, most cities now require detailed information about the location of impervious surfaces (B2 in Table 6-3). This information is used by hydrologic engineers to mitigate the impact of storm water runoff (Hodgson et al. 2003b). It is also used to levy watershed runoff taxes associated with urban/suburban development. An impervious surface map of an area near Charlotte, North Carolina is shown in Figure 6-12. It was derived using object-oriented classification techniques applied to USGS NAPP color-infrared 1 × 1 m orthophotography.

5.5. Land Cover Information Derived from Remotely Sensed Data

High spatial resolution imagery and a combination of visual and digital image processing techniques may be used to extract detailed urban/suburban land cover information (Jensen 2000). But what if we require regional, national, and even global land cover information for our models? Land cover, and human and natural alteration of land cover, play a major role in global-scale patterns of the climate and biogeochemistry of the earth system. Until recently, land-cover datasets used within models of global climate and biogeochemistry were derived from preexisting maps and atlases. These had serious problems as summarized in Friedl et al. (2002). Fortunately, a tremendous amount of

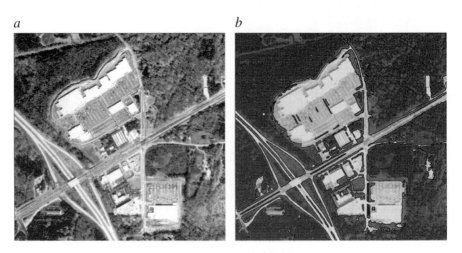

Figure 6-12. An impervious surface map derived using object-oriented classification techniques applied to USGS NAPP color-infrared orthophotography: a–USGS NAPP digital orthophotography (1 × 1 m; red band); b–extraction of impervious surface materials (courtesy Tom Tribble and Frank Obusek; North Carolina Center for Geographic Information and Analysis).

Table 6-3. Urban and suburban applications and the minimum remote-sensing resolutions required to obtain such information.

Attributes	Minimum Resolution Requirements		
	Temporal	Spatial	Spectral*
Land Use/Land Cover			
L1-USGS Level I	5-10 yrs	20-100 m	V-NIR-MIR-Radar
L2-USGS Level II	5-10 yrs	5-20 m	V-NIR-MIR-Radar
L3-USGS Level III	3-5 yrs	1-5 m	Pan-V-NIR-MIR
L4-USGS Level IV	1-3 yrs	0.25-1 m	Panchromatic
Building and Property Infrastructure			
B1-Building perimeter, area, height, and cadastral information (property lines)	1-5 yrs	0.25-0.5 m	Pan-Visible
B2-Impervious surface	1-5 yrs	0.25-0.5 m	Pan-V-NIR-TIR
Transportation Infrastructure			
T1-General road centerline	1-5 yrs	1-30 m	Pan-V-NIR
T2-Precise road width	1-2 yrs	0.25-0.5 m	Pan-V
T3-Traffic count studies (cars, airplanes)	5-10 min	0.25-0.5 m	Pan-V
T4-Parking studies	10-60 min	0.25-0.5 m	Pan-V
Utility Infrastructure			
U1-General utility line mapping, routing	1-5 yrs	1-30 m	Pan-V-NIR
U2-Precise utility line width, right-of-way	1-2 yrs	0.25-0.6 m	Pan-Visible
U3-Location of poles, manholes, substations	1-2 yrs	0.25-0.6 m	Panchromatic
Digital Elevation Model (DEM) Creation			
D1-Large scale DEM	5-10 yrs	0.25-0.5 m	Pan-Visible
D2-Large scale slope map	5-10 yrs	0.25-0.5 m	Pan-Visible
Socioeconomic Characteristics			
S1-Local population estimation	5-7 yrs	0.25-5 m	Pan-V-NIR
S2-Regional/national population estimates	5-15 yrs	5-20 m	Pan-V-NIR
S3-Quality of life indicators	5-10 yrs	0.25-30 m	Pan-V-NIR
Energy Demand and Conservation			
E1-Energy demand and production potential	1-5 yrs	0.25-1 m	Pan-V-NIR
E2-Building insulation surveys	1-5 yrs	1-5 m	TIR
Critical Environmental Area Assessment			
C1-Stable sensitive environments	1-2 yrs	1-10 m	V-NIR-MIR
C2-Dynamic sensitive environments	1-6 mos	0.25-2 m	V-NIR-MIR-TIR
Disaster Emergency Response			
DE1-Pre-emergency imagery	1-5 yrs	1-5 m	Pan-V-NIR
DE2-Post-emergency imagery	12-24 hrs	0.25-2 m	Pan-V-NIR-Radar
DE3-Damaged housing stock	1-2 days	0.25-1 m	Pan-V-NIR
DE4-Damaged transportation	1-2 days	0.25-1 m	Pan-V-NIR
DE5-Damaged utilities, services	1-2 days	0.25-1 m	Pan-V-NIR
Meteorological Data			
M1-Weather prediction	3-25 min	1-8 km	V-NIR-TIR
M2-Current temperature	3-25 min	1-8 km	TIR
M3-Clear air and precipitation mode	6-10 min	1 km	WSR-88D Radar
M4-Severe weather mode	5 min	1 km	WSR-88D Radar
M5-Monitoring urban heat island effect	12-24 hr	5-30 m	TIR

* Spectral resolution is the extent to which an application requires detection of light within narrow bands of the electromagnetic spectrum such as visible blue, green, and red light (V), a single broad band of visible light (e.g., encompassing both green and red light (Pan), near-infrared (NIR) energy, middle-infrared (MIR), and thermal-infrared (TIR).   *Source: updated from Jensen 2000.*

research has gone into the development of remote sensing systems and classification algorithms to extract regional to global land cover information. This section discusses how various remote sensing systems continue to be used to obtain regional and global land cover information.

5.5.1. Land Cover Derived from Coarse Resolution Imagery - Advanced Very High Resolution Radiometer (AVHRR)

NOAA's AVHRR sensor system has been the source of low-cost 1 × 1 km land cover data for more than two decades. The AVHRR record will be continued until the National Polar Operational Environmental Satellite System (NPOESS) is launched in 2009 (Townshend and Justice 2002). The first global land cover map compiled from remote sensing was produced by DeFries and Townsend (1994) using maximum likelihood classification of monthly composite AVHRR NDVI data at 1 degree spatial resolution (Justice 2003; Justice et al. 1994; Friedl et al. 2002). Defries et al. (1998) used a decision tree classifier to map global land cover at 8-km spatial resolution using AVHRR data. Loveland et al. (2000) used supervised classification of monthly composite AVHRR NDVI data acquired between April 1992 and March 1993 to produce a Global Land Cover dataset at 1 × 1 km spatial resolution (Belward et al. 1999). The Global Land Cover dataset is available by continent and includes seven global datasets, each based on a different landscape classification system (USGS 2002a). Tropical forest extent and areas with rapid forest-cover change have been mapped using AVHRR data by the Tropical Ecosystem Environment Observations by Satellite (TREES) project (TREES 2002).

5.5.2. Land Cover Derived from the Moderate Resolution Imaging Spectrometer (MODIS)

The Moderate Resolution Imaging Spectrometer (MODIS) was launched on the *Terra* satellite December 18, 1999. The second MODIS was launched on the *Aqua* satellite on May 4, 2002. MODIS's mission is to monitor and document global climate change, land use, land cover, and other factors affecting human habitability (Justice et al., 1998; 2003). The MODIS sensor obtains imagery at 250 × 250 m (bands 1 and 2), 500 × 500 m (bands 3-7), and 1 × 1 km (bands 8-36). The MODIS global land cover product suite includes two main parameters. The land-cover product contains 17 classes of land cover in the IGBP global vegetation classification scheme with a spatial resolution of 1 × 1 km. It is updated every quarter (i.e., every 96 days) (Strahler et al. 1999; Friedl et al. 2000). It is produced using a supervised decision tree classifier and an artificial neural network (Friedl et al. 2002). An example of

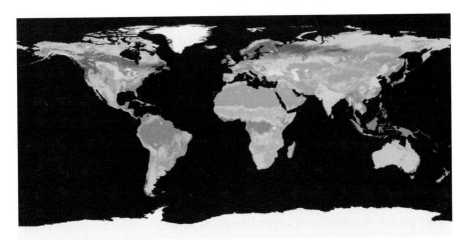

Figure 6-13. Global land cover 1 x 1 km dataset derived from MODIS data (courtesy NASA Goddard Space Flight Center and Boston University).

the MODIS-derived global land cover dataset is shown in Figure 6-13. This product is also provided at a spatial resolution of approximately 28-km for global modeling purposes. The second major product is a change detection map compiled using a change-vector algorithm. The product requires two full years of land cover information to generate the change product at 1-km spatial resolution.

5.5.3. Land-Cover Derived from Landsat Imagery

The Landsat MSS launched in 1972 had four bands at 79×79 m. Landsats 4 and 5 launched in 1982 and 1984 had Thematic Mapper sensors that recorded energy in six bands in the visible through mid-infrared at 30×30 m and in a single 120×120 m thermal band. The Landsat 7 Enhanced Thematic Mapper Plus (ETM+) launched in 1999 had one panchromatic band at 15×15 m, five bands in the visible through mid-infrared at 30×30 m, and one 60×60 m thermal band. Note that even the historical MSS data have higher spatial resolution than either AVHRR or MODIS data. The visible, near-infrared, and mid-infrared Landsat Thematic Mapper bands are particularly useful for many land cover mapping applications.

A tremendous amount of land cover mapping has been conducted over the last three decades using Landsat MSS, Landsat TM, Landsat ETM+ imagery, and SPOT data. One particularly noteworthy endeavor was initiated by Earth Satellite Corporation in 1999 when they began preparing a Landsat-based database called "GeoCover-Land Cover" for the National Imagery and Mapping Agency (NIMA). This 13-class global land cover dataset was produced at 30 × 30 m from EarthSat's GeoCover-Ortho imagery previously discussed. An example of a land-cover map of the area centered on Mount Kilimanjaro is shown in Figure 6-14.

5.6. Extracting Vegetation Amount and Condition Information from Remotely Sensed Data

The main goal of global agriculture and the grain sector of most economies is to feed the world's 6 billion people (Kogan 2001). One of the primary interests of earth observing systems is to study the role of terrestrial vegetation in large-scale global processes with the goal of understanding how the Earth functions as a system. Both these endeavors require an understanding of the global distribution of vegetation types as well as their biophysical and structural properties and spatial/temporal variations (TBRS 2003).

Biophysical measurements that document vegetation type, productivity, and functional health are needed for land resource management, combating deforestation and desertification, and promoting sustainable agriculture and rural development. Table 6-1 identifies numerous variables that should be sensed if we want to monitor vegetation type, condition, and change through time. Fortunately, we now have more sensitive remote sensing systems (e.g.,

Figure 6-14. Thirteen class land cover map of Mount Kilimanjaro derived from Landsat Thematic Mapper imagery. This is part of the EarthSat GeoCover land-cover mapping project. Compare this thematic map with the original imagery found in Figure 6-7a (courtesy Earth Satellite Corporation).

MODIS) and improved vegetation indices to monitor the health and productivity of vegetated ecosystems. Scores of vegetation indices have been developed that function as surrogates for important biophysical vegetation parameters. They may be applied to local problems or to global land cover assessments. For example, Jensen et al. (2002b) used vegetation indices to map the biomass, LAI, and chlorophyll *a* and *b* content of coastal wetland habitat in South Carolina using multispectral imagery. Blackburn (2002) used hyperspectral data to inventory photosynthetic pigments in broadleaf and coniferous forest plantations. At the other end of the scale spectrum, two of the most noteworthy indices—the Normalized Difference Vegetation Index (NDVI) and the Enhanced Vegetation Index (EVI)—have been used to monitor vegetation land cover amount on a global basis. Such information is very important for assessing global vegetation dynamics and predicting future land cover change scenarios. Recently it has also become possible to map the spatial distribution of a canopy's LAI and estimate the Fraction of Photosynthetically Active Radiation (FPAR).

5.6.1. Global Vegetation Monitoring Using the Normalized Difference Vegetation Index and AVHRR Data

Vegetation indices (VIs) are spectral transformations of two or more bands designed to enhance the contribution of vegetation properties and allow reliable spatial and temporal intercomparisons of terrestrial photosynthetic activity and canopy structural variations. The Normalized Difference Vegetation Index (NDVI) has been used successfully to estimate the vigor and amount of vegetation (Tucker et al. 1991) for more than 30 years. It is a simple ratio of the amount of red and near-infrared energy reflected from the surface of a vegetated canopy:

$$NDVI = \frac{p_{nir} - p_{red}}{p_{nir} + p_{red}} \qquad (6\text{-}2)$$

Interpretation of the NDVI is based on the fact that the healthier the plant, the greater the absorption of red light for photosynthetic purposes and the greater the reflectivity of the near-infrared light. The optimum red and near-infrared bandwidths are identified in Teillet et al. (1997). A disadvantage of the NDVI is the inherent nonlinearity of ratio-based indices and problem in scaling ratios. The NDVI also exhibits saturated signals over high biomass conditions and is sensitive to atmosphere and canopy background variations.

Despite these limitations, maps of NDVI, when modeled with other agricultural and meteorological information, are quite useful for monitoring food security problems in very fragile, drought-prone ecosystems. For example, since 1982, the Famine Early Warning System Network (FEWS NET) has used an AVHRR-derived NDVI to determine the distribution of vegetation condition across Africa every 10 days. Current FEWS NET NDVI images are compared with historical images to detect anomalies that guide famine relief efforts. Kogan (2001) described how NDVI-related information is used in a new numerical method for drought detection and impact assessment based on studies in 20 countries. He suggests that using the NDVI-assisted models,

> drought can be detected 4-6 weeks earlier than before, outlined more accurately, and the impact on grain reduction can be predicted in advance of harvest, which is most vital for global food security and trade.

The planned NPOESS will carry an improved version of the AVHRR sensor that should solve some of the problems associated with NOAA AVHRR imagery (Townshend and Justice 2002).

5.6.2. Global Vegetation Monitoring Using the Enhanced Vegetation Index

The MODIS sensor system onboard *Terra* and *Aqua* were specifically designed to provide more accurate vegetation cover and condition information. Scientists at NASA Goddard and several universities are responsible for producing a number of standard, global vegetation-related products using multitemporal MODIS data (Townshend and Justice 2002). The MODIS Land Discipline Group uses the Enhanced Vegetation Index (EVI) for use with MODIS data:

$$EVI = G \cdot \frac{p_{nir} - p_{red}}{p_{nir} + C_1 p_{red} - C_2 p_{blue} + L} \qquad (6\text{-}3)$$

where p_{nir}, p_{red} and p_{blue} represent atmospherically corrected near-infrared, red, and blue reflectance,

C_1 and C_2 are atmospheric resistance correction coefficients

L is a canopy background brightness correction factor, and

G is a gain factor,

The coefficients, C_1, C_2, and L are empirically determined. The coefficients adopted in the EVI algorithm are, $L=1$, $C_1 = 6$, $C_2 = 7.5$, and $G = 2.5$ (TRBS 2003).

This algorithm processes 250 m and 500 m spatial resolution surface reflectances and produces 250 m NDVI and EVI. The blue channel is only available at 500 m, and subsequently, this channel needs to be mapped to each 250 m pixel in order to produce the proper 250 m resolution EVI.

This algorithm has improved sensitivity to high biomass regions and improved vegetation monitoring through a decoupling of the canopy background signal and a reduction in atmospheric influences (Huete and Justice 1999). Whereas the NDVI is chlorophyll sensitive, the EVI is more responsive to canopy structural variations, including LAI, canopy type, plant physiognomy, and canopy architecture (TBRS 2003). MODIS data are being used to produce both NDVI and EVI 1-km data products on a global basis every 16 days. These are aggregated to monthly products. A global mosaic of MODIS EVI composite images over the period June 1-16, 2003 is shown in Figure 6-15. There are also NDVI and EVI standard products at 500 m and 1 km spatial resolutions.

5.6.3. Estimating Leaf-Area Index and Fraction of Photosynthetically Active Radiation

A number of biophysical parameters are required for use in advanced global models of climate, hydrology, biogeochemistry, and ecology. For example, Ramakrishna et al. (2003) recently used remote sensing-derived leaf-area index (LAI) and fraction of photosynthetically active radiation

Figure 6-15. NASA Enhanced Vegetation Index (EVI) composite obtained June 1-16, 2003. EVI is a surrogate for the amount of biomass present (courtesy of NASA Earth Observatory).

(FPAR) to document a 6 percent increase in global terrestrial net primary production from 1982 to 1999. LAI describes an important structural property of a plant canopy as the one-sided leaf area per unit ground area (e.g., 1 × 1 m). LAI is a state parameter in all models describing the exchange of fluxes of energy, mass (e.g., water and CO_2), and momentum between the surface and the planetary boundary layer (Knyazikhin et al. 1999).

The problem of accurately evaluating the exchange of carbon between the atmosphere and the terrestrial vegetation is very important. FPAR absorbed by vegetation is a key state variable in global ecosystem productivity models and in global models of climate, hydrology, biogeochemestry, and ecology (Knyazikhin et al. 1999). FPAR measures the proportion of available radiation in the photosynthetically active wavelengths (400 to 700 nm) that a canopy absorbs. Knowing how much light is absorbed and distributed among the canopy, understory, and ground provides information about the functional health and productivity of forests, rangelands, and croplands.

Thus, LAI and FPAR describe vegetation canopy structure and its energy absorption capacity. Both indexes can be derived from MODIS imagery. Daily measurements can be combined at weekly intervals into maps that show leaf area and absorbed sunlight for every square kilometer of Earth's land surface updated every 8 days. For example, LAI and FPAR maps of the U.S. derived from MODIS hyperspectral imagery are shown in Figure 6-16.

5.6.4. Remote Sensing Canopy Height/Structure

The visual analysis of stereo aerial photography has been used for some time to map canopy heights. Semiautomated approaches to extracting object heights from stereoscopic imagery or neighborhood analysis of IFSAR returns has been conducted more recently (Miller et al. 2000). Vegetation height may also be modeled through analyzing single or multiple return LIDAR data. For example, considerable work has been done in mapping canopy structure using large footprint LIDAR sensors (Drake et al. 2002; Lefsky et al. 2002). Analysis of ranges through multiple returns from the same laser pulse in vegetated areas provides estimates for canopy height or structure (Blair and Hofton 1999). Neighborhood spatial analysis of single LIDAR returns or treating multiple returns as independent estimates can also provide maps of vegetation height (Persson et al. 2002; Hodgson et al. 2003b).

6. IMPROVEMENTS IN INFORMATION EXTRACTION FROM REMOTELY SENSED IMAGERY

There have been dramatic improvements in our ability to extract information from remotely sensed data in the last 10 years. These

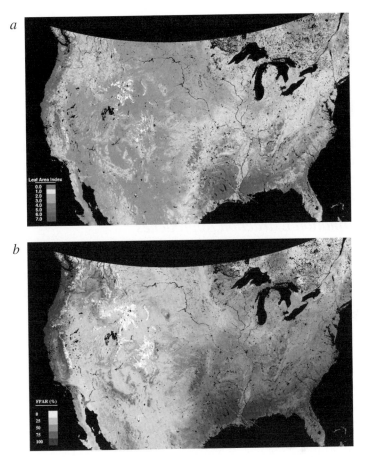

Figure 6-16. Leaf Area Index (a), and Fraction of Photosynthetically Active Radiation (b) maps of the U.S. derived from MODIS hyperspectral imagery (courtesy of NASA Earth Observatory).

improvements are fundamentally due to advances in computing technology, research in information extraction approaches, and integration of GIS and GPS.

6.1. Computational Advances

Digital remote sensor data have always required working with voluminous datasets. Multiple date scenes required for change detection studies exasperates the data storage/processing requirements. The higher spatial resolution imagery (e.g., 1 × 1 m IKONOS versus the 30 × 30 m Landsat TM) or larger number of image bands (e.g., hyperspectral sensors such as AVIRIS) magnify these requirements. Technological advances in computer processing speeds have continued to double every 18 months during the last decade (i.e., Moore's Law). The "workstations" required in the early 1990s for large-study analysis have been replaced by personal computers. Data storage advances do not follow Moore's Law but have evolved at impressive rates and have even changed forms. In the early 1990s, magnetic tapes (20 mb–120 mb capacities) were the standard. Today, red and green laser optical discs (640 mb–4,700 mb) are the standard with blue-laser disk drives already on the market with a 22,000 mb (22 gb) capacity. The concept of a local network of computers and distributed data access/storage has evolved into an Internet and a worldwide distributed data system.

6.2. Information Extraction Approaches

Significant advances in information extraction technology are having profound effects on the approaches currently used in geographic applications of remote sensing science. The selected advances discussed here are:

- Soft-copy photogrammetry
- Analog image processing
- Digital image processing:
  - Hard vs. fuzzy classification
  - Per pixel vs. object classification
  - Nonparametric information extraction techniques,
  - Parametric information extraction techniques
  - Nonmetric information extraction techniques
  - Change Detection

6.2.1. Soft-Copy Photogrammetry

At one time, most stereoscopic aerial photography was analyzed in analog (hard-copy) format using analog/analytical stereoplotters to extract urban infrastructure and topographic information. In fact, most 7.5-minute 1:24,000-scale topographic maps of the U.S. were produced in this manner. Many photogrammetric engineering firms are now migrating to the collection of digital metric photography. In addition, the photogrammetric software has been ported to work on relatively inexpensive PCs using friendly interfaces and digital (soft-copy) imagery. Advancements in soft-copy photogrammetry allow a nonphotogrammetrist to (a) perform interior and exterior orientation to scale and level a stereoscopic model, (b) extract their own DEM of the study area, (c) use a digital elevation model to produce an orthophoto of the study area, (d) view the stereomodel in three dimensions and extract true planimetric information about buildings, roads, drainage networks, and such, and (e) place all of this geographic information into a GIS as the information are compiled. Ten years ago almost all photogrammetric compilation was conducted by commercial photogrammetric engineering firms. Now many scientists are extracting the topographic and urban/suburban infrastructure information they desire using soft-copy photogrammetry algorithms.

6.2.2. Analog Image Processing

There has recently been a resurgence in the teaching of visual image-intepretation skills based on the fundamental elements of image interpretation, i.e., size, shape, shadow, texture, pattern, site, association, arrangement, etc. (Kelly et al. 1999; Jensen 2000). This is primarily a function of the availability of satellite high spatial resolution data (e.g., IKONOS, Quickbird, OrbView-3 at resolutions <1 x 1 m) which in effect look like aerial photography. For many applications, scientists and lay persons have discovered that sometimes it is not necessary to conduct elaborate digital image processing when the information desired can be extracted by simply putting the high spatial resolution image on the screen, photointerpreting the image using the elements of image interpretation, drawing a polygon around the feature of interest, and then labeling it. Remote sensing practitioners are learning that the ability to visually interpret an image is a very valuable skill. Research into how trained interpreters cognize information from imagery helps in the development of automated approaches to information extraction (Lloyd et al. 2002).

6.3.3. Digital Image Processing

Scientists and lay persons desiring to extract information from digital imagery now have a powerful suite of techniques available to them which are constantly being improved. With this improvement, however, have also come new decisions that must be made. For example, analysts must now determine at the very beginning of an information extraction project (a) if the project is to be a hard or soft (fuzzy) classification, and (b) if the classification is to be per-pixel or object-oriented.

Hard versus Fuzzy Classification: If a hard classification is selected then it is a straightforward process to utilize a standard classification scheme and identify mutually exclusive classes of interest. Conversely, scientists know that real-world landscapes often transition gradually into one another (Jensen 1996). For this reason, we may have pixels (or polygons) in the scene that contain proportions of materials rather than just a single class. If the desire is to obtain information about the relative abundance of the material(s) within a pixel or polygon then a fuzzy classification scheme and algorithm should be implemented (Foody 1996; Seong and Usery 2001).

Per-Pixel vs. Object-Oriented Classification: In the past, almost all digital image classification was performed on a pixel by pixel basis based primarily on the spectral properties of the pixel. This has changed. New object-oriented classification algorithms take into account not only the spectral characteristics of a pixel, but the spectral characteristics of surrounding pixels (Hodgson et al. 2003b). Thus, the algorithms take into account spectral and spatial information. The spatial information is based on the use of landscape ecology structure metrics such as degree of spatial compaction or smoothness (Herold et al. 2003). Object-oriented classification algorithms produce polygons that contain relatively homogeneous materials. They have been shown to be especially useful when analyzing high spatial resolution imagery in urban/ suburban environments where spatial elements of image interpretation are important (Tullis and Jensen 2003).

Parametric Information Extraction Techniques: The venerable parallepiped and minimum distance to means classification algorithms are still valuable for extracting land cover information when the classes are well separated in multispectral feature space (e.g., red and near-infrared). These algorithms only require the computation of the mean and standard deviation of training data in *n* bands. It is ideal but not crucial that the training data be normally distributed. Until recently, the maximum likelihood (ML) classification algorithm was the most widely adopted parametric classification algorithm. Unfortunately, the ML algorithm requires normally distributed training data in *n* bands (rarely the case) for computing the class variance and covariance matrices. It is difficult to incorporate nonimage categorical data into a ML classification. Fortunately, fuzzy maximum likelihood classification algorithms are now available (e.g., Foody 1996).

Nonparametric Information Extraction Techniques: Clustering continues to be a powerful nonparametric classification technique. Algorithms such as ISODATA continue to be used heavily. Unfortunately, such algorithms are extremely dependent upon how the seed training data are extracted, and it is often difficult to label the clusters to turn them into information classes. For these reasons, there has been a significant increase in the development and use of artificial neural networks (ANN) for remote sensing applications (e.g., Jensen et al. 2001). The ANN does not require normally distributed training data. ANN may incorporate virtually any type of spatially distributed data into the classification. They learn. The only drawback is that sometimes it is difficult to understand exactly how the ANN came up with a certain conclusion because the information is locked within the weights in the hidden layer. Scientists are working on ways to extract this hidden information so that the rules used can be more formally stated. The ability of an ANN to learn, however, should not be underestimated.

Nonmetric Information Extraction Techniques: Duda et al. (2001) describe various types of decision-tree classifiers as being nonmetric. There has been tremendous development in decision-tree classifiers or rule-based classifiers (e.g., Muchoney et al. 2000) because unlike parametric techniques, which are based primarily on summary inferential statistics such as the mean, variance, and covariance matrices, decision-trees and rule-based classifiers are not based on inferential statistics, but instead "let the data speak for itself" (Gahegan 2003). In other words, the data retains its precision and is not dumbed-down by summarizing through means, etc. Decision-tree classifiers can process virtually any type of spatially distributed data and can incorporate prior probabilities (McIver and Friedl 2002). There are basically three approaches to rule-creation: (1) explicitly eliciting knowledge and subsequent rules from experts, (2) implicitly extracting variables and rules using cognitive methods (Hodgson 1998; Lloyd et al. 2002), and (3) empirically generating rules from observed data and automatic induction methods (Tullis and Jensen 2003). The development of a decision-tree using human-specified rules is a time-consuming and exacting process. Nevertheless, it rewards the user with detailed information about how individual classification decisions were made (Zhang and Wang 2003). Recently, there has also been great interest in having the computer derive the rules from the training data without human intervention. This is referred to as machine-learning and is one of the most promising new digital image processing research areas (Huang and Jensen 1997). The analyst identifies quality representative training areas. The machine learns the patterns from these training data, creates the rules, and applies them to classify the remotely sensed data. The rules are available to document how classifications were made.

Hyperspectral Data Analysis: Analysis of hyperspectral data requires the use of sophisticated digital image processing software. This is because it is usually necessary to calibrate (convert) the raw hyperspectral radiance data to "apparent reflectance" before it can be properly interpreted. This necessitates the removal of atmospheric attenuation, topographic effects (slope, aspect), and any sensor system electronic anomalies. In addition, to maximize hyperspectral information extraction, it is usually necessary to use digital image processing algorithms that (1) allow analysis of a typical spectra to determine its subpixel constituent materials, and/or (2) compare the remote sensing-derived spectra with a library of spectra obtained using handheld spectroradiometers such as that provided by the USGS or NASA's Jet Propulsion Laboratory.

Change Detection: Numerous change detection algorithms are available but still require refinement (Jensen 1996; Friedl et al. 2002; Zhan et al. 2002). They can be used on per-pixel and object-oriented (polygon) classifications.

Unfortunately, there is still no universally accepted method of assessing the accuracy of change detection map products. This is an area that needs considerable attention.

6.3. Integration with GIS

Some have naively suggested that remote sensing science simply provides data and/or information as input to GIS models. Nothing could be further from the truth. It is important to remember that in many instances 1) remote sensing science can be used to extract fundamental biophysical and/or categorical information (e.g., temperature, water vapor, x,y-location, elevation, biomass) without input from the other mapping sciences, and 2) many of the most important spatial data infrastructure (SDI) datasets described are derived most efficiently and accurately using remote sensing science. Thus, while the remote sensing-derived data are a common input to GIS-based models, the relationship between these sciences is much more symbiotic. In fact, it is now quite common for GIS-derived data and/or information to be used extensively in remote sensing-related research. For example, ancillary data collected, processed, and generated during GIS processing are becoming increasingly important in the remote sensing information extraction process. GIS visualization techniques are also now used routinely in the remote sensing information extraction processes. Finally, complimentary research in the GIS and remote sensing communities has blurred the distinction between GIS and remote sensing models. Below are some integration observations.

6.3.1. Ancillary Data

The use of ancillary spatial data created in a GIS context in an image classification procedure has been with us for more than two decades (Hutchinson 1982). Although originally used as simply another "band" or "channel" of information in the image classification process, like a per-pixel classification, ancillary data are now being widely used throughout the remote sensing process. Such data are used in data normalization (Colby 1991), classification (Brown et al. 1998), accuracy assessment (Jensen 2000), and presentation (Gopal et al. 2000). Advances in GIS in the last decade for data handling and processing have allowed for improved creation of the appropriate ancillary data as input to the remote sensing process.

6.3.2. Visualization

Much of the fundamental developments in the map communication paradigm in the cartographic process in the 1970s and 1980s have evolved into what is now called visualization. More recent research has led to new techniques for the display and query of information in 2- and 3-dimensions. The on-screen digitizing and information extraction using imagery has been greatly facilitated by these improvements using simultaneous image and ancillary GIS data products. For example, synthetic 3-dimensional displays and stereo displays of LIDAR points with imagery and ancillary data are a key component in the process used to label ground, vegetation, and other laser returns (Hodgson et al. 2003a).

6.3.3. Modeling

Modeling has long been an active area of research within the GIS and remote sensing communities. Such research has led to improvements in data models, modeling language frameworks, and numerous modeling application contexts. The triangulated irregular network (TIN) developed in the late 1970s (Peucker et al. 1978) has evolved into the widely accepted terrain modeling foundation in GIS and remote sensing. The construction of DEMs from TINs of remote sensing-derived surface observations is now a transformation of surface observations integrated with ancillary data (e.g., streamlines or soft/ hard breaklines). The widely accepted cartographic modeling language is now a fundamental construct in both GIS and image processing programs (Tomlin 1990). Research into uncertainty, data types (e.g, fuzzy, vector, etc.), and spatial classification approaches with both GIS and remote sensing contexts has resulted in new approaches for developing information extraction models. The integration will continue at an increased pace.

7. CONCLUSION

This is an exciting time for remote sensing science. As predicted by Estes et al. (1980), remote sensing's real potential for providing biophysical and urban/suburban infrastructure information for spatially distributed models is beginning to be realized. The current generation of scientists has been trained in the use of GIScience including remote sensing, and they now routinely use it as a standard method of obtaining the information they require. In addition, significant advances in remote sensing technology now allow the user access to extremely high spatial resolution imagery (e.g., $<1 \times 1$ m) from several sources, as well as hyperspectral imagery for intensive spectral-dependent applications. Analog and digital image processing methods used to extract useful information from remotely sensed data continue to improve.

REFERENCES

Adams, J.C. and Chandler, J.G. (2002). Evaluation of Lidar and Medium Scale Photogrammetry for Detecting Soft-cliff Coastal Change, Photogrammetric Record 17: 405-18.

Anderson, J.R., Hardy, E.E. Roach, J.T., and Witmer, R.E. (1976). A Land Use and Land Cover Classification System for Use With Remote Sensor Data. U.S. Geological Survey Professional Paper 964.

Barnsley, M. (1999). Digital Remotely-Sensed Data and Their Characteristics. In Longley, P.E., Goodchild, M.F., Maguire, D. J., and Rhind, D.W. (Eds.) Geographical Information Systems, 451-66. New York: John Wiley.

Belward, A.S., Estes, J.E., and Kline, K.D. (1999). The IGBP-DIS Global 1-km Land-Cover Dataset GISCover: A Project Overview, Photogrammetric Engineering and Remote Sensing 5: 1013-20.

Blackburn, G.A. (2002). Remote Sensing of Forest Pigments Using Airborne Imaging Spectrometer and LIDAR Imagery, Remote Sensing of Environment 82: 311-21.

Blair, J.B. and Hofton, M.A. (1999). Modeling Laser Altimeter Return Waveforms Over Complex Vegetation Using High-resolution Elevation Data, Geophysical Research Letters 26: 2509-12.

Bowen, Z.H. and Waltermire, R.G. (2002). Evaluation of Light Detection and Ranging (LIDAR) for Measuring River Corridor Topography, Journal of the American Water Resources Association 38: 33-41.

Breaker, L.C, Krasnopolsky, V.M., Rao, D.B., and Yan, X.H. (1994). The Feasibility of Estimating Ocean Surface Currents On An Operational Basis Using Satellite Feature Tracking Methods, Bulletin of the American Meteorological Society 75: 2085-95.

Brown, D.G., Lusch, D.P., and Duda, K.A. (1998). Supervised Classification of Glaciated Landscape Types Using Digital Elevation Data, Geomorphology 21: 233-50.

Cobby, D.M., Mason, D.C., and Davenport, I.J. (2001). Image Processing of Airborne Laser Altimetry Data Improved River Modeling, ISPRS Journal of Photogrammetry and Remote Sensing 56: 121-38.

Cohen, W.B., Maiersperger, T.K., Gower, S.T., and Turner, D.P. (2003). An Improved Strategy for Regression of Biophysical Variables and Landsat ETM+ Data, Remote Sensing of Environment 84: 561-71.

Colby, J.D. (1991). Topographic Normalization in Rugged Terrain, Photogrammetric Engineering and Remote Sensing 57: 531-37.

Colwell, R.N. (1997). History and Place of Photographic Interpretation, Manual of Photographic Interpretation, Bethesda: American Society for Photogrammetry and Remote Sensing, 33-48.

Cowen, D.J., Jensen, J.R., Hendrix, C., Hodgson, M.E., and Schill, S.R. (2000). A GIS-Assisted Rail Construction Econometric Model That Incorporates LIDAR Data, Photogrammetric Engineering and Remote Sensing, 66: 1323-28.

Curran, P.J. (1987). Remote Sensing Methodologies and Geography, International Journal of Remote Sensing 8: 1255-75.

Curran, P.J., Milton, E.J., Atkinson, P.M., and Foody, G.M. (1998). Remote Sensing: From Data To Understanding. In Longley, P.E., Brooks, S.M., McDonnell, R., and Macmillan, B. (Eds.) Geocomputation: A Primer, 33-59. New York: John Wiley and Sons.

DeFries, R.S. and Townshend, J.G.R. (1994). NDVI Derived Land Cover Classifications At a Global Scale, International Journal of Remote Sensing 19: 3141-68.

Defries, R., Hansen, M., Townshend, J.G.R., and Sohlberg, R. (1998). Global Land Cover Classifications At 8 km Resolution: The Use of Training Data Derived from Landsat

Imagery in Decision Tree Classifiers, International Journal of Remote Sensing 5: 3567-86.

Drake, J.B., Dubayah, R.O., Clark, D.B., Knox, R.G., Blair, J.B., Hofton, M.A., Chazdon, R.L., Weishampel, J.F., and Prince, S.D. (2002). Estimation of Tropical Forest Structural Characteristics Using Large-Footprint Lidar, Remote Sensing of Environment 79: 305-19.

Duda, R.O., Hart, P.E., and Stork, D.G. (2001). Pattern Classification, New York: John Wiley and Sons, 394-452.

EIS-Africa (Environmental Information System-Africa) (2002). Geo-Information Supports Decision-making in Africa. http://www.eis-africa.org/DOCS/EIS-AFRICAwssd_statement-draft8.doc

Estes, J.E., Jensen, J.R., and Simonett, D.S. (1980). Impacts of Remote Sensing on U.S. Geography. Remote Sensing of Environment 10: 3-80.

FGDC (Federal Geographic Data Committee) (2002). Overview: What the Framework Approach Involves. http://www.fgdc.gov/ framework/overview.html

Foody, G.M. (1996). Approaches for the Production and Evaluation of Fuzzy Land Cover Classifications from Remotely Sensed Data, International Journal of Remote Sensing 17: 1317-40.

Friedl, M.A., Muchoney, D., McIver, D.K., Gao, F., Hodges, J.C.F., and Strahler, A.H. (2000). Characterization of North American Land Cover from NOAA-AVHRR Data Using the EOS MODIS L and Cover Classification Algorithm, Geophysical Research Letters 27: 977-80.

Friedl, M.A., McIver, D.K., Hodges, J.C.F., Zhang, X.Y., Muchoney, D., Strahler, A.H., Woodcock, C.E., Gopal, S., Schneider, A., Cooper, A., Baccini, A., Gao, F., and Scaaf, C. (2002). Global Land Cover Mapping from MODIS: Algorithms and Early Results, Remote Sensing of Environment 83: 287-302.

Gahegan, M. (2003). Is Inductive Machine Learning Just Another Wild Goose (or Might It Lay the Golden Egg)? International Journal of Geographical Information Science 17: 69-92.

Goetz, S.J. (2002). Recent Advances in Remote Sensing of Biophysical Variables: An Overview of the Special Issue, Remote Sensing of Environment 79: 145-46.

Gopal, S., Liu, W., and Woodcock, C. (2000). Visualization Based on the Fuzzy ARTMAP Neural Network for Mining Remotely Sensed Data. In Miller, H.J. and Han, Jiawei (Eds.) Discovering Geographic Knowledge in Data-rich Environments, 315-336. Heidelberg: Springer-Verlag.

Herold, M., Guenther, S., and Clarke, K. (2003). Mapping Urban Areas in the Santa Barbara South Coast Using IKONOS Data and eCognition. eCognition Application Note, Munich, Germany: Definiens Imaging GmbH, 4: 3.

Hodgson, M.E., Jensen, J.R., Schmidt, L., Schill, S., and Davis, B. (2003a). An Evaluation of LIDAR- and IFSAR-Derived Digital Elevation Models in Leaf-on Conditions with USGS Level 1 and Level 2 DEMs, Remote Sensing of Environment 84: 295-308

Hodgson, M.E., Jensen, J.R., Tullis, J.A., Riordan, K.D., and Archer, C.M. (2003b). Synergistic Use of Lidar and Color Aerial Photography for Mapping Urban Parcel Imperviousness, Photogrammetric Engineering and Remote Sensing, 69: 973-80.

Hodgson, M.E. (1998). What Size Window for Image Classification? - A Cognitive Perspective, Photogrammetric Engineering and Remote Sensing 64: 797-808.

Hodgson, M.E. and Bresnahan, P. (2004). Accuracy of Airborne Lidar Derived Elevation: Empirical Assessment and Error Budget, Photogrammetric Engineering and Remote Sensing, in press.

Huang, X. and Jensen, J.R. (1997). A Machine-Learning Approach to Automated Knowledge-base Building for Remote Sensing Image Analysis with GIS Data, Photogrammetric Engineering and Remote Sensing 63: 1185-94.

Huete, A. and Justice, C. (1999). MODIS Vegetation Index (MOD 13). Algorithm Theoretical Basis Document, Greenbelt: NASA Goddard Space Flight Center. http://modarch.gsfc. nasa.gov/MODIS/LAND/#vegetation-indices

Hutchinson, C.F. (1982). Techniques for Combining Landsat and Ancillary Data for Digital Classification Improvement, Photogrammetric Engineering and Remote Sensing 48: 123-30.

Jensen, J.R. (1995). Issues Involving the Creation of Digital Elevation Models and Terrain Corrected Orthoimagery Using Soft-copy Photogrammetry, Geocarto International: A Multidisciplinary Journal of Remote Sensing, 10: 1-17.

Jensen, J.R. (1996). Introductory Digital Image Processing: Remote Sensing Perspective. 2nd ed. Saddle River, NJ: Prentice-Hall.

Jensen, J.R. (2000). Remote Sensing of the Environment: An Earth Resource Perspective, Saddle River, NJ: Prentice-Hall.

Jensen, J.R. and Cowen, D.J. (1999). Remote Sensing of Urban/Suburban Infrastructure and Socio-Economic Attributes, Photogrammetric Engineering and Remote Sensing 65: 611-22.

Jensen, J.R. and Schill, S. (2000). Bi-directional Reflectance Distribution Function (BRDF). of Smooth Cordgrass (Spartina alterniflora), Geocarto International – A Multidisciplinary Journal of Remote Sensing and GIS 15: 21-28.

Jensen, J.R., Qiu, F., and Patterson, K. (2001). A Neural Network Image Interpretation System to Extract Rural and Urban Land Use and Land Cover Information for Remote Sensor Data, Geocarto International 16: 19-28.

Jensen, J.R., Botchway, K., Brennan-Galvin, E., Johannsen, C., Juma, C., Mabogunje, A., Miller, R., Price, K., Reining, P., Skole, D., Stancioff, A., and Taylor, D.R.F. (2002a). Down to Earth: Geographic Information for Sustainable Development in Africa, Washington: National Research Council.

Jensen, J.R., Olson, G., Schill, S.R., Porter, D.E. and Morris, J. (2002b). Remote Sensing of Biomass, Leaf-Area-Index, and Chlorophyll a and b Content in the ACE Basin National Estuarine Research Reserve Using Sub-meter Digital Camera Imagery, Geocarto International 17: 25-34.

Johannsen, C.J., Petersen, G.W., Carter, P.G., and Morgan, M.T. (2003). Remote Sensing: Changing Natural Resource Management, Journal of Soil and Water Conservation 58: 42-45.

Justice, C.O. (2003). MODIS Land Overview. http://modarch.gsfc.nasa.gov/MODIS/LAND/ #biophysical.

Justice, C., Vermote, O. Vermote, Townshend, J., Defries, R., Roy, D., Hall, D., Salomonson, V., Privette, J., Riggs, G., Strahler, A., Lucht, W., Myneni, R. Knjazihhin, Y., Running, S., Nemani, R., Wan, Z., Huete, A., van Leeuwen, W., Wolfe, R., Giglio, L., Muller, J., Lewis, P., and Barnsley, J. (1998). The Moderate Resolution Imaging Spectroradiometer (MODIS): Land remote sensing for Global Change Research, IEEE Transactions Geoscience. Remote Sensing 36: 1228-49.

Kelly, M., Estes, J.E. and Knight, K.A. (1999). Image Interpretation Keys for Validation of Global Land-Cover Data Sets, Photogrammetric Engineering and Remote Sensing 65: 1041-49.

Kerr, J.T. and Ostrovsky, M. (2003). From Space to Species: Ecological Applications for Remote Sensing, Trends in Ecology and Evolution 18: 299-305.

Knyazikhin, Y., Glassy, J., Privette, J.L., Tian, Y., Lotsch, A., Zhang, Y., Wang, Y., Morisette, J. T. , Votava, P., Myneni, R.B., Nemani, R.R., and Running, S.W. (1999). MODIS Leaf Area Index (LAI). and Fraction of Photosynthetically Active Radiation Absorbed by Vegetation (FPAR). Product (MOD15). Algorithm Theoretical Basis Document. http://eospso.gsfc.nasa.gov/atbd /modistables.html

Kogan, F.N. (2001). Operational Space Technology for Global Vegetation Assessment, Bulletin of the American Meteorological Society 82: 1949-64.

Lefsky, M.A., Cohen, W.B., Parker, G.G., and Harding, D.J. (2002). LIDAR Remote Sensing for Ecosystem Studies, Bioscience 52: 19-30.

Liverman, D., Moran, E.F., Rindfuss, R.R., and Stern, P.C. (1998). People and Pixels: Linking Remote Sensing and Social Science. Washington: National Academy Press.

Lloyd, R., Hodgson, M.E., and Stokes, A. (2002). Visual Categorization with Aerial Photographs, Annals of the Association of American Geographers 92: 241-66.

Loveland, T.R., Reed, B.C., Brown, J.F., Ohlen, D.O., Zhu, Z., Yang, L., and Merchant, J.W. (2000). Development of A Global Land Cover Characteristics Database and IGBP DISCover from 1-km AVHRR Data, International Journal of Remote Sensing 21: 1303-65.

McIver, D.K. and Friedl, M.A. (2002). Using Prior Probabilities in Decision-Tree Classification of Remotely Sensed Data, Remote Sensing of Environment 81: 253-61.

Miller, D.R., Quine, C.P., and Hadley, W. (2000). An Investigation of the Potential of Digital Photogrammetry to Provide Measurements of Forest Characteristics and Abiotic Damage, Forest Ecology and Management 135: 279-88.

Muchoney, D., Borak, J., Chi, H., Friedl, M., Gopal, S., Hodges, J., Morrow, N., and Strahler, A. (2000). Application of the MODIS Global Supervised Classification Model to Vegetation and Land Cover mapping of Central America, International Journal of Remote Sensing 21: 1115-38.

Nemani, R.R., Keeling, C.D., Hashimoto, H., Jolly, W.M., Piper, S.C., Tucker, C.J., Myneni, R.B., and Running, S.W. (2003). Climate-Driven Increases in Global Terrestrial Net Primary Production from 1982 to 1999, Science 300: 1560-63.

Ninnis, R.M., Emery, W.J., and Collins, M J. (1986). Automated Extraction of Pack Ice Motion from Advanced Very High Resolution Radiometer Imagery. Journal of Geophysical Research 91: 10725-34.

North Carolina Flood Plain Mapping Program (2002). http://www.ncfloodmaps.com

Persson, A., Holmgren, J., and Soderman, U. (2002). Detecting and Measuring Individual Trees Using An Airborne Laser Scanner, Photogrammetric Engineering and Remote Sensing 68: 925-32.

Peucker, T.K., Fowler, R.J., Little, J.J., and Mark, D.M. (1978). Digital Representation of Three-Dimensional Surfaces by Triangulated Irregular Networks (TIN). Proceedings, Digital Terrain Model Symposium, May 1978, ASP, 516-40.

Plummer, S.E. (2002). Perspectives on Combining Ecological Process Models and Remotely Sensed Data, Ecological Modelling 129: 169-86.

Raber, G.T., Jensen, J.R., Schill, S.R., and Schuckman, K. (2002). Creation of Digital Terrain Models Using An Adaptive LIDAR Vegetation Point Removal Process, Photogrammetric Engineering and Remote Sensing 68: 1307-15.

Ramakrishna, R.N., Keeling, C.D., Hashimoto, H., Jolly, W.M., Piper, S.C., Tucker, C.J., Myneni, R.G., and Running, S.W. (2003). Climate-Driven Increases in Global Terrestrial Net Primary Production from 1982 to 1999, Science 300: 1560-63.

Rizos, C. (2002). Introducing the Global Positioning System. In Bossler, J.D., Jensen, J.R., McMaster, R.B., and Rizos, C. (Eds.) Manual of Geospatial Science and Technology, 77-94. London: Taylor and Francis.

Seong, J.C. and Usery, E.L. (2001). Fuzzy Image Classification for Continental-Scale Multitemporal NDVI Images Using Invariant Pixels and An Image Stratification Method, Photogrammetric Engineering and Remote Sensing 67: 287294.

SRTM (2003). Shuttle Radar Topography Mission. http://www.jpl.nasa.gov/srtm/

Strahler, A., Muchoney, D., Borak, J., Gao, F., Friedl, M., Gopal, S., Hodges, J., Lambin, E., McIver, D., Moody, A., Schaaf, C., and Woodcock, C. (1999). MODIS Land Cover Product Algorithm Theoretical Basis Documentation, Version 5.0, Boston, MA: Center for Remote Sensing, Department of Geography, Boston University.

Terrestrial Biophysics and Remote Sensing Laboratory (TBRS) (2003). Enhanced Vegetation Index, University of Arizona. http://tbrs.arizona.edu/project/MODIS/evi.php

Teillet, P.M., Staenz, K., and Williams, D.J. (1997). Effects of Spectral, Spatial, and Radiometric Characteristics on Remote Sensing Vegetation Indices of Forested Regions, Remote Sensing of Environment 61: 139-49.

Tomlin, C.D. (1990). Geographic Information Systems and Cartographic Modeling, Englewood Cliffs, NJ: Prentice-Hall.

Townshend, J.R.G., Justice, C.O., Skole, D., Malingreau, J.P., Chilar, J., Teillet, P., Sadowski, F., and Ruttenberg, S. (1994). The 1 km AVHRR Global Data Set: Needs of the International Geosphere Biosphere Programme, International Journal of Remote Sensing 15: 3417-41.

Townshend, J.R.G and Justice, C.O. (2002). Towards Operational Monitoring of Terrestrial Systems by Moderate-resolution Remote Sensing, Remote Sensing of Environment 83: 351-59.

TREES (2002). Tropical Ecosystem Environment Observations by Satellite. http://www.gvm.sai.jrc.it/Forest/defaultForest.htm

Tucker, C., Dregne, H., and W. Newcomb (1991). Expansion and Contraction of the Saharan Desert from 1980 to 1990, Science 253: 299-301.

Tullis, J.A. and Jensen, J.R. (2003). Expert System House Detection in High Spatial Resolution Imagery Using Size, Shape, and Context, Geocarto International 18: 5-15.

USGS (United States Geological Survey) (2002). Global Land Cover Characterization Background, Sioux Fall, SD: EROS Data Center. http://edcdaac.usgs.gov/glcc/background.html

Walsh, S.J., Butler, D.R., and Malanson, G.P. (1998). An Overview of Scale, Pattern, Process Relationships in Geomorphology: A Remote Sensing and GIS Perspective, Geomorphology 21: 183-205.

Zhan, X., Sohlberg, R.A., Townshend, J.R.G., DiMiceli, C., Carrol, M.L., Eastman, J.C., Hansen, M.C., and De Fries, R.S. (2002). Detection of Land Cover Changes using MODIS 250 m Data, Remote Sensing of Environment 83: 336-50.

Zhang, Q. and Wang, J. (2003). A Rule-Based Urban Land Use Inferring Method for Fine-Resolution Multispectral Imagery, Canadian Journal of Remote Sensing 29: 1-13.

CHAPTER 7

MATTHEW ZOOK, MARTIN DODGE
YUKO AOYAMA, ANTHONY TOWNSEND

NEW DIGITAL GEOGRAPHIES: INFORMATION, COMMUNICATION, AND PLACE

Abstract   This chapter provides an overview of contemporary trends relevant to the development of geographies based on new digital technologies such as the Internet and mobile phones. Visions of utopian and ubiquitous information superhighways and placeless commerce are clearly passé, yet privileged individuals and places are ever more embedded in these new digital geographies while private and state entities are increasingly embedding these digital geographies in all of us. First is a discussion of the centrality of geographical metaphors to the way in which we imagine and visualize the new digital geographies. Then the example of the commercial Internet (e-commerce) is used to demonstrate the continued central role of place in new digital geography both in terms of where activities cluster and how they vary over space. The transformation of digital connections from fixed (i.e., wired) to untethered (i.e., wireless) connections is explored as to its significance in the way we interact with information and the built environment. Finally is an examination of the troubling issue of the long data shadows cast by all individuals as they negotiate their own digital geographies vis-à-vis larger state and private entities.

Keywords   Technology, telecommunications, social users of technology, e-commerce, mobile phones, privacy

1. INTRODUCTION

Digital communications technologies are creating complex arrays of new geographies through which we view, interact, and connect to the world. It is now possible to view live shots of the Eiffel tower, chat with a colleague in South Africa, or read a local newspaper from the comfort of your home, from an Internet café in Chiang Mai, Thailand, or via mobile phone in Helsinki. While this capability provides an image (heavily promoted by the advertising of telecommunications companies) of uniform and utopian connectivity, the reality of new digital geographies is much more complex. The cost of these technologies, the ease, availability, reliability, and portability of their use,

Stanley D. Brunn, Susan L. Cutter, and J.W. Harrington, Jr. (Eds.), Geography and Technology,
155-176. © 2004 Kluwer Academic Publishers. Printed in the Netherlands.

and even the functions to which they are turned vary across time and space. Images of famous landmarks may be easily available while vernacular landscapes are bypassed. Certain parts of South Africa and Thailand (especially privileged spaces such as cities and tourist destinations) are well wired while their hinterlands remain cut off, and parts of the "developed" world such as Appalachia are struggling to maintain meaningful digital connections. Even the space (private, public, or publicly private) and means of connection (wired PC or wireless phone) vary according to local availability and personal preference.

These new digital geographies (both social and economic) are by no means technologically determined. Rather, the way in which places and people become "wired" (or remain "unwired") still depends upon historically layered patterns of financial constraint and cultural and social variation. The geographic and technological evolution of this digital infrastructure can therefore be understood as a process of social construction of new (and often personal) digital geographies. These new geographies are both immensely empowering (for the people and places able to construct and consume them) and potentially overpowering as institutional and state forces are able to better harness information with growing personal and spatial specificity.

At the heart of new digital geographies is the ability to represent text, sound, still, and moving images in digital formats which can then be transmitted across common networks. This potentiality of shared transmission and consumption via some digital receiver, be it a wired PC or a wireless phone, is central to the geographical impact of information and communication technologies (ICTs). The exemplar of this interoperability is the Internet, which pioneered digital packet switching between disparate hardware and software systems via a standardized set of protocols (Abbate 1999). Although in existence for decades, mainstream Western society adopted the Internet during the 1990s, often with unrealistic expectations that the technology would simply substitute for geography in social and economic relations. Utopian visions of a "digital and spaceless society" abounded, and thousands of so called dot-com firms sprang up overnight intent on changing the structure of the economy. Ironically, although the rhetoric often proclaimed an end of geography, ICTs were and continue to be routinely imagined and understood through geographic metaphors.

Metaphors of information superhighways and wired cities are useful in imagining a world in which data is created, shared, accessed, and cross-checked in historically unprecedented volumes. While primarily employed to emphasize the exponential growth of "digital geographies," metaphors were also central in understanding the unevenness of these new landscapes. Some countries, particularly relatively small ones such as Singapore or Finland, emerged as so-called "cyberstates," while others such as Estonia, Qatar, and Slovenia are making considerable progress to this same goal although often with radically different forms. Larger and wealthy countries such as the U.S. with developed high technology industries, were also able to quickly expand their presence in digital geographies, albeit with significant digital divides. At the same time, much of the developing world was limited to a few "digital islands" located in capital cities and/or expatriate populations. Clearly a ubiquitous and uniform global digital geography is more rhetoric than reality (Warf 2001).

Moreover, even in the densest parts of these new digital geographies, ICTs are accessed, adapted, and appropriated differently depending on individual and societal imagination, culture, and history. This is particularly pronounced within the economic sphere, as legacies of earlier, firm, technological and cultural structures create forms of electronic or mobile commerce (e-commerce or m-commerce) that vary considerably between places. While dot-com companies envisioned a single predetermined global digital geography, evidence suggests that there are multiple digital geographies interconnected but situated in places that are instrumental in shaping any interaction. The introduction of mobile connections further amplifies the dynamic complexities of contemporary digital geographies. These technologies enable an entirely new type of interaction, whether through peer-to-peer communications or new uses of the built environment, melding access to information and instant worldwide communications in portable and personal packages. We are becoming unique and powerful digital individuals within multiple digital societies.

Yet even as people are using information in new ways and places, it is being distributed and made available in greater quantity and with unprecedented details. An ever growing number of personalized records are collected, and at times disseminated in the databases and customer management systems of businesses, organizations, and government agencies that service modern living, thereby connecting the world via a complex and ever-changing array of digitized transactions of ever more personal records. The Orwellian notion of "Big Brother" is now distributed and multiple but is indeed watching, and the implications can be simultaneously reassuring (monitoring individual health) and terrifying (tracking what is being read and with whom it is being discussed).

This chapter provides an overview of various contemporary trends of significant relevance in developing a theoretical framework for digital geographies. First, we discuss the centrality of geographical metaphors to the way in which we imagine and visualize the new digital geographies. We then use the example of the commercial Internet (e-commerce) to demonstrate the continued central role of place in new digital geography both in terms of where activities cluster, and also in how they vary over space. We explore the transformation of digital connections from fixed (i.e., wired) to untethered (i.e., wireless) connections, which has significance in the way we interact with information and the built environment. Finally, we examine the troubling issue of the long data shadows cast by every individual as they negotiate their own digital geographies vis-à-vis larger state and private entities.

2. IMAGINING DIGITAL GEOGRAPHIES

In both popular and academic discussions of digital communication technologies and their possible socioeconomic implications, a large panoply of metaphors has been coined with a premise on geographic place, such as superhighways, teleports, server farms, homepages, and so on (Adams 1997). Likewise, the transactions and data exchanges at the heart of the so-called Network Society (Castells 1996), are also frequently imagined and envisioned in terms of "spaces"—hyperspace, dataspace, netspace and, of course, cyberspace (Dodge and Kitchin 2001; Thrift 1996). The result, according to Graham (1998), is digital geographies that are made tangible and knowable through familiar territorial analogy.

Although useful for imaging new social spaces, the metaphors and geographic analogies used are rarely neutral; rather they are active, ideological constructs often deployed purposefully to hide underlying realities. The implication at the heart of many of these visions of digital geographies (particularly prevalent during the dot-com boom) is that "something new, different, and (usually) better is happening" (Woolgar 2002, 3), the rhetoric often supporting a deterministic and utopian viewpoint, with new spaces creating opportunities for free-market exploitation. There are few studies, however, that explicitly aim at "getting behind" these spatial metaphors, to begin describing how digital communication technologies actually do their "work" at the level of individual, everyday performances of space (notable exceptions include Adams 2000; Kwan 2002a; Valentine et al. 2002). Such "invisibility" of analysis of communication within the geography discipline (Hillis 1998) in part derives from the fact that, unlike transportation networks, much of the telecommunications and network infrastructures supporting cyberspace are small in scale and often remain hidden from the public view, such as the case for fiber-optic cables that carry many gigabytes of information, anonymous servers rooms, and secure, windowless buildings, with cables buried under roads and running through walls and under floors (Hayes 1997). Such invisibility may in part have led to the erroneous assumption that cyberspace is somehow immaterial, aspatial, and nongeographic. "The net cannot float free of conventional geography" (Hayes 1997, 214), however, even though most users of the Internet may be oblivious about practicalities of "where" and "how" data flows to successfully send e-mail.

> Current technology requires information to be served from somewhere and delivered to somewhere. Heisenberg's uncertainty principle not withstanding, at geographic scales a bit always has an associated location in real geographic space." (Goodchild 1997, 383-84)

The "where" and "how" of the physical embeddedness of data networks is important, first, because of their highly uneven geographical distribution and the consequent sociospatial implications in terms of access and inequalities. Second, it is important because of the increasing concern for the physical vulnerability of cyber infrastructure to terrorist attack, with damage to nodal points potentially causing major economic and social impacts for technologically dependent nations. Cartographic visualization provides one useful way to envision and begin to analyze the "where" and "how" of these digital geographies.

3. VISUALIZING DIGITAL GEOGRAPHIES

Efforts have already been made to map the material, economic, and social geographies of cyberspace (Dodge and Kitchin 2001). They range from those with a relatively basic cartographic design, such as the geographic layout of cable infrastructures from the very local scale of city streets up to global scale interconnections (as shown in Figure 7-1), to more sophisticated representations. Cheswick and Burch created a visualization of the structure of the core of the Internet using a graph representation (Figure 7-2). The "map" shows the Internet's topology as of December 11, 2000, representing over 75,000 network nodes, color coded according to the ISP, seeking to highlight who "owns" the largest sections of Internet. The choice of mapping through abstract graphs also prompts one to think about the types of visual representations, locational grids, and projections that are most effective to map new digital geographies. For geographers, it raises fundamental questions about how far Euclidean geography is useful or relevant to the analysis of these digital spaces.

One key aspect that is missing from most current work on mapping network infrastructures, including Cheswick and Burch's, is information on the nature of traffic flows and for what people are actually using the networks (see Kellerman 2003). Yet mapping can reveal the nature of information archives and social interaction by exposing their latent spatial structures (see Skupin's 2002 work on AAG abstracts). Online interaction is currently dominated by visual interfaces, rather than aural, tactile, or olfactory interfaces, which suggests that cartographic approaches are particularly apposite for

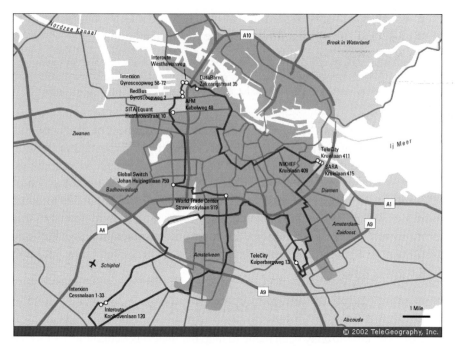

Figure 7-1. An example of how cartography can be used to envision infrastructure of digital geographies. This map shows Interoute's fibre network ring around the city of Amsterdam. (Source: TeleGeography's Metropolitan Area Networks Report 2003, http://www.telegeography.com.)

representing information spaces and providing novel tools for their navigation (Dodge and Kitchin 2001). Many of the most interesting cyberspace mapping efforts produce nongeographic visualizations of information structures using innovative processes of spatialization (Couclelis 1998).

Spatialization can be considered a subset of information visualization and information retrieval, and is defined by Fabrikant (2000, 67) as the processes of visualization that "rely on the use of spatial metaphors to represent data that are not necessarily spatial." The aim of spatialization is to render large amounts of abstract data into a more comprehensible and compact visual form by generating meaningful synthetic spatial structure (e.g., distance based on lexical similarity) and applying cartographic-like representations, for example by borrowing design concepts from terrain and thematic mapping. Innovative developments in spatialization, information visualization, and geovisualization are altering the nature of the map. Within geography, digital

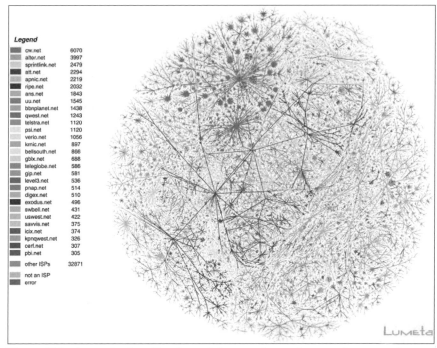

Legend	
cw.net	6070
alter.net	3997
sprintlink.net	2479
att.net	2294
apnic.net	2219
ripe.net	2032
ans.net	1843
uu.net	1545
bbnplanet.net	1438
qwest.net	1243
telstra.net	1120
psi.net	1120
verio.net	1056
krnic.net	897
bellsouth.net	866
gblx.net	688
teleglobe.net	586
gip.net	581
level3.net	536
pnap.net	514
digex.net	510
exodus.net	496
swbell.net	431
uswest.net	422
savvis.net	375
icix.net	374
kpnqwest.net	326
cerf.net	307
pbi.net	305
other ISPs	32871
not an ISP	
error	

Figure 7-2. A graph visualization of the topology of network connections of the core of the Internet, December 11, 2000. (Source: Bill Cheswick, http:// www.lumeta.com)

maps have increasingly become more tools for exploring data than static representations for communicating results. Looking forward, developments in digital communications technologies are likely to accelerate this as maps become interactive and transitory, generated "on-the-fly" to meet particular needs (e.g., web mapping services, in-car navigation, and on-demand mapping to mobile devices). As maps become "intelligent" to some degree (i.e., aware of the location of the user), we will see mobile maps that provide an individually tailored view of the world centered around the person's location and themed to match their interests. But as new modes of individualized and context aware mapping are developed at fine scales and in real time, there will be corresponding ethical and privacy implications to grapple with. These new issues will involve politics just as in other forms of cartography, and therefore their partiality and subjectivity should be taken into consideration (Dodge and Kitchin 2000; Harpold 1999).

4. COMMERCIALIZING DIGITAL GEOGRAPHIES

Arguably, one of the most powerful visions about new digital geographies was that distance would compress to nothing and physical location would become irrelevant. The initial public offering (IPO) of Netscape Communications in August 1995, through the market downturn in April 2000, marked a boom in dot-com companies that strove to change the way businesses and consumers conducted transactions. The dot-com boom was a historically unprecedented effort to define and commercialize what hitherto had been a fiercely noncommercial digital space and ushered in an extraordinary moment in the economy.

The vision of this new commercial digital geography was so compelling that the closing years of the twentieth century saw a tremendous expansion of risk capital, media attention, and stock market growth based on dot-com companies (Zook 2002). The subsequent bursting of the bubble in April 2000 and rapid retreat from Internet companies (see Kaplan 2002 and Cassidy 2002) has resulted in a marked decrease in rhetoric on the ability of the Internet and e-commerce to transform the economy. Subsequent evidence has shown that "spacelessness" was increasingly a mere product of imagination as the twentieth century came to a close. Even at the height of the boom, dot-com firms were overwhelmingly concentrated in major metropolitan and technology centers (such as the San Francisco Bay area and New York City), belying the very rhetoric they espoused (Zook 2000).

Despite this decline in visibility, e-commerce nevertheless persists and affects the way all companies conduct business, albeit with significant variation across sectors and business structures. For every spectacular dot-com flame-out, there are examples of companies using the web in new and innovative ways. This trend, however, does not mark a return to the economic system of the pre-Internet era but is simply the next step in emerging commercial digital geographies. These companies often have no formal risk capital, employ small numbers of people, and receive little in the way of media attention. One of the best known examples of this phenomenon is the listings site called Craigslist (www.craigslist.org). Founded by Craig Newmark in 1995 as an e-mail listing service, this no-frills website leverages what the Internet does best, i.e., aggregating relevant information from scattered sources (such as individuals) in an easily accessible format. In the case of Craigslist, it catalogs subjects such as apartment listings, garage sales, and job listings for a local area. It is this simplicity and the low cost of use that makes Craigslist such a success and a marked contrast to the dot-com hoopla of the 1990s. It will not turn its employees into overnight millionaires but it does have the potential of growing steadily while providing sufficient revenues and profits to continue. In fact, it is because Craigslist was not conceived as a big moneymaker that it still exists. Ironically, precisely because Craiglist did not go all out to capture a market or build a community of users, it has been able to become an important node in emerging visions of the commercial Internet.

Commercial digital geographies also manifest themselves differently across space and cultures. The way in which consumers adopt e-commerce varies greatly between societies, and such variations are directly and indirectly linked with construction of space in each society (Aoyama 2001a, b, 2003). The causes behind such variations are cultural (e.g., "keyboard allergy" or the lack of familiarity with using the standard Western keyboard for typing and entry), institutional and regulatory (i.e., structures of retail/wholesale sectors, consumer behavior), and spatial (urban form and settlement patterns). Consumer behavior is governed by convenience, familiarity, and social habits, which in turn are shaped by the historical evolution of retail trade. Hence, even societies with similar income and education levels vary in the manner and the speed of technological adoption. For example, e-commerce developments in Japan and Germany have each shown unique historical trajectories and have followed their own distinctive paths.

The notable aspects of the evolution of e-commerce in Japan include (1) the lack of widespread and historical use of long-distance, nonstore retailing, (2) the early adoption of e-commerce by neighborhood convenience stores, and (3) the widespread popularity of Internet-abled mobile telephones. The lack of popularity for nonstore retailing in Japan is attributed to its densely populated urban structure (which gave advantages to ubiquitous storefront retailers), low reputation (proliferation of fraudulent nonstore merchants), tight regulations (as a response to fraudulent merchants), and strategic blunder on the part of retailers who, in the interest of protecting and increasing profitability, did not implement generous return/exchange policies (Aoyama 2001b).

The adoption of e-commerce in Japan was initially associated with an innovative strategy of neighborhood convenience stores. Japan's convenience store franchises, such as Seven Eleven Japan, used their ubiquity and preexisting information network infrastructure (used for point-of-sale inventory reduction) to bring e-commerce into their storefronts by setting up dedicated terminals that sold game software, concert tickets, and travel packages. While the U.S.'s e-commerce sector innovated by bringing virtual storefronts to every home, Japan's e-commerce sector attempted to achieve ubiquity by bringing e-commerce to every storefront. This strategy was later superceded by the emerging Internet-abled mobile telephone, which then brought portable virtual storefronts to every consumer.

In contrast, long-distance retailing was an accepted, legitimate, and established medium of consumption in Germany for over a century before e-commerce emerged. German nonstore retailers established an early reputation for convenience, quality, and affordability, some with well-known brands that were sold exclusively via their nonstore operations. Two World Wars caused significant disruption of consumer activities, and nonstore retailers served important purposes during the time storefront retailing was being restored. E-commerce merchants benefited from this well-established practice of nonstore retailing and accumulated significant business know-how. Germany today is the largest e-commerce market in Europe, and is the home of two of the top 10 mail order businesses in the world. Furthermore, eight out of the top 10 German e-commerce websites are operated by longstanding catalog houses. This is in stark contrast to Japan's e-commerce market, where top websites are dominated by new, exclusively e-tailer merchants, with the exception of two that are run by traditional mail order firms.

Germany's retail sector has been governed by Europe's most stringent set of regulations, and they played a major role in shaping competition between storefront and nonstore retailers, the predecessors of e-commerce merchants. Regulations that controlled store closing hours, competition, and spatial planning severely restricted the use of marketing and locational strategies of storefront retailers, thereby giving opportunities for nonstore retailers to grow. In-store impromptu discounts and sales were against the law, and spatial planning policy practically eliminated opportunities for storefront retailers to use strategies such as those used by Wal-Mart in the U.S. (locating on the edge of town to reduce property cost), or those by Seven Eleven Japan (locating a small urban store near public transportation to capture commuters). Thus, not only did nonstore retailers have lower operating costs than storefront counterparts, storefront retailers could not exercise many of the conventional retail strategies to out-compete nonstore retailers.

E-commerce also reaches beyond the realm of mainstream consumption to underground and "gray" economies with decidedly geographic implications. Today, the transfer of digital products and services such as online gambling and pornography is a sizeable business generating significant revenue. These activities have been shown to often locate outside the centers of the mainstream Internet activities in more hospitable regulatory and labor regimes such as the Caribbean and Eastern Europe (Wilson 2003; Zook 2003). The technology of Internet does not itself determine the structure and role of these participating places but offers new possibilities for participation, interaction and exploitation based on existing historical and cultural attributes.

Thus, as the case of electronic commerce demonstrates, the form and function of new digital geographies varies significantly across sector, place, and culture. Moreover, any changes engendered by the use of digital communications technologies generally take much longer than technological visionaries hope. For example, David (1990) shows how a pivotal technological innovation, the electrical motor, first introduced in the 1880s, took well over four decades before altering the face of industry and fundamentally changing the production process with distinct spatial forms, e.g., from compact vertical to low-rise manufacturing factories. Likewise, the new digital geographies that are emerging based on the commercial use of the Internet are still in the state of dynamic flux—not the least because the technology that supports these activities is also continuing to evolve, as the shift towards wireless digital communications illustrates.

5. UNTETHERING DIGITAL GEOGRAPHIES

Although the dot-com boom and bust largely revolved around technologies associated with wired PCs (particularly in the U.S. context), wireless technologies (and the visions and metaphors associated with them) are emerging as an increasingly important component of digital geographies. Geographic research has only just begun to recognize the existence of this new digital infrastructure despite the fact that mobile subscribers has *always* outnumbered Internet users.

5.1. Wireless Technology, Local Variation

While mobile phone use enjoys worldwide popularity, it does so for a variety of reasons. In countries with relatively undeveloped telecommunications infrastructures, it represents an opportunity to "leap-frog" over older landline technology in an efficient and economical manner and, in many places, is fast replacing wired versions of telephony. Users in wealthier countries are often attracted to the advanced data features, i.e., e-mail, photos, and web access, that are increasingly common in mobile phones. Of course, the precise combination of factors varies with the place, while industry and regulatory issues shape the type of technology available.

The proliferation of the wireless web in Japan is an informative example of how a particular sociospatial condition which allocates premiums on space, portability, and ease of use, results in a specific digital geography. The popularity of Internet-abled mobile telephones in Japan is the result of combined technological and marketing schemes designed to provide an affordable and user-friendly alternative, to those who did not have the money, space, and computer literacy to handle PC-based Internet access (Aoyama 2003). NTT DoCoMo put together a project group in early 1997 to develop the first Internet-abled mobile telephone service with deliberately limited contents to avoid direct competition from PC-based Internet access (Matsunaga 2000; Natsuno 2001). The team conceived the service to function much like that of a hotel concierge or Japan's convenience stores, providing a limited yet essential array of services and products with instant access.

One can further speculate several society-specific factors that contributed to a wide acceptance of wireless web in Japan: market positioning, portability, urban spatial structure, and socially embedded user friendliness. Japan's sociospatial conditions accord high premiums for portability and space-saving equipment, thereby creating a market for a service that provides portability of access. Portability enabled through wireless web not only reduces cost and time of communications, it also expands the timing and location of communications, and enriches it through transfer of increasingly complex multimedia features (Sanwa Research Institute 2001). Portability is particularly attractive to residents in large metropolitan areas where commutes are long and public transit is used heavily for travel, creating idle time. Over half of the commuters in Tokyo Prefecture use public transit to get to work, with an average commuting time of 56 minutes (Ministry of Construction 2000; Japan Statistics Bureau 2002).

The popularity of Internet-abled mobile telephones as a medium of communication, particularly among the young, is not unique to Japan, however. The wireless web, although to a slightly lesser extent, has been actively adopted in Western European markets. The explanation is likely to lie in the success of implementing common compatible technical standards (GSM/TDMA), while the U.S. market is fractured between incompatible analog protocols. Another explanation may be technological leapfrogging, which tends to occur in the technological backwater areas. Unlike the U.S. market, where mass home-ownership of Internet-abled PCs was achieved early, many Japanese and Western European households lagged behind, which resulted in the absence of significant competition against wireless web adoption.

In Germany, where mobile telephone ownership is actually higher than Japan, usage (especially among the young) exploded with the introduction of prepaid cards (Koenig et al. 2003). Much like the case of Japan, the early adopters in Germany were teenagers, and today three-quarters of those aged between 12 and 19 have a mobile telephone. The killer applications (the uses of a new technology that drives its adoption) are similar across Japan and Germany, and include downloading ring-tones or display logos, SMS greetings (text messages), and simple games. Particularly for the youth, wireless web is simultaneously a critical means of freedom as well as a means of ensuring connection and mobility in an increasingly geographically and socially dispersed world (Goban-Klaz 2002).

5.2. Changing Patterns of Mobility and Social Interaction

The widespread use of wireless technologies by teenagers illustrates the way in which digital technologies are allowing people to interact with information and the built environment in new ways. In the late 1990s, observers noted changes in the mobility patterns of teens in countries around the world that exhibited high levels of mobile phone ownership. Rather than meeting at landmarks in public locations like plazas or street corners, youths tended to loosely coordinate movements and meetings through constant communications via mobile phone (Townsend 2000). Repeatedly and independently in various cities, this pattern of coordinated mobility was understood via the metaphor of flocking.

The flock-like behavior of teens using mobile phones was neither unique nor representative of a limited phenomenon. This behavior was merely the most visible manifestation of a widespread new type of emergent behavior in the untethered digital geographies, the microcoordination of daily activities. Put simply, the mobile phone permitted a much freer flow of information within social and professional networks. Operating at a highly decentralized level, these untethered networks carried the viral-like flow of information first observed in e-mail usage on the Internet, into streets, cafes, offices, and homes. In these intimate, everyday locales, untethered digital networks became far more essential and intricately interwoven into human society than any wired network ever was.

By the first decade of the 21st century, mobile communications technology had led to the creation of a massively hypercoordinated urban civilization in the world's cities. These flows had remarkably destabilizing impacts on existing social and economic structures. Employed by smart mobs, these new patterns of communication were successfully used in massive *ad hoc* antiestablishment political demonstrations and actions from Manila to Manhattan (Rheingold 2002).

While changes in the social networks of these untethered digital geographies are now well documented, there was little research to help geographers and urban planners understand the complex impacts on the physical forms of the city. The urban environment generates an enormous amount of information that needs to be anticipated, reacted to, and incorporated into everyday decisionmaking. Information about constantly changing traffic, weather, and economic conditions could be better transmitted through mobile phones and other wireless digital media. Traditionally, cities had functioned on a daily cycle of information flow with mass media like newspapers, third spaces like bars and cafes, and family conversations at the dinner table as the primary means of information exchange. With ubiquitous untethered communications, this old cycle was dramatically speeded up. As the information cycle sped up, there was a corresponding increase in the rate of urban metabolism–the pace at which urban economic and social life consumed information and materiel–and the potential number of places where interaction could occur. In effect, instead of the synchronous daily rhythm of the industrial city coordinated by standardized time and place, untethered communications were leading to a city coordinated on the fly in real time.

Untethered communications also provided more flexibility in travel, supported higher levels of mobility among certain classes and places, and helped increase the pace of all types of transactions, from making a business deal to making a date. With the ability to rapidly get information to and from the people who mattered most in any decision, the efficiency and flexibility of entities (from the corporation to the family) to deal with changing conditions was greatly enhanced. From this perspective, the use of systems such as mobile telephony can be seen as a parallel globalization process, whereby individuals may achieve the same flexible manipulation of space and time locally as corporations have globally for many decades. In short, untethered digital geographies are allowing individuals more freedom and control of the process of constructing new (and often highly personal) geographies of how and where they create and consume information. While this freedom and control is by no means equally available (relatively wealthy and urban populations are at a distinct advantage), and it does not dissolve other social divisions of gender, age and race, it does suggest the potentially liberating aspects of these new and diffusing geographies.

6. PANOPTIC DIGITAL GEOGRAPHIES

The individual empowerment afforded by the untethering of digital technologies, however, is accompanied by an increased ability of businesses, governments and other institutions to create panoptic geographies of people's lives, with important implications for individual privacy. Digital technologies create many new types of records that entangle the daily life of each person into a dense web of threads, across time and space, and easily give rise to Orwellian visions. These threads are created through routine daily electronic transactions (e.g., automatic bill payment or mobile phone calls) and interactions (e.g., companies setting cookies to track individual surfing patterns through the web). Each single item of transaction-generated information is accumulated, byte by byte, to form an ever more panoptic picture of a person's life. People are also increasingly leaving digital tracks in noncommercial arenas as more and more personal interactions are undertaken via computer-mediated communication. E-mail logs and web surfing histories can be just as revealing of a person's daily behavior patterns and lifestyle as their bank statement, medical records, and tax returns.

6.1. Digital Geographies of the Self: Data Shadows and Tracking

The result is that we have all become "digital individuals" (Curry 1997) and are represented by a parallel "data shadow." The data shadow is partial and ever changing, it represents us in transactions where we are not bodily present (e.g., authorizing an online purchase) and also identifies us to strangers (e.g., the shop assistant). We produce our own data shadow, but do not have full control over what it contains or how it is used to represent us. It has become a valuable, tradeable commodity, as evidenced by the growth of credit reference agencies, lifestyle profiling, and geodemographics systems (Goss 1995). A data shadow is inevitable in contemporary society and also necessary if we wish to enjoy many modern conveniences; we can no more be separated from it than we could be separated from the physical shadow cast by our body on a sunny day.

This is not a wholly new concern as the threats to individual privacy posed by digital "databanks" have been analyzed by commentators at least since the mid-1960s when some of the first large-scale computerized systems for storing and processing individual records were instigated (e.g., Vance Packard's 1964 book *The Naked Society*). Yet it is apparent that the growth in both the extent and level of detail of people's data shadows has inexorably and dramatically accelerated in the last decade as digital geographies are mediating more daily transactions.

The growth of our data shadows should not be viewed solely in dystopian terms, as it is an ambiguous process, with varying levels of individual concern and the voluntarily trading of privacy for convenience in many cases. But much of the data captured through routine surveillance are hidden in the background, easily accepted as part of everyday activities (Lyon 2003). Those who try to opt out of using digital technologies as far as possible in their personal life are hard pressed to avoid their routine inclusion in government, business, and medical databases. Even technologically sophisticated people often focus on the benefits that flow through their data shadow and give little thought to the type and amounts of personal data that are captured every time a card is swiped or a pin number entered. It is likely to become even harder to undertake routine daily transactions in an anonymous fashion over the next few years. Increasingly, developments in sociotechnical systems are able to personally identify and track people through the objects people use. Examples include smartcard tickets, electronic road pricing, new intelligent postal systems that can track the sender of all letters via "personalized stamps" (using 2D barcodes), and the likely deployment of radio frequency identification (RFID) labels in retail goods. The security paranoia, post-9/11, is making it much easier for governments and corporations to justify the introduction of new layers of tracking, facilitated in large part by digital geographies.

The data shadow is undergoing changes as it becomes mobile, continuous across time and space, longer lasting, and more widely accessible. The very wireless technologies that afford us new flexibility in constructing our personal daily geographies are simultaneously providing the means for our tracking. The ability to determine the location of individuals at all times via cheap and accurate Global Positioning Systems (GPS) and wireless tracking systems will become commonplace in the next few years, encouraged in large part by the development of novel location based services. There will be significant implications for the nature of individual privacy as the data shadow can be tied to place on an almost continuous basis. Conventionally, locational coordinates for a person have only been recorded sporadically in their data shadow at certain points of interactions (e.g., using an ATM). As yet, few people have really begun to think through the consequences of the fact that their movements through space are being tracked by mobile phones and recorded by the phone company (see Phillips 2003 for an in-depth discussion).

6.2. Forever Storage and Bottom Up Surveillance

While people's data shadows are becoming mobile and continuous, they are also developing much longer memories, potentially holding digital records forever. The capability to log, process and permanently store streams of transaction-generated information about individuals (e.g., time, place of a mobile phone call) has become feasible for most businesses and organizations as the cost of computer storage has tumbled. Hard disks, in particular, have become orders of magnitude bigger in the last decade, which has driven down the cost per megabyte of storage at a rapid pace. This has effectively removed the technical and cost barriers to storing the complete data shadow for the whole life of the subject, with the consequence that what you say or do today (e.g., posting a message to a listserv, paying in a store with a credit card, speaking to a friend on the phone) may well be logged and kept for the rest of your life, with the potential to be recalled and analyzed at any point in the future. (Note, there are, of course, still major technical challenges in the intelligent summarization and analysis of such huge and detailed peta-byte databases.)

Long-term data retention is also being encouraged by the realization of the commercial value, or likely future potential value, locked up in individual-level transaction data. Also, governments are mandating long term data retention on certain service providers (especially banking, health, and telecommunications) because of its perceived rich evidentiary potential for law enforcement. Like India ink, the marks people leave in cyberspace may remain indelibly with them forever. The costs for individuals and society as a whole of data shadows that never forget are manifold (see Blanchette and Johnson 2002). When we all live in place with no fading memory, will this give rise to a much more intolerant and self-policed society?

The emergence of "forever storage" of personal data is not only available to institutions with large IT budgets. Access to massive digital storage is within range of most anyone, with average retail PCs having ballooning hard disks. Many academics and other professionals, for example, can store all their e-mail communications in searchable archives, realizing their value as information repositories. Many will have experienced the fact that each new generation of PC they purchase has a storage capacity several times larger than the old one, and they can simply copy over their entire digital store of work. It becomes easier to keep documents and data just in case they might be needed in the future as there is no longer a constraint of digital space for the majority of users. This may well give rise to new forms of individual memory keeping, the creation of permanent digitized scrapbooks and a virtual diary of life events (an experimental example of this is the *MyLifeBits* project, see Gemmell et al. 2002).

The beginnings of wholesale storage of your own copy of your data shadow can be interpreted as part of a larger societal shift from a centralized "Big Brother" to distributed "bottom-up" surveillance (Batty 2003). Digital tools and software are providing many individuals with new opportunities for detailed surveillance of physical and virtual spaces. Examples include networked webcams, picture phones, and location tags for tracking loved ones. Also, many large and detailed information archives and databases are now online and available to individuals and can be quickly and easily searched. Indeed, a surprising amount of personal information is seeping onto the web and can be freely accessed directly through search engines. There is a noticeable "Google effect" as information can be tracked down much more effectively, fast dissolving the accepted notion of "privacy through obscurity."

> Twenty-somethings are going to search engines to check out people they meet at parties. Neighbors are profiling neighbors. Amateur genealogists are researching distant family members. Workers are screening co-workers. (Lee 2002, 1).

In the realm of academic research, the *JSTOR* archive (www.jstor.org) has provided access to the full content of many major journals, through a simple, searchable web interface. It has unlocked a wealth of previously hard-to-access material and added real value to the journal articles. Another example, is the

Internet Archive's *Wayback Machine*, a serial archive of the web that provides a powerful illustration of the potential of "forever storage." The *Wayback Machine* makes it possible to travel in virtual time to surf websites as they looked in the past (Figure 7-3). It enables everyone to access web pages that have long been deleted.

7. FUTURE RESEARCH AGENDA

Newly emerged digital geographies have already had pervasive effects on the contemporary economy and society, from businesses to peer-to-peer communications, from consumption to urban space, and from network infrastructure to the digital self. Digital geographies entail ever more complex webs of infrastructural, social, cultural, and economic interactions that are increasingly supported by, and are, digitized information. While the interactions among technology, society, and geography are by no means unique to the Internet or mobile phones, this chapter identified a set of processes and emerging trends which are particular to digital communication technology. Visions of utopian and ubiquitous information superhighways and placeless commerce are passé, yet individuals and places are increasingly embedded in new digital geographies while private and state entities are increasingly embedding these digital geographies in all people and places.

Digital geographies are "democratizing" spatial data, making it less and less the preserve of the professional geographer, cartographer, and surveyor. As GPS and other spatially aware digital tools become more affordable and widely used, the unique power of the locational key becomes available to any business or individual, as easily as reading the time. Just as the marine chronometer of the 18th century allowed ships to more efficiently locate themselves and thereby facilitate the expansion of global trade, digital and mobile GPS systems are ushering in a new era in the use of locational knowledge, only at a vastly finer social and geographical scale.

Continuous and real-time knowledge of location is also inspiring a raft of innovative new projects, particularly from computer-savvy artists and community activist groups. Notable examples include GPS drawing, virtual treasure hunts and games of hide and seek played in real places, and the posting of geo-notes, so-called "mid-air messaging." (This is a piece of information assigned to geographic location that can then be read by people at the place). These are speculative projects at the moment to a large degree but focus attention on the ways that new digital geographies of wireless communications, information sharing, and ubiquitous location data will be able to "remake" material geographic environments in surprising, playful, and useful ways. There is also hope that these types of efforts can lead to the creation of fine-

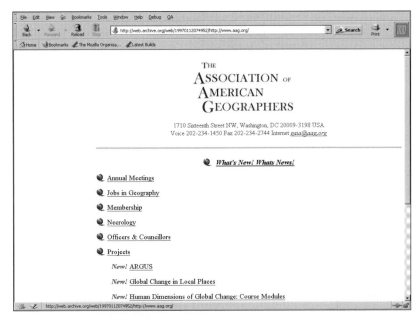

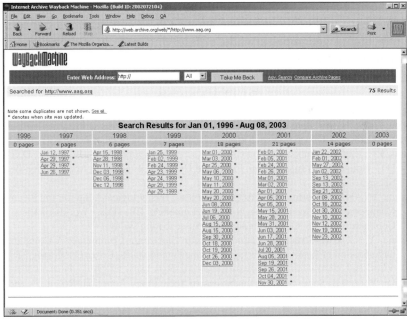

Figure 7-3. Sample of Wayback Machine webpages: top–a screenshot of the AAG homepage as it was in January 1997; bottom–a listing of all available views of the AAG web pages from different dates stored in the Wayback Machine. (Source: http://www.archive.org/)

scale spatial data that is gathered in a participatory fashion, a kind of ground-up, open-source digital map that is richer and more diverse in themes than the conventional topographic data of government and commercial mapping concerns.

In the realm of academic geography research, these digital geographies potentially open new avenues as they make available unique new data sources and quantitative methods of analysis. To give just one example, drawing on the themes of visualization and individual tracking through the mobile data shadow, we believe there is scope for a new type of real-time social geography. The fusion of fine-scale individual activities patterns that are automatically logged and novel forms of geovisualization could give rise to fully dynamic time-space diagrams. (The work visualization in Kwan 2003b is clearly pointing in this direction.) When many individual diagrams are aggregated to the level of cities and regions, these visualizations may provide geographers, for the first time, with truly dynamic maps of dynamic human processes. One might imagine them as twenty-first century "weather maps" of social processes. Yet, at the same time as the digital geographies give us new means to observe and model society, they will also challenge current notions of privacy and make the object of study that much more fragmented, dynamic, and chaotic. The challenge will be to appreciate and use the complexity and richness of the new digital geographies without dissolving into chaos or crystallizing into authoritarian structures.

REFERENCES

Abbate, J. (1999). Inventing the Internet. Cambridge, MA: MIT Press.

Adams, P.C. (1997). Cyberspace and Virtual Places, The Geographical Review 87: 155-71.

Adams, P.C. (2000). Application of a CAD-Based Accessibility Model. In Janelle, D.G. and D.C. Hodge (Eds.) Information, Place, and Cyberspace: Issues in Accessibility. Berlin: Springer-Verlag.

Aoyama, Y. (2001a). Information Society of the East?: Technological Adoption and Electronic Commerce in Japan. In Brunn, S.D. and Leinbach, T.R. (Eds.) Worlds of Electronic Commerce, 109-28. Chichester and West Sussex: John Wiley.

Aoyama, Y. (2001b). Structural Foundations for Electronic Commerce: A Comparative Organization of Retail Trade in Japan and the United States, Urban Geography 22: 130-53.

Aoyama, Y. (2003). Socio-Spatial Dimensions of Technology Adoption: Recent E- and M-Commerce Developments Environment and Planning A 35: 1201-21.

Batty, M. (2003). The Next Big Thing: Surveillance from the Ground Up, Environment and Planning B: Planning and Design 30: 325-26.

Blanchette, J.F. and Johnson, D.G. (2002). Data Retention and the Panoptic Society: The Social Benefits of Forgetfulness, The Information Society 18: 33-45.

Cassidy, J. (2002). Dot.com. New York: Harper Collins.

Castells, M. (1996). The Rise of the Network Society. Oxford: Blackwell.

Couclelis, H. (1998). Worlds of Information: The Geographic Metaphor in the Visualization of Complex Information, Cartography and Geographic Information Systems 25: 209-20.

Curry, M.R. (1997). The Digital Individual and the Private Realm, Annals of the Association of American Geographers 87: 681-99.

David, P. (1990). The Dynamo and the Computer: An Historical Perspective on the Modern Productivity Paradox, American Economic Review 90:355-61.

Dodge, M. and Kitchin, R. (2000). Exposing the Second Text of Maps of the Net, Journal of Computer-Mediated Communication 5. http://www.ascusc.org/jcmc/vol5/issue4/dodge_kitchin.htm

Dodge, M. and Kitchin, R. (2001). Mapping Cyberspace London: Routledge.

Fabrikant, S.I. (2000). Spatialized Browsing in Large Data Archives, Transaction in GIS 4: 65-78.

Gemmell, J., Bell, G., Lueder, R., Drucker, S., and Wong C. (2002). MyLifeBits: Fulfilling the Memex Vision ACM Multimedia'02 Juan Les Pins, France, December 1-6. http://research.microsoft.com/~jgemmell/pubs/MyLifeBitsMM02.pdf

Goban-Klaz, T. (2002). SMS Gener@tion or the Emergence of Mobile Media People. In Pantzar, E. (Ed.) Perspectives on the Age of the Information Society, 217-27. Tampere, Finland: Tampere University Press.

Goodchild, M.F. (1997). Towards a Geography of Geographic Information in a Digital World, Computers, Environments and Urban Systems 21: 377-91.

Goss, J. (1995). We Know Who You Are and We Know Where You Live: The Instrumental Rationality of Geodemographics Systems, Economic Geography 71: 171-98.

Graham, S. (1998). The End of Geography or the Explosion of Place? Conceptualizing Space, Place and InformationTechnology, Progress in Human Geography 22: 165-85.

Harpold, T. (1999). Dark Continents: Critique of Internet Metageographies, Postmodern Culture (on-line journal) 9.

Hayes, B. (1997). The Infrastructure of the Information Infrastructure, American Scientist May-June 85: 214-18.

Hillis K. (1998). On the Margins: The Invisibility of Communications in Geography, Progress in Human Geography 22: 543-66.

Japan Statistics Bureau (2002). Commuting Employed Persons and Persons Attending School 15 Years of Age and Over by Means of Transport, 2000 Census Preliminary Results. http://www.stat.go.jp/data/kokusei/2000/sokuhou/zuhyou/a029.xls

Kaplan, P.J. (2002). F'd Companies: Spectacular Dot-Com Flameouts. New York: Simon and Schuster.

Kellerman A. (2003). Internet on Earth: The Geography of Information. New York: Wiley.

Koenig, W., Wigand, R.T., and Beck, V.R. (2003). Globalization and E-Commerce; Policy and Environment in Germany. Working Paper. University of California at Irvine.

Kwan M.P. (2002a). Time, Information Technologies and the Geographies of Everyday Life, Urban Geography 23: 471-82.

Kwan M.P. (2002b). Feminist Visualization: Re-envisioning GIS as a Method in Feminist Geographic Research, Annals of the Association of American Geographers 92: 645-61.

Lee, J. (2002). Trying to Elude the Google Grasp, New York Times, July 25, G, 1.

Lyon, D. (2003). Surveillance as Social Sorting: Computer Codes and Mobile Bodies. In Lyon, D. (Ed.) Surveillance as Social Sorting: Privacy, Risk and Digital Discrimination London: Routledge, forthcoming.

Matsunaga, M. (2000). I-mode Jiken. Tokyo: Kadokawa Shoten.

Ministry of Construction (2000). Tokyo Toshi-ken No Sogoteki Na Kotsu Jittai Chosa No Kekka Gaiyo (Summary Results of the Survey of Comprehensive Travel Patterns in Greater Tokyo Region). http://www.iijnet.or.jp/tokyopt/masmedia1.html

Natsuno, T. (2001). I-mode Sutorateji: Sekai Wha naze Oitsukenaika. Tokyo: Nikkei BP Shuppan.

Packard, V. (1964). The Naked Society. London: Longmans.

Phillips, D.J. (2003). Beyond Privacy: Confronting Locational Surveillance in Wireless Communication, Communication Law and Policy 8: 1-23.

Rheingold, H. (2002). Smart Mobs: The Next Social Revolution New York: Perseus Books.

Sanwa Research Institute (Ed.) (2001). Jisedai Keitai Bijinesu No Kachigumi No Hosoku. Tokyo: Kosaido.

Skupin, A. (2002). A Cartographic Approach to Visualizing Conference Abstracts, IEEE Computer Graphics and Application 22: 50-58.

Thrift, N. (1996). New Urban Eras and Old Technological Fears: Reconfiguring the Goodwill of Electronic Things, Urban Studies 33: 1463-93.

Townsend, A.M. (2000). Life in the Real-Time City: MobileTelephones and Urban Metabolism, Journal of Urban Technology 7:85-104.

Valentine, G., Holloway, S.L., and Bingham, N. (2002). The Digital Generation? Children, ICT and the Everyday Nature of Social Exclusion, Antipode 34: 296-315.

Warf, B. (2001). Segueways into Cyberspace: Multiple Geographies of the Digital Divide, Environment and Planning B: Planning and Design 28: 3-19.

Wilson, M. (2003). Chips, Bits, and the Law: An Economic Geography of Internet Gambling, Environment and Planning A 35: 1245-60.

Woolgar, S. (2002). Five Rules of Virtuality. In Woolgar S. (Ed.) Virtual Society? Technology, Cyberbole, Reality. Oxford: Oxford University Press.

Zook, M.A. (2000). The Web of Production: The Economic Geography of Commercial Internet Content Production in the United States, Environment and Planning A 32: 411-26.

Zook, M.A. (2002). Grounded Capital: Venture Financing and the Geography of the Internet Industry, 1994-2000, Journal of Economic Geography, 2: 151-77 .

Zook, M.A. (2003). Underground Globalization: Mapping the Space of Flows of the Internet Adult Industry, Environment and Planning A 35: 1261-86.

PART III

NEW GEOGRAPHIES WITH NEW TECHNOLOGIES

CHAPTER 8

ROGER M. DOWNS

FROM GLOBES TO GIS: THE PARADOXICAL ROLE OF TOOLS IN SCHOOL GEOGRAPHY

Abstract   Computing and GIS appear to have the potential for revolutionizing geographic learning, offering students access to the "real world" and powerful tools for thinking geographically. They also appear to have the potential to reestablish a significant role for geography in American education. The former may be correct; the latter is unlikely. To understand the impact of computing and GIS, we must set them into the context of prior tools and technologies for geographic learning. Tools have played a double and paradoxical role in the history of school geography. They have been necessary for teaching geography, and they have epitomized the failure of the subject. Using textbooks as a case study, this chapter shows how school geography has squandered the original social mandate to learn geography. Tools were instrumental in this demise of school geography. Geography went from being indispensable to marginal, from intellectually challenging to dull and boring. It is the lack of a compelling social mandate to learn geography that prevents the reestablishment of geography as a school subject in K–12 education in America.

Keywords   GIS, textbooks, geographic education, social mandate, K–12 education

1. A GLOBE IN EVERY CLASSROOM

Histories of American education often contain a grainy black-and-white photograph of a typical late nineteenth-century classroom. As an educator looking at such a photograph, I am struck by the parallel rows of desks, occupied by neatly, if not uniformly dressed students sitting attentively and segregated on the basis of sex. In front of the classroom stands a teacher, most often a woman, holding a pointer in the military at-ease position. In the corner of the room hangs the American flag.

As a geographer, I am struck by two other objects in this stereotypical depiction of a classroom: a globe and a wall map of the world.[1] These teaching tools are a mark, literally and symbolically, of the significant role of geography in American K–12 education. As I will argue, unfortunately the presence of

Stanley D. Brunn, Susan L. Cutter, and J.W. Harrington, Jr. (Eds.), Geography and Technology,
177-200. © 2004 Kluwer Academic Publishers. Printed in the Netherlands.

this pair of tools also represents the high-water mark–and a distantly receding one–of school geography in American K–12 education.

Today, geographers see school geography as an endangered species in the educational ecosystem, a subject that is under-supported and under-appreciated. We have become inured to frequent laments about the persistence of geographic ignorance among students and the general public. We see geography as having lost ground in American schools. It is no longer a stand-alone subject, with the teaching of human geography having been subsumed within the social studies curriculum (especially in grades K–6) and physical geography being replaced by Earth science. Stand-alone or infused, geography is no longer offered at most grades (Bednarz, Downs, and Vender in press). Indeed, if present in classrooms today, the globe and the wall map are not functional representations but decorative symbols of a well-furnished classroom. They have gone from being essential working tools to vestigial remains without any significant role in the process of education.

From our perspective today, therefore, it is impressive that during the nineteenth century, scarce resources were spent on providing tools for geography education. It is remarkable when we realize that "(t)he introduction of globes and maps marked the first appearance of scientific demonstration apparatus in the American common school" (Rosen 1957, 406). It is even more remarkable when we appreciate the rapid and widespread commitment to such expenditures:

> On April 12, 1837, the Massachusetts Legislature authorized school districts to raise money for the purchase of globes and maps. This action, first of its kind in the United States, was hailed by Horace Mann as "hardly second in importance to any law passed since 1647, when the common schools were established." By 1860, geographical globes were standard equipment in practically all school systems in the country. (Rosen 1957, 406)

Even allowing for the hyperbole in Horace Mann's assertion, how can we account for the authorization of such significant expenditures on tools for geography? Why would globes so rapidly become standard equipment in the American classroom?

In fact, the material support for geography education has extended far beyond the purchase of globes and maps. In sharp contrast to its current lowly status and impoverished condition, school geography has been richly supported in the past. A diverse range of tools has been brought to bear in support of geography education. There are geography-specific tools: maps, atlases, gazetteers, globes (terrestrial and celestial), orreries, and working models (e.g., stream tables). There are tools adapted to geography education: hornbooks, textbooks ("Geographies"), dictionaries, encyclopedias, and geography-based board and card games (Shefrin 1999a). General-purpose tools have been used in geography teaching: blackboards, pointers, and chalk and slates. Students have created maps and landscape models out of an ingenious range of everyday materials from papier maché to pasta (Mitchell 1932).

2. THE PARADOXICAL ROLE OF TOOLS IN SCHOOL GEOGRAPHY

The classroom photograph, therefore, prompts us to explore the historical links between tools and the place of geography in K–12 schools. I want to argue that tools have played a double role in the history of school geography. Paradoxically, tools were necessary in making the teaching of geography possible, and yet those same tools epitomized the problematic nature of school geography. They symbolized a subject that became less interesting, less challenging, and less relevant to students and to society. They presented an image of geography that seemed out-dated, out-moded, and eventually, out of favor. While I do not want necessarily to argue that tools for teaching geography caused *all* of these problems, they most certainly personified them and became emblematic of them. Given their central role in this paradox, I will develop this argument by focusing on one tool, geography textbooks.

But before we can address the paradox itself, we have to understand why school districts would authorize expenditures on geography instruction in the first place. Why invest in teaching geography to students? In order to answer that question, we have to recognize the remarkable value that American society placed on an understanding of geography: by the turn of the eighteenth century, there was a powerful social mandate for geography. To appreciate the basis for that mandate, we have to understand what exactly geography meant to people in past centuries, why it was valued, and how school geography was constructed to meet the mandate for knowledge about geography.

The development of school geography has been shaped by four major factors: the emerging educational infrastructure (specifically types of schools and their associated curricula); the abilities and needs of teachers (relatively untrained and certainly not expert in geography); the changing beliefs about what students needed to learn (the nature of geography); and how students could learn and teachers could teach that subject matter (particular systems of pedagogy). As we understand the interactions among these factors, we can see why tools were necessary for teaching school geography.

The geography that emerged from this shaping process was simple to teach and simple to learn but also simplistic in conception. As we see how tools–especially textbooks–met the challenges of simplification only too well, we can understand why the classroom photograph represents the high-water mark for school geography. In effect, school geography squandered the opportunity presented by the social mandate to teach geography in a way that was both responsive to society's needs and yet intellectually challenging and responsible.

If tools have played a central role in the growth and subsequent demise of school geography, what role can they play in the future of school geography? The question takes on great import today because of the high hopes that many people have for the latest set of geography teaching tools: computers and geographic information systems (GIS). Can these new tools help to reestablish a significant role for geography education in American schools? My answer is no. Tools are a means, not an end in themselves; no tool is powerful enough to reestablish the necessary social mandate for geography instruction.

3. THE NEED FOR GEOGRAPHY AND FOR THE TOOLS TO TEACH IT

3.1. Jigsaws and GIS

In the 1670s, John Spilsbury, a London mapmaker and engraver, advertised for sale a "dissected map." As his business card stated, it was intended "to facilitate the Teaching of Geography" (Shefrin 1999b). Dissected maps are better known today as jigsaw puzzles after a technological change in the mode of production (Williams 1990). Spilsbury's first dissected map, the kingdoms of Europe, was hand-colored and hand-cut from a piece of mahogany. Practically, there is no great trick to pasting a printed paper map onto wood and then carving it into interlocking pieces. Conceptually, however, there is something provocative about this intersection between what has become an artifact of popular culture and the process of geography education. Why the need for a dissected map? Or to put it another way, why was there a market for such an expensive thing made of hand-tooled mahogany?

We can see a similar intersection–but one that seems much more readily explicable–between another cultural artifact and the process of geography education in the latest suite of teaching tools: computers and GIS. It is argued that GIS has the potential to revolutionize the teaching and learning of geography (Audet and Ludwig 2000). Computers and GIS offer students direct access to geospatial data. They are powerful tools for the support of thinking spatially. They have the potential for both independent and collaborative learning. They offer students the opportunity to solve problems with genuine real-world applications (Malone, Palmer, and Voight 2002). Computers and GIS promise authenticity and relevance, empowerment and excitement, and access to job skills and careers. The results of the 2001 National Assessment of Educational Progress in Geography indicate that higher rates of use of CD-ROMS and the Internet correlate with higher performance on the NAEP geography test (for grades 4, 8, and 12) (National Center for Education Statistics 2002). The need for and market potential of an expensive computer and a GIS seem self-evident.

3.2. Forms and Functions of Tools

At first glance, the humble dissected map seems far removed in form and function from the computer and the sophisticated GIS software package. But if we look carefully, we can see that these two tools have much in common as they meet the fundamental challenges of teaching geography in schools. They offer knowledge and intellectual satisfaction but they do so in an appealing way. Jigsaw puzzles and GIS, especially through the parallel to computer games, are examples of "edutainment," something that makes learning pleasurable (although not all tools are or should be edutaining).

Both tools reflect clever marketing: the redesign and repositioning of an existing product to create and exploit a new market in education. They require the active manipulation of representations of the world around us. In the case of the jigsaw puzzle, the world is predetermined and therefore fixed: the intellectual challenge is to recreate it correctly and quickly. In the case of GIS, the world is to be constructed and therefore represented creatively. Both tools require the exercise of spatial skills (operations such as rotation in 2- or 3-dimensional space or translation, and shape, pattern, and color matching in the case of dissected maps, and far more sophisticated skills such as overlay, buffering, and multiple classification in the case of GIS). In assembling a hard-copy or creating a virtual representation, the challenge for the student is to understand the context of geographic phenomena, to see patterns of spatial relationships, and to manipulate spatial representations.

There are, however, important conceptual transitions from the process of assembling the pieces of a dissected map to the process of creating and analyzing representations in a GIS. For example, there is a shift from a single and fixed view of the world to multiple and flexible views, from demonstrating learning by recreating to displaying comprehension by creating, from place learning to learning about places, from knowledge about the world to skills in thinking about the world, from pattern recognition to pattern analysis. Despite these significant differences in intellectual demands, both tools meet a fundamental challenge of geography education: they actively engage students in doing geography. Students are manipulating geospatial information in order to solve a geographic problem. The tools enable the process of thinking geographically and learning about the world.

3.3. The Value of Learning about Geography

But why *should* anyone want to learn to do geography? To a professional geographer the answer is obvious, but then, unfortunately, we are not the ultimate arbiters of what is taught in school. The reason for the rich material support of geography was simple: geography mattered. Until the early part of the twentieth century, *all* students were expected to learn geography and therefore geography was central to the curriculum of the American school system at all levels, elementary through high school. We need to understand the basis for this expectation and the role of tools in meeting it.

The roots of this societal expectation are deep and wide. The project to teach geography began as an attempt to create and instill a national identity that was rooted in a knowledge of place, American place. On August 13, 1776, John Adams wrote in one of his many remarkable letters to Abigail, his wife, that:

> Geography is a Branch of Knowledge, not only very useful, but absolutely necessary, to every Person of public Character whether in civil or military Life. Nay it is equally necessary for Merchants...America is our Country, and therefore a minute Knowledge of its Geography, is most important to Us and our Children. (Butterfield, Garrett, and Sprague 1963, 90)

The letter contained a detailed account of maps collected by the Board of War for display in the War Office. After describing seven maps, Adams wrote:

> You will ask me why I trouble you with all these dry Titles, and Dedications of Maps.–I answer, that I may turn the attention of the Family to the Subject of American geography.–Really there ought not to be a State, a City, a Promontory, a River, an Harbour, an Inlett, or a Mountain in all America, but what should be intimately known to every Youth who has any Pretensions to liberal Education. (Butterfield et al. 1963, 91)

Adams's rationale for geography education is captured in a famous passage from a letter of May 12, 1780:

> I must study Politicks and War that my sons may have liberty to study Mathematicks and Philosophy. My sons ought to study Mathematicks and Philosophy, Geography, natural History, Naval Architecture, navigation, Commerce and Agriculture, in order to give their Children a right to study Painting, Poetry, Musick, Architecture, Statuary, Tapestry, and Porcelaine. (Butterfield and Friedlaender 1973, 342)

Adams's view of geography, eloquently and powerfully expressed, was typical of the time. An educated citizen needed to know geography in order to understand his nation and its place in the world *and* to appreciate his own place in that nation and world.[2] Of equal significance to us is geography's central place in the hierarchy of knowledge–it is listed alongside mathematics, philosophy, natural history, and other major intellectual domains. The elements of a basic rationale for geography are present: it offers practical, useful and "absolutely necessary" knowledge that ought to be studied to ensure a liberal education for all Americans.

In one form or another, this citizenship model dominated the case for geography education through the nineteenth and into the beginning of the twentieth centuries (Libbee and Stoltman 1988). Thus Thomas Jefferson put geography firmly into his model for the curriculum of an age-based education system:

> In the first [elementary schools] will be taught reading, writing, common arithmetic, and general notions of geography. In the second [district colleges], ancient and modern languages, geography fully, a higher degree of numerical arithmetic, mensuration, and the elementary principles of navigation. In the third [a university], all the useful sciences in their highest degree. (Bergh 1907, 155-56)

In Jefferson's view, geography was sufficiently rich and challenging that it required an age-based, cumulative progression from "general" to "full" understanding, and it was suitable for a university education. And ironically, it is striking that geography was listed alongside two school subjects—reading and arithmetic—that are now the most valued and therefore the most comprehensively assessed in current educational philosophies.[3]

Adams and Jefferson are chosen to make the point that geography was valued by opinion leaders in American society. Geography was seen as developing necessary skills in using knowledge and in instilling a sense of citizenship at the state and national levels (see Antonelli 1970). As Rumble (1946, 266) argued: "By the early nineteenth century, the question became not *whether* geography should be taught, but *how* it should be taught." Knowledge of geography was essential to the making of an American citizen.

3.4. The Value of School Geography

But what exactly was school geography valued for and how was it defined? Before answering these questions, we must acknowledge a telling observation by William Warntz (1964, 7):

> Who are we to look back upon an early "geography" course and label it as "not geography"? If one is searching out the development of a specific concept or idea then one can expect to cross disciplinary boundaries. But if one is searching for the continuity of a name associated with an entire discipline, then changes in emphasis and method from time to time must be expected within that discipline. Therefore, no one has the right to rule out any part of the subject's record as not geography because it fails to conform to his own parochial notions of the subject or his rationalization of his own efforts in the field of geography.

Within the broad domain of geography, there are multiple forms of geography–popular, academic, school, college, applied–and each has its particular purposes and audiences. As a consequence, we must not privilege any one form of geography as representative of the "true" geography although equally well, we do not have to accept uncritically all versions of all geographies. That said, school geography is *not* some "reduced" or "diluted" (and implicitly "inferior") version of "real" geography, where the latter is defined in terms of the latest fruits of academic scholarship. Indeed, during the late eighteenth and for much of the nineteenth centuries, there was no formal program of academic geography in America. For all intents and purposes, geography *was* school geography and school geography was developed in order to meet society's needs for liberally educated American citizens.

By the end of the nineteenth century, the educational value of school geography was typically captured under four rubrics. It was seen as (1) an "information subject" providing valuable general knowledge; (2) an "aid to other subjects" (especially history); (3) a "disciplinary subject" fostering mental training in memory, imagination, and reason; and (4) a "culture subject" substituting for travel in the pursuit of "breadth of mind and liberality" (excerpts from Miller 1900). The third rubric is incidental to geography itself except inasmuch as instruction in geography emphasizes mental training. The exercise of memory, however, in the process of geography education has always far exceeded the exercise of either imagination or reason, with the unfortunate consequences that have led to the popular conception of geography as the memorization of boring facts about the world. The second rubric provides a justification for geography only inasmuch as history itself is valued and given the subservience of geography to history in the twentieth-century concept of social studies, we know all too well the dangers of such a derivative justification built on the primacy of another subject.

It is in the first and fourth rubrics that school geography found its distinctive rationale. As an information subject and a culture subject, geography was expected to provide the broad general knowledge of the world that was essential to the development of American citizenship. The idea of substituting for travel is particularly significant: if you cannot go there in person, geography can take you there. For the vast majority of American citizens at this time, international travel was nonexistent (except perhaps during their arrival as immigrants) and even travel within the U.S. was highly restricted. General knowledge is that which is neither specialized nor limited, something that concerns all or most people and that involves only main features. Geography offered general knowledge about places in America and the world. Taken together, these two rubrics shaped what society wanted and expected from school geography in the nineteenth century. Ironically, school geography met them so well that it laid the groundwork for its own demise.

While the general purposes of geography education remained the same, the nature of the geography taught changed significantly during the nineteenth century (see the excellent history of geography education by Schulten 2001). The purposes centered on providing general knowledge of the world and more particular knowledge about the U.S. Initially this knowledge was primarily knowledge of places, as suggested by John Adams, but during the latter part of the nineteenth century, it gradually expanded to encompass physical geography (or physiography), and human geography (as commercial geography, and eventually as political geography).

3.5. The Need for Tools

Tools were essential to geography instruction for two linked reasons, both of which derived from the nature of the developing education system. During the late eighteenth and early nineteenth centuries, the educational infrastructure emerged in a complex series of developments. There were different types of age-graded schools–common schools, Latin grammar schools, private academies, primary and then high schools–organized at local and state scales and run as public, private, or religious institutions. There was a burgeoning support industry, producing textbooks and other materials. The Normal School system was developed to produce teachers, particularly for common and elementary schools (Cremin 1970, 1980, 1988). During the nineteenth century, a remarkable variety of educational philosophies affected the pedagogy of geography: the catechetical method, with its emphasis on questions and answers; the Pestalozzian approach, with its emphasis of the child's immediate experiential world (the home) and the process of contrast and comparison; the geographical reasoning approach of Guyot; and the child study approach (Koelsch 2002).

It is the conjunction between teachers and pedagogy that explains the central role of tools in teaching geography. Formal training of teachers was a late development, emerging in the latter part of the nineteenth century. Most teachers were generalists, expected to teach a variety of school subjects to students varying widely in age and ability (with the classic one-room school being the extreme case of such variation). While geography was important and mandated, teachers could not be expected to know much, if anything about it. They were expected to teach it, however, and therefore materials, especially textbooks, were essential. Those materials in effect defined what was to be known, how it was to be taught, and how it was to be assessed.

Pedagogy, especially early in the nineteenth century, emphasized the learning of factual knowledge: memorization and rote learning were the dominant, though not exclusive pedagogical strategies. The catechetical method (Q and A) presented what was to be known in simple declarative-statement form and then assessed the student's knowledge. Failure to answer the question correctly required additional study of the question and the answer. Blank globes allowed students to demonstrate their knowledge of coordinate systems, land and water masses, and significant places by drawing the requisite patterns in chalk on the globe. In a variant of the Q and A approach, students learned their geographic knowledge from song or rhyming verse (Ravitch 2000). Textbooks were equipped with place name pronunciation guides to enable students to respond orally to questions.

School geography was constructed to fit into these pedagogical forms, and as I will argue in the next section, there was a mutually successful adaptation among the nature of school geography, its pedagogy, and its material support system. In the long term, the adaptation was, unfortunately, counter-productive; it led to a school subject that no longer possessed the challenge and relevance sufficient to ensure its continued prominence and eventually its presence in American schools.

3.6. The Presence of Geography

By the end of the nineteenth century, geography was present throughout the American school system. The penetration is readily documented. Rumble (1946) lists states that required some form of geography instruction: Massachusetts (1827), Vermont (1827), New York (1827), Ohio (1848-1849), Pennsylvania (1854), Indiana (1855), Illinois (1857), Washington Territory (1859), and Mississippi (1873). Warntz (1964, 60) lists universities, including Harvard, that set geography knowledge as an entrance requirement: "geography soon came to be almost universally demanded among the American colleges as a requirement for admission and entrance examinations in geography became the general rule."

The list of successes could be extended to include a journal devoted to geography pedagogy, numerous books on the same subject, and a burgeoning cadre of trained and enthusiastic teachers of geography at all levels, elementary to college.

At exactly the same time, however, there was widespread debate about the nature and value of geography as a school subject. This debate is exemplified in the discussions of the role of geography in American schools that was the result of the work of the various committees of the National Education Association. For example, the famous 1894 Committee of Ten report on the future of high school education in America contained only one disciplinary subcommittee report for which there was a minority report: that for geography. Legitimate though the debate might have been, it suggested a subject at odds with itself. Geography was seen as challenging, but for the wrong reasons: "(t)he subject of geography presents many difficulties not only because of its vastness, but also through the indefiniteness of its limits" (Salmon quoted in Anonymous 1902). Its public image was now less than positive: "In general, it may be said that inert, lifeless things, such as lists of capes, headlands, islands, etcetera, called by the Committee of Ten 'Sailor Geography,' will not do much to train the mind" (Miller 1900, 10). It was losing ground in the high school system, with science replacing physical geography as the preferred required course of study. Colleges and universities no longer required geography as part of their entrance requirements (Warntz 1964).

To add to the irony of the situation, this was also the time during which forms of geography other than school geography were beginning to flourish in America. Popular geography was being supported by a range of local and national organizations, most particularly the National Geographic Society (founded in 1888) (Bryan 1987). Academic geography was being organized under the aegis of the Association of American Geographers (founded in 1903) (James and Martin 1978). Universities were beginning to offer undergraduate and graduate degrees in geography (Rugg 1981).

The obvious question is: Why and how did school geography fail to capitalize on the opportunity to remain an integral and essential part of the American school curriculum? To paraphrase Rumble (1946), by the early twentieth century, the question became not *how* geography should be taught but *whether* it should be taught. Tools played an enabling, if not causal role in the demise of school geography.

4. THE ROLE OF TOOLS IN GEOGRAPHY EDUCATION

4.1. The Patrimony of Jedidiah Morse

Jedidiah Morse is widely labeled as the "father of geography."[4] He earned this soubriquet by creating a new market for an existing product. While Morse did not invent the geography textbook, he was a prolific writer of books and a relentless promoter, publishing a series of best-selling geographies that were reprinted, abridged, and revised in a series of editions between 1784 and 1828. He created the market for and, of equal importance, the model for American geography textbooks, and in that respect Morse deserves his title. While his patrimony did much to establish geography as a part of American education, however, it also had the unintended effect of helping to establish many of the conditions that led to its gradual demise during the subsequent century. To understand this paradox, it is necessary to explore the structural links between the market for geography textbooks and the types of geography represented in those textbooks.

Morse identified an unmet need for American geography textbooks in the immediate post-Revolutionary War period:

> So imperfect are all the accounts of America hitherto published, even by those who once exclusively possessed the best means of information, that from them little knowledge of this country can be acquired. Europeans have been the sole writers of American Geography, and have too often suffered fancy to supply the place of facts, and thus have led their readers into errors, while they have professed to aim at removing their ignorance. But since the United States have become an independent nation, and have risen into Empire, it would be reproachful for them to suffer this ignorance to continue; and the rest of the world have a right now to expect authentic information. To furnish this has been the design of the author of the following work. (Morse 1789, v)

The result, *The American Geography*, was a nationalistic paean written by an American, for Americans, and largely about America:

> Every citizen of the United States ought to be thoroughly acquainted with the Geography of his own country, and to have some idea, at least, of the other parts of the world; but as many of them cannot afford the time and expence, necessary to acquire a compleat knowledge of the several parts of the Globe, this book offers them such information as their situation in life may require; and while it is calculated early to impress the minds of American Youth with an idea of the superior importance of their own country, as well as to attach them to its interests, it furnishes a simplified account of other countries, calculated for their juvenile capacities, and to serve as an introduction to their future improvement in Geography. (Morse 1789, vii)

Thus America accounted for 438 of 532 pages in *The American Geography* of 1789 and 210 of 323 pages in *Geography Made Easy* of 1791.

Morse carefully matched his books and their intended audience as evidenced in the remarkable dedication for the 1791 edition of *Geography Made Easy*:

> To the young masters and misses throughout the United States, The following Easy Introduction to the Useful and Entertaining Science of Geography, Compiled particularly for their Use, is dedicated, With his warmest Wishes For their Early Improvement In every Thing that shall make them truely happy, By their sincere Friend, Jedidiah Morse. (Morse 1791, iv)

Morse could write such an easy introduction because "(h)appily, there is no science better adapted to the capacities of youth, and more apt to captivate their attention, than Geography." Unhappily for the future of geography in America, school geography was indeed adapted to what were perceived to be the needs and capacities of students. Those perceptions understated the capacities of students and as a consequence, school geography neither captivated their attention in the way that Morse might have wished and nor in the end did it meet American society's expectations.

4.2. The Characteristics of School Geography in the Nineteenth Century

I use the word "unhappily" to characterize this adaptation because the consequence was a form of school geography that during the nineteenth century became increasingly problematic in its intellectual character. The resulting geography, though having the virtues of being simple, instructive, easy to teach, and easy to understand, was neither useful nor entertaining. Morse lost sight of the intellectual challenge of a science, substituting description for explanation, and he inundated the readers, overwhelming them in accurate, complete, but uninteresting detail.

The geography in Morse's textbooks had two characteristics that, when imitated throughout the nineteenth century, led to a version of geography that was derisively called "sailor geography" and that we might now call "the principal products of Peru" approach. The characteristics are (1) an imbalance of emphasis between two approaches to geography, general and special, and (2) an overemphasis on facts arranged within simplistic conceptual structures. While Morse cannot be blamed for all of the problems of school geography during the nineteenth century, we can see how and why they arose by looking at the way in which he fitted geography into a school textbook.

4.3. General and Special Geography

In exploring the first characteristic, the imbalance between the two approaches to geography, I draw extensively on the work of William Warntz (1964), a brilliant and underappreciated scholar of geography. Warntz distinguished between two complementary approaches to geography, general and special, or as Morse (1789, 11) put it: "Geography is either universal, as it relates to the earth in general; or particular, as it relates to any single part."

General geography, as the name suggests, dealt with the fundamental structures and dynamics of Earth: through astronomy, it made links to the structure of the solar system; through geology, it offered accounts of the origin of Earth; through spherical trigonometry, it dealt with what we would now call geodesy and map projections; through hydrography, it explained ocean currents. It was a rigorous, systematic, and intellectually challenging approach to an explanatory science.

Special geography dealt descriptively with the nature and character of particular places. It responded to John Adams's call for "a minute Knowledge of its [U.S.] Geography," providing the detailed factual information that could be "intimately known." The vast bulk of Morse's books dealt with special geography and with that of the U.S. in particular. He acknowledged, however, the importance of general geography:

> A complete knowledge of geography cannot be obtained without some acquaintance with Astronomy. This Compendium, therefore, will be introduced with a short account of that Science. (Morse 1791, 9)

It was indeed short, accounting for only 10 of 532 pages in *The American Geography* of 1789 and 11 of 323 pages in the *Geography Made Easy* of 1791.

General and special geography coexisted in early geography instruction, but gradually, during the nineteenth century, special geography became dominant for a variety of reasons. In part, general geography was seen as too challenging for the minds of students. Other disciplines such as science, geology, and astronomy were organizing themselves as independent subjects and taught—and owned—ideas that might earlier have been presented in a geography class. General geography was also linked to cosmic—and sometimes religious—explanations of the origins of systems, both physical and human. As education became increasingly secular in orientation, these explanations were downplayed. One consequence of the shifting balance was that "a part of geography expanded so much as to appear to the casual academic observer and especially to the general public as the entire subject" (Warntz 1964, 26). Therefore, a

> perfectly justifiable, desirable, and for many purposes extremely valuable turning to special geography as a remedy for the lack of adequately detailed knowledge of our own country brought with it in time an unfortunate side effect, the academic demotion of the subject. (Warntz 1964, 27)

Warntz's conclusion remains true today:

> geography acquired its reputation as a school subject, a tag still firmly fixed in the minds of many modern educators as well as those of the general public. Despite the fact that geography subsequently enjoyed intellectual and academic success for a time in the nation's colleges, the image of the subject as a "dry as dust" set of facts to be memorized remains vivid. (1964, 138)

4.4. Geography as Factual Memorization

The memorization of facts lies at the heart of the second characteristic that Morse's books bequeathed to school geography. Morse established many of the intellectual structures through which special geography was presented to students and by which geography was defined for American society in general. Using *The American Geography* as a model, we can see how these structures shaped what was to be known about geography. "Geography is a science describing the surface of the earth divided into land and water" (Morse 1789, 11). Water is then divided into oceans, lakes, seas, straits, bays, and rivers, and land into continents, islands, peninsulas, isthmuses, promontories, and mountains, hills, etc. In many ways, this is an obvious and meaningful characterization of physical features, reflecting the eye-level experience of travelers and the dominant transportation systems. The description of America brings to bear another set of nested categories: climate, soils, productions, rivers, gulfs, gulf stream, mountains, number of inhabitants, aborigines, and the first peopling of America. The chapter for Pennsylvania presents a more finely subdivided set of categories: length and breadth in miles; latitude and longitude; boundaries; mines and minerals; civil divisions; rivers; swamps; mountains, face of the country, soil, and productions; climate, diseases, longevity, etc.; population, character, manners, etc.; religion; literary, humane, and other useful societies; colleges, academies and schools; chief towns; trade, manufactures and agriculture; curious springs; remarkable caves; antiquities; constitution; new inventions; and history. Not only does the order and number of categories vary between states but there is little sense of organization within a category. The reader is presented with a deluge of detailed information about the nature of the place. The result was indeed a compendium but one lacking in a compelling internal logic and interest. It was literally a compilation, and that had two unfortunate effects on the development of school geography.

First, it established a tradition of borrowing, imitation, and what nowadays might well be called plagiarism:

> In the prosecution of the work, he has aimed at utility rather than originality, and of course, when he has met with publications suited to his purpose, he has made a free use of them; and he thinks it proper here to observe, that, to avoid unnecessary trouble, he has frequently used the words as well as the ideas of the writers, although the reader has not been particularly apprized of it. (Morse 1789, vi)

Intellectual structures and detailed information show remarkable similarities across authors. Despite superficial differences, the form and content of a geography textbook became formulaic. Teachers and students knew what to expect, a virtue that was gradually transformed into a liability as familiarity bred indifference, if not contempt.

Second, the process of compilation was unsystematic. In Morse's case, the process was initially based on four years of work that involved visits to states and extensive correspondence with other scholars: "he has had to collect a vast variety of materials—that these have been widely scattered—and that he could derive but little assistance from books already published" (Morse 1789, v). Morse correctly recognized the possibility of inaccuracies but he saw an organic parallel between the book and the Nation, both of which would grow in parallel, the former becoming a larger and more perfect reflection of the latter. Thus Morse set in motion a process of continual revision and updating to match the changing world depicted in the geography textbook. There was a fetish for detail and accuracy that became the driving force for new, if not better editions and that led to intense competition between marginally different books produced by authors and publishers.

4.5. The Nature of School Geography

School geography was, therefore, established as a derivative, self-referential and largely descriptive factual account of the special geography of America and a simplified account of the rest of the world, all of which was matched to 'juvenile capacities.' It was easy to teach and easy to learn. In this way, Morse set the stage for the vibrant and highly competitive geography textbook industry that developed to serve American schools. With the development of printing technologies, textbooks incorporated numerous black-and-white and then colored maps, drawings, and eventually black-and-white photographs; these images met the needs of a school audience that could not experience other places through travel. Textbooks were packaged in a variety of ways: as age-graded series, to match the changing intellectual capacities and needs of students; as local editions (by state and even city), to meet the demands for locally specific information as part of the expanding environments curriculum; and as sets with atlases, wall maps, and globes, to provide teachers with a complete and integrated set of classroom tools. Textbooks claimed to provide accurate, complete, and authentic information. Because of the changing nature of the world, textbooks enjoyed a built-in obsolescence, requiring frequent revisions and new editions and considerable expenditures by schools.

Textbooks, therefore, played a dual and paradoxical role in the history of geography. During the nineteenth and early twentieth centuries, they were the predominant tool for teaching school geography. They presented what students should know and be able to do in geography. They were a window on a larger world that could not be experienced directly. At the same time, they became a, if not the, public face for the discipline of geography as a whole. They represented what the general public came to understand as geography. Morse, and the subsequent generation of textbook writers, did succeed in making geography easy, simple, accurate, and complete. Unfortunately, they did not make geography useful and entertaining. Sadly, they did not make it intellectually challenging. If geography was easy to learn, it was equally easy to belittle, to forget, and ultimately to dismiss.

5. RE-PLACING GEOGRAPHY IN THE CURRICULUM?

5.1. A Role for GIS?

The late twentieth century saw a concerted effort to reestablish geography in American schools (Bednarz et al. in press). One part of that effort, the creation of the National Geography Standards, offers an interesting insight into the complex and still paradoxical links between school geography and teaching tools. *Geography for Life* (Geography Education Standards Project 1994) is marked by what now appears to be a surprising omission. GIS appears only as a two-page Appendix to the main text, and it does not figure prominently in the detailed specifications of what students should know and be able to do in geography: "the standards were written with geographic information systems in mind but not immediately in sight" (Geography Education Standards Project 1994, 257).

As the Writing Coordinator, I well recall the collective editorial decision to remove any significant substantive references to GIS. The decision was motivated by a review of a draft of the Standards offered by a senior education official from one of the Southern states. The official argued against the inclusion of GIS not on intellectual grounds but on grounds of fairness. The power of GIS was accepted, as was its potential role in geography. The argument, however, was that widespread access to computing equipment and to trained teachers would not be available to students in that state for at least the next few years. Therefore students in that *state* would be unfairly disadvantaged if GIS figured prominently in the *National* Geography Standards. We were persuaded by the equity argument, by the need to balance state with national expectations, and by the need to heed the advice of prominent stakeholders.

Looking back over the ten years, that decision now seems to be a mistake, excusable and understandable though it might be. The National Geography Standards will be around for longer than the digital divide. They were intended to be the goal for the future, something to which we—society, the education community, geographers—could aspire. GIS should have been part of that goal but it is difficult to resist an argument based on fairness and equity, especially in the context of educational opportunity.

Technological change—qualitative and quantitative—has been such that computers now seem to be integral to American education. The nature and degree of access, however, especially in terms of geography education, are still not clear. Despite the laudable efforts of ESRI to distribute its educational GIS software to schools, access to software, to computers, to high-speed internet connections, to trained instructors, and to computer support staff seems spotty to say the least. While there are no reliable data, we can ask the question: Can computing and GIS help to make room for geography in an already overcrowded curriculum by reestablishing the intellectual credibility and therefore educational necessity of school geography?

Without question, GIS can reflect the best of geography. It is interesting, challenging, and powerful. It is engaging and entertaining. It does much to support spatial thinking and to enhance geographic literacy. Despite all of these undeniable positives, there is a major problem. While academic critics in geography are perfectly correct in arguing that GIS does not equal geography, it is equally true that GIS does not necessitate geography. To the extent that GIS is a general purpose decision-support system, then it has innumerable applications across the range of school subjects. It is as valuable in biology and anthropology as it is in geography. GIS is as likely to make an impact across the curriculum as it is to foster the development of one school subject, geography. Given the current pattern of GIS adoptions by nongeography versus geography teachers, the former seems more likely than the latter.

5.2. A Role for Tools?

As we know all too well, having access to a map is not the same as doing geography. And so I would argue that having access to computers and GIS is not the same as doing geography. Schools will only make room for geography, that is teaching students to think geographically, *if* there is a compelling case for doing so. Cliché though it is, tools are only a means to an end. While it is true that tools inadvertently led to the demise of school geography during the nineteenth century, there is no reason to believe that a tool—no matter how powerful—can reestablish a role for geography in American schools in the twenty-first century. For that to happen would require the inspired advocacy of latter-day public intellectuals such as John Adams or Thomas Jefferson. Unfortunately, geographers squandered what was seen as an indispensable link between the necessity of knowing geography and the goals of a liberal education. While tools aided in that loss, I do not believe that any tool, no matter how powerful, can be the way back into schools for geography. If there is a convincing and compelling rationale for the need for geography, then the education system will have to find the time—and the tools—to meet it. The rationale-tool nexus existed early in the history of American geography but the rationale came first. We cannot magically reverse that order two centuries later.

Reestablishing a place for geography is difficult. The solution is not to be found in GIS or any other technology. Other school subjects are neither defined by tools nor validated by particular technologies. Those subjects find their basis in a compelling argument that students, and therefore American citizens, need to know that subject. The best indicator of the magnitude of the problem facing school geography is the current "No Child Left Behind Legislation." Geography is listed in the preamble as one of the core academic subjects. Sadly, it is the only subject for which neither programs nor funds are explicitly mandated. The case for geography is simply not compelling enough for the Federal government to find a place for geography. Mahogany jigsaw puzzles in 17th-century England and globes and maps in early 19th-century America were necessary to teach a necessary subject. Until we can make the case for the necessity of geography, we will remain an endangered species. Until we can confront and deal with the tension between academic and popular visions of geography, we will stay misunderstood, the intellectual power of academic geography unable to compete with what Warntz called "the image of the subject as a 'dry as dust' set of facts to be memorized."

5.3. A Globe In Every Classroom, Still?

Ironically, however, the globe remains a potent symbol. The *Washington Post* participates in the "Newspaper in Education" program, offering low-cost classroom subscriptions and curriculum support materials. A recent advertisement for the program has a photograph of an earnest young boy intently reading a newspaper, with a smiling woman teacher standing behind him. As the advert notes, "It's hard to master a subject when you don't understand it." They are posed in front of a blackboard and to the far right of the image is a globe showing the Americas. It is hard to master geography if you do not use–or have–the tools at hand. It is even harder if you and society at large do not understand and value the subject. School geography needs another social mandate whereby our children ought to study geography in order to aspire to a liberal education.

ACKNOWLEDGMENTS

Without the critical eye of Lynn Liben and the library research skills of Jennifer Adams, this chapter would not have been possible. I thank them both for their efforts.

NOTES

1.   Bruckner (1999) points to the frequent use of the discursive materials of geography–maps, atlases, etcetera–in the background to American portraits of the late eighteenth century.

2.   Despite Adams's reference to his sons, the expectations of geography knowledge applied to girls and boys. Significant numbers of early textbooks were written by women.

3.   The 2002 Elementary and Secondary School Education Act (the so-called Leave No Child Behind Act) requires assessment of reading, mathematics, and science every year during grades 3 through 8.

4. Given Morse's role in religious and intellectual life, there is an extensive literature on his life and contributions. There are two excellent introductions from historians: Moss (1995) and Phillips (1983). The major discussion in geography remains the article by Brown (1941).

REFERENCES

Anonymous (1902). Geography Teaching: The Relation of Geography and History, Journal of Geography 1: 333–36.

Antonelli, M.F. (1970). Nationalism in Early American Geographies, Journal of Geography 69: 301–05.

Audet, R. and Ludwig, G. (2000). GIS in Schools. Redlands, CA: ESRI Press.

Bednarz, S.B., Downs, R.M., and Vender, J.C. (in press). Geography Education. In Gaile, G. and Wilmott, C. (Eds.) Geography in America. 2nd ed.

Bergh, A.E. (Ed.) (1907). The Writings of Thomas Jefferson. Washington, DC: Issued Under the Auspices of the Thomas Jefferson Memorial Association of the U.S.

Brown, R.H. (1941). The American Geographies of Jedidiah Morse, Annals of the Association of American Geographers 31: 145-217.

Bruckner, M. (1999). Lessons in Geography: Maps, Spellers, and Other Grammars of Nationalism in the Early Republic, American Quarterly 51: 311-43.

Bryan, C.D.B. (1987). The National Geographic Society: 100 Years of Adventure and Discovery. Washington, DC: National Geographic Society.

Butterfield, L.H., Garrett, W.D., and Sprague, M.E. (Eds.) (1963). Adams Family Correspondence, vol. 2, June 1776-March 1778. Cambridge, MA: Belknap Press of Harvard University Press.

Butterfield, L.H. and Friedlaender, M. (Eds.) (1973). Adams Family Correspondence, vol. 3, April 1778-September 1780. Cambridge, MA: Belknap Press of Harvard University Press.

Cremin, L.A. (1970). American Education: The Colonial Experience, 1607–1783. New York: Harper and Row.

Cremin, L.A. (1980). American Education: The National Experience, 1783–1876. New York: Harper and Row.

Cremin, L.A. (1988). American Education: The Metropolitan Experience, 1876–1980. New York: Harper and Row.

Geography Education Standards Project (1994). Geography for Life: National Geography Standards. Washington, DC: National Geographic Society Committee on Research and Exploration.

James, P.E. and Martin, G.J. (1978). The Association of American Geographers: the First Seventy-Five Years 1904–1979. Washington, DC: The Association of American Geographers.

Koeslch, W.A. (2002). G. Stanley Hall, Child Study, and the Teaching of Geography, Journal of Geography 101: 3–9.

Libbee, M. and Stoltman, J. (1988). Geography in the Social Studies Scope and Sequence. In Natoli, S.J. (Ed.). Strengthening Geography in the Social Studies, 22–43. Washington, DC: National Council for the Social Studies Bulletin 81.

Malone, L., Palmer, A.M., and Voigt, C.L. (2002). Mapping Our World: GIS Lessons for Educators. Redlands, CA: ESRI Press.

Miller, E.I. (1900). Educational Value of Geography Study, Bulletin of the American Bureau of Geography 1: 5–10.

Mitchell, L.S. (1932). Young Geographers: How They Explore the World and How They Map the World. No. 5 The Cooperating School Pamphlets. New York: John Day Company.

Morse, J. (1789). The American Geography: Or, a View of the Present Situation of the United States of America. Elizabeth Town: printed by Shepard Kollock for the author (Reprinted by the Arno Press, Inc., New York, 1970).

Morse, J. (1791). Geography Made Easy: Being an Abridgement of the American Geography. (Third Edition, Corrected). Boston: Samuel Hall.

Moss, R.J. (1995). The Life of Jedidiah Morse: A Station of Peculiar Exposure. Knoxville, TN: University of Tennessee Press.

National Center for Education Statistics (2002). Geography Highlights 2001: The Nation's Report Card. Washington, DC: U.S. Department of Education, Office of Educational Research and Improvement, NCES 2002-485.

Phillips, J.W. (1983). Jedidiah Morse and New England Congregationalism. New Brunswick, NJ: Rutgers University Press.

Ravitch, D. (2000). Left Back: A Century of Battles Over School Reform. New York: Simon and Schuster.

Rosen, S. (1957). A Short History of High School Geography (to 1936), Journal of Geography 56: 405–13.

Rugg, D.S. (1981). The Midwest as a Hearth Area in American Academic Geography. In Blouet, B.W. (Ed.). The Origins of Academic Geography in the United States, 175-91. Hampden, CT: Shoe String Press.

Rumble, H.E. (1946). Early Geography Instruction in America, The Social Studies 37: 266–68.

Schulten, S. (2001). The Geographical Imagination in America, 1880–1950. Chicago: University of Chicago Press.

Shefrin, J. (1999a). Make it a Pleasure and Not a Task: Educational Games for Children in Georgian England, Princeton University Library Chronicle 60: 251–75.

Shefrin, J. (1999b). Neatly Dissected for the Instruction of Young Ladies and Gentlemen in the Knowledge of Geography. Los Angeles: Cotsen Occasional Press.

Warntz, W. (1964). Geography Then and Now. New York: American Geographical Society Research Series 25.

Williams, A.D. (1990). Jigsaw Puzzles: An Illustrated History and Price Guide. Radnor, PA: Wallace-Homestead Book Company.

CHAPTER 9

PHILIP W. PORTER
LAWRENCE S. GROSSMAN

FIELDWORK IN NONWESTERN CONTEXTS: CONTINUITY AND CHANGE

Abstract   Fieldwork has a long and varied history in geography, from being employed in the service of empires in the past to its use in participatory research for social advocacy and empowerment of the poor and the powerless today. This paper focuses specifically on continuity and change in the nature of fieldwork in nonwestern settings over the last half century. Geographical questions originate, are documented, and are tested in the field. Although technological innovation has benefited some aspects of fieldwork, many other dimensions of fieldwork remain unchanged. The advance of technology is no guarantee that the quality of fieldwork will improve. The character of fieldwork has also been influenced by the changing nature of societies themselves. The past fifty years have seen the end of formal European colonies. Finally, discussions of fieldwork in geography have been influenced by the growing interest in poststructuralism and feminism. Geographers are increasingly aware of their situatedness or positionality as researchers and of their responsibilities toward those whom they study as well as toward scholarship itself. We discuss the practice of fieldwork informed by our experiences in three different parts of the world—Africa (Porter), Papua New Guinea (Grossman), and the Eastern Caribbean (Grossman)—representing research at different times beginning in the 1950s. We consider more general discussions of fieldwork in the literature that address issues of the politics and ethics of knowledge production. While the ways of doing fieldwork change and even purposes change, the task ultimately is to understand our world and help make it a good, continuing place for humankind.

Keywords   fieldwork—methods, technology, situatedness, gender, ethics, developing countries

1. INTRODUCTION

This chapter characterizes the various methods and goals of field research in developing countries. We provide a sketch of the changing social context in which fieldwork has been conducted over the past fifty years, and we discuss the changing technology and techniques of fieldwork, as well as features that are changeless. We consider relations among researchers and the researched, and the consequences of an increased awareness of the researcher's

Stanley D. Brunn, Susan L. Cutter, and J.W. Harrington, Jr. (Eds.), Geography and Technology,
201-220. © 2004 Kluwer Academic Publishers. Printed in the Netherlands.

situatedness or positionality. The chapter next discusses interviews, questionnaires, and sampling. We conclude with an exploration of gender and ethics in fieldwork.

Before we begin, have a moment of respite; a poem you may not have read since high school:

I wandered lonely as a cloud

I wandered lonely as a cloud
That floats on high o'er vales and hills,
When all at once I saw a crowd,
A host, of golden daffodils;
Beside the lake, beneath the trees,
Fluttering and dancing in the breeze.

Continuous as the stars that shine
And twinkle on the milky-way,
They stretched in never-ending line
Along the margins of the bay:
Ten thousand saw I at a glance,
Tossing their heads in sprightly dance.

The waves beside them danced; but they
Out-did the sparkling waves in glee:
A poet could not but be gay,
In such a jocund company:
I gazed—and gazed—but little thought
What wealth the show to me had brought:

For oft, when on my couch I lie,
In vacant or in pensive mood,
They flash upon that inward eye
Which is the bliss of solitude;
And then my heart with pleasure fills,
And dances with the daffodils.

William Wordsworth

The creation of knowledge is an iterative process. Geographical knowledge is created through a dialectic that involves shuttling back and forth between the observable world and the quiet place where one can think about what one has seen and experienced (the field of daffodils recollected in tranquility). For example, it is easy to look at a Michelin map of west Africa and see that a road connects Ouagadougou in Burkina Faso, with Mopti in Mali, and that on the journey of 449 kilometers, one will pass through Yako, Ouahigouya, and Bankass (Michelin 1970). Yet there is nothing like actually driving that road to get a sense of terrain and landscape, as well as the people one has visited with along the way. Forever after, once the trip is completed, one looks at the map and at those place names in a different way. Thus, the field plays an obvious role in the creation of geographical knowledge.

Fieldwork involves both quantitative and qualitative techniques. Mapping, sampling, counting, and measuring are important. Even more significant in research are qualitative methods, especially participant observation (Kitchin and Tate 2000). What makes fieldwork unique as a methodology is the key role of establishing personal relationships with the subjects of research. Informal social visits to the homes of informants, sharing personal stories while consuming alcohol in various forms and in various contexts, and attending and participating in social and ceremonial events are just as central to the endeavor as are counting and measuring.

There are many aspects of such fieldwork that contrast with research in western contexts. For example, the now ubiquitous Research on Human Subjects Committees at universities have to be satisfied that one will not harm those interviewed, either physically or psychologically. Cross-culturally, such committees have tin ears, and the stipulation that those interviewed must "sign a release" on the first page of an interview form, while routinely accepted in the U.S., can generate incalculable fear and reluctance when pressed on someone in a nonwestern setting. Similarly, a basic tenet of survey research in western countries is the importance of not interviewing people with whom one has a preexisting, personal relationship. In contrast, researchers in nonwestern countries must establish such relationships before hoping to obtain meaningful information. Also, those conducting fieldwork in western countries have a much more extensive supporting structure of federal, state, and local governments to aid them in data collection, including relatively accurate and frequently updated censuses, maps at a variety of scales, widespread airphoto coverage, detailed soil surveys, and numerous official publications. Language and cultural barriers also may not exist, or at least they can be more easily crossed.

Geographers have been more reticent in writing about their fieldwork experiences in nonwestern contexts compared to anthropologists (see Wolf 1996; Marcus 1998; DeWalt and Dewalt 2002). Recent collections of essays by geographers about fieldwork, however, include discussions of such research in nonwestern countries, thus helping to correct somewhat this gap in the literature.[1]

2. A BRIEF SOCIAL HISTORY OF GEOGRAPHY AS CONTEXT FOR FIELDWORK

The first decades of the latter half of the 20th century were a time of geography's quantitative revolution and a call to make the discipline a "normal" science (Barnes 2001). Theory and method were important. There was a prescriptive, normative character to research as well—for planning, for development, for rational use of resources. It was a time of white shirts (open collar, sleeves rolled up), punch cards, and printouts. The middle decades of the latter half of the 20th century brought geography's politicization—a time of Marx, political economy, social and racial justice, and change. (Geography, of course, was equally political before this time. It is just that the fact lay unacknowledged and unexamined.) Research displayed new forms of the prescriptive and the normative, only the enemies were different. It was the time of the red bandana. The most recent decades of the latter half of the 20th century have witnessed a reordering of field and mind. Like the book and the map, the landscape—where fieldwork occurs—has become text. The task of the scholar is the deconstruction of these various texts to reveal the relationships of power and oppression that underlie the texts. It is the time of play, pastiche, simulacrum, local knowledge, and the wry postcolonial gesture (Geertz 1983; Clifford and Marcus 1986; Wolf 1996; Marcus 1998).

All these changes in thought are reflected in the questions geographers have asked as they embarked upon fieldwork. They also changed the way geographers have asked them. In response to developments in the literatures on poststructuralism, postcolonial studies, and feminism, geographers are increasingly aware of their positionality as researchers. Fieldworkers have become self-conscious, introspective and self-examining—as to their goals, their methods, their motives, their gender, and their situatedness as agents of query among those queried. There was a time when researchers took many of these considerations as givens, needing no explanation.

During the latter half of the 20th century, there has also been a decline in fieldwork in nonwestern countries as well as in geography generally (Rundstrom and Kenzer 1989; Zelinsky 2001). Several trends help account for this pattern. A major force was the rise of the quantitative revolution in the 1960s and 1970s and its emphasis on spatial modeling and abstract theories of human spatial behavior. Another was increasing interest in computer-based analytical techniques, especially GIS (geographic information systems), in the last two decades, and the explosion in availability of secondary data for analysis.

Over the period 1950-2000, a large part of the world stopped being colonies and became nominally or actually independent states. This development brought another set of previously unexamined relationships under scrutiny—namely, the western researcher, the newly independent state, and rural people. No longer did the researcher check in pro forma with the colonial office and the local district officer before beginning fieldwork. New gatekeepers and new goals for research came into being, ones that furthered the developmental priorities of the new country. In addition, there were new requirements on local collaboration, sharing of data, and reporting of findings.

At the same time, the issue of "development" in nonwestern societies became a central theme in research. Politically committed activists were and are combining fieldwork with participatory research to mobilize disadvantaged communities. The goal is to induce social change resulting in the alleviation of inequalities based on race, gender, and class (Katz 1994; Wolf 1996).

Another pattern during this period is also noteworthy. Fieldwork in nonwestern settings benefited from collaborations between anthropologists and geographers. Such efforts have helped meld traditional anthropological concerns about culture, kinship, and social organization with traditional geographic interests in environment, settlement, and land use (see Porter 1965; Brookfield and Brown 1963; Rappaport 1968; Clarke 1971).

As an indication of the extent of change in perceptions about fieldwork in geography, it is instructive to compare the content and message of an earlier discussion of fieldwork in *American Geography: Inventory and Prospect*, the magisterial survey of the state of geography at the time of the AAG's semicentennial year, 1954. Charles Davis contributed a chapter called "Field Techniques" (Davis 1954; see also Sauer 1956 and Platt 1959), which covered methods of recording field observations, field mapping and traverse, methods of mapping single phenomena and associations of phenomena, mapping phenomena of spatial interchange, methods of note taking, including sketching, sketch maps, field diagrams (all seemingly lost arts today), taking photographs, and interviewing by questionnaire and informal conversation. The chapter's appeal for objectivity and impartiality and its apolitical stance contrast markedly with discussions of fieldwork today.

3. CONTINUITY AND CHANGE IN FIELDWORK

One of the unique enduring aspects of fieldwork is that many researchers have received so little formal training in preparation for it (DeWalt and DeWalt 2002). Walter Goldschmidt, UCLA anthropologist, took his Ph.D. at Berkeley in the 1930s. Before going to do his fieldwork among the Yurok of northern California, he asked A.L. Kroeber if he had any advice on doing fieldwork. "Take plenty of pencils," was all that Kroeber suggested. Similarly, few geographers receive adequate formal training in the issues involved in fieldwork (DeLyser and Starrs 2001). Thus we discuss in the following section some of the experiences and techniques related to fieldwork in nonwestern settings. While some have changed over time, many others are as valid today as they were in the past.

3.1. Impressions and Context

For many people in the nonwestern world, having foreigners wanting to conduct research among them is a novel and sometimes disconcerting prospect. Indeed, the notion of collecting information for the purpose of research (and, of course, for the researcher's personal academic advancement), in itself, is completely alien. Even assertions that fieldwork is intended to improve people's living conditions and/or raise their consciousness in relation to their exploitation may be met with considerable skepticism.

Fieldworkers may arouse suspicions in a variety of ways, which can limit their ability to obtain meaningful information. They sometimes need to contact local government officials to obtain permission to travel and conduct research in certain areas, to obtain access to documents, and for interviewing. Being seen with such government officials can lead community members to experience doubt, skepticism, and distrust as to the fieldworker's real purpose. Surveying garden plots with compass and tape for land use studies may create the false impression that the researcher is prospecting for valuable minerals or is interested in purchasing or expropriating the surveyed plots.

There is usually an accepted hierarchy of authority that affects the nature of fieldwork. In a community, there is always a power structure and to ignore it is to risk getting off on the wrong foot, and perhaps ruining all chance of effective fieldwork. By going through "proper" channels, however, one runs several risks, such as cooptation and being guided, channeled, and staffed in ways that do not serve the goals of one's research. The local power structure may attempt to determine for a researcher where to live, whom to interview, whom to employ, and so on. Patriarchal authority structures may create particular problems for females, whose research activities may be circumscribed by preconceived notions about appropriate gender roles (Wolf 1996).

Shakespeare wrote of "the law's delays." Fieldwork is another activity subject to delays, disruptions, and reconfigurations, and the researcher should always be flexible and have a "Plan B" in readiness when the unexpected occurs, as will inevitably happen. Sometimes it is weather, sometimes lack of official permissions, sometimes it is personal tragedy. In 1993, Porter's research in Tanzania experienced delay when Pitio Ndyeshumba, who was conducting interviews, learned that his sister had died. He had to leave the base at Muheza, Tanga Region, travel to Dar es Salaam, and then accompany his sister's body to Bukoba for the funeral, being absent for ten days in all. Connie Weil, who works in Latin America, proposes a *rule of three*; that is, estimates of the time it will take to do something in the field or the amount it will cost should be multiplied by at least three (Connie Weil, personal communication, 2001).

Fieldwork can be a lonely enterprise for a single researcher, but such a condition also has certain advantages. Loneliness can spur a researcher to make sustained efforts to complete fieldwork so as to return sooner to family, friends, and peanut butter; it can also motivate a researcher to interact more intensively on a social basis with those in the local community.

At the same time, the presence in the field of a spouse and children has many benefits (see DeWalt and DeWalt 2002, 61-64). There is fellowship and sharing of laughter and pain as the family reviews and interprets the events of the day. Gender-related features of research may be accomplished more effectively with the presence of a spouse. Children are wonderful ambassadors, often of great interest to those in the local community, helping to establish closer social relations and a sense of shared life experiences. But childcare can also require considerable time, limiting that available for field research.

3.2. Technology and the Fieldworker

In Porter's first research in Liberia (1955-56), he took into the field: eight big rolls of Gateway Natural British 60-gramme Tracing Paper (Wiggins Teape, fine paper makers since 1761). He also took thin leather mosquito boots, an umbrella, two cameras (a Voigtländer Bessa II and a Kodak Retina IIa), a pocket stereoscope, a linen prover (magnifying glass), a typewriter, a large supply of carbon paper, and an eight month supply of Aralen (anti-malarial drug). The main equipment in Grossman's first research in the highlands of Papua New Guinea (1976-77) involved a camera, tape recorder, simple surveying equipment (Suunto compass, clinometer, staff leveler, and tape measure), a soil corer, and a simple hollow steel drum for measuring soil water infiltration rates.

To document changes that have taken place, we can examine the budgets in recent dissertation proposals sent to funding agencies. Increasingly, research proposals request funding for battery-operated laptop computers with solar arrays to recharge batteries; portable printers; digital cameras; video equipment; global positioning system (GPS) devices for field and settlement mapping; and if the question warrants it, satellite imagery to be taken at specific times.[2]

Laptops are a valuable innovation in fieldwork. They help resolve two major concerns of fieldworkers: losing field notes, and the need to organize data for subsequent access and analysis while still in the field. It is essential to make duplicates of field notes, periodically sending the second set to a relative or colleague. (One should also keep a journal or diary, recording what one does as well as what one thinks. Such a record is invaluable later in interpreting the notes collected in the field.) Until laptops became widely employed in fieldwork, carbon paper for making duplicate notes was an essential element in fieldwork. With the advent of laptops, researchers can print duplicate copies of notes, save them on a variety of storage media, and, increasingly, send them to home base as e-mail attachments. Also, fieldwork in nonwestern contexts often involves extended stays, sometimes lasting one or two years. Living in a community for such an extended period enables the accumulation of hundreds of pages of field notes. Searching for particular information stored in handwritten notes, even if carefully filed into useful categories, can be very time consuming. In contrast, storing information in computer files makes it easy to retrieve data collected weeks or months earlier and analyze them while still in the field.

Tape recorders have long been employed by fieldworkers. Certainly from the standpoint of accuracy, use of tape recorders is beneficial, especially when long or complex narratives are involved. The respondent does not have to be continually interrupted while the researcher records his or her comments. Taping is more readily done after a degree of rapport has been established between the researcher and the person interviewed. But there are certain problems associated with using tape recorders: those interviewed may feel uncomfortable, making their answers less spontaneous and forthcoming. Also it takes a tremendous amount of time and attention to transcribe taped interviews.

Cameras also remain a valuable tool in fieldwork. Sometimes a photograph truly is worth a thousand words. Photographs enter our mind, as do maps, as gestalts rather than linear narrative arguments. They can support text in incomparable ways. Personal taste governs how one goes about taking photographs. Some researchers wait a long time before bringing out the camera, until the researcher is well known and a familiar sight in the community. It is common courtesy to ask permission to take a photo, especially of a person, and to send such pictures back to those photographed after leaving the field.

If one is in a research setting for the first time and there is the possibility that this may become a place to which one will return time and again, leading to useful longitudinal research, one may wish to consider what Robert Huke calls "saturation photography" (R. Huke, personal communication June 12, 2002). In saturation photography one takes many pictures of anything and everything—people, buildings, fields, trees, rivers, etc. One great example of the value of "then and now" photography is richly illustrated by B.L. Turner (father of B.L. Turner II) in the book *Vegetational Changes in Africa Over a Third of a Century*. This book contains a comparison of photos taken from

Shantz and Marbut, The Vegetation and Soils of Africa (1923) and photos from a return visit a third of a century later (Shantz and Turner 1958).

Other new technologies have benefited fieldworkers. Years ago, Porter explored the possibility of using Landsat satellite imagery to figure out which areas in east Africa had been planted to crop, a rather basic geographic question. It proved impossible for two reasons: (1) the planting took place during the rains, a time of almost continuous cloud cover, and (2) the fields on average were smaller than the pixel size of the thematic mapper then in use. Today there are imaging systems that can see through clouds and the resolution is much greater, making it possible to distinguish among small fields and adjacent land.

Although we do not consider physical geography fieldwork in this essay, we note that there have been major technical developments in climatology and biogeography (field sensors to measure photosynthesis, albedo, wind speed, etc. in the atmosphere, and heat transfer, wetness, carbon dioxide levels, etc. in soils and plant tissues). The most striking development in physical geography has been the astonishing proliferation of ways of dating things (land surfaces and deposits in strata) and of measuring change through time. These ways include cosmogenic nuclides and isotope analysis, thermochronology, rock coatings, and thermoluminescence (Ronald Dorn, personal communication, May 28, 2003; Singhvi and Wintle 1999). Such advances in the tools used in fieldwork have enabled researchers to obtain a more sophisticated understanding of environmental change (Whitlock 2001; Kennedy et al. 2003).

3.3. The Researcher and the Researched

When one conducts research in a culture different from one's own, one faces many challenges to understanding what is going on and why. This statement is true even if one is fluent in the local language. In Porter's cultural-ecological research, he has attempted to understand local knowledge and local livelihood systems. In the anthropological parlance of the 1960s, this was known as an emic approach. The terms derives from a distinction made by Kenneth Pike, an anthropological linguist, who contrasted how a word sounds (phon*etic*), with what a word means (phon*emic*). Universal international standards have been developed that allow anyone to know how a word sounds, but word meanings are contingent and particular (Pike 1966, 152ff.). If one attempts to understand local knowledge, the best approach is to assume the role of student or apprentice, which calls for an attitude of respect, deference, and patience on the part of the learner. One does not thereby abandon standards of judgment and scholarship. One still gets the story from independent sources, as a means of cross-checking what one is told.

The length of fieldwork is often directly related to the quality of data obtained and the ability of the fieldworker to assist and benefit those who are the subject of research. Learning to ask culturally appropriate and meaningful questions takes time. Perhaps the most important method in this regard is known as "participant observation" pioneered by anthropologist Malinowski (1922) in his classic study of Trobriand Islanders in the Pacific Ocean during his internment there during World War I. Participant observation can be defined as a

> method in which a researcher takes part in the daily activities, rituals, interactions, and events of a group of people as one of the means of learning the explicit and tacit aspects of their life routines and their culture. (DeWalt and DeWalt 2002, 1)

Although participant observation has been employed by geographers throughout the period under consideration here, it is only during the last twenty years that fieldworkers have become more sensitive to the fact that they do not record objective observations of people and the "real" world (e.g., Sundberg 2003). As anthropologists DeWalt and DeWalt (2002, 143) note:

> The researcher decides what goes into the field notes, the level of detail to include, how much context to include, whether exact conversations are recorded or just summaries.

What a fieldworker records reflects his or her own biases, perceptions, personality, gender, age, ethnic background, and class. Fieldnotes and interpretations of them are thus social constructions. Given the importance of the uniqueness of each fieldworker in the process of research, some decide to include discussions of their own influence on the fieldwork process in their publications, though such reflexive approaches are more typical in anthropology.

Issues of reflexivity, intersubjectivity, and positionality in fieldwork have become key issues in ethnographic research since the 1980s, reflecting the influence of writings on poststructuralism and feminism (see Wolf 1996). Some quotes from Richa Nagar's dissertation shed light on these issues (Nagar 1995: 52-53, 58-59; see also Nagar 1997, 2002).[3] They reflect the more introspective and sensitive ways in which the puzzles, contradictions, and dilemmas of fieldwork are approached today in comparison with earlier times:

> The nature of my research topic and methodology, makes reflexivity and intersubjectivity central to my work. Reflexivity refers to the capacity of the self to reflect upon itself as well as on the underlying systems that create it (Prell 1989, p. 251). Intersubjectivity can be defined as the shared perceptions and conceptions of the world held by interacting groups of people (Johnston et al., 1986, p. 236). At the heart of both reflexivity and intersubjectivity lies the issue of positionality. The process of positioning, as Kamala Visweswaran has aptly noted, itself becomes an epistemological act. "The relationship of the knower to the known is constituted

by the process of knowing. Conversely, the process of knowing is itself determined by the relationship of the knower to the known" (Visweswaran [1994], p. 48). ...

In my research, I have employed the ethnographic method with a sensitivity to the role of place and space in constituting identities and communities as well as in the production of knowledge itself. Between 1991 and 1993, I spent a total of twelve months in Dar es Salaam conducting 58 life histories and 150 shorter interviews/ conversations with Goan, Sikh, Ithna Asheri and Hindu men and women from different backgrounds.... Additionally, I also collected information from newspapers, community records, and personal collections of individuals....

Participant observation formed both the core of my research and the heart of my experience in Dar es Salaam. There was no clear line of separation between my personal life and my research. Most of my time was spent with friends and acquaintances from different communities in temples, mosques, clubs, halls, playgrounds, beaches, religious classes, and community houses; in weekly communal gatherings; at celebrations of secular/religious festivals and weddings; and in people's homes where I frequently spent time or lived as a guest, friend, researcher, or "adopted" family member. There was hardly any street in the Asian section that I did not know intimately and on which I did not know at least a few faces....

Dar es Salaam became a kaleidoscope of social sites for me, as I traversed the segregated, gendered, classed, raced and communalized spaces in the course of my daily life. With every turn of the kaleidoscope, I was conscious of my changed position, both geographically and socially. Not only did I behave differently in each situation, I was also "textualized" differently by people in each social site, and this dialogical process between me and my informants continuously shaped the structure and interpretation of the narratives that were produced in the course of my work. I often tried to make sense of these shifting sites, positions and interactions by writing about my personal experiences, observations and feelings in letters and journals....

For me, fieldwork in Dar es Salaam was, in many ways, like knitting a large familial net. Differences, whether religious, political or ideological, were part of the same multi-textured, multi-colored net where threads did not always match perfectly. Amidst my many friends, "mothers," "aunts," "uncles," "sisters," "brothers," "sisters-in-law," and "grandparents," I always felt at home in Dar es Salaam. At the same time, however, some of these very individuals who trustfully told me their stories, fed me regularly, showered affection on me and received my affection in return, might disapprove of the manner in which I have used their words in this dissertation. The dilemma that this situation has posed for me has been clearly articulated by Lila Abu-Lughod (1993, p. 41) in her introduction to *Writing Women's Worlds*:

> Does using my knowledge of individuals for purposes beyond friendship and shared memories by fixing their words and lives for disclosure to a world beyond the one they live in constitute some sort of betrayal? As someone who moves between worlds, I feel that confronting the negative images I know to exist in the United States toward Arabs is one way to honor the kindness they have shown me. So is challenging stereotypical generalizations that ultimately make

them seem more "other." Yet how will my critical ethnography be received? This is the dilemma all those of us who move back and forth between worlds must face as we juggle speaking for, speaking to, and ... speaking from.

3.4. Interviews and Questionnaires

In the western world, researchers use both interviews and questionnaires. In contrast, researchers in nonwestern settings tend to eschew employing questionnaires for a variety of reasons, especially in rural areas. The degree of literacy may be a constraint in people's understanding and ability to fill in responses in questionnaires. In addition, the creation and utilization of questionnaires must be understood as a product of a particular western cultural context in which people have, throughout most of their lives, been exposed to such instruments and have learned to appreciate the role of such tools. In contrast, such instruments are culturally alien in much of the nonwestern world.

Interviewing can involve a variety of formats. If an interview can be a conversation, rather than a "fill in the blanks" exercise, the results will be more meaningful (Sheskin 1985). It is useful to have in mind (or to unobtrusively peek at in a field notebook) a list, a telegraphic aide-mémoire, of the topics one wants to cover. In this way, the conversation can go off on different paths, and one may learn new and surprising things. Yet the list is there to jog one's memory on topics that are to be covered. In contrast, following a rigid, preset interview schedule with closed-ended questions involving multiple-choice and yes-no options, which are often employed in western countries, is not only culturally inappropriate, but will also provide bits of information divorced from their real-world contexts, a point made in Waddell's (1977) valuable critique of methodologies employed in natural hazards research in the 1970s.

3.5. Sampling

Sampling is often a key feature of fieldwork. The purpose is to ensure that a representative segment of the population of interest is included in the study so that research results can be generalized to the larger population. Sampling in nonwestern contexts, however, can present a variety of problems. An accurate sampling frame from which to draw a sample is rarely available, whether one is attempting to select a representative sample of communities in a region or a representative sample of people within a particular community. In addition, established political hierarchies may attempt to alter researchers' sampling plans to ensure that favored clients are more intensively sampled, or to ensure that those with unfavorable viewpoints or characteristics are not included or are underrepresented in the selected samples.

Although sampling is most often employed to select a segment of the population for intensive study, it can also be used effectively to examine time-allocation patterns among members of a community. In his research in both Papua New Guinea (1984) and on St. Vincent (1993), Grossman adapted a method devised by the anthropologist Johnson (1975) to determine the percentage of time allocated to various activities by members of households in his sample. The method involved making random spot-check observations of the activities of members of sample households on randomly selected days each month during the year. In data analysis, the percentage of time spent on any particular activity is determined by dividing the number of observations made of that activity by the total number of observations of all activities.

The method has several advantages: it does not interfere with the daily patterns of community members because the researcher only has to locate the sample member once in a day; several thousand observations of activities can be obtained during the year for analysis; and it forces the researcher to travel throughout the community territory in search of sample members at random times, thus making unplanned, chance discoveries about community life likely (Johnson 1975). This method is particularly useful for comparing differences in the percentage of time allocated to key activities, such as subsistence production and commodity production, and differences in the activity patterns of those in contrasting social categories, such as men and women or male-headed and female-headed households (e.g., Grossman 2000).

4. GENDER AND FIELDWORK

Gender is highly relevant to fieldwork. Women may experience constraints that men never encounter in fieldwork. Women may be treated differently because of gender-stereotypes in strongly patriarchal societies and are more likely than men to be pressured to adhere to traditional expectations about gender roles (e.g., Berik 1996). Unwanted sexual advances or pressures to marry are other potential obstacles to research.

Feminist scholars have been especially concerned about the issue of gender in geographic inquiry and increasingly about "intersectionality" among gender, class, race, sexuality, and location (Sundberg 2003). Discussions about feminism and fieldwork in geography have developed mostly since the 1990s (Nast 1994). Feminist researchers hold the view that

> methodologies that promote mutual respect and identification of commonalities between researcher and researched in non-authoritative ways are deemed preferable in that they allow for "others" to be heard and empowered. (Nast 1994, 58)

A major goal of such fieldwork is to induce change that will improve the lives of women in relation to equity, access to resources, social justice, and cultural freedom (Katz 1994, 70).

The literature on feminism also has raised several critical issues in relation to fieldwork. Many feminist scholars assert that women are better able than men to study women in the field for several reasons: they have experienced sexist oppression and inequality at home and thus are more able to understand the experiences of women in nonwestern contexts (Nast 1994), and women are more oriented toward empathy and sharing, making them better fieldworkers (Wolf 1996). Others have questioned such assumptions, raising the issue of whether they create an essentialist view of women. They note that being a woman, in itself, does not confer insider status in relation to understanding women in nonwestern contexts; other forms of difference between researchers and the subjects of research—race, income, westernness, and personality—can also limit understanding (Wolf 1996; Staeheli and Lawson 1994).

Also, feminist critiques of positivism and associated assumptions of a value-free approach, objectivity, and distance and noninvolvement between the researcher and the researched (Wolf 1996, 4) have led to a reevaluation of the role of quantitative methods in fieldwork. Some view quantitative methods in fieldwork with suspicion because of the perceived link of such methods to positivism and claims about objectivity, and because the categories employed in quantitative research may fail to reflect the complexities of women's experiences (Kwan 2002). Nonetheless, many feminist researchers find value in some aspects of quantitative field methods (Mattingly and Falconer-Al-Hindi 1995; Rocheleau 1995). The distinction between qualitative and quantitative methods in field research should not be overemphasized; quantitative studies involve some degree of interpretive acts, and qualitative methods involve some aspects of counting (Mattingly and Falconer-Al-Hindi 1995). Quantitative techniques can be useful in characterizing differences between men and women and among different groups of women (Mattingly and Falconer-Al-Hindi 1995), and for documenting the pervasiveness and distribution of gender-related problems (Rocheleau 1995). But issues remain in relation to quantitative studies and gender, such as "who does the counting, whose realities are counted, and which social and institutional context constitutes the sampled universe for a given group" (Rocheleau 1995, 459).

5. ETHICS AND FIELDWORK

Fieldwork raises numerous ethical concerns. Fieldworkers have long been sensitive to many of the ethical issues involved in research, but discussions of them, especially those related to asymmetries in power, have become much more prominent in the literature during the last two decades.

Several ethical concerns are longstanding. The primary one is to ensure that no harm comes to those who are the subject of research. Fieldworkers thus have an obligation to maintain the confidentiality of their informants, and not to publish any information that would result in harm to those who are the subject of research. Another such ethical concern relates to informed consent (DeWalt and DeWalt 2002). Researchers have the responsibility to tell people about the nature of their research, obtain their permission to conduct research, and inform people about their right to choose whether or not to participate in the fieldworker's endeavors.

Another enduring issue that presents an ethical dilemma relates to honesty. Fieldworkers expect complete honesty from informants because accurate information is essential for analysis, but they are sometimes less than honest about their own lives back home when discussing their own life situation with informants. Fieldworkers may offer somewhat distorted portrayals about their own incomes, marital status, or religion to facilitate research (Katz 1996; Wolf 1996).

Concerns about power relations in fieldwork are more recent. Researchers—by virtue of their foreign origin, class, race, and education—often have a considerable degree of influence in relation to those who are the subjects of research. Informants can sometimes be intimidated by the presence of rich foreigners asking questions. Certainly, every fieldworker can recall instances of people who were unwilling to cooperate in research, but ultimately the researcher is the one who has the ability to determine the nature of the research project, pay for assistance, leave the field and return home, and decide what and where to publish. Feminist scholars, in particular, have been sensitive to the asymmetrical power relations in fieldwork and have sought to develop more egalitarian and reciprocal relationships with those who are the subjects of their research (Wolf 1996; Achebe 2002).

In nonwestern fieldwork, knowledge is never created by a sole author. Its creation is inevitably a collaboration, transnational and transcultural. Sometimes fieldwork can be done only through an interpreter or by employing field assistants. Again, common sense, the golden rule, and generally accepted rules of ethical behavior come into play. Yet the circumstances of employee and employer are fraught with special potential problems. There is differential power (the researcher drives up in a Landrover full of expensive equipment), perhaps different goals, and different expectations. The researcher may say, "We are going to share in this enterprise as equals, and you will also share in the credit for the outcome equally;" but these rarely eventuate. Edgerton (1965) called such relationships pseudo-friendships. The limits of friendship are undefined at the outset, and sometimes the employee tests them, such as when repeatedly asking the employer to lend money. Relations with those with whom one works need constant care and thought.

Another concern is to ensure that the research process benefits both the researcher and the researched. Although the fieldworker is almost always the prime beneficiary of the research process, given the benefits to his/her academic career, fieldworkers also attempt to benefit those studied in a variety of ways, both while in the field and after leaving the field. That is, reciprocity is a moral obligation.

Researchers attempt to offer assistance in numerous ways. Providing limited medical care or transportation services to those without regular access to such basic services is one example. Some fieldworkers attempt to aid people by raising their consciousness in relation to their exploitation based on income, gender, race, and ethnicity (Wolf 1996). In other cases, community members appreciate the role of researchers as recorders of local oral traditions and oral histories to ensure their preservation for future generations (Stevens 2001). Perhaps one of the prime examples of a researcher attempting to aid those whom he studied is the well-known case of Bernard Nietschmann, who worked with the Miskito Indians of Eastern Nicaragua. Nietschmann literally risked his life during repeated return visits to the area in his efforts to provide continuing support to the Miskito Indians in their struggle against the Sandinista government.

In the past, local researchers in the third world would sometimes clam up or become vague and general when asked by the visitor from the West: "Well, what research are you up to these days?" They may have had painful experience of sharing their data and insights with Western visitors, only to see them published in refereed journals by the visitor, with at best a cheerful acknowledgement of the help of the local researcher. The external advantages have lain with the visiting researcher—access to funding, networks of academics, familiarity with publishing. The internal advantages of the local researcher include understanding the local context much better and being able to use preexisting social networks. Increasingly, scholars from nonwestern and Western countries come together as colleagues in the joint enterprise of research, not as competitors for knowledge.

The production of knowledge in the field is almost always a joint effort. Informants have as important a role in creating knowledge as the researcher. The researcher, however, is in the privileged position of sifting and selecting among the mass of data fieldwork created, the task of analysis, and the responsibility to generalize data in a truthful way. Researchers who work in nonwestern areas have a dual responsibility: (1) toward their professional colleagues, and (2) toward those whose world they are exploring. The goal is to have research pass a dual test: does it make sense to geographers and to other academics who read it, and does it make sense and is it useful to those whose world the researcher is attempting to describe?

6. SHARING DATA AND RESULTS

The responsibilities of the researcher toward those studied do not end with the termination of fieldwork. Sharing data collected and the results of analysis with communities and relevant organizations and government agencies (while keeping sources of information confidential) is now an accepted dimension of the post-fieldwork experience. Grossman used his knowledge of banana production in the Eastern Caribbean to lobby British government officials to continue providing assistance to Windward Islands banana growers. He also provided information and advice to Vincentian agricultural officials in an effort to make agrochemical use safer. Porter has always sent draft manuscripts (and later on, reprints) back to those who helped undertake the study. Upon completion of his eight months of fieldwork in Tanga Region, Tanzania (1992-1993), he left disks containing agrometeorological data and analytical crop simulation computer programs with governmental, U.N., and academic institutions. This included a laboriously created set of meteorological data (daily rainfall values for 24 stations in Tanga Region from 1926 -1991, plus other information on evaporation and soils).

7. FIELDWORK IN GEOGRAPHY

Fieldwork is always difficult and almost always rewarding. It is the only way certain questions can be answered. Not everything can be found in a census volume, a map, or an air photo. It is only through careful fieldwork that we can challenge the conventional wisdom about pastoralism, deforestation, and dessication (Bassett and Zueli 2000), appreciate the role of agency and environment in the persistence of the peasantry (Zimmerer 1991), or understand the importance of gender in agricultural and forestry schemes (Rocheleau 1995).

Good fieldwork, like good geographical writing, has been done in every era. Technological changes may gain us new efficiencies in the research process, but finally, it is the quality of exchanges between individuals in the field that makes scholarship possible and worthwhile.

NOTES

1.   For example, see The Professional Geographer 1994, vol. 46, number 1, which explores feminist perspectives on theory and methodology in fieldwork, including the politics of fieldwork and the politics of representation; The Professional Geographer 1995, vol. 47, number 4, which examines debates concerning the role of quantitative methods in feminist fieldwork and research; Geographical Review 2001, vol. 91, numbers 1 and 2, which provide a collection of personal essays highlighting the broad diversity of fieldwork experiences in geography; and Gender, Place and Culture, 2002, vol. 9, number 2, which discusses issues associated with transnational/transborder feminist praxis. The book titled Fieldwork in Geography: Reflections, Perspectives, and Actions (Gerber and Chuan 2000) is essentially about using the field for student instruction, viz., field trips and student field exercises. Its focus thus is different from that of our essay.

2.   Apropos of global positioning systems (GPS), we can place Porter's pre-GPS dream for field mapping in east Africa in 1961 in the "I was young and naïve" category. It was to place an inflatable tethered balloon carrying a six-inch Polaroid camera over the area to be mapped. He corresponded with the Schjeldal Company (maker of the early Telstar satellite balloon) and even got a balloon and a camera. The project, however, never got off the ground.

3.   We thank Richa Nagar for advice in writing this essay and for permission to quote from her work.

REFERENCES

Abu-Lughod, L. (1993). Writing Womens Worlds. Berkeley: University of California Press.

Achebe, N. (2002). Getting to the Source: Nwando Achebe-Daughter, Wife, and Guest-A Researcher at the Crossroads, Journal of Women's History 14: 9-31.

Barnes, T. (2001). Retheorizing Economic Geography: From the Quantitative Revolution to the "Cultural Turn," Annals of the Association of American Geographers 91: 546-65.

Bassett, T. and Zueli, K.B. (2000). Environmental Discourses and the Ivorian Savanna, Annals of the Association of American Geographers 90: 67-95.

Berik, G. (1996). Understanding the Gender System in Rural Turkey: Fieldwork Dilemmas of Conformity and Intervention. In Wolf, D.L. (Ed.) Feminist Dilemmas in Fieldwork, 56-71. Boulder, CO: Westview Press.

Brookfield, H.C. and Brown, P. (1963). Struggle for Land: Agriculture and Group Territories Among the Chimbu of the New Guinea Highlands. Melbourne: Oxford University Press.

Clarke, W.C. (1971). Place and People: An Ecology of a New Guinean Community. Berkeley: University of California Press.

Clifford, J. and Marcus, G. (1986). Writing Culture: The Poetics and Politics of Ethnography. Berkeley: University of California Press.

Davis, C.M. (1954). Field Techniques. In James, P.E. and Jones, C.F. (Eds.) American Geography: Inventory and Prospect, 496-529. Syracuse, NY: Syracuse University Press.

DeLyser, D. and Starrs, P.F. (2001). Doing Fieldwork: Editors' Introduction, Geographical Review 91: iv-viii.

DeWalt, K.M. and DeWalt, B.R. (2002). Participant Observation: A Guide for Fieldworkers. Walnut Creek, CA: Altamira Press.

Edgerton, R.E. (1965). Some Dimensions of Disillusionment in Culture Contact, Southwestern Journal of Anthropology 21: 231-43.

Geertz, C. (1983). Local Knowledge. New York: Basic Books.

Gerber R. and Chuan, G.K. (Eds.) (2000). Fieldwork in Geography: Reflections, Perspectives, and Actions. Dordrecht: Kluwer Academic Publishers.

Grossman, L. (1984). Peasants, Subsistence Ecology, and Development in the Highlands of Papua New Guinea. Princeton, NJ: Princeton University Press.

Grossman, L. (1993). The Political Ecology of Banana Exports and Local Food Production in St. Vincent, Eastern Caribbean, Annals of the Association of American Geographers 83: 347-67.

Grossman, L. (2000). Women and Export Agriculture: The Case of Banana Production on St. Vincent in the Eastern Caribbean. In Spring, A. (Ed.) Women Farmers and Commercial Ventures: Increasing Food Security in Developing Countries, 295-316. Boulder, CO: Lynne Rienner Publishers.

Johnson, A. (1975). Time Allocation in a Machiguenga Community, Ethnology 14: 301-10.

Johnston, R.J., Gregory, D. and Smith, D.M. (1986). The Dictionary of Human Geography. Oxford: Blackwell.

Katz, C. (1994). Playing the Field: Questions of Fieldwork in Geography, The Professional Geographer 46: 67-72.

Katz, C. (1996). The Expeditions of Conjurers: Ethnography, Power, and Pretense. In Wolf, D.L. (Ed.) Feminist Dilemmas in Fieldwork, 170-184. Boulder, CO: Westview Press.

Kennedy, L., Horn, S. and Orvis, K. (2003). A 4000-Year Sediment Record of Fire and Forest History from Valle de Bao, Dominican Republic. Paper presented at the Annual Meeting of the Association of American Geographers, New Orleans.

Kitchin, R. and Tate, N. (2000). Conducting Research into Human Geography. London: Prentice Hall.

Kwan, M. (2002). Quantitative Methods and Feminist Geographic Research. In Moss, P. (Ed.) Feminist Geography in Practice: Research and Methods, 160-72. Oxford: Blackwell.

Malinowski, B. (1922). Argonauts of the Western Pacific. London: G. Routledge and Sons, Ltd.

Marcus, G. (1998). Ethnography Through Thick and Thin. Princeton, NJ: Princeton University Press.

Mattingly, D.J. and Falconer-Al-Hindi, K. (1995). Should Women Count? A Context for the Debate, The Professional Geographer 47: 427-35.

Michelin #153 (1970). Afrique Nord et Ouest. 1: 4,000,000.

Nagar, R. (1995). Making and Breaking Boundaries: Identity Politics Among South Asians in Postcolonial Dar es Salaam, Ph.D. dissertation, University of Minnesota.

Nagar, R. (1997). Reflections on Methodology and the Puzzles of "Fieldwork." In Jones, J.P. III, Nast, H.J., and Roberts, S.M. (Eds.) Thresholds in Feminist Geography: Difference, Methodology, Representation, 203-24. Lanham: Rowan and Littlefield.

Nagar, R. (2002). Footloose Researchers, "Traveling Theories," and the Politics of Transnational Feminist Praxis, Gender, Place, and Culture 9: 179-86.

Nast, H.J. (1994). Opening Remarks on "Women in the Field," The Professional Geographer 46: 54-66.

Pike, K.L. (1966). Emic and Etic Standpoints for the Description of Behavior. In Smith, A.G. (Ed.) Communication and Culture, 152-63. New York: Holt, Rinehart and Winston.

Platt, R.S. (1959). Field Study in American Geography. Chicago: University of Chicago, Department of Geography Research Paper No. 61.

Porter, P.W. (1965). Environmental Potentials and Economic Opportunities—A Background for Cultural Adaptation, American Anthropologist 67: 409-20.

Prell, R.-E. (1989). The Double Frame of Life History in the Work of Barbara Myerhoff. In Personal Narratives Group (Eds.) Interpreting Womens Lives, 251. Bloomington and Indianapolis: Indiana University Press.

Rappaport, R. (1968). Pigs for the Ancestors: Ritual in the Ecology of a New Guinea People. New Haven: Yale University Press.

Rocheleau, D. (1995). Maps, Numbers, Text, and Context: Mixing Methods in Feminist Political Ecology, The Professional Geographer 47: 458-66.

Rundstrom, R.A. and Kenzer, M.S. (1989). The Decline of Fieldwork in Human Geography, The Professional Geographer 41: 294-303.

Sauer, C.O. (1956). The Education of a Geographer, Annals of the Association of American Geographers 46: 287-99.

Shantz, H.L. and Marbut, C.F. (1923). The Vegetation and Soils of Africa. New York: National Research Council and the American Geographical Society.

Shantz, H.L. and Turner, B.L. (1958). Vegetational Changes in Africa over a Third of a Century. Report 169, Tucson, AZ: University of Arizona.

Sheskin, I.M. (1985). Survey Research for Geographers. Washington, DC: Association of American Geographers, Resource Publications in Geography.

Singhvi, A.K. and Wintle, A.G. (1999). Luminescence Dating of Aeolian and Coastal Sand and Silt Deposits: Applications and Implications. In Goudie, A.S., Stokes, S., and Livingstone, I. (Eds.) Aeolian Environments, Sediments and Landforms, 293-317. Chichester, UK: Wiley.

Staeheli, L.A. and Lawson, V.A. (1994). A Discussion of "Women in the Field": The Politics of Feminist Fieldwork, The Professional Geographer 46: 96-102.

Stevens, S. (2001). Fieldwork as Commitment, Geographical Review 91: 66-73.

Sundberg, J. (2003). Masculinist Epistemologies and the Politics of Fieldwork in Latin Americanist Geography, The Professional Geographer 55: 180-90.

Visweswaran, K. (1994). Fictions of Feminist Ethnography. Minneapolis: University of Minnesota Press.

Waddell, E. (1977). The Hazards of Scientism: A Review Article, Human Ecology 5: 69-76.

Whitlock, C. (2001). Doing Fieldwork in the Mind, Geographical Review 91: 19-25.

Wolf, D.L. (1996). Situating Feminist Dilemmas in Fieldwork. In Wolf, D.L. (Ed.) Feminist Dilemmas in Fieldwork, 1-55. Boulder, CO: Westview Press.

Wordsworth, W. (1951). The Works of William Wordsworth. Roslyn, NY: Black's Readers Service.

Zelinsky, W. (2001). The Geographer as Voyeur, Geographical Review 91: 1-8.

Zimmerer, K. (1991). Wetland Production and Smallholder Persistence: Agricultural Change in a Highland Peruvian Region, Annals of the Association of American Geographers 81: 443-63.

CHAPTER 10

JOHN A. JAKLE

THE CAMERA AND GEOGRAPHICAL INQUIRY

Abstract   Geographers have long made use of cameras in producing photographs useful in teaching, in the conduct of research, and in publishing. Like lay people generally, they are also consumers of the massive amount of photographic imagery daily encountered on television, in the cinema, in the "pictorial" print media, in outdoor advertising, etc. Explored in this essay is academic geography's embrace of still photography taken at or near ground level. (Other essays in this anthology treat aerial photography, including remote sensing and geographical information systems, and motion pictures). Credit is due those geographers who have made camera use and/or the analysis of photographs central to their work. The fundamental centrality of visual imagery in contemporary society suggests that much more remains to be accomplished, however. This essay seeks to offer appropriate conceptual focus encouraging to a fuller and more sophisticated embrace of photography in geography.

Keywords   cameras, photography, research, learning, visuality

Never before has visual imagery, especially that of photography, been more influential than in today's highly interconnected world. Pictures (in many instances more so than the spoken or written word) substantially influence comprehension and understanding, including the search for geographical meaning. Geographers have long employed photographs in classroom teaching: from use of two-by-two slide transparencies to computer-generated imagery. Handheld cameras have long facilitated field work in the discipline. Cameras help geographers orient to new areas, record data in the field, and, with photographs taken in sequence, record change. Physical geographers, for example, have used historical photos, viewed as realistic documentation, to measure landform, vegetational, and other change. Photography has been widely used as a complement to verbal and statistical reporting in professional journals and books.

Human geographers have begun to analyze photographs as they constitute a form of visual culture: subjective social representation clearly value laden. Their concern, of course, centers on the spatial distributional aspects of photo-

Stanley D. Brunn, Susan L. Cutter, and J.W. Harrington, Jr. (Eds.), Geography and Technology,
221-242. © 2004 Kluwer Academic Publishers. Printed in the Netherlands.

depiction, especially that which reflects the human creation of, and use of, landscapes and places. Produced by mechanical and chemical means, photographs are easily thought of as realistic—truthful as visual likenesses— and thus fully useful in scholarly inquiry. But today, geographers, as other scholars, appraise photographic images more as "narratives" influenced by what photographers choose to picture, and by what editors, layout artists, printers, and other media people decide to emphasize. Well recognized is the fact that every geographer operates within, and thus confronts in every endeavor, social realities substantially constructed through visual representation.

1. PHOTO TECHNOLOGY

Technology amplifies humankind's ability to do things. The development of photography amplified the power to accurately fix visual images on surfaces: images that closely approximated the seeing of things first hand. Photographs could be held in the hand ready for close examination: every photograph an invitation to spectate. Photographs could be stored, retrieved, and used over and over again as a kind of permanent record of things seen and seeable. The camera, for its part, was perfected from the *camera obscura*, a device, widely used by artists through the early nineteenth century, fundamental to the rise of Cartesian perspectivism: a model for conceptualizing human vision based on geometrical optics. The aperture of the camera was seen as corresponding to a single mathematically definable point from which the visual world could be logically deduced and represented (Mitchell 1994, 282). In early cameras, light was admitted into a darkened chamber through a converging lens that focused an image onto a surface opposite there, to be recorded on a light-sensitive emulsion spread on metal, glass or, later, on a transparent film of cellulose or acetate. Traditional photographic processes depended on the light-sensitivity of silver halide crystals.

Photographs could replicate scenes even in fine detail. Consequently, unlike the work of painters and other visual artists, photographs were considered to be exact visual representations that spoke without apparent codes of insinuated meaning, the mind and the hand of the artist being subsumed by photo technology. Was not a camera a mechanical device? Was not the photograph the result of chemical reactions? One did not create photographs so much as "take" them. Photographs were not art, but science, it was argued. Photographs spoke with decided objectivity, that pictured by a camera being realistically represented. Photographs seemed to be co-natural with nature. Asserted was a sense of overwhelming truth (Tagg 1988, 1).

But in fact photography was (and is) very much an art form. Photographs very definitely involve coded meaning. What the photographer chooses to picture and how he or she chooses to compose a scene are very much a matter of human decision making. Indeed, the camera is both a means of experiencing the world and a means of denying it. Photographs can lull viewers into knowing the world only through its surficial features. Photographs of scenes (or kinds of scenes) seen over and over again, can deaden awareness and, thereby, diminish knowing rather than enhance it, viewers becoming complacent and thus emotionally disengaged (Lowe 1982, 135). Additionally, photographs can dislocate viewers from original contexts. Presented are worlds in fragmentation, photographic images being detached; set adrift not only geographically, but temporally as well. Essayist Susan Sontag (1977) saw in photography not knowledge so much as a semblance of knowledge: a semblance of wisdom that affected not only the consumers of photographs but the producers as well. "Ultimately, having an experience," she wrote, "becomes identical with taking a photograph of it, and participating in a public event comes more and more to be equivalent to looking at it in photographed form" (p. 24).

Improvement came steadily to photographic art, making the creation of photographs, both behind the lens and in the darkroom, increasingly faster and convenient. Direct-positive photo processes produced single images that could not be replicated in kind (daguerreotypes for example). Negative-to-positive processes enabled replication of numerous positive prints from a master negative (calotypes for example). Initially, photographic plates had to be coated with a light-sensitive wet-emulsion immediately before their exposure. When photographing away from their studios, photographers were required to carry small portable "darkrooms" for this purpose. Perfection of gelatin dry-plates in the 1870s greatly streamlined the act of photographing, especially when conducted in the field. Thus amateurs, and not just professionals, were encouraged in the picturing of things.

Producing traditional black-and-white photographs involves highly standardized darkroom procedure. After negatives are exposed, they are immersed in three solutions: a developer to bring out an image, a stop bath to halt its development, and a fixer to make the image permanent. A water wash then removes residual fixer. In black and white negatives, tonal values are recorded opposite those actually in the pictured scene: light areas appearing dark and dark areas appearing light. Positive prints are obtained when light-sensitive paper is exposed to a negative and then similarly "developed." Today, such darkroom procedure seems all too commonplace. But it required decades to perfect such simplicity after photography was first introduced in 1839.

The earliest cameras were large in format, and thus not only bulky but very heavy. They required the anchoring of a tripod or other device. An unexposed plate was inserted by hand each time a photo was taken. Beginning in the 1880s, cellulose nitrate film (replaced in the 1930s by nonflammable cellulose acetate film) was packaged to be inserted into cameras in rolls, thus cutting down substantially a photographer's time and motion. Amateur photography, including that of geographers, was much encouraged by two camera innovations. First was the inexpensive handheld snapshot camera of the 1890s developed by Eastman Kodak–a kind of "point and shoot" in today's parlance that required little if any skill. Second was the equally portable single-lens reflex (or mirror) camera of the 1930s. With its through-the-lens viewfinder, and later on its through-the-lens light-metering, serious photography was possible with only a modicum of training. Single-lens reflex cameras were the result of several technological improvements: camera miniaturization through use of light metals, development of optically fast lenses, and the development of fast films. Especially important was color film.

Before 1900, color in photographs resulted from hand tinting. Then came the autochrome process that used grains of starch (microscopic in size and colored orange, green, and violet) dispersed on a glass plate to be developed as a positive. In 1935, Eastman Kodak introduced a film built of three separate layers each sensitive to a different light spectrum. The film was developed as a direct-positive transparency. Popularized, thereby, was the two-by-two color slide.

Today, camera use is made further convenient through computer automation, electric flash strobes, and ultra-fast films, both black-and-white and color. But even more facilitating today is electronic film recording and electro-optic cameras that are fully digitized. Implicit is a shift in the location of final photographic production from the "chemical darkroom" to the "electronic darkroom" of the computer (Wells 1997, 251). Electronic impulses are readily passed from digital cameras over telephone lines and satellite links long distance. Electronic images are readily manipulated digitally: forever dispelling (one would think) the notion that "photographs never lie," the myth of photo realism. There is today an unprecedented convergence of the still photograph with other, previously distinct imagery, including that produced by the video camera and through computer animation. It is inappropriate, however, to assume that computer science has come lately to photo technology. Modern computers were made possible through emergence in the 1970s of microprocessors–minuscule solid-state structures that depended upon the microfabrication of computer chips, a photo-lithographic technology (Maynard 1997, x).

Important in the promotion of photography as a medium of communication were advances that enabled photos to be reproduced in newspapers, magazines, and books, or printed en mass, for example, as postcards. Lithography was very much at the heart of this enterprise. To be printed, photos initially had to be copied onto wood or metal blocks, at first freehand by engravers, and then with the help of chemical etching. Produced were simplified images removed from the mimetic specificity of original photos: photography's apparent factuality substantially muted. Then came perfection of various photo-mechanical printing technologies in the 1890s, prime among them halftone printing that used dot-screen overlays to break photographic images into printable patterns of black, white, and gray. Especially revolutionary was development of four-color half-tone. Magazines like *National Geographic* reproduced images obtained directly from photographs, their graphic resolution and coloration substantially preserved.

2. PHOTOGRAPHY'S SOCIAL IMPACT

Through the popularity of photographic images, Americans, like the readers of *National Geographic*, were instructed in the art of spectating. Photos consumed not only suggested what was worth seeing, but how that seen, might be looked at. Increasingly, Americans came to live in a world of visual symbolism, a world promoted most intensely, perhaps, through a deluge of advertising art not only in the print media but out-of-doors on billboards and other hoardings. Ways of looking evolved, governed by largely unconscious epistemic rules or presuppositions (Lowe 1982, 9). Increasingly, the appearance of things did, in fact, come to represent knowledge.

Philosophers were challenged to understand the relationship between sensory appreciation and acknowledged reality (Tuan 1979). From emphasis on things, philosophers had turned in the seventeenth century to emphasize ideas, and in the early twentieth century to emphasize words–the so-called "linguistic turn" (Rorty 1979, 263). For the late twentieth century, however, W.J.T. Mitchell (1994, 12) identified still another reemphasis: the "pictorial turn." Concepts appropriate to understanding written discourse, Mitchell noted, were inadequate to understanding visual display, and to comprehending the relationships between the sayable and the seeable. Guy Debord (1994) had written of the "age of spectacle" and Michel Foucault (1972) of the "age of surveillance," the roles of seeing and of being seen in human experience outlined. But, Mitchell (1994, 6) argued,

> we still do not know exactly what pictures are, what their relations to language is, how they operate in observers and on the world, how their history is to be understood, and what is to be done with or about them.

In English, "looking," "seeing," and "knowing" became substantially intertwined. The conflation between "seen" and "known" was reinforced by modern science, a product of the Enlightenment's embrace of visual observation as the essential means of understanding. Science strove to bring the outside world inside. Through careful observation, the order of the outside world was made knowable and predictable. Positivist revelation was achieved through seeing, visual observation made, as well, the basis for believing. Of course, positivist science does not embrace the whole of what potentially can be known. Science offers only "partial sight." It does so through selection (focusing sight on particular aspects of reality), abstraction (altering visualization from one level of generalization to another), and transformation (reordering visual content) (Jenks 1886, 3). Science, as a means of imagining the world, is a cultural practice with deep visual implications.

Scholarly concern with human vision has been joined by concern for human visuality, the latter fully culturally inflected. In this regard, emphasis has come to be placed on spectatorship. In a figurative sense, to spectate is like looking through a window. It frames what is seen, mediating the possibilities of vision (Burnett 1995, 4). Implied is a boundary between the perceiver and the perceived, a separation between one who sees into the world and that which is seeable. Implicit is a kind of retreat or withdrawal of the self from its surroundings. One is not in the world so much as surrounded by it. "Ensconced behind the window the self become an observing subject, a spectator, as against the world which becomes a spectacle, an object of vision" (Romayshyn 1989, 42).

The spectator gazes through a "prism of sympathetic imagination" (deBolla 1996, 75). People read both others and themselves as symbols, such "readings" always contextualized geographically as to place. There are different kinds of gaze (Rogoff 1996, 189). There is the gaze of cognition (surveying, investigating, verifying) and the gaze of desire (wanting, needing). Indeed, it has been the latter that has engaged social theorists most profoundly, especially the masculine objectification of women in public places. The masculine gaze, of course, is but part of a much broader social discourse pitting various classes of people in often antagonistic relationship. Thus are people, and the places that they routinely occupy, stereotyped along gender, class, race, ethnic, and other lines of social reference. Gazes vary place to place, a connection always operating between legitimate ownership of a designated locale—the sense of belonging there—and the right to look. The propensity to look away, hazard a look, look on, and look straight back is substantially a function of one's sense of place appropriateness.

The flanneur was a kind of nineteenth-century tourist for whom the gaze was all important. He moved on foot through city streets observing its spectacles, a would-be connoisseur of human nature (Tester 1994). He was a seeker of and an interpreter of visual images held in the head. Attention to the flanneur was focused initially by Walter Benjamin (1973) in his consideration of the French poet Charles Baudelaire. The flanneur was male. He took pleasure from being hidden in the crowd, protected in his obscurity from the return gaze of those othered. He walked at will almost without purpose, but was always inquisitive, wondering at that seen. Late in the nineteenth century, mass-produced photography substantially appropriated "the gaze." Especially did scenic photography direct it toward the appreciation of landscape both natural and human built. Emphasized was scenery as spectacle.

Invited was not just seeing, but different kinds of seeing. Catherine Lutz and Jane Collins (1993), in their assessment of *National Geographic*, and the social codes implicit in the photography that the magazine used, differentiated the following (p. 364). The photographer's gaze involved the actual look through the viewfinder of the camera. The institutional gaze of the editor/publisher involved all manner of photo manipulations: the selecting, cropping, framing, captioning, and arraying of images. The viewer's gaze was one of selecting; selection based on personal experience both with photo depiction in general and with the category of thing specifically depicted. Important also, where people were pictured, was the pictured gaze back through the camera lens. As a rule, photographers, in the act of "taking" pictures, tend to be divorced from, if not alienated from, the people and places that they photograph, their primary role being quite literally one of extracting or expropriating images. The institutional gaze likewise tends to be fully opportunistic. Photographs are selected, modified, and combined with other photographs, or with written or other "texts," for a purpose, usually a commercial purpose in the American experience. At hand is the production of saleable product: the photograph as commodity.

Technological advance and social embrace underpinned a tightly knit photography industry. There were the commercial photographers themselves, those who actually made photographs either in the field or in the studio. There were the studio assistants who prepared, developed, and printed negatives and prints, and the technicians who mounted and otherwise finished photographs for customers. There were the salespeople who sold to wholesalers and to retailers (for example, mass-produced stereographs, postcards, and other products for resale) (Jakle 2003). They, in turn, were supported by a large cadre of stock manufacturers who perfected and supplied films and photo papers to meet increasingly specialized needs. Additionally, there were the many photo operatives employed by magazine and book publishers.

Amateur photography was greatly encouraged by a maturing photography industry with its increasingly cheaper supplies, and cheaper and more easily operated cameras. Indeed, amateur purchase of cameras and film today anchors a rapidly growing multi-billion dollar industry. In the U.S., camera sales increased from $1.6 billion in 1998 to $3.5 billion in 2001, and film sales from $3.4 billion to $3.9 billion. In total (including film processing and the buying of darkroom and all other supplies), U.S. amateur photography represented an estimated $16.6 billion market (Photo Marketing Association International 2001).

Today, ours is a world saturated with visual images, most of which are photographic in origin. Indeed, visual images have become critical in the cultural construction of social life. Americans consume images as well as things, making it often impossible to separate satisfactions derived from real things from satisfactions derived from their representations. Sustained, largely through the mass consumption of photographic art, is "life as a simulacrum," in the words of Jean Baudrillard (1988). Ours is a "postmodern" world in which seeing, looking, and representing things visually has assumed important centrality, photography being at the hub of the wheel of social change. In the future, digital cameras, therein direct interface with computer storage and retrieval laser printers, real-time electronic display surfaces, and sure-to-come technical innovations yet unimagined will only reinforce contemporary society's substantive orientation to visuality.

3. GEOGRAPHY AND VISUALITY

Geographers have by no means been immune from the social effects of photography's rising import. Sight, and the use of photographic images, has impacted substantially what we think of today as geographical knowledge. Nonetheless, it can be argued that, from a scholarly point of view, geographers (human geographers most critically) have tended to ignore the importance of "the gaze." "Perhaps geographers," wrote Yi-Fu Tuan (1979, 413) "take the supremacy of the eye for granted." Geographers, like most physical and social scientists, and like most humanists as well, do use photographs, but rarely in ways fundamental. Use of photographs, whether in the classroom or in published articles and books, has tended to be supportive rather than creative. Visuals are used to reinforce other narratives. They supplement. They enliven. But they usually do not fundamentally express. According to geographer Peter Goin (2000, 367) the myth of objective realism has been little challenged by geographers in their use of photographs.

> Photographs *are* documents at one level of the visual spectrum, yet the photograph is a constructed illusion woven within a cultural frame of reference and point of view. Within any photograph the subject may become a symbol that can transcend its own appearance...At the same time, the politics of representation cannot be

ignored. *How* the subject is represented is as important as the subject itself. *In what relationship* with other photographs was the appropriated image intended? For whom was the photograph intended?

Geographers are not visually illiterate. As a class of scholars, however, they have substantially favored the map as a device for both conceptualizing and communicating geographical relationships visually. As Peirce Lewis observed (1985, 467) geographers tend to suffer from "'cartopohia': the visceral love of maps." The turn from largely descriptive regional geography toward a more analytical, quantitative geography, beginning in the 1950s, also privileged use of diagrams, especially graphs, that illustrated spatial association. Figure 10-1 shows that in research articles published in the *Annals of the Association of American Geographers*, the use of photographs declined across the 1940, 1970, and 2000 volumes: from 17.7 to 8.9 to 5.5 photos per hundred pages of text, respectively. Map use increased from 2.9 to 25.0 per hundred pages, and then declined to 11.3. Use of diagrams increased from 11.0 to 19.0 per hundred pages, and then declined to 8.5. Decreasing use of photographs was even more exaggerated in *Economic Geography*, a journal perhaps more fully impacted by the discipline's so-called "quantitative revolution" and its embrace of positivism. In the 1940 volume of *Economic Geography*, each hundred pages contained on average 38.6 photos. But neither the 1970 nor the 2000 volume contained any photographs whatsoever. Use of maps declined from 48.7 per hundred pages in 1940 to 16.6 in 1970 to 5.7 in 2000. Use of diagrams changed from 6.9 per hundred pages in 1940 to 21.8 in 1970 to 9.8 in 2000. These data seem to suggest that geographers are not only less reliant on photographs as a research reporting tool, but seem to be less inclined to use any kind of visual illustration, even maps.

4. PHOTOGRAPHY AS A TEACHING AID

Geography teachers were encouraged to use photographs in the classroom as a means of enabling students to better "picture" the world (for example, see Dexheimer 1929; Thomas and Raup 1956; Griffen 1970). Pictures of landscape and place found a market in the U.S. first through landscape paintings, affordable only by the well-to-do, and then through mass-produced engravings and lithographic prints (including map-like "bird's eye views") that were affordable by the middling classes as well. Actual photographs were popularized commercially first as so-called "cabinet cards" (collected and usually displayed and stored in albums), and then as stereographs (paired photos which when properly viewed gave 3-D effect: looking in full amplification of Cartesian perspective). Then came photo-mechanically reproduced postcards, which after 1900, were marketed annually by the tens of millions. The western world became thus saturated with cheap "views" to be collected and/or sent through the mails (Jakle 2003).

Emphasized in such photography in the U.S. were the nation's rapidly growing cities, and its rapidly disappearing frontier expanses. As historian Peter Hales (1991, 207) observed:

Graphics in the *Annals of the Association of American Geographers*

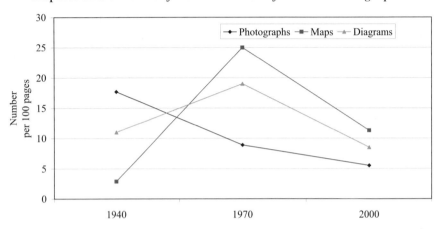

Graphics in *Economic Geography*

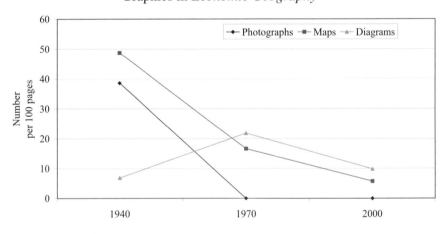

Figure 10-1. Comparison of the number of graphics included in the Annals of the Association of American Geographers and in Economic Geography, 1940-2000.

The colonization of these new landscapes, both urban and wilderness, required more than simple expropriation: they had to be ordered, made sensible, judged, and then inserted in their proper places in the dominant visions of American life and purpose.

So also did photos of foreign lands serve to place life in America more fully in context. Photography became an important means of bringing the distant near, and of making the strange comprehensible if not familiar. Americans could appraise their progress as a society by witnessing through photographs what other people and other places looked like.

Firms, like Keystone, Underwood and Underwood, and the American Stereoscopic Company, packaged stereographs especially for classroom use (Earle 1979; Batzli 1997). Valued were photographs that, standing alone, told important stories–photos that distilled life into "quintessential moments" of understanding such that painters sought to achieve—"tableaux" that not only reported factually, but carried clear metaphorical meaning. But very few single photographs spoke with such force. Photographs could communicate powerfully, but usually only when arrayed with other photos or grouped, in other words, as visual essay. Popularization of the "magic lantern," which could project images from glass slides onto screens or other surfaces, and, later, the slide projector for 2x2 color transparencies, fostered the art of the slide show. Geographers were early and eager adopters (Thompson 1954; Gregor 1956; Miller 1965). Geography was a subject ideally taught out-of-doors through the venue of the "field trip," but, lacking the resources to do so, slide shows stood as important surrogate, a means of "bringing the world into the classroom" (see Teske 1959).

5. THE RESEARCH USES OF PHOTOGRAPHY

For their portability and ease of operation, single-lens reflex cameras quickly impacted geographic field research. Use of the camera was a way of orienting to new areas. The act of taking pictures was a means of "reading" landscape for commonalities, place to place, as well as that which made each place unique. Photographs could stand as a permanent record of that observed, both intended and unintended. Underpinning all was continued faith in photography's realism as a means of documenting truth. "The photograph is non-reflective," wrote geographer Charles M. Davis (1954, 520), and in this way serves as a good check on the subjective interpretation of the facts and conditions of an area as portrayed in the symbolism of the map. But in reviewing the camera's place in field work for *Inventory and Prospect* (1954, 522), Davis outlined what did and did not constitute photographic excellence, as if to admit that even photographs, especially poor ones, could misdirect if not lie.

Good geographic pictures should be able to withstand certain critical challenges. Does the picture document a geographical idea regarding a characteristic phenomenon or association, or is it unique and random? Can the picture be defended as a representative sample? Is the picture properly composed; the exposure technically correct? Can the view be precisely located on a map?

Anthropologists, rather than geographers, developed a sophisticated literature focused on photography's "field" uses, what came to be called "visual anthropology" (Banta and Hinsley 1986; Collier and Collier 1986; Norman 1991). (In recent years, however, there has been a decided shift toward something quite broader: the rise of an anthropology of visual communication.) More than a few geographers have proven quite imaginative in using photography in field-based research. For example, some have deliberately inserted themselves into research locales not for purposes of picture-taking per se, but to record the reactions that locals make to their photo-taking "intrusions" (for example, see Kinsman 1995). If the act of photographing is a form of extractive "taking," then how do those photographed respond? And what do their responses mean regarding sense of self, sense of place, etc. Geographers have used photographs in interviewing respondents, asking them, for example, to look at pictures and tell stories about what it is they see. Thus Thomas Saarinen (1966) used photographs of storm cloud formations to tease out information on farming strategies on the American Great Plains. Geographers have asked respondents to create and interpret photo diaries and then used those diaries to assess landscape meaning (for example, see Young and Barrett 2001).

Most physical geographers routinely use photos in describing data collection procedures, especially the set up of field instrumentation. Photography, however, can also play a fundamental and quite essential role in physical-geographical research (for example, as previously mentioned, when so-called "repeat photography" is used) (Rogers, Malde, and Turner 1984). Conrad Bahre and David Bradbury (1978), and Robert Humphrey (1987), studied vegetational change along the Mexican-American border, replicating, through contemporary use of a camera, extant historical photography from the past. Also, by way of example, Thomas Vale (1987) used vintage photographs in relating vegetational change to changing management practice in the higher elevations of Yosemite National Park.

The vintage photograph has much to offer human geographers also (Rundell 1978; Schlereth 1980; Leary 1985). A dated photo invites the viewer to "enter" a scene and look around–to visually engage a past time (and place) divorced from present-day experience. It is also invitation to understand historical codes of visual representation. Time eventually positions most photographs, even the most amateurish, at the level of art, Susan Sontag (1977,

21) observed, rarity, if nothing else, bestowing value. Of course, intellectual value depends mostly on what the viewer brings to the viewing. It is the knowledge brought to observing old photos that make them truly significant as "history." Recently historical and cultural geographers have begun to explore the pitfalls of using vintage photos "as evidence" (Schwartz 1995; Rose 2000; Ryan 2000). For example, Thomas Vale and Steven Helscher contributed centrally to the Wisconsin Sesquicentennial Rephotography Project, an examination of landscape change as a kind of historical process (Bromberg 2001).

Numerous scholars are presently focused on how the world in the past was pictured photographically (see Schwartz and Ryan 2002). Assessing how photographs were used in sustaining various "world views" lies very much at the core of this effort. Examples include Joan Schwartz's (1996; 1998) concern with nineteenth-century American "imaginative geographies" as constructed through photography, and James Ryan's (1997) consideration of the kinds of photos used to "picture" the British Empire in the nineteenth century. Geographers have joined anthropologists, sociologists, and others, in emphasizing the camera as an instrument of "othering." As Schwartz (1996, 30-31) observed:

> The camera distanced viewer and object, elevated observer over observed, and maintained power relations that privileged the position behind the lens over that in front of the lens...Not only as a pool of visual facts, [but also as] symbols of imperial expansion, colonial development, commercial enterprise, military might, and scientific knowledge, these mutually held visual images contributed to national identity, stimulated patriotic effort and reinforced one's sense of place in the world.

Sociologists Catherine Lutz and Jane Collins (1993) assessed the photography used in *National Geographic*, long the most widely-circulated "geography" magazine in the world. People of the third and fourth worlds, they demonstrate, were consistently portrayed as exotic or, as they phrased it, "idealized," "naturalized," and "sexualized" (p. 89). Historian Susan Schulten (2001), in her assessment of America's collective "geographical imagination," as it changed from the mid-nineteenth through the mid-twentieth centuries, also focused on *National Geographic*, in assessing both its photography and its cartography: visuals calculated to make "Americans feel at home in the world by transforming controversy into wholesome fare for popular consumption" (p. 175).

More than a few geographers contribute to what has become an important new interdisciplinary research emphasis–the study of visual culture. They have used photographs to measure not only change, but velocity of change in the built environment (for example, see: Vale and Vale 1983; Klett et al. 1984; Foote 1985). They have begun to assess the work of specific photographers (for examples, see Shortridge 2000; Watts 2000). They have joined other scholars in asking just how photographs might be analyzed for social meaning, especially for power relationships (Rose 2001). At the core of this effort is the study of "iconology": the study of images and their relation to social discourse (Mitchell 1994, 3). It is an emphasis that embraces modern social theory through, among other approaches, concern for "socio-semiotics": the study of symbols (Gottdiener 1995, 25; Rose 2001, 69).

6. LANDSCAPE PHOTOGRAPHY

Human geographers are especially attracted to pictures of "place"—places being centers of social meaning (and thus behavioral expectation) defined at various scales and nested or embedded in landscape. Geographers have considered how specific places (and, more importantly perhaps, kinds of places) have been "framed" photographically (for example, see Mayer and Wade 1969; Shortridge 2000). The landscape concept, today used mainly in assessing the material culture of built environment, is fully rooted in the act of visualization (Cosgrove 1985; Palka 1995). Landscape photography, of course, constitutes an important photographic genre in the U.S. Ever since photography's development, landscape imagery has been fully implicated in the sustaining of a distinctive American identity (Taft 1964; Davis 1989; Davis 1992). Every nation produces a distinctive iconography in self-representation. For the U.S., it is an iconography that emphasizes idealized scenes of wild nature, the tamed "middle landscape" of farming, small town and big city urbanization (including industrialization), and sectional or regional difference and similarity. Emphasized are such values as progress, modernity, and democracy. Closely related to the study of landscape and landscape representation is concern with travel and tourism. Much landscape photography in the past was generated both in promoting travel and in validating travel experience. The "gaze" of the tourist was an important means of imagining geography for nation building among other purposes (Albers and James 1988; Crang 1997; Hoeslscher 1999).

How people look at and how they conceptualize their visual surroundings—the visual landscape, if you will—has only recently come to be emphasized in academic geography (Appleton 1975; Jakle 1987). Photographs have been used as landscape surrogates, especially in "user-perception" studies designed to understand landscape preferences (see Porteous 1996, 139-43). Geographers, among others, have been not been totally remiss in recognizing one important fact about environmental perception and cognition. People do "picture" their surroundings in the "mind's eye": a picturing that use of actual pictures, like photographs, substantially influences.

Mental "pictures" do influence imagined geography–even geography imaged in map form. Indeed, the geographer's concern with landscape visualization remains substantially wed to map use and map interpretation as reflected, for example, in the behavioral geographer's preoccupation with "mental maps."

That things "pictured" in the mind do get related map-like, is probably a very safe assumption. But how? How do people picture their surroundings? And how does this picturing influence a sense of geographical reality? To what extent do picture-like schemata (cognitive structures or coding systems) enter into the creation of imaginative geography? Although sight is not the only means by which people directly sense the environment, it is primary. Place meanings are usually cued in the first instance through sight recognition. Even those who are not sighted operate in built environments substantially configured by (and for) those who are. How we picture the world in the mind, and how that picturing relates to our ability to map the spatial environment (not just as location, but as place) remains an important research agenda.

Behavioral geographers, in their attending to environmental cognition, were quick to adopt the analogy of the map. "Mental maps" (or cognitive maps) were made a primary focus. Analysis of sketch maps solicited from respondents persists as a means of understanding the cognitive aspects of location decisionmaking among spatial behaviors (for example, see Downs 1981, 111; Kitchen and Bladen 2002, 141). But to what extent are "maps in the head" reliant upon, indeed constructed from, what we might call photographic thinking? How is observational information summarized, reconstituted and fore-grounded in the mind in order to facilitate geographical decision making? In other words, how do individuals transform impressions from one kind of visual thinking to another (Golledge 2002, 4)? As geographer Denis Wood (1993) has pointed out, a map is the product of a spectrum of codes: iconic, linguistic, tectonic, and temporal. "Iconic codes govern the manner in which graphic expressions correspond with geographic items, concrete or abstract," he wrote (p. 117). How do photographs figure in such coding? What do photographs teach us to see and remember about landscape and place? And how does that, in turn, influence human spatially?

Landscape photography, for its part, is quintessentially geographical and, accordingly, should be valued by every geographer. Five photos illustrate. The view through a camera lens can be a way of exploring how people experience space. In motoring, highway landscapes rapidly loom ahead as they rapidly occlude behind (Figure 10-2). Photos can amplify a sense of spatial relationship as when environmental factors (wind and topographic relief) are seen to be conflated through human technology in the generation of electricity (Figure 10-3). Places, of course, are bounded spaces (often also enclosed spaces) that are readily understood in terms of behavioral expectation.

Figure 10-2. View through a rearview mirror on U.S. 30 west of Toledo, Iowa, 2002.

Photos can sometimes capture not only how places are created, but how they are valued. Children, stopped in their play, are pictured helping to restore a dated family home (Figure 10-4). "Sense of place" involves strong personal attachment to, if not a sense of ownership of, a place. As pictured, the stoic countenance to defend the territory of a big city neighborhood (Figure 10-5). In today's world, visual images reverberate in the creation of and use of places, especially in cities. Through photographic imagery in advertising and other signs, places are related symbolically: for example, in the enticing of city people out into nature, albeit nature commodified (Figure 10-6).

All of these photos communicate understandings about landscape that are difficult to fully express in words alone, and certainly to express through statistical tabulations. They are realizations not easily graphed or mapped. In order to communicate well, landscape photos should picture things in ways readily recognizable, but also in ways not fully anticipated. Therein lies the power of landscape photography to excite as well as inform. Implicit is invitation to geographers to be not only scientists and humanists, but to be artists as well.

Figure 10-3. Windmill farm northwest of Palm Springs, California, 2002.

Figure 10-4. Log house north of Andover, Pennsylvania, 1991.

Figure 10-5. The image of former Mayor Frank Rizzo on South Ninth Street in largely Italian-American South Philadelphia, 1995.

Figure 10-6. Tour bus on "The Strip" in Las Vegas, 2002.

7. CODA

Geographers have learned and accomplished much through embrace of cameras and photography. But there is much more that remains to be done. Visual literacy in academic geography remains tied largely to cartography (and through maps to map-like aerial photos and satellite images). Geographers, as a group, do use cameras, but in ways mainly complementary rather than fundamental. They generate images primarily in bolstering teaching and/or strengthening research reporting. In this regard, too few geographers work diligently to raise the quality of their photography to high skill levels thus to better appreciate the power of photography as a means of communication. Indeed, many geographers never develop any competency with cameras at all and, unfortunately, stand proud of it. The art of photography is neither taught nor especially encouraged in most programs of graduate study in Geography. Perhaps the growing embrace of computer graphics will become an avenue for constructive change in this regard?

A few geographers do make use of photographs in ways fully central to their research. But, even so, one wonders if the amount of effort expended to date in any way matches the profound social importance that photographic imagery has assumed as an instrument of social power in contemporary society. Indeed, one could easily argue that geographers have been little concerned to understand human visuality. One could argue that they have even been reticent to understand how visual representation operates in people's comprehending the world as geography. It seems that geographers, like most people, generally have tended to take visualization very much for granted. Things are seen and identified. Only then does serious thinking begin. For many geographers, search for geographical significance begins by drawing a map. Perhaps the future will bring endeavor more fully cognizant of the visual environment as a starting place for geographical inquiry?

REFERENCES

Albers, P.C. and James, W.R. (1988). Travel Photography: A Methodological Approach, Annals of Tourism Research 15: 134-58.

Appleton, J. (1975). The Experience of Landscape. Chichester, UK: John Wiley and Sons.

Bahre, C.J. and Bradbury, D.E. (1978). Vegetation Change along the Arizona-Sonora Boundary, Annals of the Association of American Geographers 68: 145-65.

Banta, M. and Hinsley, C.M. (1986). From Site to Sight: Anthropology, Photography and the Power of Imagery. Cambridge, MA: Harvard University Press.

Batzli, S. (1997). The Visual Voice: Armchair Tourism, Cultural Authority, and the Depiction of the United States in Early Twentieth-Century Stereographs, Ph.D. dissertation, University of Illinois at Urbana-Champaign.

Baudrillard, J. (1988). Selected Writing (M. Poster, trans). Cambridge, UK: Polity Press.

Benjamin, W. (1973). Charles Baudelaire: A Lyric Poet in the Era of High Capitalism, (H. Zohn, trans). London: NLB.

Bromberg, N. (2001). Wisconsin Then and Now: The Wisconsin Sesquicentennial Rephotography Project. Madison: University of Wisconsin Press.

Burnett, R. (1995). Cultures of Vision: Images, Media, and the Imaginary. Bloomington: Indiana University Press.

Collier Jr., J. and Collier, M. (1986). Visual Anthropology: Photography as a Research Method. Albuquerque: University of New Mexico Press.

Cosgrove, D. (1985). Prospect, Perspective and the Evolution of the Landscape Idea, Transactions of the Institute of British Geographers n.s. 10: 45-62.

Crang, M. (1997). Picturing Practices: Research through the Tourist Gaze, Progress in Human Geography 21: 359-73.

Davis, T. (1989). Photography and Landscape Studies, Landscape Journal 8: 1-12.

Davis, T. (1992). Beyond the Sacred and the Profane: Cultural Landscape Photography in America, 1930-1990. In Franklin, W. and Steiner, M. (Eds.) Mapping American Culture, 191-230. Iowa City: University of Iowa Press.

de Bolla, P. (1996). The Visibility of Visuality. In Brennan, T. and Joy, M. (Eds.) Vision in Context, 70-77. New York: Routledge.

Debord, G. (1994). The Society of the Spectacle [1967], (D. Nicholson-Smith, trans.). New York: Zone Books.

Dexheimer, L.M. (1929). Picture Study in Geography, Journal of Geography 28: 334-49.

Downs, R. (1981). Cognitive Mapping: A Thematic Analysis, In Cox, K.R. and Golledge, Reginald G. (Eds.) Behavioral Problems in Geography Revisited, 101-22. New York: Methuen.

Earle, E. (1979). Points of View: The Stereograph in America: A Cultural History. New York: Visual Studies Workshop Press.

Foote, K.E. (1985). Velocities of Change of a Built Environment, 1880-1980, Urban Geography 6: 220-45.

Foucault, M. (1972). Discipline and Punish: Birth of the Prison (A.S. Smith, trans.). New York: Pantheon.

Goin, P. (2001). Visual Literacy, Geographical Review 91: 363-69.

Golledge, RG. (2002). The Nature of Geographic Knowledge, Annals of the Association of American Geographers 92: 1-14.

Gottdiener, M. (1995). Postmodern Semiotics: Material Culture and the Forms of Postmodern Life. Oxford, UK: Blackwell.

Gregor, H.F. (1956). Slide Projection Techniques in the Geography Class, Journal of Geography 55: 298-303.

Griffen, P.F. (1970). Photographs in the Classroom, Journal of Geography 69: 291-98.

Hales, P.B. (1991). American Views and the Romance of Modernization. In Sandweiss, M.A. (Ed.) Photography in Nineteenth-Century America, 205-57. Fort Worth: Amon Carter Museum; New York: Harry N. Abrams.

Hoelscher, S. (1998). The Photographic Construction of Tourist Space in Victorian America, Geographical Review 88: 548-70.

Humphrey, R.R. (1987). 90 Years and 535 Miles: Vegetation Change Along the Mexican Border. Albuquerque: University of New Mexico Press.

Jakle, J.A. (1987). The Visual Elements of Landscape. Amherst, MA: University of Massachusetts Press.

Jakle, J.A. (2003). Postcards of the Night: Views of American Cities. Santa Fe, NM: Museum Press of New Mexico.

Jenks, C. (1995). The Centrality of the Eye in Western Culture. In Jenks, C. (Ed.) Visual Culture. London: Routledge.

Kinsman, P. (1995). Landscape, Race and National Identity: The Photography of Ingrid Pollard, Area 27: 300-10.

Kitchin, R. and Bladen, M. (2002). The Cognition of Geographic Space. London: I.B. Tauris.

Klett, M. et al. 1(984). Second View: The Rephotographic Survey Project. Albuquerque: University of New Mexico Press.

Leary, W.H. (1985). The Archival Appraisal of Photographs: A Ramp [Records and Archives Management Programme] Study with Guidelines. Paris: Unesco.

Lewis, P. (1985). Beyond Description, Annals of the Association of American Geographers 75: 465-78.

Lowe, D.M. (1982). History of Bourgeois Perception. Chicago: University of Chicago Press.

Lutz, C. and Collins, J. (1993). Reading National Geographic. Chicago: University of Chicago Press.

Mayer, H.M. and Wade, R.C. (1969). Chicago: Growth of a Metropolis. Chicago: University of Chicago Press.

Maynard, P. (1997). The Engine of Visualization: Thinking Through Photography. Ithaca, NY: Cornell University Press.

Miller, E.W. (1965). The Use of Color Slides as a Geographic Teaching Aid, Journal of Geography 64: 304-07.

Mitchell, W.J.T. (1994). Picture Theory. Chicago: University of Chicago Press.

Norman, Jr., W.R. (1991). Photography as a Research Tool, Visual Anthropology 4: 193-216.

Palka, E.J. (1995). Coming to Grips with the Concept of Landscape, Landscape Journal 14: 63-73.

Photo Marketing Association International (2001). PMA U.S. Consumer Photo Buying Report. http://www.FirstSearch.oclc.org

Porteous, J.D. (1996). Environmental Aesthetics: Ideas, Politics and Planning. London: Routledge.

Rogers, G.F., Malde, H.E. and Turner, R.M. (1984). Bibliography of Repeat Photography for Evaluating Landscape Change. Salt Lake City: University of Utah Press.

Rogoff, I. (1996). Others Others: Spectatorship and Difference. In Brennan, T. and Jay, M. (Eds.) Vision in Context. New York: Routledge.

Romanyshyn, R.D. (1989). Technology as Symptom and Dream. New York: Routledge.

Rorty, R. (1979). Philosophy and the Mirror of Nature. Princeton, NJ: Princeton University Press.

Rose, G. (2000). Practicing Photography: An Archive, a Study, Some Photographs, and a Researcher, Journal of Historical Geography 26: 555-71.

Rose, G. (2001). Visual Methodologies: An Introduction to the Interpretation of Visual Materials. London: Sage Publications.

Rundell, W. (1978). Photographs as Historical Evidence, American Archivist 41: 379-98.

Ryan, J.R. (1997). Picturing Empire: Photography and the Visualization of the British Empire. Chicago: University of Chicago Press.

Ryan, J.R. (2000). Photos and Frames: Toward an Historical Geography of Photography, Journal of Historical Geography 26: 119-24.

Saarinen, T.F. (1966). Perception of the Drought Hazard on the Great Plains. Chicago: University of Chicago, Department of Geography Research Paper No. 106.

Schlereth, T.J. (1980). Mirrors of the Past: Historical Photography and American History. In Schlereth, T.J. (Ed.) Artifacts and the American Past, 11-47. Nashville: American Association of Local History.

Schulten, Susan (2001). The Geographical Imagination in America, 1880-1950. Chicago: University of Chicago Press.

Schwartz, J.M. (1995). We Make Our Tools and Our Tools Make Us: Lessons from Photographs for the Practice, Politics, and Poetics of Diplomatics, Archivaria 40: 40-74.

Schwartz, J.M. (1996). The Geography Lesson: Photographs and the Construction of Imaginative Geographies, Journal of Historical Geography 22: 16-45.

Schwartz, J.M. (1998). Agent of Site, Site of Agency: The Photograph in the Geographical Imagination, Ph.D. dissertation. London, Ontario: Queens University.

Schwartz, J.M. and Ryan, J.R. (Eds.) (2002). Picturing Place: Photography and the Geographical Imagination. London: I.B. Tauris.

Shortridge, J.R. (2000). Our Town on the Plains: J.J. Pennell's Photographs of Junction City, Kansas, 1893-1922. Lawrence: University Press of Kansas.

Sontag, S. (1977). On Photography. New York: Farrar, Straus and Giroux.

Taft, R. (1964). Photography and the American Scene [1938]. New York: Dover.

Tagg, J. (1988). The Burden of Representation: Essays on Photographies and Histories. Amherst: University of Massachusetts Press.

Teske, A.E. (1959). Geography Field Trips by Colored Slides, Journal of Geography 58: 334-40.

Tester, K. (Ed.) (1994). The Flanneur. London: Routledge.

Thomas, A.K. and Raup, H.F. (1956). Photography for the Geography Teacher, Journal of Geography 55: 243-47.

Thompson, J.H. (1954). A Colored Slide Collection for Geography Teachers, Journal of Geography 53: 117-23.

Tuan, Y. (1979). Sight and Pictures, Geographical Review 69: 413-22.

Vale, T.R. (1987). Vegetation Change and Park Purposes in the High Elevations of Yosemite National Park, Annals of the Association of American Geographers 77: 1-18.

Vale, T.R. and Vale, G. (1983). U.S. 40 Today: Thirty Years of Landscape Change in America. Madison: University of Wisconsin Press.

Watts, M.J. (2000). Master of Harlem: Roy DeCarava and the Post-War American City. In Watts, M.J. (Ed.) Struggles Over Geography: Violence, Freedom and Development at the Millenium, 75-109. Heidelberg: University of Heidelberg.

Wells, L. (1997). Photography: A Critical Introduction. London: Routledge.

Young, L. and Barrett, H. (2001). Adapting Visual Methods: Action and Research with Kampala Street Children, Area 33: 141-52.

CHAPTER 11

DEBORAH P. DIXON
LEO E. ZONN

FILM NETWORKS AND THE PLACE(S)
OF TECHNOLOGY

Abstract   Interrogation of the complex links between film networks and technology is a recent and promising trend within cinematic geography that moves beyond concerns with relative accuracies of place representation, reel-real distinctions, and singularly focused textual readings associated with constructions of place. This chapter contributes to the emergent dialogue by regarding film and technology as part of a broad, relational network comprised of diverse objects and knowledges, such that they can be read both as a product of processes and as being productive of other processes. The nature of this complex and multiscale network is examined by providing an assessment of interrelations that bind film financiers, producers, distributors, personnel, viewers, and public institutions into a series of smaller and still-complex networks, all of which are deeply embedded within a varied array of economic, political, and cultural settings. Three examples within this larger field are explored within such a frame: the early years of film, the Hollywood System, and "global" cinema and its discontents. The key features of each are considered in terms of its relationality between the components of each network to the larger system, and in terms of the respective technological apparatus brought to bear in their respective realizations.

Keywords   film networks, global cinema, national cinema, Third Cinema

1. INTRODUCTION

During the course of the 20th century, history rather than geography has mattered in the telling of film's relationship with technology. Hagiographies, composed for the most part by writers working within the industry itself, have noted the emergence, development, and eclipse of particular technologies and genres (see Fielding 1967; Grau 1914; Hampton 1931; Macgowan 1965; and Ramsaye 1926). The emergence of historical materialist accounts in the 1970s broadened the field of inquiry by revealing the socioeconomic context within which key technological innovations were embedded (for example, Allen 1977; Branigan 1979; Buscombe 1978; Commolli 1971; Spellerberg 1979). In such analyses, technology and film were regarded as part of a

Stanley D. Brunn, Susan L. Cutter, and J.W. Harrington, Jr. (eds.), Geography and Technology,
243-266. © 2004 Kluwer Academic Publishers. Printed in the Netherlands.

"network" or "field" of causal elements within a specific mode of production, such that they can be read as a product of other processes as well as productive of other processes. This reassessment of the film industry was also premised on the rejection of a teleological approach to the technologies used in the industry. The pejorative description of early film as "primitive" was replaced by an appreciation of the particular set of technologies used prior to the emergence of the classic Hollywood production system which is described below.

Despite this significance, geographers have tended to eschew the promising terminology of networks and fields in favor of analyses that dwell on the accuracy (or not) of filmic representations of place. As Aitken and Zonn (1994: 9-15) explain, geographers have traditionally presumed a simple distinction between the "real" and the "reel," whereby the latter, as a mere mechanical process, allows for a necessarily inferior representation of the former. While some contemporary geographers have retained this notion of film as a mimetic—most notably Harvey's (1989) dismissal of film as a soporific ideological form that serves to obscure the real-life process of exploitation—others have deliberately blurred the boundary between the "real" and the "reel" by focusing on the processes by which particular meanings are produced, distributed, and consumed (see Benton 1995; Hanna 1996; Gold 2002; Strohmeyer 2002). Cinema is considered a popular medium through which spatial images are filtered for a specific audience (Zonn 1984, 1985; Aitken and Zonn 1993), as an assemblage of technological practices that allows for the unsettling of taken-for-granted discourses on people in place (Aitken 1991; Natter and Jones 1993), and as a modernist, urban-based display of alienated spectatorship (see Clarke's 1997 introduction). In so doing, geographers have begun to flesh out the complex interrelations that bind film-financers, film-makers, film-personnel, film-distributors, and film-viewers into a spatially and temporally specific "network," and have embedded cinema within broad economic, political, and cultural settings (Cresswell and Dixon 2002).

This chapter builds on these analyses in its assertion that technology is more than simply "social." Technology is part and parcel of a broader network, in that it is produced, deployed, and made redundant through its connections to a host of other phenomena. Strathern (1996) raises an important question, viz., where does one "cut" the network? Clearly, it is not possible to tease out all of the phenomena that form a collective for action through space or through time. Instead, we must select particular moments of stability, or what Law (1992) has termed "puntualisation," wherein networks revolving around film pursue: *longevity*, in the sense that ideas and actions are inscribed in various texts, such as film prints, books, and memoirs, as well as various types of apparatus including cameras, editing suites, and exhibit displays; *mobility*, in which texts and these apparatus forms circulate through space and thus maintain the ordering of the network at a distance; and as a *centre of calculation*, whereby the reception of particular ideas and techniques is monitored through audience surveys, profit measures, and so on.

Below we discuss three such "puntualisations" that have underlain the development of the film industry: the early years of film, the Hollywood System, and "global" cinema and its discontents. In all of these film networks, technology has played a role in the pursuit of longevity and mobility, as well as the monitoring and adjustment of the network itself.

1.1. Film Networks

The elements of a film network can be thought of simply in terms of the flow of elements from place to place over time, such as the transfer of money from one group to another, the training of an apprentice in film lighting techniques, the hiring of an established "star" from one vehicle to the next, traveling to a movie palace, or the movement of a rental videotape from store to house and back again. But it is more useful, we would argue, to acknowledge how such flows are shaped by, and in turn help shape what we might term the "identity" of those peoples, places, and things. Such flows are made possible by the particular form and character of phenomena; they also can alter the form and character of phenomena. For example, a "star" performer is constructed from a range of knowledges; she/he has been trained, styled, lighted, choreographed, and filmed in a series of prior productions, and it is this particular package that is called upon when a star is placed in a new production. In turn, the arrival of such a star can transform an everyday B movie into a classic. Investment capital has its own particular baggage in that it too is produced under a series of money-making knowledges and techniques. Funneling such money into what is seen as a profitable cinema industry can change the character of place from a rural backwater to a thriving economic powerhouse. Even sitting in a cinema, watching and listening to film, can have a transformative effect, offering opportunities for escapism, voyeurism, and so on, as well as an exposure to a host of ideas, concerns, and even emotions.

From our perspective, the networks within which films are embedded are constituted from a range of ontological phenomena and epistemologies; moreover, these are differentially related through power. Here we draw a simple analytic distinction between the power to achieve something and power as the sum of the opportunities and constraints a person is afforded by virtue of the particular position she/he holds within a network. The reader already may have discerned a weak form of Actor-Network-Theory as the framework for this chapter. Latour (1997) finds the notion of networks useful in the sense that they do away with "inside" and "outside," asking only if and how a connection is made. But we prefer a broader emphasis on how people and things are placed in relation to one another (following Bingham 1996; Escobar 1994; Massey 1993), such that issues of inclusion and exclusion, hegemony and dependence, remain at the forefront of analysis. All phenomena in such a network, therefore, whether they be human or nonhuman, can be considered powerful in the sense that they operate as part of a collective to allow for a particular event to occur or entity to perform; in this sense, all such phenomena are of explanatory significance. Thinking of power as the consequence of one's positionality, however, lends an extra dimension to the analysis that we can differentiate those who can pursue a variety of avenues of behavior and thought in terms of who can lay claim to greater access to people, things, and ideas, and who can more easily transform a given network from within versus from those who cannot.

Our aim in the following sections is to outline the key features of each of these networks, focusing on the relationality between the components of each network, as well as the technological apparatus brought to bear in the realization of that network.

2. NETWORK ONE: THE EARLY YEARS OF FILM

Development of the film industry depended upon the assemblage of knowledge concerning the physiology of the senses, the finance and skill of inventors who produced several devices for the display of moving images, the entrepreneurial spirit of film exhibitors, and the receptive curiosity of spectators from all stations and walks of life. Initially it was the technology itself that was the focus of interest as inventors competed with one another to produce devices that would convey a sense of mobility to an audience, while the actual content was derivative of established forms of entertainment, such as vaudeville and literature. As the number of films produced increased exponentially and the venues at which films could be viewed continued to grow in number, this "wonder" was displaced by an appreciation of film as an entertainment medium. Movie viewing became a popular pasttime and audiences grew accustomed to the form and function of particular *kinds* of films, such as comedy, fantasy, and documentary. Below we outline the emergence of this network of people, things, and knowledges, noting in particular the varied flows and transformations that characterized this phase of the film industry, as well as the positionalities that various groups held in relation to one another.

2.1. Experiments with Optics

The unique ability of the "motion picture" to represent movement is contingent upon two complementary perceptual processes that involve interaction between the observer and the action–one being optical and the other psychological. The optical phenomenon, known as the persistence of vision, has been recognized for many centuries, but it was not documented until 1824, when Peter Mark Roget found that the eye had the ability to retain an image of an object after its removal, which he estimated to be from 1/5 to 1/20 of a second. This discovery means that if a series of pictures is presented, the eye retains the vision of one frame until the next arrives. In filmic terms, when a series of stills is projected at the rate of 24 frames per second (fps), the currently standard rate of film projection, the eye does not distinguish the pause between each frame, which helps to create a sense of continuity (Parkinson 1996). While studying the persistence of vision, Max Wertheimer (1912) discerned the phi phenomenon, which Hugo Munsterberg (1916) later documented in the context of film. This phenomenon is the psychological means whereby an individual connects one still of a quickly shown series to the next as if it were a continuous whole. This bridging ability creates a meaningful continuity on the part of the viewer, rather than a perception of the discrete fragments that actually exist.

The persistence of vision was much used in the development of optical toys such as the *Thaumatrope*, created by John Paris in 1824, which had a parrot on one side of a disk and a cage on the other. When the disk was quickly spun over and over, the parrot appeared to be in the cage. In 1832 two inventors, Joseph Plateau and Simon Stampfer, simultaneously and independently used this fundamental approach to develop a device (called the *Phenakistiscope* by Plateau and the *Stroboscope* by Stampfer) that was comprised of a circular board with a variety of drawings that was held to a mirror. When the board was spun the drawings displayed continuous action. William Horner improved the toy in 1834 when he placed the images in a circular drum, whereupon the viewer would look from the side and through slots to watch the animated actions. The device was labeled and patented by William Lincoln in the U.S. as a *Zoetrope* in 1867 and was patented in Europe the same year by Milton Bradley. The *Praxinoscope*, developed by Emile Reynaud in 1877, contained a series of pictures inside an outer cylinder which was spun around a post that was covered by a set of small mirrors. When the cylinder was spun, the illusion of moving action was created. Reynaud took the process yet one step further by creating the *Theatre Optique*, whereby images were produced on a small screen.

By the 1880s, Edward Muybridge, an Englishman living in California, had developed the *Zoopraxiscope*. In order to settle a bet for the governor of California, Leland Stanford, that indeed all of the horse's hooves left the ground when it ran, Muybridge posted 24 cameras in a row along a racetrack. The horse tripped each wire as it moved down the track, providing a series of images. Muybridge developed the process over the next few years and combined the results with a projection device to create the *Zoopraxiscope*. Another inventor, Etienne-Jules Marey, who had visited with Muybridge during the latter's tour of Europe with his new device, created in 1882 the first camera to shoot multiple pictures in a series. These were shot at the rate of 12 photos per second onto a photographic plate.

The pace of development began to increase in the late 1880s and early 1890s as inventors continued to collaborate and compete with one another. Much of the credit for the development of early cinema goes to Thomas Edison, although most of the achievements were by William Kennedy-Laurie Dickson, his employee in Edison Studios, including the earliest film on record, *Fred Ott's Sneeze* (1891). Months later, the studio gave the first public showing of a projected, moving film—of a man who bowed, smiled, moved his hands, and removed his hat—to the National Federation of Women's Clubs (Mast and Kawin 2002).

2.2. Moving Pictures

Edison believed the future of film was in terms of the single viewer, so much of his cinematic contribution was toward the technological and commercial ends of this approach. As a result, he first introduced the *Kinetoscope*, a machine that allowed the viewer to look down through a window to watch the film, at the Chicago World's Fair in 1893. By the next year, he was marketing the product throughout the country and eventually into Europe, which resulted in the proliferation of the *Kinetoscope* Parlors. The shows were of people actions, such as dancing and juggling, and were made in the world's first film studio, which was built in New Jersey in 1893. It should be noted that a problem with the *Kinetoscope* was that each show lasted only 20 seconds, because the films were shown at more than 40 fps and the size of the machine did not allow for any more film.

At roughly the same time, Dickson left Edison's studios and created his own machine, the *Mutoscope*, which was hand-cranked. The new machine eventually dominated the market and made the Edison machine passé. But soon Edison bought the rights to a recently invented machine that projected the film onto a screen, named it the *Vitascope*, and showed the first film to a paying audience as part of a Vaudeville act in New York in 1896. Nonetheless, he was a year late to be able to claim the title as the founder of cinema; that distinction belonged to the Lumière brothers.

August and Louis Lumière worked at their father's photography equipment factory in Lyon, France and were inspired directly by Edison's *Kinetoscope*. They were able to develop a markedly improved product within a short time. By 1895 they had developed the *Cinematographe*, which was portable, developed film as it shot, and was able to project the image onto a screen. They also established 35 mm as the standard for film, which exists to this day, and the rate of 16 fps, which was the standard until sound film.

Soon thereafter, the brothers used the new device to film *Workers Leaving the Lumière Factory*, and opened, also in 1895, the first theater for a paying public showing this film and several others. The Lumière brothers were also the first filmmakers to record geographic settings throughout the world and to bring these images back to the general public. They were the first to make a narrative film (a brief comedy), developed the first cinematic catalogue (1200 films for purchase), and can claim, through the naming of their device, to be the source of the word "cinema." Despite their premier standing within the film community, their status and influence soon faded, at least partially because of the intense competition and rapid changes in the nature of the industry (Mast and Kawin 2002).

According to Thompson (1985), the actual film content of these early products, as well as the filming technique itself, was highly derivative of existing entertainment media. Action was noncentered as on a stage and was shot from the front by a distanced camera, such that the camera replicated a stationary audience perspective. Films were regarded, he argues, as "spectacles," hailing the attention of the spectator by showing an "attraction," such as a series of novel views including nonfictional (current events, natural wonders) vaudeville acts, or famous fragments (realizations of well-known paintings or moments from novels).

By the turn of the century, audiences were becoming bored with the standard material and, in fact, there was a brief lull in demand for film. But a number of creative and pioneering technological advances changed the nature and character of the demand. Examples include the work of Georges Melies, a stage magician, who moved his efforts to the screen, beginning with the *Dreyfuss Affair* (1899), which was shown in eleven scenes on eleven reels. This film, which was the first multiscene production and the first to be censored for political reasons, was followed by his *A Trip to the Moon* (1902), which included the first screen illusions, and, it should be noted, the first parody of academics. Another example is the work of Edward S. Porter, who worked in Edison's studios. He produced two influential works, *Life of an American Fireman* (1903) and *The Great Train Robbery* (1903), which included the first efforts to follow the action by camera movement, to produce more than one shot for a scene, the fundamental practice of "ellipsis" (which refers to shooting and editing that collapse time and space, leaving the viewer to fill in the blanks), and the "cross cut," which involves the interweaving of two scenes.

2.3. Exhibiting Film

Toward the end of the 19th century, inventors and producers such as Edison had depended upon existing entertainment industries and venues, including vaudeville palaces, traveling tent shows, amusement parks, fairground exhibitions, and educational exhibitions in schools and churches, to sell their products. From this diverse assortment of venues, films were exhibited in towns and cities across Europe and the U.S., in middle and working-class, white and black neighborhoods (Allen 1977, 1982; Musser 1991; Waller 1995).

As the new century began, however, the dominance of the film producer was challenged. Attempts by Edison and others to control the emergent industry by selling films and projection machines as complete packages failed, as exhibitors insisted on buying machines and films separately. Middle-level individuals appeared, buying films from several producers and renting them to exhibitors. In the U.S., there was an explosive growth in the number of nickelodeons (converted store-front theaters), each able to show new films on a daily basis (Merritt 1976).

For Musser (1991), exhibitors held a powerful position by virtue of the fact that they could endow meaning on the film-watching experience by showing the film as part of an entertainment package that could include the linking of contrasting films, interspersing slides or recitations, adding music or other sound effects, or a narration of events. Smith (2002) provides a fascinating case study of how exhibitors did indeed provide an ambience for Flaherty's *Nanook of the North*, selling snow cones in the foyers and special music sheets. Hansen (1991) makes the significant point, however, that the nickelodeons also performed a transformative role within the traditional public sphere, which was otherwise masculinized, white, and middle-class. In these new gathering places, previously excluded groups could participate as spectators on an equal footing with other members of society. They, too, could view film as part of a collective, as well as construct their own personal meanings from the events, peoples, and places shown on screen. Films, therefore, have an inherently democratic potential. As Benjamin stated so well in regard to early film:

> Our taverns and our metropolitan streets, our offices and furnished rooms, our railroad stations and our factories appear to have us locked up hopelessly. Then came the film and burst this prison asunder by the dynamite of the tenth of a

second, so that now, in the midst of its far-flung ruins and debris, we calmly and adventurously go traveling. (cited in Cresswell and Dixon 2002: 4-5)

In the years just before World War I, several film companies were formed, with most eventually being characterized by a significant degree of vertical integration of production. Each company was manufacturing cinematic equipment, creating movies, and building its own theatres to show its own creations; the ensuing competition was chaotic and frenzied. George Melies, Leon Gaumont, and Charles Pathé of France, and R.W. Paul of the U.K., all of whom produced films in addition to creating companies, were in competition with one another and with Edison and Biograph Studios in the U.S. Unable to compete, the Lumière brothers had gone out of business by 1902. Nonetheless, by 1908 most of these companies had formed a consortium, closer in fact to a cartel, which by World War I dominated most of the essential aspects of the industry production. By 1914 French companies were selling a dozen films a week to U.S. exhibitors, while French and Italian films dominated the burgeoning Latin American market (Balio 1993).

This European dominance was to be short-lived, as by 1914 a new group of companies, Universal, Famous Players, and Mutual, had arrived on the scene. Located in Hollywood, California, they initiated the development of this small town into an economic powerhouse that today provides not only an almost unrivaled number of films onto the market, but also a particular style of film that has become for some a global-scale, hegemonic cultural force.

3. NETWORK TWO: THE HOLLYWOOD STUDIO SYSTEM

The Hollywood studio system is recognized as one of the most potent economic and cultural forces of the 20th century. Below we outline the key components of this network from production to distribution to exhibition. We note how the studio system was predicated on the efficient deployment of people and machines, such that a flexible yet regular production system was established. The key to the success of the studio system was a monopoly hold on the distribution and exhibition of film in nickelodeons (and later picture palaces) across the U.S. As more and more "classic" Hollywood films made their way onto the screen, viewers became enmeshed in what some commentators labeled a hegemonic form of ideology.

We conclude this section by noting how Hollywood has adapted to changing political and economic conditions. In the search for alternate venues for the exhibition of film, Hollywood studios have taken advantage of diverse media such as TV and the Internet, each of which demands a particular technology such as cable, video, and DVD. While some commentators refer to this series of technological shifts as the "end" of the Hollywood system (see Jenkins 1995), we note how film studios have become part and parcel of entertainment conglomerates, such that "global cinema" has become both the model and the bête noir of film-makers everywhere.

3.1. The Rise of the Hollywood Studio System

The consolidation process apparent in the European film industry outlined above was to be repeated in the American context. Prior to the 1920s, films were being produced by a variety of "independents" in and around Los Angeles, a setting that provided a sunny climate for outside shooting and a low unionization rate (Wasko 1982). Despite this latter condition, for some commentators, this was a "golden age," in that conditions were in place for the development of a "progressive" industry geared towards filming working-class issues for a working-class audience (Guback 1969). Moreover, the independents were selling to the international as well as the domestic market. By 1916, U.S. firms were exporting more films to Europe than were being imported (Balio 1993).

In the 1920s, the film industry took on an industrial mode of production. New York financial capital made its way into Hollywood, allowing some companies to develop a flexible labor system for the regular production and distribution of feature-length films as well as short subjects and newsreels (Staiger 1982; Wasko 1982). It was the introduction of sound technology that set the seal on Hollywood's success as a production center for film. English was the language of rule in many parts of the globe, thanks to British and American imperialist expansion; the new Hollywood talkies, therefore, had an enormous market potential (Acheson and Maule 1994). Musicals were produced for non-English markets, and by the end of the decade Hollywood films were selling across Asia and Latin America, as well as to Britain and Australia (Armes 1987). The technology used to produce these films was also proving to be an export success (Hay 1987).

By the 1930s, Hollywood was ruled by the "big five," Warner Brothers, Paramount, Loews (parent of MGM), RKO, and Fox Films (later 20th Century Fox), in a consolidated studio system, while smaller-scale companies such as Universal Pictures, Columbia Pictures, United Artists, Republic Pictures, and Monogram imitated their production model as best they could. Hollywood, as the locus of production headquarters and studio sets, had become an established learning region in regard to the making of film (Bordwell, et al. 1985).

3.2. Classic Hollywood Style

During this era of consolidation, camera techniques, narrative structures, and production qualities were rapidly changing. These developments led not to an increasingly diverse array of products, but toward the emergence of a particular, hegemonic film form, the "classic" Hollywood movie. The emergence of this model is closely tied to the work of director D.W. Griffith, who, arguably, brought more changes to the cinema than any other director before or since. He is often remembered for his racist views, of course, most obviously in *Birth of a Nation* (1915), but Griffith was also known for his portrayal of the rapidly changing urban environment of the day and the new technologies, rail, automobile, telegraph, and telephone, as they appeared in this landscape (Mast and Kawin 2002). In films such as *The Lonely Villa* (1909) and *The Lonedale Operator* (1911), Griffith portrayed these technologies as perilous and miraculous, emblematic of a new modern era.[1]

For the influential French film theorist Bazin (1951), the Hollywood classic is characterized by a particular mode of filmmaking, one that relies on long takes, deep-focus cinematography, staging in-depth, and continuous editing. These particular techniques allow for the regeneration of realism in storytelling and are the hallmark of such directors as John Ford and Orson Wells. While Bazin is working within an Althusserian Marxism, it is Commolli (1971), Metz (1974), and MacCabe (1974) who emphasize the ideological function of such techniques, arguing that they allow for bourgeois conceptions of the "real" to be transmitted with crystal clarity. For these critics, classic Hollywood film is akin to propaganda. Unlike the agitprop of Sovfilms, which deployed montage and surrealism to convey the alienation induced by capitalism (see Staddon et al. 2002), classic Hollywood film deliberately worked to a form of realism to convey the naturalness of this mode of production. As Ray (1985) argues, what makes Hollywood cinema so successful is its power to hide the processes behind its own production, and to make the audience forget that they are indeed watching a film. This critique of Hollywood cinematography, as well as actual film content, has clearly influenced the work of geographers such as Harvey (1989).

Such an argument is, perhaps, something of a monolithic one in its view of the production and reception of film.[2] For Bordwell et al. (1985), cognitive psychology provides a more nuanced means of understanding how viewers are predisposed to look for cues in a film's narrative and form, such that the information provided is processed and organized into a discernable pattern. Technology plays a crucial role here, in that camera placement, lighting, set-design, and framing are all used to focus the viewer's attention on the central narrative. Successive analyses have drawn out how various social positionalities, provided by gender, race, and sexuality as well as class, shape this cognitive process (including Kaplan 1992 and Mulvey 1977). De Lauretis (1984, 1987), for example, inaugured a feminist appraisal of spectatorship by pointing out that female viewers must necessarily watch "against the grain" of masculinist films that portray women as either housewives or femme fatales.

3.3. Distributing the Dream

The development of a flexible system for the regular production of films, shorts, and newsreels is not enough to ensure the economic and cultural dominance of the Hollywood film. The key to the success of this particular film industry lies with the monopoly control each of the big five companies developed in regard to distribution and exhibition. The number of U.S. venues had grown from 10,000 nickelodeons in 1910 to 28,000 picture palaces by 1928 (May 1983). Three-quarters of these were owned by the big five studios (Gomery 2000).

The vertical integration of the U.S. film industry ensured not only a continuous stream of income for the major companies and the establishment of the classic Hollywood movie as the filmic "norm," but it also set in motion a revolution in spectatorship. In the 1920s, the average audience was 25-30 million people a week, and by the 1930s, this increased to an estimated 85-110 million people each week (Dieterle 1941). Building on the work of Bazin noted above, Ewen and Ewen (1982) point out that spectating became an ordered and disciplined activity in these special venues. This immensely popular mode of entertainment, they argue, became the means of stimulating the desires of spectators; in the process, pleasure was disassociated from the "real" world of work and was invested instead in the on-screen world of the signifier. On a more prosaic note, they suggest that immersion in the classic Hollywood film served to socialize the burgeoning number of immigrants, providing role models for dress, work, love, and marriage. Censorship at the local, regional, and national level also ensured that films played to what were posited as "community" standards. Right-wing lobbyists feared what they saw as the potentially immoral and subversive impact of Hollywood film on the masses. The nation-wide Hays Production Code, introduced in the 1930s and still operative in the 1960s, outlawed nudity and explicit sexuality as well as prostitution and drugs.

During the next few decades, Hollywood was to decrease its dependence on domestic picture palaces for distribution and exhibition and turned instead to its international sales as well as alternative distribution systems. Partly this shift was forced upon the studios because a series of antitrust legislation packages had been introduced during the Depression. As an outcome of the Paramount case of 1948, the big five studios were forced to sell off their "first show" movie theaters. They retained, however, their control of picture palaces in a host of other countries and proceeded to build on this power base. In Brazil the Hollywood studios had long since displaced local producers by buying out local distributors (Shohat and Stam 1994); this practice continued unabated across the globe. In the 1970s and 80s, for example, hyperinflation in Argentina and Mexico all but destroyed local film production centers, and in the aftermath, Hollywood film corporations consolidated their hold on Latin America cinema (Himpele 1996). Moreover, as Harley (1940) and Elsaesser (1989) note, the U.S. film industry had been helped enormously by tax-credit schemes and film commissions, as well as State and Commerce Department representation; all of these practices served to increase its international market share. For example, Hollywood's African Motion Picture Export Company has dominated sales to the former British colonies since the 1960s (Diawara 1992). In the aftermath of World War II, and particularly during the Cold War years, this political support was predicated on Hollywood presenting a positive picture of the American way of life. This portrayal was understood to serve as a dual purpose, combating Socialist ideology and fueling international demand for American goods (Schatz 1988). Of course, support is also provided at the state and local level, as Swann (2001) demonstrates in the case of Philadelphia.

Meanwhile, the studios had become increasingly flexible in regard to locating and funding alternative distribution outlets, such as drive-in movies and multiscreen cinemas in shopping malls.[3] In the 1970s, Time Inc. developed Home Box Office for cable TV in order to screen Hollywood films. They were swiftly followed by Ted Turner's Superstation. By the 1980s the studios had even begun to make made-for-TV films, expanding into miniseries and novels adapted for television. The same decade saw a revolution in home-based technologies as sales of video recorders increased dramatically. Hollywood capitalized on this by distributing videos for rent, facilitating the emergence of the neighborhood video store as well as chains such as Blockbuster Video.[4] As Wasko (1995) illustrates, the adoption of these particular technologies and forms of distribution across the globe allowed Hollywood film unprecedented access to individuals, families, and communities.

The arrival of digital technologies in the 1990s has had an interesting impact on the distribution and exhibition of Hollywood film. Digital technology has allowed film to be presented via computer-based systems as well as DVD, opening up the Internet as a venue for film. It has also allowed for the form and content of films to be digitally remastered, leading some commentators to debate the "democratic" potential of these new information technologies. Echoing the comments of Benjamin, noted earlier in this chapter, several authors point to the constant reproduction of meaning through the viewing process, such that diverse articulations of people and place can be produced (see Clarke 1997). Others have pointed to the emergence of a "digital divide" as the skills and knowledges, as well as the finance necessary for this practice remain the property of the minority (see Wasko 1995). And yet, as we note in the example of Nigerian cinema below, these new media technologies do allow for films to be produced at a fraction of the cost demanded by Hollywood Cinema.

All of these changes had led commentators to talk about the "end" of the Hollywood studio system (Gomery 2000; Jenkins 1995; Neale and Smith 1998). Certainly in economic terms, the corporate structure of the film industry has changed considerably, as production studios have been dismantled and then refitted into global-scale entertainment conglomerates. In the process, the traditional in-house production format has been replaced by the assemblage of "packages," whereby independent producers are teamed with stars, directors, and vehicles. These conglomerates are just as interested in pushing international sales of film as their predecessors. Whereas the Motion Picture Association of America lobbied on behalf of MGM and Paramount Studios in the 1930s, it now lobbies on behalf of the "big six," namely Disney, Twentieth Century Fox, Seagram's Universal, Viacom's Paramount, Sony's Columbia and Time Warner's Warner Brothers.[5] It is worth noting that ownership of these conglomerates has been variously Australian, French, Japanese, and Canadian, a fact that draws further attention to the "global" character of the Hollywood film industry.

4. NETWORK THREE: "GLOBAL" CINEMA AND ITS DISCONTENTS

As intimated above, it would be unwise to conflate "Hollywood film" with "American film," as the conglomerates that currently dominate film production are not limited in their ownership, labor, or studio sets to the U.S. In producing, distributing, and exhibiting film, each conglomerate cuts across political, economic, and cultural borders. As a major example, among other holdings, Viacom owns Paramount pictures, as well as Blockbuster Video, a variety of cable networks (including Showtime, The Movie Channel, MTV, etc.), and Simon and Schuster publishers. All of these divisions have international offices, franchises, and distributors and are actively engaged with international markets. Further, Paramount is a primary owner of United International Pictures, which distributes American and international films and encourages and supports "indigenous producers." It would seem appropriate, therefore, to talk not of American cinematic hegemony, but rather of "global" cinema and its discontents.

The economies of scale that can be achieved by conglomerates, and the sheer political clout of organizations such as Disney, have fueled debate over the hegemonic impact of films produced under the auspices of these giants.

In this section, we draw out the key features of this debate. In order to do so, we must first outline more carefully what we mean by "global," a term which has been laden with a series of negative connotations through the deployment of the associated term "globalization." We suggest that globalization is not a set of abstract flows or homogenizing processes, but is instead a particularly complex and dynamic series of networks that produce and cut across scale to produce difference and diversity. In this sense, global cinema is a network of people, things, and ideas that stand in numerous, power-laden relationships to one another, albeit some more distanced than others.

4.1. Global Cinema and Globalization

From the details noted above, a case can readily be made for a global cinematic hegemony. Initially, this dominance accrued from the market leverage of the film studios based in and around Los Angeles. In recent years, economic, political, and cultural power has been invested in the products of a few major entertainment conglomerates, each responsible for producing commodities designed from a Hollywood-based template. By 1994, revenue from the international market for these productions was higher than that generated within the U.S., with 88 of the world's top 100 grossing films being Hollywood productions (Rockwell 1994). As Miller et al. (2000) summarized the situation,

> people of color are the majority filmmakers in the world, with much more diverse
> ideological projects and patterns of distribution than Hollywood. But Los Angeles
> culture and New York commerce dominate screen entertainment around the globe
> (p. 145)

It is the nature of this "domination" that has fueled debate. According to Miller et al. (2000),

> while its (global cinema) mounting dominance is there for all to see, U.S.
> governments and businesses continue to assault other countries' attempts to assert
> rights to national self-determination on-screen via barriers to imports. (p. 9)

For Miller et al. (2000), global cinema is akin to Americanization, as alternative forms of identity are aggressively eroded. Such an argument is itself predicated on a particular conception of globalization, defined in general terms as the "rapidly developing and ever increasing network of interconnections and interdependences that characterize modern social life" (Tomlinson 1999:1-2) throughout the world. And yet, two contestable issues relevant to this discussion are the presumption of increased cultural homogenization and the role of the state in this process.

First, the discourse of cultural homogenization considers globalization as a process that standardizes the demands and products of consumer cultures, creating an increased sense of uniformity, beginning at the level of popular culture. This presumption is too facile and superficial to account for enormous variations of daily-lived experiences that have hardly changed with globalization and in fact may have become enriched and emboldened in their practices and representations.[6] The many and often cited cases in which features of dominant ideologies are being accepted, an anecdotal study being the rise of McDonald's in Beijing (Yan 2000), are more likely to reflect national and local adaptations rather than wholesale acceptances of global-inspired practices. Sen (1994), for example, notes how the Hollywood model has been imitated and adapted by Indonesian national cinema in its production of the Si Boy cycle of films.

Moreover, such "other" cinemas not only respond to screened American culture in diverse ways, they also contribute to what we conceive of as products of the "Hollywood" studio system. For example, recent Hollywood blockbusters, such as *Tomb Raider*, *Rush Hour*, *Shanghai Noon,* and *The Matrix*, have borrowed heavily from the immensely popular martial arts genre, produced within the film industries of Japan, China, and Korea. Stars, directors, and technicians have all been exported to the U.S., alongside the narrative and graphic norms that animate this genre. Such cinemas also influence one another–one example being when Hollywood similarly adopts the martial arts genre in films (see Anderson 2001).

The second contestable notion involves the nature of the state. It is contended that the state is being "hollowed out," as global-scale economic and political entities wield greater and greater influence. The dismissal of the state in terms of the broader discourses of globalization is, however, somewhat naïve. Smith (2001) offers instead the concept of transnationalism. While the globalization discourse draws attention to social processes that are "largely decentered from specific national territories, as in the case of Manuel Castells's (1997) discussion of globalization(s) as taking place in a "space of flows," research on transnational processes depicts social relations as "anchored in" while also transcending one or more nation states" (Smith 2001: 3). For example, some writers feel that the concept of "national cinema" is now defunct in light of globalization (Miller et al. 2000; Shiel and Fitzmaurice 2001), but for others, national cinema remains a significant category because of the "embeddedness" of particular film-making norms and technologies (Hjort and Mackenzie 2000). Recent studies of film-making in Indonesia (Heider 1991; Sen 1994), India (Chakravarty 1998), Subsaharan Africa (Barlet 2001), and Thailand, Vietnam, and Sri Lanka (Dissanayake 1994) point to this embeddedness and the contingencies that arise from the interplay of American and indigenous production and viewing practices. The links between cinema and the promotion of a national identity are particularly pertinent for those "new" states that continue to emerge. For example, the nascent Canadian Territorial Government of Nanuvut has expended considerable time, effort, and money into the promotion of an Inuit-led "national" film industry that can stand in opposition to global cinema (Igloolik Asuma Productions 2000).

4.2. Films of Third Cinema

The relationality of global cinema and its discontents is usefully illustrated through the case of Third Cinema which, as Macdonald (1994) makes clear, is a label given on the basis of a film's intent to contest the prevailing "norm." First Cinema refers to "the cinema of profit," normally characterized by large-scale productions; Second Cinema consists of "art cinema;" Third Cinema is judged to be "the cinema of subversion." For Wayne (2001), whose work is influenced by Gabriel's (1991) *Third Cinema in the Third World*, the latter category can be described as:

> a cinema of social and cultural emancipation…[these films] are political in the sense that they in one way or another address unequal access to a distribution of material and cultural resources, and the hierarchies of legitimacy and status accorded those differentials. (p. 1)

One prominent example of Third Cinema is the Nigerian video industry. Nigeria has recently seen the emergence of a local film industry, affectionately and confidently referred to by some as Nollywood, that produces cheaply made video (VHS) films for markets within Nigeria and in a few surrounding countries. Made in English, a local pidgin (of Lagos), and indigenous languages, the films have become exceptionally popular, are widely advertised and anticipated, and have become an integral part of popular culture throughout Lagos, the countryside, and even in other parts of West Africa. The films are normally made on budgets of around $15,000, each being filmed with one digital camera over the span of a few days, and are usually based in the Surulere district of Lagos, which the *New York Times* calls "West Africa's new movie making capital" (Onishi 2002). The salary for actors is minimal, so most participants are part-time, even though they are often recognized throughout the country and beyond. The films are distributed by rental shops and are often viewed in private homes operating as movie centers. A recent estimate places the annual revenues at $45,000,000 a year (Haynes 1997).

Nigerian video films often focus on the daily street life of Lagos. These are aimed primarily at the youth market; violence, poverty, homelessness, disintegration of the family, and a variety of other social issues are treated as the foci of the urban narratives, with an implicit focus on the collapse of social order in the city. Most productions clearly portray an antiurban and provillage subtext in which the latter represents the past when/where community and other positive virtues prevailed before Lagos began its struggles to integrate into the world economy. The city is often signified as a jungle, while issues of the supernatural and customs of local cultures occasionally become a part of the plot (Oha 2001).

The prevailing discourse within this urban-rural frame centers on the experiences of the individual in a chaotic and unruly postcolonial city and era where the country is considering (intentionally and not) western values of individualism and achievement at the expense of all else. Thus these films undermine the interpretation of the city as a civilizing center in postcolonial Nigeria by drawing attention to the "signs" (symptoms) of the postcolonial city, in fact to the sickness that *is* the postcolonial city (Haynes 1997; Okome and Haynes 1995) These films also provide a strongly subversive critique of the prevailing order, particularly of a Nigeria struggling to establish itself within a globalizing economy, despite the fact that the films are neither sanctioned nor opposed by the government.

5. CONCLUSION

The capacity of cinema to represent the world's cultures, spaces, and places is enormous. While we can use the concept of the "network" to discuss how this capacity has been produced via the assemblage of people, things, and knowledge, in time and through space, it is important to bear in mind the varied power relations that knit people together in these networks. In the early years of the film industry, inventors and producers of the new cinematic technologies played a key role in determining the content, form, and distribution of film. As the individual gave way to the film company, which was geared towards the establishment of a consumer market for its products, the network within which films circulated began to transform. Exhibitors and middle-level persons made their mark not only on which films were to be shown where, but also on how those films were to be interpreted by an audience of paying spectators. As Benjamin (1969) remarked, cinematic technologies opened up a series of new conceptual and experiential horizons for those film audiences.

With the consolidation of the studio system in the 1920s, the Hollywood "classic" film became the "norm." Produced under a mode of technological production that specified particular lighting, camera, sound, and acting techniques, the Hollywood picture was exported around the world, displacing local production centers as it sold the American way of life. Under the pressure of declining domestic sales, the mobility of this form increased apace in the second half of the 20th century. Taking advantage of the arrival of new media technologies, the producers of Hollywood film distributed their product via video, DVD, and the Internet, increasing not only their audience figures, but also the variety of sites wherein film could be experienced.

By the 1980s when cinema had reached across the globe, the industry could no longer be described as either "Hollywood" or "American." The entertainment conglomerates that house film-production facilities cut across national boundaries, ensuring the constant flow of capital, people, and ideas. This has not meant the end of "national" interest in the film industry, as such territorial means of identity formation continue to play an important role in the everyday lives of people. Moreover, the flow of people, things, and knowledges does not imply the abstraction of each from local conditions or processes. The term transnationalism attempts to capture this notion of the embeddedness of these conditions and responses within a series of contexts.

One final point is that film networks do not exist in isolation, but rather intersect with one another through space and over time, as well as with a host of other networks. The phases of "puntualisation" we have drawn merely serve to highlight the extensive, often spatially disparate, relations that enable film to exist and act as an economic commodity, as an agent for cultural change, and even as a form of political leverage. As the director John Ford once remarked, "Hollywood is a place you can't geographically define" (cited in Miller et al. 2000, 151); it is precisely this indefiniteness, perhaps, that affords such networks their longevity.

NOTES

1. This relationship between film and modernity is elaborated by Benjamin (1969) who draws attention to how filming techniques, such as rapid montage, slow and fast motion, and huge close-ups, were both symptomatic of, and reflective considerations of, the shock of modernism's forms and experiences. (See also Friedberg 1993; Kracauer 1995; Natter 1994.)

2. This is a view that dismisses the prominence given to those same productive practices in film festivals, and in particular, in film award ceremonies. At the Oscars, for example, numerous awards are provided for lighting, camerawork, editing and so on. Also various trade and popular magazines, including Premier, Hotdog, and SRX, also contribute to the diffusion of information regarding the technologies by which films are produced. And new media technologies have provided a wealth of background material on film-making to audiences across the globe. DVDs, for example, now include sections on "The Making Of ...," while the Internet allows access to reviews, commentaries, and other opinion pieces. Such events and texts point to the wider content within which films are discussed and assessed by "experts" and public alike.

3. In subsequent years, the studios were hit by declining domestic audiences. Numerous reasons have been posited for this, including the arrival of TV and the increasing costs of suburban living (see Gomery 1992).

4. There have been the more eccentric experiences such as 3-D (see Neumann 2001), but few of these have had a lasting impact.

5.   Piracy of theatrical films, which costs the industry tens of billions of dollars a year, occurs throughout the world by copying CDs directly by downloading from the Internet and by retransmitting satellite programming. Given expanding digital technologies, there is little generational loss from one copy to the next, as in the case of analogue forms, and the costs of reproduction are a small fraction of the legitimate market costs. The U.S. government and film corporations have taken a variety of legislative and direct actions to try and curb piracy, usually under the guise of Intellectual Property Rights. Within the corporate realm, two examples are the MPAA (Motion Picture Association of America), which is an aggressive lobbyist on behalf of the major U.S. corporations, and the International Intellectual Property Alliance (IIPA), which is a private-sector coalition that represents U.S. IPR interests, including films, software, music, and videogames sectors; it pursues a variety of public relations and legal actions. Governmental actions include American actions within the World Trade Organization (WTO), the associated General Agreement on Trade in Services (GATS), and the Global Business Dialogue on Electronic Commerce, which works parallel with the WTO and the "Special 301" of the Omnibus Trade and Competitive Act of 1988 which mandates the U.S. Trade Representative to take relevant actions against those countries that are not in compliance with IPR protections. In a rather aggressive move, a bill proposed to the U.S. Congress in 2003 would allow the industry to "hack" into the computers of those individuals and companies suspected of downloading and trading films.

6.   The presumption of cultural homogenization is "a little like arriving by plane but never leaving the terminal, spending all one's time browsing among the global brands of the duty-free shops" (Tomlinson 1999, 6).

REFERENCES

Acheson, K. and Maule, C.J. (1994). Understanding Hollywoods Organization and Continuing Success, Journal of Cultural Economics 18: 271-300.

Aitken, S.C. (1991). A Transactional Geography of the Image-Event: The Films of Scottish Director Bill Forsyth, Transactions of the Institute of British Geographers 16: 105-18.

Aitken, S.C. and Zonn, L.E. (1994). Re-presenting the Place Pastiche. In Aitken, S.C. and Zonn L.E. (Eds.) Place, Power, Situation and Spectacle: A Geography of Film, 3-25. Lanham, MD: Rowman and Littlefield.

Aitken, S.C. and Zonn, L.E. (1993). Weir(d). Sex: Representation of Gender-Environment Relations in Peter WeirsPicnic at Hanging Rock and Gallipoli, Environment and Planning D: Society and Space 11: 191-212.

Allen, R.C. (1977). Film History: The Narrow Discourse, Film Studies Annual: Part II: Film Historical-Theoretical Speculations, 9-17.

Allen, R.C. (1982). Motion Picture Exhibition in Manhattan, 1906-1912: Beyond the Nickelodeon. In Kindem, G. (Ed.) The American Movie Industry: The Business of Motion Pictures, 12-24. Carbondale: Southern Illinois University Press.

Anderson, A.D. (2001). Asian Martial Arts Cinema, Dance, and the Cultural Languages of Gender, Asian Journal of Communication 11: 58-78.

Armes, R. (1987). Third World Film Making and the West. Berkeley: University of California Press.

Balio, T. (1993). Grand Design: Hollywood as a Modern Business Enterprise 1930-1939. New York: Scribners.

Barlet, O. (2001). African Cinemas: Decolonizing the Gaze (Turner, C., trans.). London: Zed Books.

Bazin, A. (1951). What is Cinema? Berkeley: University of California Press.

Benjamin, W. (1969). Illuminations: Essays and Reflections. (Arendt, H. Ed.; Zohn, H. trans.). New York: Schocken Books.

Benton, L. (1995). Will the Real/Reel Los Angeles Please Stand Up? Urban Geography 16: 144-64.

Bingham, N. (1996). Objections: From Technological Determinism Towards Geographies of Relations, Environment and Planning D: Society and Space 14: 635-57.

Bordwell, D., Staiger, J., and Thompson, K. (1985). The Classical Hollywood Cinema: Film Style and Mode of Production. London: Routledge.

Branigan, E. (1979). Color and Cinema: Problems in the Writing of History, Film Reader 4: 16-34.

Buscombe, E. (1978). Sound and Color, Jump Cut 17: 23-25.

Castells, M. (1997). The Rise of the Network Society. New York: Blackwell.

Chakravarty, S. (1998). National Identity in Indian Popular Cinema, 1947-1987. Austin: University of Texas Press.

Clarke, D. (Ed.) (1997). The Cinematic City. London: Routledge.

Commolli, J.L. (1971). Technique et idéologie, Cahiers du Cinéma 229 (May-June): 4-21.

Cresswell, T. and Dixon, D. (Eds.) (2002). Engaging Film: Mobility, Identity, Pedagogy. Lanham, MD: Rowman and Littlefield.

De Lauretis, T. (1987). Technologies of Gender: Essays on Theory, Film, and Fiction. Bloomington: Indiana University Press.

De Lauretis, T. (1984). Alice Doesnt: Feminism, Semiotics, Cinema. Bloomington: Indiana University Press.

Diawara, M. (1992). African Cinema: Politics and Culture. Bloomington: University of Indiana Press.

Dieterle, W. (1941). Hollywood and the European Crisis, Studies in Philosophy and Social Science 9: 96-103.

Dissanayake, W. (1994). Colonialism and Nationalism in Asian Cinema. Bloomington: University of Indiana Press.

Elsaesser, T. (1989). New German Cinema: A History. London: British Film Institute.

Escobar, A. (1994). Welcome to Cyberia: Notes on the Anthropology of Cyberculture, Current Anthropology 35: 211-31.

Ewen, S. and Ewen, E. (1982). Channels of Desire. New York: McGraw-Hill.

Fielding, R. (Ed.) (1967). A Technological History of Motion Pictures and Television. Berkeley and Los Angeles: University of California Press.

Friedberg, A. (1993). Window Shopping: Cinema and the Postmodern. Berkeley: University of California Press.

Gabriel, T.H. (1991). Third Cinema in the Third World: The Aesthetic of Liberation. London: Umi Research Press.

Gold, J. (2002). The Real Thing? Contesting the Myth of Documentary Realism Through Classroom Analysis of Films on Planning and Reconstruction. In Cresswell, T. and Dixon, D. (Eds.) Engaging Film: Mobility, Identity, Pedagogy, 209-25. Lanham, MD: Rowman and Littlefield.

Gomery, D. (1992). Shared Pleasures: A History of Movie Presentation in the United States. Madison: University of Wisconsin Press.

Gomery, D. (2000). Hollywood as Industry. In Hill, J. and Church Gibson: (Eds.) American Cinema and Hollywood: Critical Approaches, 19-28. Oxford: Oxford University Press.

Grau, R. (1914). Theater of Science: A Volume of Progress and Achievement in the Motion Picture Industry. New York: Broadway.

Guback, T.H. (1969). The International Film Industry. Bloomington: University of Indiana Press.

Hampton, B. (1931). [Repr. 1970] History of the American Film Industry from its Beginnings to 1931. New York: Dover.

Hanna, S.P. (1996). Is it Roslyn or is it Cicely?: Representations and the Ambiguity of Place, Urban Geography 17: 633-49.

Hansen, M. (1991). Babel and Babylon: Spectatorship in American Silent Film. Cambridge, MA: Harvard University Press.

Harley, J.E. (1940). World-Wide Influences of the Cinema: A Study of Official Censorship and the International Cultural Aspects of Motion Pictures. Los Angeles: University of Southern California Press.

Harvey, D. (1989). The Condition of Postmodernity. London: Blackwell.

Hay, J. (1987). Popular Film Culture in Fascist Italy: The Passing of the Rex. Bloomington: Indiana University Press.

Haynes, J. (Ed.) (1997). Nigerian Video Films. Ibadan: Kraft Books for the Nigerian Film Corporation.

Heider, K.G. (1991). Indonesian Cinema: National Culture on Screen. Honolulu: University of Hawaii Press.

Himpele, J.D. (1996). Film Distribution as Media: Mapping Difference in the Bolivian Cinemascape, Visual Anthropology Review 12: 47-66.

Hjort, M. and Mackenzie, S. (Eds.) (2000). Cinema and Nation. New York: Routledge.

Igloolik Asuma Productions (2000). Briefing Paper on Development of a Nunavut Film Industry. http://www.isuma.ca/news/correspondance/film_industry_pdf. PDF accessed April 6, 2003.

Jenkins, H. (1995). Historical Poetics. In Hollows, J. and Jancovich, M. (Eds.) Approaches to Popular Film, 99-122. Manchester: Manchester University Press.

Kaplan, E.A. (1992). Motherhood and Representation: The Mother in Popular Culture and Melodrama. New York: Routledge.

Kracauer, S. (1995). The Mass Ornament: Weimar Essays (Levin, T.Y., Ed. and trans.). Cambridge, MA: Harvard University Press.

Latour, B. (1997). On Actor-Network Theory: A Few Clarifications. Keele, UK: STOT Resources, Center for Social Theory and Technology, Keele University.

Law, J. (1992). Notes on the Theory of Actor Network: Ordering, Strategy and Heterogeneity, Systems Practice 5: 379-93.

MacCabe, C. (1974). Realism and Cinema: Notes on Some Brechtian Theses, Screen 15/2 (Autumn): 7-32.

Macdonald, G.M. (1994). Third Cinema and the Third World. In Aitken, S.C. and Zonn, L.E. (Eds.) Place, Power, Situation and Spectacle: A Geography of Film 27-46. Lanham, MD: Rowman and Littlefield.

Macgowan, K. (1965). Behind the Screen. New York: Delta.

Massey, D. (1993). Power Geometry and a Progressive Sense of Place. In Bird, J., Curtis, B., Putman, T., Robertson, G., and Tickner, L. (Eds.) Mapping the Futures, 59-69. London: Routledge.

Mast, G. and Kawin, B.F. (2002). A Short History of the Movies. London: Longman.

May, L. (1983). Screening Out the Past: The Birth of Mass Culture and the Motion Picture Industry. Chicago: University of Chicago Press.

Merritt, R. (1976). Nickelodeon Theaters 1905-1914: Building an Audience for the Movies. In Balio, T. (Ed.) The American Film Industry, 59-79. Madison: University of Wisconsin Press.

Metz, C. (1974). Film Language: A Semiotics of the Cinema. New York: Oxford University Press.

Miller, T. (2000). Hollywood and the World. In Hill, J. and Church Gibson (Eds.) American Cinema and Hollywood: Critical Approaches, 145-55. Oxford: Oxford University Press.

Miller, T., Govil, N., McMurria, J., and Maxwel, R. (2001). Global Hollywood. London: British Film Institute.

Mulvey, L. (1977). Visual and Other Pleasures. London: Macmillan.

Musser, C. (1991). Before the Nickelodeon: Edwin S. Porter and the Edison Manufacturing Company. Berkeley: University of California Press.

Natter, W. (1994). The City as Cinematic Space: Modernism and Place in Berlin: Symphony of a City. In Aitken, S.C. and Zonn, L.E. (Eds.) Place, Power Situation and Spectacle; A Geography of Film, 203-28. Lanham, MD: Rowman and Littlefield.

Natter, W. and Jones II, J.P. (1993). Pets or Meat: Class, Ideology and Space in Roger and Me, Antipode 25: 140-58.

Neale, S. and Smith, M. (Eds.) (1998). Contemporary Hollywood Cinema. London: Routledge.

Neumann, M. (2001). Emigrating to New York in 3-D: Stereoscopic Vision in IMAXs Cinematic City. In Shiel, M. and Fitzmaurice, T. (Eds.) Cinema and the City: Film and Urban Societies in A Global Context, 109-21. Oxford: Blackwell.

Oha, O. (2001). The Visual Rhetoric of the Ambivalent City in Nigerian Video Films. In Shiel, M. and Fitzmaurice, T. (Eds.) Cinema and the City: Film and Urban Societies in A Global Context, 195-205. Oxford: Blackwell.

Okome, O. and Haynes, J. (Eds.) (1995). Cinema and Social Change in West Africa. Jos, Nigeria: Nigerian Film Corporation.

Onishi, N. (2002). Step Aside, L.A. and Bombay, for Nollywood, New York Times, September 16, A1.

Parkinson, D. (1996). The History of Film. London: Thames and Hudson.

Ramsaye, T. (1926). A Million and One Nights (2 vols). New York: Simon and Shuster.

Ray, R. (1985). A Certain Tendency of the Hollywood Cinema, 1930-1980. Princeton, NJ: Princeton University Press.

Rockwell, J. (1994). The New Colossus: American Culture as Power Export, New York Times January 30: H1 and H30.

Schatz, T. (1988). The Genius of the System: Hollywood Filmmaking in the Studio Era. New York: Pantheon.

Sen, K. (1994). Indonesian Cinema: Framing the New Order. London: Zed.

Shiel, M. and Fitzmaurice, T. (Eds.) (2001). Cinema and the City: Film and Urban Societies in A Global Context. Oxford: Blackwell.

Shohat, E. and Stam, R. (1994). Unthinking Eurocentrism: Multiculturalism and the Media. London: Routledge.

Smith, L. (2002). Chips Off the Old Ice Block: Nanook of the North and the Relocation of Cultural Identity. In Cresswell, T. and Dixon, D. (Eds.) Engaging Film: Mobility, Identity, Pedagogy, 94-122. Lanham, MD: Rowman and Littlefield.

Smith, M.P. (2001). Transnational Urbanism: Locating Globalization. New York: Blackwell.

Spellerberg, J. (1979). Technology and Ideology in the Cinema, Quarterly Review of Film Studies 2/3 (August): 288-301.

Staddon, C. et al. (2002). Using Film as a Tool in Critical Pedagogy: References on the Experience of Students and Lecturer. In Cresswell, T. and Dixon, D. (Eds.) Engaging Film: Mobility, Identity, Pedagogy, 271-95. Lanham, MD: Rowman and Littlefield.

Staiger, J. (1982). Dividing Labor for Production Control: Thomas Ince and the Rise of the Studio System. In Kindem, G. (Ed.) The American Movie Industry: The Business of Motion Pictures, 94-103. Carbondale: Southern Illinois University Press.

Strathern. M. (1996). Cutting the Network, Journal of the Royal Anthropological Institute 2: 517-35.

Strohmeyer, U. (2002). Practising Film: The Autonomy of Images in Les Amants du Pont-Neuf. In Cresswell, T. and Dixon, D. (Eds.) Engaging Film: Mobility, Identity, Pedagogy, 193-208. Lanham, MD: Rowman and Littlefield.

Swann: (2001). From Workshop to Backlot: The Greater Philadelphia Film Office. In Shiel, M. and Fitzmaurice, T. (Eds.) Cinema and the City: Film and Urban Societies in a Global Context, 88-98. Oxford: Blackwell.

Thompson, K. (1985). Exporting Entertainment: America in the World Film Market, 1907-34. London: British Film Institute.

Tomlinson, J. (1999). Globalization and Culture. Chicago: University of Chicago Press.

Waller, G. (1995). Main Street Amusements: Movies and Commercial Entertainment in a Southern City, 1896-1930. Washington: Smithsonian Institution Press.

Wasko, J. (1982). Movies and Money. Norwood, NJ: Ablex.

Wasko, J. (1995). Hollywood in the Information Age. Austin: University of Texas Press.

Wayne, M. (2001). Political Film: The Dialectics of Third Cinema. London: Pluto Press.

Yan, Y. (2000). Of Hamburger and Social Space: Consuming McDonalds in Beijing. In Davis, D.S. (Ed.) The Consumer Revolution in Urban China, 201-25. Berkeley: University of California Press.

Zonn, L.E. (1985). Images of Place: A Geography of Media, Proceedings of the Royal Geographical Society of Australasia (South Australian Branch). 84: 34-45.

Zonn, L.E. (1984). Landscape Depiction and Perception: A Transactional Approach, Landscape Journal 3: 144-50.

CHAPTER 12

JAMES M. RUBENSTEIN

MOTOR VEHICLES ON THE AMERICAN LANDSCAPE

Abstract   Geographers have studied the impacts of automobiles and related economies at various scales during the past century. Among the major topics investigated are industrial sites, influences on urban and suburban morphology, and the automobile in American culture and society. The automobile has been a catalyst for changing class, race, and gender relations in rural and urban settings. Entire new service economics, including leisure, entertainment, lodging, and fast foods, have emerged, all reflective of a machine-dependent technology. The increased reliance on motor vehicles has raised social and environmental awareness, especially on pollution, dependency on international producers of fossil fuels, traffic congestion, and urban sprawl. Many of the features of the American motor vehicle landscape are diffusing to other world regions and major cities.

Keywords   motor vehicles, landscape, highways, urbanization, Fordist production, pollution

1. INTRODUCTION

When the Association of American Geographers met for the first time in 1904 in Philadelphia, the motor vehicle was a novelty toy for the very rich. The twenty-six geographers at the AAG organizational meeting undoubtedly arrived in Philadelphia by rail from New York, Boston, Chicago, or elsewhere, and transferred from the train station to the meeting site by horse-drawn carriage or streetcar.

Geographers in 1904 would not have appreciated the central role about to be played by the motor-vehicle industry in restructuring the distribution and organization of U.S. manufacturing. The Ford Motor Co., which would dominate U.S. motor-vehicle production for the first quarter of the twentieth century, had been incorporated in 1903, only one year earlier than the AAG. Also incorporated in 1903 were the Buick Motor Co. and the Cadillac Motor Car Co., combined in 1908 to form General Motors, the dominant motor-vehicle producer of the second and third quarters of the twentieth century. In 1904, Michigan had just surpassed Connecticut as the leading car-producing

Stanley D. Brunn, Susan L. Cutter, and J.W. Harrington, Jr. (Eds.), Geography and Technology,
267-284. © 2004 Kluwer Academic Publishers. Printed in the Netherlands.

state, and Ford's mass production innovations were still a decade in the future (Rubenstein 1992).

Physical geography topics dominated the twenty papers at the first AAG gathering. None of the papers concerned motor vehicles, although a physical geographer, Albert Perry Brigham of Colgate University, delivered one of the four nonphysical geography lectures at the first AAG meeting, "The Development of the Great Roads across the Appalachians," a historical preautomotive survey. Brigham earlier in 1904 published the first article in a major geographic journal containing the word *automobile*, "Good Roads in the United States," in the *Bulletin of the American Geographical Society*.

The condition of early twentieth-century U.S. roads was "terrible" (Davies 2002: 11). Brigham's "Good Roads" paper argued that improving roads in the U.S. would bring economic and social benefits. Economically, Brigham found that the cost per ton-mile of hauling crops from farm to market was 25¢ by road, much higher than 3/4¢ by rail, and less than 1/4¢ by water, but the benefits of higher farm property values and market prices would more than offset the cost of constructing roads. Socially, Brigham claimed that good roads would improve rural life in such areas as education (better access to consolidated schools), morals (better access to churches), and citizenship (better access to polling stations). Good roads would encourage rural families, especially young people, to remain on the farm, and urban families to take more vacations.

Perhaps a few of the two dozen geographers arrived at the initial AAG meeting in one of Philadelphia's electric taxis, then beginning to compete for space on the crowded streets of Philadelphia and other large cities. Most of the 45,000 motor vehicles in the U.S. in 1904, about half electric-powered, were confined to large cities because of a lack of paved roads elsewhere.

1.1. The Motor Vehicle at the AAG's 25th and 50th Anniversaries

At the AAG's 25th anniversary meeting in 1929, a geographer could have argued with considerable justification that the single most distinctive feature of the U.S. landscape, compared with that of the rest of the world, was the large number of motor vehicles, but none did. The motor vehicle was not a major concern of academic geographers during the early twentieth century despite, or perhaps because of, its extremely rapid diffusion in the U.S. During the 1920s, the world had about 20 million motor vehicles, of which about 17 million were registered in the U.S. and another half-million were in Canada (Rubenstein 2001).

Industrial geography attracted the interest of a number of geographers at the 1929 AAG meeting. The U.S. then had nearly as many motor vehicles as households and was in process of getting the good roads advocated by Brigham a quarter-century earlier. But the railroad and shipping industries were cited as principal transportation influences in studies of Cleveland, St. Louis, and the Twin Cities (the latter by Richard Hartshorne). Not mentioned was the presence in all three communities of Ford Motor Co. assembly plants, let alone Ford's impact on mass production, such as the moving assembly line, sequencing of work tasks, and deskilling of the work force. Nor were industrial geographers in 1929 discussing the implications of Ford's recently constructed River Rouge complex outside Detroit, where 100,000 people worked in 11 million square feet of buildings spread over 2,000 acres, the world's largest manufacturing facility.

John Orchard's "Can Japan Develop Industrially?" also presented at the 1929 AAG meeting, ignored Japan's motor vehicle industry, then dominated by Ford and GM as in nearly every country prior to World War II. Ford produced one-half of Japan's cars in the 1920s and 1930s, GM one-third. While access to motor vehicles was nearly universal in the U.S. by 1929, it would remain rare in the rest of the world for another three decades.

As late as 1950, 60 percent of the world's motor vehicles were registered in the U.S. and 80 percent were manufactured there. Even in Europe, where late nineteenth-century producers like Daimler, Benz, Renault, and Peugeot were key motor-vehicle innovators, car ownership rates remained low until the 1950s, restricted by high price and limited production primarily designed for aristocrats. Hitler's vaunted autobahns were used primarily by the military, and Volkswagen produced only prototypes of its "People's Car" until 1949.

The rapid diffusion of the motor vehicle into U.S. society and economy did not go unnoticed by other social scientists during the 1920s. Robert and Helen Lynd's *Middletown: A Study in American Culture* (1929), which compared Muncie, Indiana, in 1890 and 1924, described the centrality of the motor vehicle in daily life, which was neatly summarized by one interviewee: "We'd rather do without clothes than give up the car." A U.S. Department of Agriculture study during the 1920s found that many farm families acquired motor vehicles before installing indoor plumbing or electricity. In the words of another interviewee: "You can't go to town in a bathtub."

Nearly universal access to motor vehicles transformed gender relations in the U.S. during the first quarter of the twentieth century. When the AAG first met in 1904, female motorists were nearly as rare as female geographers. A woman driver was regarded as somewhere between undignified and scandalous. Obstacles were partly practical: starting the vehicle by turning a crank in front and then racing into the driver's seat demanded strength, agility, and a sense of danger that few men, let alone women, possessed in 1904, and the open-carriage body style offered no protection from mud, soot, and rain.

The self-starter and closed passenger compartment introduced during the 1910s made driving safer and more pleasant for women. To appeal to women, exteriors were brightly painted and interiors were comfortably appointed during the 1920s. With nearly all middle-class households in possession of one by the 1920s in the U.S., the car was turned over to the housewife during the week to chauffeur the children and collect groceries, while the husband took the train or streetcar to work. The husband took back the wheel on the weekend for leisure trips. The husband paid for the car, but because the wife told him which one to buy, carmakers started pitching their advertisements to women. Many Americans during the 1920s regarded the eagerness of women to drive or be driven in a car as both evidence and a result of "loose" behavior. The car enticed a woman to be courted by a man, whisked away to wild parties, and "ravished" by a "masher" in the language of the day (Rubenstein 2001).

The car was providing African-Americans with the most visible expression of personal freedom in the segregated America of the 1920s. Especially satisfying for African-Americans, who had migrated from the rural south to Detroit and other northern cities during the 1920s, was to return home for a visit driving a new car. Friends and family would turn out in the streets of an impoverished community, not even electrified, to greet the motorist as a hero who had overcome "the back of the bus." Michigan's carmakers hired especially large numbers of African-Americans. One-third of Ford's workforce was African-American, and Henry Ford constructed the city of Inkster for his African-American workers who were prohibited from living in all-white Dearborn (Rubenstein 2001).

Economic studies of the impact of the motor vehicle also abounded during the 1920s. The U.S. Department of Agriculture found that 40 percent of U.S. farmers were shipping crops to market by truck during the 1920s, allowing them to double shipping distances and halve shipping costs per ton-mile. One-half million children in rural areas were bused to schools in 1926, sparking rapid consolidation of rural schools in small towns. The 1930 U.S. Census found that 222 of 982 cities with at least 10,000 inhabitants had no public transit so were already entirely dependent on the private motor vehicle for transportation.

Many of the papers at the 50th annual meeting of the AAG in 1954 in Philadelphia purported to offer overviews of academic geography's first half-century. Motor vehicles remained of limited interest to the fifty-plus presenters, although the influences of the railroad and shipping industries were again extensively documented. One exception, Chauncy D. Harris's "The Last Fifty Years and the Next" (1954), noted that motor vehicles had caused major modifications in the layout of cities, that their manufacturers constituted one of a number of "new industrial giants," and that they added "diversity" to the nation's transportation network.

Harris's 50th anniversary "diversity" observation was noticeably ill-timed: streetcar tracks were being dismantled, rail companies were falling into bankruptcy, and the 50,000-mile (80,000-km) interstate highway system was being launched. On hindsight, the appearance of "diversity" in the U.S. transportation system during the 1950s was in reality a snapshot of rail and other public transit approaching death, and the motor vehicle approaching complete triumph.

1.2. The Motor Vehicle at the AAG's 100th Anniversary

Geographers during the AAG's second half-century no longer ignored the role of the motor vehicle in shaping the American landscape. Geographers during the 1950s began to document the distinctive spatial distribution of the U.S. motor-vehicle production, in which most parts were made in Michigan and shipped to branch assembly plants near major population centers (Boas 1961; Henrickson 1951; Hurley 1959).

Ron Horvath in 1974 called the automobile "the single most significant innovation in American culture during the twentieth century." The automobile, he wrote, "is so shrouded in myth and has so much symbolic meaning that we could call it our 'sacred cow'" (Horvath 1974, 168).

For the most part, rather than being the principal object of study, the motor vehicle invariably lurked behind the scenes of geographic inquiry as the explanatory agent (or culprit) underlying patterns and processes observed in the landscape and organization of space. Given its near-universal ownership and dominance of the nation's transportation system, the motor vehicle was frequently called upon by geographers to serve as key independent variables in the analysis of some other culture, physical, economic, or environmental phenomenon, but rarely as the dependent variable.

Geographers have observed the impact of motor vehicles on the landscape at four scales: local, regional, national, and international. At the local scale, Americans place on the landscape objects such as wide streets, multicar garages, and drive-through restaurants that accommodate and respond to the characteristics of the motor vehicle. At the regional scale, near-universal ownership of motor vehicles has been both a cause and an effect of suburban sprawl in urban areas, and tourism and depopulation in selected rural areas.

At the national scale, the motor vehicle is one of the most important icons of U.S. popular culture, and an agent of transforming social class, gender, and race relations. Motor vehicles came to dominate the U.S. landscape and culture during the first half of the twentieth century while still rare elsewhere, and during the 1990s, the U.S. became the first populous country with more motor vehicles than licensed drivers.

At the international scale, motor vehicles have long since diffused from the U.S. to other world regions. Increasing motor-vehicle usage is also central to understanding global-scale issues of concern to geographers such as air pollution and depletion of fossil fuel resources.

2. THE MOTOR VEHICLE AT THE LOCAL SCALE

The motor vehicle was nearly but not completely invisible in the AAG's seventy-fifth anniversary retrospective, published in the March 1979 *Annals*. In a photograph accompanying a description of his pioneering work on folk housing, Fred B. Kniffen was shown (p. 59) in 1965 utilizing the most important yet least appreciated tool for geographic inquiry of local-scale cultural landscape: a car.

In John Fraser Hart's review of geography in the 1950s, Hart (1979) commented that "Highways [in the 1950s] were fairly slow two-lane affairs, for the most part, and they gave you a chance to get the feel of the country" (p. 111). Most geographers, according to Hart, still traveled to national meetings and other distant locations by train during the 1950s. The role of motor vehicles in revolutionizing fieldwork for cultural geographers, such as Kniffen and Sauer, let alone physical geographers, was mentioned only in passing in the *Annals'* other essays.

The car enabled geographers to observe the landscape, record its features, and ultimately analyze its distinctive patterns. The car promoted efficient fieldwork: more territory could be covered more quickly, so more samples could be collected in less time. The car promoted diverse fieldwork: remote locations not served by railroads were now accessible by car, so fieldwork could be done in a wider variety of places. The car promoted scientific fieldwork: bulky or heavy samples could now be transported from the field to laboratory in sufficiently large quantities to make valid generalizations. The car promoted comfortable fieldwork: enough camping gear or personal belongings could be hauled into the field to eliminate or at least reduce the hazards and discomforts of fieldwork.

Geographers have come to recognize that the organization of particular landscapes in the U.S. can be attributed in large measure to universal usage of motor vehicles (Jakle and Sculle 1994, 1999; Jakle et al. 1996). Ironically, of particular interest to geographers has been the automotive-influenced landscape of the early twentieth century, the distinctiveness of which earlier generations of geographers somehow didn't appreciate. Inside cities, elaborate palaces were built as showrooms, clustered along "Automobile Row," such as South

Michigan Avenue in Chicago, Broadway in Midtown Manhattan, North Broad Street in Philadelphia, and East Jefferson Street in Detroit.

What Horvath called "automobile territory" accounted for two-thirds of the land area in downtown East Lansing and Detroit in 1971. Automobile-related "machine space" included streets and other road surfaces, parking spaces, garages, and automobile-related retailers such as gasoline stations and supply stores (Horvath 1974).

Drive-through off-street gasoline stations became common in 1920s cities, and there were substantial ornate structures designed to convey confidence in the product, much as banks were designed like Greek temples. Uniformed attendants pumped the gasoline, checked the tire pressure, filled the radiator with water, cleaned the windshield, and emptied the ashtray. Off-street stations replaced curbside pumps that had become hazards because of proximity of sparks from speeding cars to flammable gasoline and blockage of the street as cars waited in line.

Historical and cultural geographers have been especially interested in the rural and intercity automotive landscapes. Lincoln Highway (later U.S. 30) was built in sections as an act of patriotism (Davies 2002). The National Road (later U.S. 40) incorporated portions of the early nineteenth-century engineering marvel (Raitz 1996). The National Old Trails Roads that had carried pioneers to southern California in the late nineteenth century were strung together in 1926 as a U.S. highway, paved during the 1930s by the Works Progress Administration, and immortalized in popular songs and television programs during the 1950s under its more prosaic official name, Route 66. Advertisements aimed at motorists were painted on barns and planted alongside the roads, most famously Burma-Shave's sequences of five signs— four rhyming lines of a poem, then a wrap-up—beginning in Minnesota in 1925.

Three types of services sprouted along the two-lane intercity roads to serve motorists: gasoline stations, lodging houses, and restaurants. Gas stations proliferated the earliest because they were essential to long-distance travel: motorists could not carry along their own supplies. After obtaining the breakup of the Standard Oil Trust in 1911, the U.S. government demanded competition in gasoline retailing, thereby assuring that the highway landscape would feature a multiplicity of gasoline stations next door to each other at regular intervals.

Roadside campgrounds evolved into motor hotels or motels, locally owned and operated. Through the decades, these lodgings attracted motorists through such amenities as cleanliness and radios during the 1920s, swimming pools during the 1930s, air conditioning during the 1940s, and television during the 1950s.

Restaurants catering primarily to motorists were also owned by local families. At first, the traditional style of eating prevailed, in which the motorist parked the vehicle and ate in the restaurant. Distinctively automotive forms of serving the food soon emerged. During the 1950s, waitresses came to the car, took the order, and served the food on a tray attached to the car. A generation later, the drive-in restaurant gave way to the drive-through restaurant, in which the motorist collected the food at a window. The first franchised restaurant chain, Steak 'n Shake, originated in St. Louis during the 1930s. Fanciful architecture was especially popular during the 1950s, such as a hot-dog stand shaped like a giant hot dog.

With construction of interstate highways, services were clustered at interchanges rather than along the side of the road. The gasoline, lodging, and restaurant retailers were organized into what Jakle and Sculle called "placed-product-packaging." National retail chains provided local franchises with standardized architecture, décor, product, service, and operating routine in a deliberate attempt to make the landscape familiar and predictable to passing motorists. Gasoline was refined to the same formula, motels were equipped with the same furniture, and restaurants offered the same menu. National chains of retailers emerged first in gasoline during the 1910s and became important in lodging in the 1950s and in fast-food in the 1960s. Corporations awarded exclusive franchises for local trade territories to promote competition with other brands and minimize cannibalizing other franchises of the same brand.

3. THE MOTOR VEHICLE AT THE REGIONAL SCALE

Academic geographers entered the second half-century complacent about the regional-scale effects of motor vehicles. Jean Gottmann's 1961 landmark *Megalopolis* helped to focus geographic concepts on the patterns and processes of suburbanization; it also contributed a compelling term to the geographic vocabulary. The 800-page book meticulously documented the coalescence of northeastern urban areas through spread of land area and population, but curiously downplayed the role of motor vehicles. Gottmann believed that motor vehicles had peaked in importance in 1960 and usage would decline because people regarded operating costs as too high, road congestion as intolerable, and their use in leisure activities as less alluring (Gottmann 1961, 684).

Outside academic geography, critics had already opened fire on the motor vehicle. The same year that *Megalopolis* was published, Jane Jacobs wrote in *Death and Life of Great American Cities*:

> Today everyone who values cities is disturbed by automobiles. Traffic arteries, along with parking lots, gas stations and drive-ins, are powerful and insistent instruments of city destruction. To accommodate them, city streets are broken down

into loose sprawls, incoherent and vacuous for anyone afoot. Downtowns and other neighborhoods that are marvels of close-grained intricacy and compact mutual support are casually disemboweled. (Jacobs 1961, 338)

Lewis Mumford was harsher in his essay *The Highway and the City*:

When the American people through their Congress, voted [in 1956] for a twenty-six-billion-dollar highway program, the most charitable thing to assume about this action is that they hadn't the faintest notion of what they were doing. Within the next fifteen years they will doubtless find out; but by that time it will be too late to correct all the damage to our cities and our countryside, not least to the efficient organization of industry and transportation, that this ill-conceived and preposterously unbalanced program will have wrought...The fatal mistake we have been making is to sacrifice every other form of transportation to the private motorcar. (Mumford 1963, 244, 247)

As Jacobs pointed out, however, "we blame automobiles for too much" (Jacobs 1961, 338). The motor vehicle was not responsible for luring Americans from rural to urban areas and from cities to suburbs, rather it was the means to a desired end.

The motor vehicle was actually more of a savior than a destroyer of cities during the first half of the twentieth century. The streets of U.S. cities were congested with horse-drawn and electric vehicles in 1900, and teeming throngs of humanity jostled on sidewalks.

The first impression which a stranger receives on arriving in Chicago is that of the dirt, the danger and the inconvenience of the streets. (Cook, et al. 1973, 62; Stead 1894)

The confusing rattle of busses and wagons over the granite pavement in Broadway almost drowns his own thoughts, and if he should desire to cross the street a thousand misgivings will assail him...although he sees scores of men and women constantly passing through the moving line of vehicles. (Buel 1882, 26; Schlesinger 1933, 87)

As motor vehicles replaced horses and streetcars in U.S. cities during the first half of the twentieth century, people were able to travel faster. During rush hour, average traffic speed in cities increased from less than 5 miles per hour in 1900 to 20 mph in the 1950s and 30 mph in the 1960s. Even in Manhattan, rush-hour speeds increased during the first half of the twentieth century from less than 2 mph to more than 11 mph. Outside rush hour and in smaller cities, speeds increased more dramatically (Meyer, et al. 1965, 68).

The U.S. population increased by about 125 million in the second half of the twentieth century, and all of that growth went to suburbs, while population remained about the same in central cities and nonmetropolitan areas. Americans were about evenly divided, living in central cities, suburbs, and nonmetropolitan areas in 1950, compared to 60 percent in suburbs and only

20 percent each in central cities and nonmetropolitan areas in 2000. When Chauncy Harris updated his multiple nuclei model in 1997, forty-two years after first describing it with Edward L. Ullman, urban geographers gave the motor vehicle its due. Surrounding the old central city in the multiple nuclei model was now a suburban circumferential beltway (Harris 1997; Harris and Ullman 1945).

Taking advantage of increased speeds afforded by motor vehicles and highways, people chose to make more trips of longer duration rather than reduce travel time. In 1950, before construction of the interstate highways, 150 million Americans drove 48 million vehicles a total of 458 billion miles on 2 million miles of paved roads. In 2000, 275 million Americans drove 220 million vehicles a total of 2.5 trillion miles on 4 million miles of paved roads. Thus, between 1950 and 2000, the number of Americans nearly doubled, the number of roads doubled, the number of vehicles more than quadrupled, and the number of miles driven more than quintupled (Rubenstein 2001, 315).

Americans coped with congestion not by altering their driving patterns but by regarding their motor vehicles as an important expression of personal space. Increasingly elaborate entertainment systems were placed in the vehicles: AM radios (popular since the 1930s) were supplemented with FM radios, tape decks, CD players, satellite entertainment services, and Internet access. Interior space was designed to accommodate beverages, and drive-through restaurants packaged food for convenient in-car consumption.

According to one prediction,

> by the end of the 21st Century...several generations of drivers will perceive the automobile not as a mechanical device that transports them to and from work. Rather, they will see the automobile as a bundle of complicated electronic, cellular and satellite gadgetry that keeps them—for better or worse—in communication with bosses, friends and even government authorities. (Konrad 1999)

4. THE MOTOR VEHICLE AT THE NATIONAL SCALE

The most visible contribution of the motor vehicle at the national scale in the U.S. has been the creation of a "car culture." Motor vehicles have figured prominently in popular culture, including television, film, and sport. In the 1950s, G.M. extolled Americans to purchase a Chevrolet as a patriotic act. Dinah Shore and Pat Boone sang:

> See the U.S.A. in your Chevrolet,
> America is asking you to call.
> Drive your Chevrolet through the U.S.A.,
> America's the greatest land of all.

Stock car racing, including the National Association for Stock Car Racing (NASCAR), Championship Auto Racing Teams (CART), and the Indy Racing League (IRL) together attracted more than $1 billion in sponsorship revenue in 2000 and drew higher television ratings in the U.S. than any other sport with the exception of football. Popularity is especially high in the U.S. for racing of vehicles that appear similar to mass produced models.

Most significant to national-scale geographic inquiry has been the term "Fordism" or "Fordist production" as shorthand for the distinctive characteristics of industrial production prevailing in the U.S. for much of the twentieth century. Widespread use of the term recognizes the central role of motor vehicle producers, especially the Ford Motor Co., in creating the twentieth century's dominant mode of industrial production (Clark 1986; Dicken 1998).

Mass production was not invented by the motor vehicle industry, but the U.S. motor vehicle industry made three key contributions. The first was the invention of methods for making large quantities of essentially identical products efficiently and inexpensively. Most famously was the moving assembly line which Ford Motor Co. installed at its Highland Park, Michigan, plant in 1914. Several years before, Ford had also pioneered arranging work tasks in a logical sequence and placing specialized machine tools next to the workers.

Second, the U.S. motor vehicle industry also pioneered vertical integration, the creation of corporations that maintained tight control over all phases of a highly complex production process, from initial research to a final sale. General Motors acquired partsmakers scattered through the Great Lakes region, whereas Ford clustered all phases of production at its enormous River Rouge complex.

Third was the attraction, retention, and fashioning of a large supply of workers who were minimally skilled yet highly productive. Deskilling the workforce took away the control over the workplace exercised by early auto workers who were skilled craftspeople with a scarce talent. Some automotive executives, although not Henry Ford, were familiar with the work of Frederick Taylor in fashioning work tasks.

The term "Fordism" was actually first used during the 1920s in the Soviet Union and Germany to describe both the Ford-inspired mass production methods as well as the cult of personality of Henry Ford, who established the company in 1903 and ran it until his death in 1947. Ford was a hero in the Soviet Union, where he was a major investor soon after the Russian Revolution. Nazi Germany awarded Henry Ford the Grand Cross of the German Eagle, the highest possible honor for a foreigner, in 1938, five months after the Austrian anschluss and one month before the Munich pact (Rubenstein 2001).

When "Fordism" crept into U.S. geography literature around 1980, the connection to Henry Ford's personality conveyed by the Russian and German language terms was gone. Instead, "Fordism" was contrasted with "Post-Fordism," a Japanese-inspired "flexible" or "lean" mode of production whose ascendancy in the 1970s was a cause and consequence of Western capitalism's crisis in the 1970s.

The automotive industry favored the term Toyota Production System, in honor of the company credited as the earliest and best practitioner of lean production. Toyota pioneered producing a large variety of products in small batches with machines that could be retooled quickly to meet a changing market. Key tasks were outsourced to independent suppliers, and workers were organized into teams with a variety of tasks and responsibilities. The result was better quality products built more efficiently.

Geographers noted spatial implications of the motor-vehicle industry's restructuring (Bloomfield 1978, 1981; Glasmeier and McCluskey 1988; Hoffman and Kaplinsky 1988; Holmes 1983; Rubenstein 1992, 2001). Final assembly plants were closed along the East and West Coasts and constructed in the interior, mostly in rural and Southern communities not historically associated with motor vehicle manufacturing. Partsmakers also opened new facilities in the U.S. interior to assure just-in-time delivery to the assembly plants.

As Fordist and post-Fordist production became an accepted dichotomy in geographic literature, inside the motor vehicle industry itself, differences between the two modes of production were blurring into what was called in the industry "optimum lean production." U.S. and European companies adopted optimum lean production by blending key elements of the Toyota Production System with mass production, thereby narrowing the productivity and quality gaps.

Japanese companies moved toward optimum lean production for a different reason; they were building high-quality vehicles efficiently but were making lower profits during the 1990s than their American and European competitors, and they were losing market share among younger customers. Key mass-production practices were incorporated, such as building larger batches and offering fewer model choices. To raise profits, Toyota started reducing the quality of its vehicles in places, where the company hoped, consumers would not notice, such as by no longer painting the inside of the bumper. Honda survived by selling more cars in North America than in its home market of Japan. The other Japanese companies all sold controlling interests to European and American companies.

Optimum lean production encouraged design of truly "global" vehicles. Under flexible production, particular vehicles were most efficiently built in batches of one-quarter million per year—the capacity of a single assembly plant—so major differences in vehicles sold in North America, Europe, and Japan were tolerated. Under optimum lean production, economies of scale were maximized by manufacturing vehicles in batches of one million per year. In most cases, one million could be sold only by building and offering vehicles in Europe and Japan as well as North America. Most efficient was to operate assembly plants in two or all three of the major market areas, and to purchase most of the parts from a handful of major international suppliers also operating in more than one of the major market areas.

At the same time they pursue the "global car," manufacturers have also accommodated increasingly diverse consumer tastes and preferences. As all vehicles offer reliable transport, sales are increasingly differentiated on the basis of style. Individual brands have widely varying market shares by gender, ethnicity, and age. Market shares also vary between urban and rural areas and between interior and coastal regions.

5. THE MOTOR VEHICLE AT THE GLOBAL SCALE

The motor vehicle figured prominently in global-scale geographic analysis during the late twentieth century, especially around resource issues such as air pollution and petroleum dependency. Geographers fretted that consumer preferences in engineering (internal combustion engine rather than electric power) and styling (large trucks rather than small cars) reflected limited concern for global resource issues.

At the AAG's first meeting in Southern California in 1958, the region's air pollution was already "infamous," according to one of the hosts, UCLA's Howard Nelson. People in Los Angeles were calling the brown haze *smog*, a combination of "smoke" and "fog," although the term was not yet being used in geographic writing. Photochemical smog formed when sunlight mixed with hydrocarbons and nitrogen oxides, although in the 1950s, hydrocarbons received most of the attention. Southern California's three million motor vehicles were cited at the AAG meeting as the principal source of the offending hydrocarbons (Nelson 1959, 98).

Nationally, motor vehicles were held accountable for 50 percent of nitrogen oxides and 60 percent of hydrocarbons emitted into the air in 1960. A pound of fuel burned in a car resulted in discharges of 0.2 pounds of nitrogen oxides and 0.1 pounds of hydrocarbons. Motor-vehicle emissions also generated more than two-thirds of the national total of carbon monoxide. Breathing carbon monoxide reduces the oxygen level in the blood, impairs vision and alertness, and threatens breathing problems.

Nelson told the 1958 AAG meeting that "No immediate solution seems in sight" to reducing hydrocarbon emissions. But three years later, in 1961, the California Motor Vehicle Pollution Board (renamed California Air Resources Board in 1968) started placing emission controls on motor vehicles. CARB required crankcase blowby devices beginning in 1963 and exhaust-control systems to reduce hydrocarbons and carbon monoxide beginning in 1966. The 1970 Clean Air Act followed with nationwide air quality standards and mandated reductions of nitrogen oxide, hydrocarbon, and carbon monoxide emissions in motor vehicles.

Carmakers led by then dominant GM successfully fought most efforts to make vehicles cleaner and safer during the 1960s and 1970s (Nader 1965; Wright 1979). Rather than redesign the engine, carmakers met the Clean Air Act standards through installing catalytic converters and lead-free gasoline. Catalytic converters were first installed on cars sold in California in 1975 and in the rest of the country two years later. Nitrogen oxide and hydrocarbon emissions declined by more than 95 percent between 1970 and 2000, and carbon monoxide emissions declined more than 75 percent. Most gains were realized between the mid-1980s and mid-1990s, once vehicles lacking catalytic converters became scarce.

While addressing one problem, the catalytic converter unintentionally contributed to another environmental problem of concern to the next generation of geographers: global warming. Increased concentrations of trace gases in the atmosphere contribute to global warming by blocking or delaying the return of heat from Earth to space. Concentrations of two trace gases, carbon dioxide and nitrous oxide, increased during the twentieth century. Both gases were being discharged into the atmosphere as a result of chemical processes inside catalytic converters.

Progress on reducing emissions was being realized just as the energy crisis hit. The close connection between the motor vehicle and petroleum industries was scrutinized by political and economic geographers as well as environmentalists. The background to the 1970s energy crisis was clear: The motor vehicle and petroleum industries had flourished together since Texas fields were first exploited in 1901. With gasoline selling for only a few cents a gallon, U.S. carmakers never had an incentive to build fuel-efficient models. The Model T, which accounted for nearly half of all cars sales during the 1910s and early 1920s, achieved less than 20 mpg, and average efficiency for all vehicles in the U.S. was only 15 mpg in 1930 and 12 mpg in 1975.

Once domestic petroleum became relatively expensive to exploit, the U.S. became a net importer of petroleum in 1947, and imports increased from 14 percent in 1954 to 40 percent in 1970. Foreign-owned petroleum fields were nationalized or more tightly controlled, and prices were set by governments, especially acting through OPEC (Organization of Petroleum Exporting Countries), formed in 1960 by the major Middle East producers Iran, Iraq, Kuwait, and Saudi Arabia, plus Venezuela.

After the 1973-74 disruption of petroleum supplies, the U.S. shifted imports to Mexico, Venezuela, and Saudi Arabia, stockpiled about 10 weeks supply in the Strategic Petroleum Reserve, and imposed fuel efficiency standards on carmakers. Each manufacturer had to meet a Corporate Average Fuel Efficiency (CAFE) standard, calculated as the efficiency of each vehicle weighted by sales. Fuel efficiency in the U.S. rose from 15.8 mpg in 1975 to 26.0 in 1982, and petroleum use declined from 7.0 to 6.5 million gallons per day.

As memories of the 1970s shortages faded, gains were wiped out in the 1990s through increased driving and purchase of less fuel-efficient trucks. A strong lobby group of motor vehicle manufacturers, energy companies, labor unions, and allied industries was able to prevent the U.S. government from mandating significant improvements in fuel efficiency or pollution.

6. CONCLUSION: THE SECOND CENTURY OF GEOGRAPHIC ANALYSIS OF MOTOR VEHICLES

The love affair with the car may have become less intense in the U.S. at the AAG's centennial, but it is alive and well in much of the rest of the world, especially less developed countries. Growth in motor vehicle production and sales was especially strong in Latin America and Asia.

Geographers interested in motor vehicles into the twenty-first century were looking across the Pacific to China. Fewer than 1 million private individuals owned cars in China in 2000, and 85 percent of the vehicles were owned by companies or the government. China's government, however, believed that the desire to own a motor vehicle was universal, and was helping producers meet that demand. The profound impacts of millions of motor vehicles on China's landscape, cities, national economy, and resource base are likely to occupy many geographers in this century.

In the U.S., little short of a national emergency is likely to pry motorists from their vehicles in the years ahead. Congestion is likely to be addressed through construction of yet more highways, though national budget limitations mean that they will be constructed as toll roads in many metropolitan areas. An increasing percentage of vehicles will be powered by hybrid or diesel engines, as the short-term bridge to alternative fuel and lower-emission vehicles. In the first decade of this century, both U.S. and European governments are banking on fuel-cell technology to wean motorists from the internal combustion engine. Whether fuel cells prove to be the transformational technology or merely a footnote is anybody's guess in 2003.

In the twenty-first century, geographers may take more notice of motor vehicles: they may attack SUVs (sports utility vehicles), push for alternate fuels, despair of congestion and deteriorating infrastructure, and applaud smart growth and traffic calming. Meanwhile, they will contain to purchase them—and conduct their fieldwork in them.

REFERENCES

Bloomfield, G.T. (1978). The World Automotive Industry. Newton Abbot, London, and North Pomfret, VT: David and Charles.

Bloomfield, G.T. (1981). The Changing Spatial Organization of Multinational Corporations in the World Automotive Industry. In Hamilton, F.E.I. and Linge, G.J.R. (Eds.) Spatial Analysis, Industry and the Industrial Environment, vol. 2: International Industrial Systems. New York: John Wiley.

Boas, C.W. (1961). Locational Patterns of American Automobile Assembly Plants, Economic Geography 37: 218-30.

Brigham, Albert Perry (1904). Good Roads in the United States, Bulletin of the American Geographical Society 36: 721-35.

Buel, James W. (1882). Metropolitan Life Unveiled. St. Louis: Historical Publishing.

Clark, G.L. (1986). The Crisis of the Midwest Auto Industry. In Scott, A.J. and Storper, M. (Eds.) Production, Work, Territory: The Geographical Anatomy of Industrial Capitalism. Boston: Allen and Unwin.

Cook, Ann, Gittell, Marilyn, and Mack, Herb (Eds.) (1973). City Life, 1865-1900. New York: Praeger.

Davies, Pete (2002). American Road: The Story of an Epic Transcontinental Journey at the Dawn of the Motor Age. New York: Henry Holt.

Dicken, P. (1998). Global Shift: Transforming the World Economy. 3rd ed. New York: Guilford.

Glasmeier, A.K. and McCluskey, R.E. (1988). U.S. Auto Parts Production: An Analysis of the Organization and Location of a Changing Industry, Economic Geography 64: 142-59.

Gottmann, Jean (1961). Megalopolis. New York: Twentieth-Century Fund.

Harris, Chauncy D. (1954). The Last Fifty Years and the Next, Annals of the Association of American Geographers 44: 211.

Harris, Chauncy D. (1997). The Nature of Cities and Urban Geography in the Last Half Century, Urban Geography 18: 15-35.

Harris, Chauncy D. and Ullman, Edward L. (1945). The Nature of Cities, Annals of the American Academy of Political and Social Science 143: 7-17.

Hart, John Fraser (1979). The 1950s, Annals of the Association of American Geographers 69: 109-14.

Henrickson, G.R. (1951). Trends in the Geographic Distribution of Suppliers of Some Basically Important Materials Used at the Buick Motor Division, Flint, Michigan. Ann Arbor: University of Michigan Institute for Human Adjustment.

Hoffman, K. and Kaplinsky, R. (1988). Driving Force: The Global Restructuring of Technology, Labor, and Investment in the Automobile and Components Industries. Boulder: Westview Press.

Holmes, J. (1983). Industrial Reorganization, Capital Restructuring and Locational Change: An Analysis of the Canadian Automobile Industry in the 1960s, Economic Geography 59: 251-71.

Horvath, Ronald J. (1974). Machine Space, Geographical Review 64: 167-88.

Hurley, N.P. (1959). The Automobile Industry: A Study in Industrial Location, Land Economics 35: 1-14.

Jacobs, Jane. (1961). Death and Life of Great American Cities. New York: Random House.

Jakle, John A. and Sculle, Keith A. (1994). The Gas Station in America. Baltimore: Johns Hopkins University Press.

Jakle, John A., and Sculle, Keith A. (1999). Fast Food. Baltimore: Johns Hopkins University Press.

Jakle, John A., Sculle, Keith A., and Rogers, Jefferson S. (1996). The Motel in America. Baltimore: Johns Hopkins University Press.

Konrad, Rachel (1999). Drive to the Future, Detroit Free Press April 28, 1999.

Lynd, Robert and Lynd, Helen (1929). Middletown: A Study in American Culture. New York: Harcourt, Brace.

Meyer, John R., Kain, J.F., and Wohl, M. (1965). The Urban Transportation Problem. Cambridge, MA: Harvard University Press.

Mumford, Lewis (1963). The Highway and the City. New York: Harcourt, Brace and World.

Nader, Ralph (1965). Unsafe at Any Speed. New York: Grossman.

Nelson, Howard J. (1959). The Spread of an Artificial Landscape over Southern California, Man, Time and Space in Southern California: A Symposium, Annals of the Association of American Geographers 49: 80-99.

Orchard, John E. (1929). Can Japan Develop Industrially?, Annals of the Association of American Geographers 19: 39-40.

Raitz, Karl (Ed.) (1996). The National Road. Baltimore: Johns Hopkins University Press.

Rubenstein, James M. (1992). The Changing U.S. Auto Industry: A Geographical Analysis. London and New York: Routledge.

Rubenstein, James M. (2001). Making and Selling Cars: Innovation and Change in the U.S. Automotive Industry. Baltimore: Johns Hopkins University Press.

Schlesinger, Arthur M. (1933). The Rise of the City 1878-1898. New York: Macmillan.

Stead, William (1894). If Christ Came to Chicago. Chicago: Laird and Lee.

Wright, J. Patrick (1979) On a Clear Day You Can See General Motors. New York: Avon Books.

CHAPTER 13

THOMAS R. LEINBACH
JOHN T. BOWEN, JR.

AIRSPACES: AIR TRANSPORT, TECHNOLOGY, AND SOCIETY

Abstract   In the course of the last century, air transport leapt from the pages of science fiction to become a relentless mechanism for economic and social change. For millions of people, the airline industry has redefined the scope and pace of everyday life. The success of air transport, both for passenger and cargo traffic, has been founded upon technological change, much of it defense-related. Commercial aircraft rapidly improved, especially after World War II, in range, capacity, speed, and safety. Combined, these changes have driven the cost of air transport steadily downwards, fostering unprecedented personal mobility for an increasing share of the world's population and helping to shape a new international division of labor. Although the airline and aircraft industries were profoundly shaken by the attacks of September 11, 2001, ongoing technological advances promise to amplify the impact of air transport still further, not only in the busiest hubs, but also in the distant places that have heretofore lain on the margins of the jet age.

Keywords   air transport, aircraft technology, airline industry, airports

1. INTRODUCTION

Despite all that has been written about a truly global system, it may be argued that such a system has really been dependent upon the rise of aviation, and this has come about only relatively recently. Civil aviation began in the U.S. in the 1920s with the movement of airmail, but had an important precursor as a people carrier in Germany before World War I. Given high costs until recently, civil aviation focused upon the movement of information and people with specialized skills. But since the late 1960s, declining costs have made recreational travel not only possible but also commonplace (Hugill 1995, 249). In addition, air freight has enjoyed explosive growth as firms move high-value, low-bulk cargoes around the globe.

Through the unrelenting advance of technology, air transport has become a potent mechanism for economic and social change over the last century.

Stanley D. Brunn, Susan L. Cutter, and J.W. Harrington, Jr. (eds.), Geography and Technology,
285-314. © 2004 Kluwer Academic Publishers. Printed in the Netherlands.

For millions of people, the airline industry defines the scope and pace of everyday life. In 2002, 1.6 billion passengers took to the air (ICAO 2003). They were joined by a large and rapidly rising share of world trade, from computer chips to orchids to running shoes, shipped in either freighter aircraft or the bellyhold of passenger aircraft. The success of air transport in the movement of both people and goods has been founded on technological change, much of it defense-related, which has made commercial aircraft larger, faster, safer, and able to fly much farther without refueling. Combined, these changes have driven the costs (whether measured in monetary or time units) of air transport steadily downwards.

The imprint of air transport upon geography has become much more pronounced in the fifty years since the publication of *American Geography: Inventory and Prospect* (James and Jones 1954). Ullman made a passing reference to the emerging importance of air transport and airport land uses in his chapter on transportation geography (Ullman 1954, 316 and 324). With the onset of the jet age, both the importance of air transport and the attention it garnered expanded rapidly. Some of the most significant early work was undertaken by geographers at the University of London in the mid-1950s, and specifically East et al. (1957, 1). In their introduction, the authors note that from the record of world exploration:

> two points seem especially important at the moment. In the first place, it is no accident that the periods of intense exploratory or colonizing activity were also periods of advance in other spheres. The importance of 'place' and, more particularly, the ease with which one can travel from one place to another, is an essential ingredient in an expanding economy. 'Accessibility' is part and parcel of man's material progress. Secondly, movement implies a means of locomotion, and each of the great historical periods of discovery or colonization is linked with a major step in transport development.

With this opening the authors go on to discuss the physical geography of aviation and its economic and technical background as well as world air routes, air transport in Europe and the United States and airport locations (East et al. 1957, 19).

Following the East et al. volume, Sealy published a single-authored volume on the topic (1966). During this same early period in the U.S. the work in air transportation by Edward Taaffe is noteworthy. Among his many publications are those dealing with air traffic patterns (Taaffe 1952), air transportation and the urban distribution (Taaffe 1956), trends in air passenger traffic (Taaffe 1959), and an air traffic-defined urban hierarchy (Taaffe 1962). In the years since, geographic research concerning air transport proliferated rapidly and in diverse directions, including the changing spatial structure of aviation, the impact of air transport upon patterns of economic development, and the effect of state regulation and deregulation on the geography of the airline industry.

Our intent in this chapter is to capture recent trends in aviation and particularly those developments which have been driven by technological changes. We first examine the application of technology on aircraft development and then, logically, the factors instrumental in producing growth in the airline industry. A section follows this on the spatial organization of air travel that has emerged in the era of liberalization. Subsequently, we examine the reasons for the current financial crises among the airlines and, amid this discussion, the impacts brought about by two major trends: the growth of the regional jet phenomenon, and "no-frills" carriers. These topics are followed by a discussion of two defining forces for the future: the impact of the Internet on air travel, and the growth of air cargo and logistical services.

2. FASTER AND FARTHER: ADVANCES IN AIRCRAFT TECHNOLOGY

Like the automobile, the airplane emerged from inventors' workshops to become an important means of transport in the early 20th century. In the years since, a cascade of innovations dramatically altered the size, range, speed, and safety of aircraft. Improvements along each of these dimensions continue to be achieved by the dwindling number of major manufacturers. The most important new aircraft in development in 2003, is the Airbus Industrie A380, an airplane that will surpass even the Boeing 747, the behemoth whose introduction in 1970 helped to dramatically reduce the cost of air travel, making feasible an unprecedented degree of globalization.

It is ironic perhaps that a technology (the airplane) whose use in the attacks of September 11, 2001 brought unprecedented distress to the airline industry is also a technology whose advance has been propelled, to an important degree, by its military value. Several of the most important innovations in airframes and aircraft engines have been defense-related, illustrating the crucial linkages between civil and military aviation. In particular, World War II accelerated the development of the jet engine and also pushed the envelope of aircraft capabilities outward along the frontiers of range and size. The first patent for a jet engine was issued in 1932, but the technology remained commercially unimportant until German drawings for a swept wing, crucial to attaining the full speed advantage of jet aircraft, were spirited out of the defeated country to Boeing's Seattle headquarters in the mid-1940s (Rodgers 1996). First bombers and then passenger aircraft were transformed by these technologies. The first workhorse of the jet age, the Boeing 707, introduced in 1958, was remarkably different from the Douglas DC-3, which had accounted for as much as 95 percent of commercial aircraft sold in the years just before the war (Table 13-1). Its productivity advantages helped to foster a rapid expansion of passenger travel. A variant of the Boeing 707 also became the first aircraft capable of nonstop transatlantic flight from London or Paris to New York, even in the face of prevailing westerly winds (Hugill 1993).

The catalytic relationship between the American military-industrial complex and the development of passenger aircraft, the technological leadership of postwar America, and the sheer size of the American air-travel market meant that American firms, especially Boeing and Douglas (later McDonnell Douglas) dominated the burgeoning aircraft industry through the

Table 13-1. Selected significant commercial passenger aircraft.

Aircraft	Year of First Commercial Service	Cruising Speed (km/hr)	Maximum Range with Full Payload (km)	Typical Seating Capacity
Douglas DC-3	1935	346	563	30
Douglas DC-7	1947	501	4,835	52
Boeing 707-100	1958	897	6,820	110
Boeing 727-100	1963	917	5,000	94
Boeing 747-100	1970	907	9,045	385
McDonnell Douglas DC-10	1971	908	7,415	260
Airbus A300	1974	847	3,420	269
Boeing 767-200	1982	854	5,855	216
Boeing 747-400	1989	939	13,444	416
Boeing 777-200ER	1995	905	13,420	305
Airbus A340-500	2003	886	15,800	313
Airbus A380	2006	930	14,800	555

Source: Derived from information available at the websites of Boeing Commercial Airplane Group (www.boeing.com) and Airbus Industrie (www.airbus.com).

1970s. The most important aircraft introduced after the 707 was another Boeing product, the gargantuan 747. The largest of 20th century passenger jets grew in part out of the design competition for a military transport aircraft (Irving 1993). With its bulbous nose (to accommodate front-loaded oversized cargo in freighter versions), the familiar form of the Boeing 747 became an icon of the jet era.

Yet development of the Boeing 747 nearly destroyed Boeing financially. The aircraft industry has been nicknamed "the sporty game" because each new aircraft requires so large a financial commitment for development costs before a single aircraft is flown that a company's financial future is imperiled (Newhouse 1982). McDonnell Douglas, for instance, lost an estimated $2.5 billion on the DC-10 and never recovered as a commercial manufacturer. Between making the decision to develop the B747 in 1965 (after receiving orders for 50 aircraft from launch customer, PanAm) (a launch customer is the first airline to commit to a new aircraft) and the first flight with paying passengers and cargo in 1971, Boeing spent a billion dollars and teetered on the brink of bankruptcy (Rodgers 1996).

The introduction of the 747 and two other new American widebody aircraft, the DC-10 and Lockheed L1011, in the late 1960s and early 1970s, reflected the American-dominated aircraft industry's confidence that the remarkable postwar growth of traffic would continue unabated (Figure 13-1). But a few years later, the 1973 Arab Oil Embargo caused fuel prices to spike and created an opening for Airbus Industrie. Formed as an alliance of German,

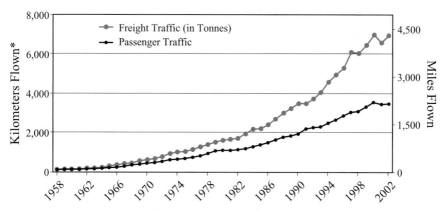

* Indexed so that 1958=100 for both cargo and passenger traffic.
Source: *U.N. Statistical Yearbook and International Civil Aviation Organization news releases.*

Figure 13-1. The rise of global air traffic.

French, British, and Spanish state-linked aerospace companies, Airbus entered the market with a relatively fuel-efficient widebody jet with just two engines, the twinjet A300. Sales accelerated sharply after 1973 (Newhouse 1982), and in a sense, the takeoff of the modern European aircraft industry, like that of the Japanese auto industry, was sparked by high oil prices that undermined the competitiveness of relatively fuel-inefficient, but previously dominant American products.

Airbus's advantages included not only a well-positioned product but also strong government backing (Fisher et al. 1992). Airbus has defended its government subsidies in part by pointing to the massive support the U.S. aircraft industry has garnered from the American military. Under a 1992 bilateral agreement, government aid to Airbus is limited to 33 percent of the development costs of new aircraft (including, for instance, $2.6 billion in government loans to defray the costs of developing the A380). The same agreement allowed Boeing to continue to receive indirect subsidies through its space and defense contracts. Airbus, which may come to more closely resemble Boeing in its relationship to the military, was partially privatized in 2000 when the multinational defense contractor European Aeronautic Defense and Space (EADS) acquired 80 percent of the plane-maker. As a result, both of the world's major commercial aircraft manufacturers are now integrated into leading defense firms (Pfleger 2002). By the end of the century, Airbus and Boeing were the only two remaining major aircraft manufacturers (the latter acquired McDonnell Douglas in 1997).

Airbus's A300 was the first of several widebody twinjets whose importance on long-haul routes has grown rapidly since the 1980s. Reflecting far greater confidence in the safety of commercial jets (Figure 13-2), the U.S. Federal Aviation Administration (FAA) and other aviation authorities began permitting airlines to operate specially equipped and maintained twinjets on routes for which alternate airports (to be used in emergencies) are up to 180 minutes away from any point along the route. As a result, twinjets came to dominate the transatlantic market by the early 1990s, with Boeing 767s and 757s and Airbus A300s and A310s displacing the Boeing 747. The transition to smaller jets enabled many long, thin point-to-point "pencil" routes to be opened. As a result, a greater number of nonstop international city-pairs came into operation, such as Charlotte-Paris and Chicago-Krakow.

Boeing is expecting a similar transformation of the transpacific market ("Place your bets..." 2002). Ironically, Boeing and Airbus are betting their respective futures in that market and on aircraft that differ from those that have accounted for much of each company's success. Boeing derived as much as 80 percent of its profits during the 1980s and early 1990s from the four-engine B747, and is betting that the transpacific market will be fragmented,

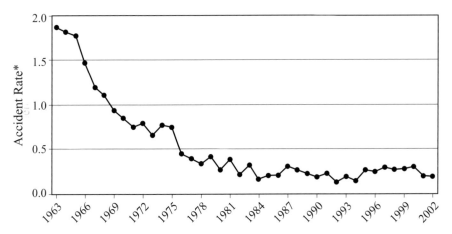

* Accidents per 100,000 hours flown by large commercial aircraft in the U.S.
Source: *Annual Review of Aircraft Accident Data.*

Figure 13-2. The falling commercial airline accident rate.

favoring its long-range B777 twinjet. Airbus, whose early success was based on twinjets, is wagering that transpacific traffic (and Europe-Asia transcontinental traffic) will continue to move via hubs, favoring the four-engine superjumbo A380.

The A380 is designed to carry a 30 percent larger payload than the largest version of the Boeing 747 and will have a range greater than almost any commercial aircraft at 9,200 mi (14,800 km). Airbus has indicated that the direct operating costs of the A380 will be 15- 30 percent lower than the B747-400 (Bartlett 2002). With its giant size, the A380 will also enable airlines to make better use of limited takeoff and landing slots, particularly at capacity-constrained airports like the Tokyo's Narita.

Together Airbus and Boeing now account for virtually all sales of commercial aircraft with at least 100 seats (as well as virtually all new freighter aircraft). The absence of any Asian manufacturer is noteworthy, but somewhat misleading. A commercial airplane incorporates literally millions of parts (e.g., three million in the case of the B777), and a significant number of those parts are sourced globally. While no large Asian aircraft assembler has emerged (nor is any likely soon given the extraordinary capital expenditure necessary to play the "sporty game"), Asian suppliers do play a critical role in the industry. Japanese heavy manufacturing firms (e.g., Mitsubishi, Kawasaki, and Fuji) produce about 20 percent of the B777's fuselage (NRC 1994).

Firms across the world are involved in the production of new aircraft. Suppliers for the Boeing 777 include not only Japanese firms, but also manufacturers in Italy (wing flaps), the U.K. (flight computers), Canada (landing gears), Singapore (landing gear doors), and many other countries. About 30 percent of the plane is made outside the U.S., much higher than for earlier Boeing models (McMillan 2003). The globalization of Boeing (and Airbus) has been driven not only by the need to lower production costs, although that has been important, but also by the need to lower labor costs. For example, labor costs at an Italian firm producing components for the B757 are less than one-third of those at the Boeing plant where the work used to be done. Boeing and Airbus have also shifted work to gain access to markets (especially the rapidly growing Chinese market) and crucial resources (including skilled labor).

Nevertheless, most aerospace jobs remain concentrated in the U.S. and Europe. The aircraft industry has been a mixed blessing for cities where it is based. The ordering of aircraft years in advance of delivery contributes to the boom and bust cycle in the airline industry, that is, airlines order too many aircraft during boom periods, exacerbating overcapacity during the lean periods, and then order too few in the lean periods, exacerbating capacity shortages during the next boom. The boom-and-bust cycle in the airline industry translates into a similar pattern in the commercial aircraft industry, especially in the cities where assembly work is carried out (Seattle, Long Beach, and Wichita in the case of Boeing; Toulouse and Hamburg in the case of Airbus). The aftermath of September 11 was particularly harsh in Seattle as Boeing eliminated 16,000 jobs in 2002, reflecting a decline in Boeing's production of commercial aircraft from 575 jets in 2001 to a planned total of just 280 in 2003 (Bowermaster 2002).

The financial crisis in the airline industry, and concomitantly, the commercial aircraft industry, will slow the advance of aircraft technology. In 2002, Boeing was forced to abandon a planned aircraft called the Sonic Cruiser that would have flown at 98 percent of the speed of sound, far faster than conventional passenger aircraft. The Sonic Cruiser elicited little industry enthusiasm after 2001. In its place, Boeing launched a new aircraft provisionally termed the 7E7 which, in recognition of the acute financial crisis in the airline industry and longer-term concerns about fuel supplies, promises no great improvement in speed, range, or capacity, but is instead aimed to dramatically lower operating costs through improved fuel efficiency (Done 2003).

Indeed, the period since 2001 has marked somewhat of a retreat from the frontier of aircraft technology. In 2003, both British Airways and Air France ended scheduled operations with the Concorde, the only supersonic passenger aircraft. Introduced in 1976, the aircraft, which cruised at twice the speed of sound, was a commercial failure because its small fuselage and prodigious thirst for fuel placed its speed beyond the means of all but the richest travelers (Odell 2003). With the withdrawal of the Concorde from service, aviation experts expect that the return to supersonic flight will not occur for several decades (U.S. House of Representatives 2000).

One of the most promising technologies to permit the return to supersonic flight is the scramjet, a version which was successfully tested in-flight for the first time in 2002. The scramjet uses oxygen directly from the atmosphere, making it far lighter and less costly than conventional rockets that require liquid oxygen. This jet is expected to initially revolutionize space launch vehicles, but its use in passenger aviation is already envisioned, perhaps beginning around 2020. With speeds of up to fourteen times the speed of sound, the scramjet would enable flights from London to Sydney in less than two hours. It is noteworthy that the first successful test of the scramjet was carried out by a team of Australian university scientists whose budget was a small fraction of that enjoyed by a U.S. National Aeronautics and Space Administration (NASA) team working on the same technology (Osborne 2002).

Nevertheless, NASA and other space agencies are expected to play a more important role in the development of aviation technology. The end of the Cold War has been associated with lower defense expenditures and a weaker synergy between the development of military and civilian aircraft. Moreover, because future gains in aircraft speed (e.g., via the scramjet) are expected to involve rocket technology employed at very high altitudes, the prospects for stronger synergies between space and aviation technologies are good (U.S. House of Representatives 2000). NASA's aviation-related research, for example, includes nanotechnology (permitting much stronger, lighter weight airframes) and biologically based systems (including research into "wing-morphing" technology enabling the wing of an aircraft to rapidly change shape like that of a bird to improve control) (U.S. House of Representatives 2000).

Although larger aircraft (like the A380) and faster aircraft (propelled perhaps by the scramjet) will play an important role in the industry's future, there is an alternative, perhaps complementary, vision of the industry based on very small jets operating personalized point-to-point routes. Both the U.S. government and the private sector are advancing technologies that will allow very inexpensive small jets to be introduced within a generation. In 2006, an Albuquerque, New Mexico-based company plans to introduce a six-passenger jet for less than $1 million per copy, less than a quarter of the cost of comparable business jets. Such aircraft are central to the alternative future that the writer and amateur pilot James Fallows mapped in his book *Free Flight*. The answer "to the inefficient hell that modern airline travel has become…" (Fallows 2001, 6), he argued, lies in small jets ferrying small numbers of passengers directly from origin to destination, circumventing hubs and reducing ground travel time by using smaller airports that are located closer to where most people live.

3. THE RISE OF THE AIRLINE INDUSTRY

Prior to 2001, the global airline industry had recorded five decades of almost uninterrupted growth (see Figure 13-1). Passenger traffic, measured in passenger-kilometers, had grown every single year, and cargo traffic, measured in freight ton-kilometers, expanded in every year but 1991 (the Gulf War and its aftermath) and 1998 (the Asian economic crisis). Over the course of the post-World War II period, the growth of traffic has been concentrated in the world's richest economies. In 2001, North American and European airports accounted for nearly three-quarters of passenger traffic and nearly two-thirds of cargo traffic (Baker and O'Toole 2002). Nevertheless, the last several decades have witnessed the rapid growth of air traffic in the industrializing economies of Asia in particular, and across the rest of the developing world to a lesser extent. In this section, we consider those factors that have fueled the industry's growth and account for its current prominence.

First, as described above, technological change has made air travel safer, faster, and much less expensive. In 1949, the lowest available fares for transatlantic travel and for travel across the U.S. were, in inflation-adjusted year-2000 U.S. dollars, approximately $4,300 (Solberg 1979, 342) and $660 (Heppenheimer 1995, 127), respectively. By 2000, these fares had fallen to as little as $225 and $160, respectively, during the off-peak season. As fares fell, air travel ceased to be the preserve of the affluent "jet set" and became a form of mass transportation, a transition that many lament. Solberg (1979, 413) wistfully ends his book on the history of the airline industry, "The miracle of flight […] had been triumphantly turned into a repetitive and unexciting routine. The adventure was over."

The success of the airline industry has come mainly at the expense of other modes of travel. The number of passengers crossing the Atlantic by air exceeded the number traveling by ship for the first time in 1957 (Hugill 1993), and within the U.S., domestic airline passengers exceeded the number of coach and Pullman rail travelers in 1956 (Heppenheimer 1995). With the introduction of widebody aircraft (the Boeing B747, McDonnell Douglas DC-10, Lockheed L1011, and Airbus A300) in the early 1970s, fares fell still further and the number of domestic and international airline passengers came to dwarf the number traveling by other long-haul modes. More recently, air travel has begun to draw travelers from short-haul modes as well. Southwest Airlines, for instance, has relied upon a strategy of offering fares so low that travel by air is less expensive than driving one's own car (Kim and Mauborgne 1999).

Second, the postwar expansion of the world's major economies drew millions more into the class of people who could afford to travel by air. In the U.S., for example, the combination of rising real incomes and falling real airfares meant that the cost of a roundtrip transatlantic fare fell from the equivalent of three months' wages for a typical manufacturing worker in 1949 to less than three days' wages by 2000. Rapid income growth has fostered the expansion of air travel from newly industrializing economies as well, especially in East Asia. In the 1990s, for example, China emerged as an increasingly important source of air travelers, with the number of international leisure travelers from that country doubling every three years in the 1990s to reach 30 percent of the corresponding U.S. figure by 2001 (Yau 2002; ITA 2002).

Third, the expansion of the airline industry took place largely within a setting made safe and predictable by the postwar *pax Americana*. After World War II, much of the world and especially its richest economies comprised a realm of increasingly open borders, facilitating both international trade and travel. Moreover, a principal foreign policy objective of the U.S. has been maintaining low oil prices, a goal whose successful pursuit has had obvious importance for the airline industry.

The success of *pax Americana* was not absolute, of course. Because airlines are quite literally "flag carriers" on their overseas routes, bearing the flag and name of their home countries, passenger aircraft became the target of many high-profile terrorist incidents. In one six-year period (1985-91), eighteen major terrorist attacks were conducted against the world's airlines, including the bombing of Pan Am 103 over Lockerbie, Scotland (Wallis 1993). Nevertheless, such events occurred infrequently enough that, at least until 2001, most air travelers confidently boarded aircraft expecting to arrive at their destination safely.

Fourth, the broader geographic scale of economic activity in the postwar era has been facilitated by the expansion of the airline industry, but has also sustained the further rapid growth of that industry. From its early days, the airline industry has played a critical role in the expanding scale of corporate operations. The first airlines sped long-distance communication through airmail, one of the most important effects of which was to accelerate the velocity of the money supply through the banking system (Ciccantell and Bunker 1998). Later, the expansion of passenger transport after World War II enabled large corporations to more easily coordinate globally dispersed operations from a single world headquarters (Hugill 1995). And like mail and passenger traffic earlier last century, freight traffic shifted away from other modes towards air transport, particularly in the twenty years after about 1980. Lower air freight rates, the adoption of just-in-time production systems, and the trend towards higher value-to-weight ratios in manufactured goods (particularly with the advent of microelectronics) produced a robust expansion of air cargo traffic. In fact, for most of the postwar period, cargo traffic grew faster than air passenger traffic (see Figure 13-1).

Fifth and finally, the liberalization of the airline industry, beginning in the 1970s, helped to sustain the long-term decline in airfares and provided airlines greater freedom to operate more frequently among more city-pairs. The airline industry emerged from the aftermath of World War II as the most politicized transport system (Glassner 1993). Most airlines were either partially or fully state-owned. Thousands of bilateral agreements were negotiated between pairs of countries; the most restrictive of these agreements named the carriers from each side that could serve a particular international route, set the capacity that each could offer, and tightly controlled fares. This heavy state intervention in the industry, both at the domestic and international levels, has been partly dismantled during the past two decades as part of a more general "hollowing out of the nation-state" (Knox et al. 2003). Many airlines have been privatized, especially in wealthy industrialized and industrializing economies, and many markets, both domestic and international, have been deregulated.

Liberalization has shifted the balance of power to shape airline networks and patterns of accessibility from national governments to the industry's leading airlines. The most powerful airlines have formed alliances to capitalize on the synergies in their networks and to overcome the remaining state-imposed strictures on the development of those networks. These alliances, such as the Star Alliance (14 carriers, including United Airlines, Lufthansa, Singapore Airlines, and Varig), align carriers operating from the industry's most important hubs. The alliances also are designed to capture more traffic, especially high-fare business traffic, and to move it via the alliances' hubs (U.S. DOT 2002).

The alliances pointedly exclude airlines from the poorest parts of the world, including Subsaharan Africa and South Asia. Although the airline industry serves to integrate the world economy, it does so unevenly. Poor regions are marginalized within the international airline industry by the weakness of their flag carriers, the relative inaccessibility of their hubs within global networks, and their limited ability to generate high-yielding traffic. Liberalization has tended to exacerbate that marginalization because privatization and deregulation have strengthened the competitive advantages of the strongest airlines and, indirectly, the strongest hubs (Bowen 2002).

And yet, it is in developing countries that the airline industry has its greatest importance as a symbol. An airline bearing the name and flag of its home country is a potent sign of national identity and technological prowess (Raguraman 1997). Not surprisingly, governments in almost every country have invested heavily in airlines and the infrastructure they require. Air Malawi, Estonian Air, and Lao Aviation are examples of the many flag carriers inaugurated almost simultaneously with the newly independent countries they serve. Many of those developing-country airlines remain heavily subsidized.

Together the economic, social, and political forces described above have propelled the expansion of the airline industry. Those same forces also fostered an unprecedented degree of time-space convergence. Until the industrial revolution, transport technology permitted only limited access to other regions of the world. Technological innovations in the domain of transportation were essentially used to increase the efficiency of advanced economies, enabling them to have access to resources and markets. This trend began with mercantilism and gradually shaped the global economic space, leaning on the transmission of information and fast and inexpensive transport systems. But it is the concepts of time-space and cost-space convergence, both well known in the geographical literature, that capture the critical contribution of air transport to social and economic development in a global context. For example, air cargo services play an integral role in holding down global production networks and enable competitive advantage as, increasingly, firms need to outsource aspects of their production and distribution operations. In a social context, the impact of time-space convergence is felt in the ordinary travel behavior of individuals at holiday seasons and during vacation periods. The European "pleasure periphery" is, for example, almost entirely dependent upon air travel. In still another perspective, the movement of export laborers and permanent migrants in a global context is closely aligned with airline networks and schedules.

In facilitating the globalization of economic activity and, for some, personal mobility, the airline industry itself has been globalized, evident in the globe-encircling networks operated by many of the world's largest carriers. The globalization of the industry has also made international coordination in the industry more important. In this regard, the two principal organizations associated with air travel are the International Civil Aviation Organization (ICAO) which deals with issues of standardization, facilitation, and technical cooperation for development in regard to air transport and a variety of international issues, and the International Air Transport Association (IATA) which has the basic mission of representing the airline industry worldwide. IATA promotes safety and security, seeks to achieve recognition of the importance of a healthy air transport industry to world-wide social and economic development.

An interesting element of the global standardization of air transport that these organizations have fostered is the recommendation that English be used as the language of air traffic control everywhere, even for domestic flights within non-English-speaking countries. The intent of this policy is to accelerate the exchange of information among pilots and ground controllers. The lack of English fluency and consequent miscommunication has been blamed for several serious accidents, including an Avianca crash near New York City in 1990 (Philips 1990). Similarly, the reluctance of some controllers, including those in France, to use only English has played a role in deadly crashes (Webster 2001).

4. WHERE THE FUTURE TOUCHES DOWN: AIRPORTS IN A GLOBALIZING WORLD

The remarkable increase in air traffic has been most concentrated in a relative handful of the thousands of airports scattered across the globe (Tables 13-2 and 13-3). Hubs are a paramount feature of the industry and have become more so as a result of airline restructuring following liberalization of the industry. Hubbing enables carriers to take advantage of economies of density (the lower per-passenger cost associated with serving many passengers on a single-network segment) by drawing traffic feed from a larger network, for example, feeding passengers from a dozen or more different origin cities through Chicago onto a single westbound flight to San Francisco. But the advantages of a hub-and-spoke system come at a cost. Such a system requires not only the spatial concentration of activity at key hubs (80 percent of airline traffic in the U.S. takes off or lands at the busiest 1 percent of airports), but also the timed concentration of activity in those hubs in several connection banks each day ("The way we fly now" 2001). The concentration in both space and time also makes such systems prone to severe congestion and being acutely vulnerable to interruptions from weather and other disruptions. Delays at a major hub ripple across the air transport system. The frustration of travel via hubs has been a major factor in the rise of air rage as a social phenomenon.

The increased importance of air transport has made major hubs, whether for passenger or cargo traffic or both, gateways articulating the relationship between regional economies and the larger global economy. For cities like London and New York, the international airport is both a vortex drawing together people, goods, information, and money from across the world and a window through which the global reach of such cities extends. For cities aspiring to this international stature, the airport is a crucial piece of infrastructure. Accordingly, governments at both the local and national level have invested heavily in new airports and airport improvements. At the end

Table 13-2. The world's busiest passenger airports, 2002.

Rank	Airport	Total Passengers (million)	Rank	Airport	Total Passengers (million)
1	Atlanta	76.9	16	Minneapolis-St Paul	32.6
2	Chicago	66.5	17	Detroit	32.4
3	London-Heathrow	63.3	18	Bangkok	32.2
4	Tokyo-Haneda	61.1	19	San Francisco	31.4
5	Los Angeles	56.2	20	Miami	30.1
6	Dallas-Forth Worth	52.8	21	London-Gatwick	29.6
7	Frankfurt/Main	48.5	22	Singapore	29.0
8	Paris-CDG	48.3	23	Newark	29.0
9	Amsterdam	40.7	24	Tokyo-Narita	28.9
10	Denver	35.7	25	New York-JFK	28.9
11	Phoenix	35.5	26	Beijing	27.2
12	Las Vegas	35.0	27	Seattle	26.7
13	Houston-Bush	33.9	28	Orlando	26.7
14	Madrid	33.9	29	Toronto	25.9
15	Hong Kong	33.9	30	St Louis	25.6

Source: ACI 2003.

of the 1990s, government investment in recently completed, in-progress, or planned airport development and redevelopment projects across the world totaled at least US$225 billion (Dempsey 2000).

The commitment of so vast a sum to airports across the world is still further testament to the importance of air travel in the international economy. The confluence of immigrants, cosmocrats, tourists, the transnational capitalist class, and all the other myriad travelers on any given day in the world's

Table 13-3. The world's busiest cargo airports, 2002.

Rank	Airport	Total Cargo (metric tons)	Rank	Airport	Total Cargo (metric tons)
1	Memphis	3,390.3	16	Chicago	1,279.2
2	Hong Kong	2,516.4	17	Bangkok	957.2
3	Anchorage	2,027.8	18	Indianapolis	866.0
4	Tokyo-Narita	2,000.8	19	Newark	821.5
5	Los Angeles	1,758.0	20	Osaka	805.4
6	Seoul	1,705.9	21	Dubai	785.0
7	Singapore	1,660.4	22	Atlanta	732.5
8	Frankfurt-Main	1,631.5	23	Tokyo-Haneda	707.1
9	Miami	1,624.2	24	Dallas-Ft. Worth	669.5
10	New York-JFK	1,574.5	25	Beijing	668.7
11	Louisville	1,523.9	26	Oakland	650.4
12	Paris-CDG	1,397.0	27	Shanghai	635.0
13	Taipei	1,380.7	28	San Francisco	593.8
14	London-Heathrow	1,310.6	29	Guangzhou	592.5
15	Amsterdam	1,288.6	30	Philadelphia	524.1

Source: ACI 2003.

dominant hubs, makes the world's leading airports places "where worlds collide" (Iyer 1995). Los Angeles International Airport (LAX), for example, contains a traveler's aid desk dispensing advice in dozens of different languages including eleven from India alone. Airports like LAX, London-Heathrow, and Tokyo-Narita are at the vanguard of globalization; these and other major hubs are where "the future touches down" (Iyer 1995, 50).

But the greater global importance of airports has also fostered local conflicts concerning noise and conflicting land uses. In contrast to other modes of transport, aviation creates relatively little aural pollution along the transport corridor; rather, noise is concentrated very sharply in the vicinity of airports (Graham 1995). In developed countries, concern about aircraft noise became a prominent issue in the 1970s, reflecting the conjunction of the rapid increase in jet travel and the nascent environmental movement. Concerns about noise severely limited the utility of the Concorde (the loudest commercial aircraft) because it could not attain supersonic speed over land. More generally, increasingly stringent standards have been placed on airlines, and noise levels for new aircraft are much lower than for their predecessors. For example, at a distance of 4 mi (or 6.5 km, the standard used for such measurements) from a first-generation Boeing 747-100 taking off, a person perceives a noise level of 100.5 decibels (dBA). For a new Boeing 747-400, the comparable measure is 90.8, a tenfold difference (U.S. FAA 2002). In developed countries, virtually all older aircraft like the 747-100 have been retired from airline fleets or fitted with hushkits to reduce their noise footprint. As a result, it has been estimated that fewer than two percent of people in most rich countries are regularly exposed to aircraft noise pollution (Graham 1995). In developing countries, the combination of dense airport-adjacent populations and a greater proportion of older, louder aircraft mean that less progress has been made in alleviating this problem.

In both developing and developed countries, new airport construction and expansion projects have been very controversial. Airports occupy relatively little land, certainly much less than consumed by highways and their interchanges. For example, Chicago-O'Hare occupies approximately 9 sq mi (23 sq km). The necessary proximity of airports to major population centers amplifies the likelihood of land-use conflicts. The construction of new runways at airports like O'Hare, London-Heathrow, and Tokyo-Narita has been greatly complicated by the need to occupy already densely developed land, a problem compounded by suburbanization. When O'Hare opened in the 1940s, much of the nearby land remained covered in farmland (the airport's code, ORD, is derived from the orchard that once stood on the airport site); but in the subsequent decades, the urban area moved rapidly towards and ultimately beyond the airport. Moreover, the imprint of an airport upon land-use patterns now extends far beyond the airport perimeter. Airport access highways, vast parking lots, hotels, logistics parks, office parks, and light industry vie for space (Sudjic 1993). Although the massive complex of activities surrounding a major airport now often rivals that of a central business district in economic importance, the expansion of airports has spawned grassroots resistance, often successful, in many countries.

5. THE REGIONAL JET PHENOMENON

One additional impact of deregulation was to allow the growth of regional and commuter carriers. Regional/commuter airline revenue enplanements in the U.S. increased from 7 million passengers in 1975 to over 72 million in 1999 (U.S. DOT 2000). Larger air carriers began using code-sharing agreements (where an airline offers services in its own name, but the transport service is provided by another carrier) and purchasing regional partners as a way of insuring a source of traffic to their major flights. Examples include United Express as a feeder to United Airlines with subsidiary operations performed by Air Wisconsin in the upper Midwest and Atlantic Coastal Airlines in the Southeast. Another model is ComAir, an independently owned feeder based in Cincinnati that is linked to Delta; it also operates in the Southeast. Serving a range of markets and utilizing a diverse fleet of aircraft has also stimulated the evolution of the commuter/regional airline industry.

As a derivative effect, another major development in airline interaction has been the growth of the regional jet (RJs) phenomenon (Taylor 2000; Yung 2000). Essentially smaller jet aircraft (50 and 70 seats) have begun to gradually replace traditional turbo-prop aircraft, and in some instances, larger jets in nodes where demand is fairly low. These RJs combine the range and speed of larger jets with the flexibility and cost-effectiveness of turboprops. The pace of this replacement will accelerate rapidly over the next twenty-five years. As a reflection of the expansion of the regional airline market over time, the length of the average trip has gradually increased from 105 mi (169 km) in 1975 to over 260 mi (418 km) in 2000 (U.S. DOT 2000). The result of the application of this innovation can be immediate access for smaller communities and improved scheduling to larger communities. The 50-seat aircraft (the Canadair Regional Jet produced by Bombardier in Toronto is one example) has many advantages over more traditional aircraft. These carriers are cheaper to fly because they cost less to purchase, they burn less fuel, and the crews are paid less. In some situations, the use of these aircraft will provide business travelers with an alternative to congested hubs and crowded airports. More important is that it allows airlines to cope more easily with fluctuating demand. For example, after September 11, United Airlines was able to develop a more frequent schedule of service and accommodate demand between Lexington, Kentucky and Chicago by using the RJs. Larger jet aircraft, the British Aerospace 146, formerly used on the same segment, were displaced to western markets where the higher capacity could be better utilized.

If the RJ has been a convenience for passengers, it has been a bonanza for the small airlines that fly them. Before the RJ, the mainstays of many regional carriers' fleets were turboprops which cruised at 300 mph (482 kph), and had a range of 300-400 mi (482-644 km) thus limiting them to ferrying passengers to hub airports. The new RJs can fly at more than 500 mph (805 kph) and as far as 1,500 mi (2,413 km), thus enabling them to link cities too far apart to be served by turboprops. The RJ has shifted the fare-splitting formula with the major airlines in the regionals' favor. This development is possible because the major airlines are willing to give a larger share of the pie to regional carriers and because RJs route so many more customers to hubs than do the turboprops. The extra volume more than makes up for the majors' smaller share of the revenue (Stein 2000).

The regional airline industry has taken over a significantly larger proportion of the U.S. air transport industry in the aftermath of September 11. The use of regional jets has grown 40 percent since that date while the use of narrow-body aircraft has dropped almost 19 percent. Revenue passenger miles (RPMs) for the regionals increased by over 20 percent compared to a 9 percent drop in RPMs for majors. In the current aviation climate, virtually every carrier is cutting back on its aircraft operations, but regional operations likely will continue to grow. The advent of both longer-range and higher-capacity regional aircraft merely underscores the magnitude of potential opportunities. That RJs have arguably become one of the more effective and visible tools of airline competition in recent years is displayed in the Boston-New York and other shuttle markets. Here the RJs have begun to insert themselves in markets which were long the exclusive preserve of Boeing and Airbus models.

6. NO-FRILLS AIRLINES

The future of the airlines is uncertain, particularly in periods when the domestic and indeed the global economy are struggling. One partial answer certainly lies in downsizing and the creation of a low-cost, no-frills model. Low-fare carriers now account for about 20 percent of the seats offered nationwide, an increase of 15 percent in 1998. The prototype for this trend in the U.S. is Southwest Airlines. Begun over thirty years ago by Herb Kelleher, the airline is based upon sound management principles, viz., labor productivity and the appreciation of customers ("Southwest Airlines..." 1995). The "no-frills" descriptor usually means that all tickets are electronic, seat assignments are on a first-come, first-serve basis, no food is served, and strict baggage restrictions apply. More importantly, the airline offers low fares and refuses to link up with computer reservation systems. Southwest also flies only one type of aircraft (Boeing 737) in order to achieve lower maintenance and crew training costs. This feature also gives them a greater ability to leverage purchases from Boeing to obtain a lower unit cost. Moreover, the airline operates in largely short haul, low-margin markets and has eschewed the hub-

and-spoke system in order to attain high aircraft and personnel utilization while at the same time achieving a very fast aircraft turnaround. Southwest Airlines has become the fourth largest airline in America, flying more than 64 million passengers a year to over 60 cities while operating 355 aircraft which have an average age of less than nine years. It also offers, as of mid-2003, a transcontinental flight from Baltimore to Los Angeles.

The success of an airline utilizing economical measures and flying from secondary airports has encouraged other similar carriers to emerge seeking niche markets throughout the U.S. Two U.S. examples are Air Tran Airways and Jet Blue. Air Tran is based in Atlanta and serves over 40 nodes ranging from Minneapolis, Dallas, Houston, Boston, and many other East Coast points to more than ten destinations in Florida. An agreement with Air Wisconsin provides a connecting service from nodes where the airline does not operate to places where it does. Operating with hubs in New York (JFK) and Long Beach, Jet Blue concentrates low-fare services in the Northeast, the West Coast, and Florida and flies only the Airbus A-320 (which reduces scheduling and maintenance problems). Jet Blue operated 36 aircraft at the end of 2002. Efficiency is also created using a largely Internet ticketing basis and only e-tickets. The airline's basic strategy for expansion is to test the market link, and when it becomes profitable, to begin adding additional flights.

In a move that is perhaps a harbinger of other industry actions, Delta Air Lines announced in late November 2002 that it would create a new low-fare subsidiary to compete with the growing number of no-frills discount carriers. The carrier will initially operate with a fleet of 36 Boeing 757 jets in order to compete head-to-head with other carriers in the low-fare market segment. It is designed to take business away from Jet Blue, Southwest, and Air Trans Airways. The new unit will initially operate its dedicated fleet in the Northeast to Florida market, with later expansion across Delta's entire network. Immediate responses from the airline industry per se were skeptical of success, with a major question being the extent to which the new carrier will cannibalize its own business.

Once a tightly regulated environment, the no-frills airline movement has spread to Europe, where carriers such as easyJet, Go, and Ryanair are thriving. They are providing stiff competition for British Airways and Lufthansa on European short hauls. In Europe, low-fare carriers account for about 10 percent of the intra-Europe traffic. As in the U.S., low fares, use of Internet ticketing (90 percent of their business), short turnaround times, high plane usage, and paperless ticket distribution have been the critical ingredients for success. British Airways, bmi British Midland, and SAS have already modified their fare structure and reduced restrictions in response to the no-frills competition.

Traditionally, low-cost entrants in Asia have been constrained by a host of restrictions including political agendas and the lean-cost structures of existing carriers. Along with the low-cost of incumbent airlines, the slowness of Asia to follow U.S. and European trends can also be explained by the difficulties in competing against sixth freedom (an airline's ability to carry traffic between two foreign countries via an airport in its own territory, e.g., Sydney to London via Singapore or Bangkok) carriers that are able to price flight legs marginally into and out of the region through their home ports. Basically low-cost carriers have been confined to the liberalized domestic markets in Malaysia (Air Asia), the Philippines (Cebu Pacific), and Japan (Skymark and Air Do) (Bowen and Leinbach 1995; Doyle 2002). New carriers in Thailand and Indonesia do not really follow the no-frills model. The most successful has been Malaysia's Air Asia, which is aligned with the low-fare model of Southwest-Ryanair. In part, this airline's success to date (350 thousand passengers per year and 80-90-percent load factors) has been associated with its ability to market holiday packages through its website. Most promising is the huge future potential as currently only 6 percent of Malaysians travel by air (Prystay 2002; Thomas 2002).

7. AIR TRAVEL, THE INTERNET, AND VIRTUAL TRAVEL AGENTS

Figures released by the U.S. Census Bureau show that Americans spent more than US$28 billion on the Internet in 2000. Some US$7.8 billion was spent on airline tickets, making them the highest-ranking consumer product, ahead of personal computers and hotel rooms. Through the third quarter of 2002, Internet sales totaled US$52.5 billion, up 41 percent compared to the same period in 2001. During this same period in 2002, airline tickets accounted for approximately US$14 billion of the total sales.

The rapid growth of Internet sites that offer airline tickets, hotel rooms, and rental cars to passengers is, essentially, a reinventing and expansion of the concept of distributed data that allowed computer reservation systems to revolutionize air travel in the 1980s. The original concept of a central repository for all computerized airline scheduling information was not achieved mainly because airlines found their own computer reservation systems could be very profitable. Selling airline tickets—especially at discounted prices—on the Internet has become a huge business and involves not only the airlines' own websites, but also a number of other enterprises including Orbitz, Cheaptickets, AirfarePlanet, Priceline, and Hotwire. The Orbitz (which is co-owned by five airlines) travel site has created a dedicated business travel site; it claims that it can save business travelers more than 90 percent on transaction fees compared with a business travel agent. This claim has brought accusations of undercutting by the National Business Travel Association and a call for investigation by the U.S. Department of Transportation (DOT). Of increased importance are those sites that provide a more comprehensive travel service to consumers, that is, more than simply airfares. Microsoft's Expedia is one such free service that allows individuals to find low fares, book flights, make hotel reservations, and rent cars. This site also allows access to a comprehensive source of travel information to allow the consumer to decide where to go and what to do once they get to a particular destination. Among the benefits of such sites is the fare-tracker service, which provides an e-mail communication at no charge that updates customers on the best fares between destinations they choose. The service also allows them to select seats and use a mammoth hotel directory. Travelocity is another site that is based upon comprehensive travel planning. Given the fragmented Internet market, the aim of these sites is to capture business by linking up a broad set of services to make them more accessible.

Airlines are finding success on their own websites. These sites are an increasingly important tool that allows cost-cutting and the development and strengthening of a customer base. For example, United Airlines expected the percentage of its online bookings to grow from 5 percent of its current passengers in 2000 to 20 percent by 2003. Such sites, in addition to purchasing tickets, allow passengers to view their itineraries, receive flight-change notices, and monitor flight delays, departure times, and progress, check their mileage program accumulation, and request upgrades and "free" mileage award travel. Airlines increasingly, despite hosting their own websites and forming shared sites with alliance partners and even competitors, are also promoting themselves on private sites. Intriguing future developments to monitor include the role of Internet database manipulation to achieve further efficiency and more personalized services built upon a customer's profile.

Evidence as of mid-2003 shows that online travel has rebounded from the September 11 incident, with major reservation gains being reported by Expedia, Travelocity, Orbitz, and other sites. This trend represents one of the few dot-com sectors to thrive despite an overall decrease in air travel due to terrorist concerns.

8. AIR CARGO AND LOGISTICS

After growing 6 percent and 7 percent in 1999 and 2000, respectively, world air cargo traffic fell a dramatic 6 percent during 2001. This decline, the worst ever in the modern air cargo industry, was the result of two concurrent factors: the U.S. economic slowdown and the collapse of the "technology bubble" (a portion of the economy that during the mid- to late-1990s resulted in inflated values relating to the proliferation of increasingly sophisticated information and telecommunication technologies). The bursting of this "bubble" (collapsing values) began with the "dot.com" sector in early 2000 (Boeing 2002a). The decline in air cargo worsened in 2001, well before the terrorist attacks of September 11, and by year-end, major air trade lanes linking North America, Europe, and Asia had contracted by nearly 10 percent. On a positive note, signs of a recovery have emerged in 2002 led by strong Asian trade lanes and the domestic U.S. market. During the next two decades, growth will likely expand at nearly 6.5 percent annually. Asian air cargo markets will continue to be the regional leaders with intra-Asian and domestic Chinese markets expanding at 8 percent and 10 percent annually respectively. Without the strong forecast growth from the Chinese market, the world air cargo market would expand significantly slower (Boeing 2002b). More than two-thirds of total traffic is carried by non-U.S. airlines, and historically, their growth has outpaced that of U.S. carriers.

Although economic activity remains the primary driver for the air cargo industry, other factors affect the development of this traffic. Some of this development is airline controlled: for example, the express and small-package market growth acquisition of aircraft and expansion of services. But factors beyond airline control have also acted favorably on growth. Developments include inventory management techniques, deregulation, market liberalization, national development programs, and a stream of new air-eligible commodities (electronics components, perishable goods). All have played major roles in air cargo growth. Of these, a primary driver of air cargo demand continues to be inventory, or more appropriately, supply-chain management. A shift in corporate strategies over the last decade to the externalization of production has made firms more and more reliant on external resources. The role of suppliers has been elevated to a level such that they are now strategically inseparable from internal aspects. A dominant proportion of competitive advantage now rests with the management of external relations of production and the flow of resources from source to consumer (Hall and Braithwaite 2001). On the other hand, forces from outside the industry have had a constraining effect. Unfavorable issues are debt burdens, high interest rates, trading blocs and protectionism, commodity price weaknesses, and political volatility. Positive influences on the other hand include expanding Asian markets, currency strength, oil marketing agreements, and Middle East stability (Boeing 2002b).

World air cargo comprises freight (that is, scheduled freight, charter freight, and express) and mail. Scheduled freight and express together constitute the largest component of total air cargo. The definitions of express versus nonexpress air cargo are increasingly blurred as traditional airlines expand their offerings of "time-definite" services, that is, cargo services with a performance guarantee based on time, which often includes a refund of all or a portion of the payment made for same service if the advertised time is not met. Government postal authorities also continue to make strides in becoming full-fledged "logistics providers" (that is, functions involving the procurement, distribution, maintenance, and replacement of material and personnel) largely through the acquisition of established firms. This area has become a lucrative source of revenue. The heightened competition among firms such as FedEx, UPS, Airborne, and DHL means that the consumer will increasingly benefit from service options and lower prices as competing products enter the market. The international express market continues to grow at an extraordinary rate of over 20 percent per year since 1991. Within the U.S., the express market constitutes about 60 percent of domestic air cargo.

One of the most important developments in the movement of freight in general has been the emergence of logistical hubs and providers. Freight traffic is concentrated at a relatively small number of points in order to achieve economies of scale. In addition, certain locations have evolved with such a hub function being critical to their development. Singapore and Hong Kong are two such nodes (see Table 13-3 for the major air cargo hubs). Whether goods travel to global markets by air or sea traffic, they either arrive from multiple locations and are repacked and then shipped to other locations, or they are stored to serve a specific region or country as growth occurs. Manufacturers that earlier set up distribution points and manufacturing plants for specific and immediate markets are now reexamining their options. Time-based competition, information technology, multifunctional outsourcing, and the globalization of manufacturing are having a profound impact on logistics, distribution, and warehousing. The trend is clearly toward indirect relationship management where third-party operators are now corporate collaborators. By using a third-party logistics service provider, the manufacturer does not have to invest in land, buildings, handling equipment, and hiring and training of logistics personnel. Therefore, the manufacturer avoids the risk of disinvestments in case of major changes in the market situation. Further, more companies are now locating near their customers, especially those in the perishables business. The expanded role between manufacturer and third-party logistics operator is also having an impact on global site selection considerations. Not only are logistics operators opening facilities around the world to accommodate their growing base of clients, logistics centers such as Amsterdam Westpoint, and free trade zones such as Jebel Ali Free Zone in Dubai, the United Arab Emirates, are being developed to attract these customers. Meanwhile, large manufacturers are opening and operating regional distribution sites in markets in which they serve (Thuermer 2000).

The competitive nature of the air cargo industry requires innovation and flexibility. The freighter "wet lease" (arrangement that includes all facets of operating an airplane on a carrier's behalf including the airframe, crew, and most, if not all of the airplane-related expense items) airline, or ACMI (aircraft, crew, maintenance, and insurance) provider, provides traditional airlines a new competitive option. Wet-lease carriers can offer airlines the flexibility to contract for air transportation services on a trial basis if demand is uncertain to augment existing markets, or to provide service in markets that are highly seasonal without the investment in dedicated equipment (Boeing 2002a).

The profit squeeze within the passenger industry has focused attention on the lower-hold revenue opportunities in the cargo market. Cargo revenue represents, on average, 13 percent of total traffic revenue, with some airlines reaching as high as 40 percent. Industry yields (airline charges as measured in units of aggregated weight and distance, e.g., revenue per ton-kilometer) for both cargo and passenger services have steadily declined since 1970. Such declines reflect airline productivity gains, technical improvements, and intensifying competition (Boeing 2002b).

9. THE FUTURE OF AIR TRANSPORT

The global airline industry suffered its worst year ever in 2001, losing an estimated US$11.6 billion before taxes (ICAO 2002). The industry's massive losses coincided with the demise of some of its most storied members. Sabena and Swissair both stopped operating shortly after September 11, 2001, though the latter reemerged as a somewhat smaller carrier called simply Swiss shortly after. In the U.S., United Airlines and USAir, two of the largest carriers in the world, were forced into bankruptcy. By early 2002, nearly 1,000 aircraft, or approximately 6 percent of the world's commercial aircraft, were idled, many of them parked in desert storage (Smith 2003).

While the scale of the industry's difficulties is without precedent, it is important to note that the airline industry has never been particularly profitable. In the past thirty years, the industry's pretax profit margin exceeded 6 percent only once. There are several reasons for the lackluster performance. First, as noted earlier, the ordering of aircraft years in advance tends to foster a boom-and-bust cycle in the airline industry. Second, the pervasive involvement of the state in air transport has diverted the industry's emphasis from profit-making. Third, some suggest that the romance of air travel accounts for the continued willingness of investors to put money into an industry with so poor a track record, resulting in persistent overinvestment. Finally, economists have contended that an "empty core" may characterize the industry, meaning that there is a structural tendency towards unprofitability (Button and Stough 2000). The marginal cost of serving an additional passenger on a flight that would otherwise not be full (an extra bag of peanuts, a miniscule amount of additional fuel, etc.) is so small that competitive pressure, particularly during cyclical downturns, compels carriers to offer fares that are well above that marginal cost, but fall below the carrier's average cost.

Nevertheless, the industry's financial problems did reach an unprecedented scale in 2001. That year was already expected to be among the worst, especially in the U.S., before the terrorist attacks of September 11th. The industry's pre-attack problems reflected a shift in the underlying geography of air travel. Point-to-point carriers, especially the no-frills airlines described earlier, captured an increasing share of traffic. Their success reflects the routinization of air travel, a trend which has made the high-cost structure of hub-and-spoke "network airlines" unsustainable (Leonhardt and Markels 2002).

After the attacks of September 11, the industry's already bad situation eroded rapidly. Demand briefly collapsed, then struggled to recover. New airport procedures mean that it takes longer to process passengers, their baggage, and cargo on the ground, making it more difficult to keep aircraft in the air (where they make money). New biometric scanning technologies are being speeded to market to alleviate this problem; one promising technology employs iris-scanners to confirm the identity of previously enrolled travelers permitting them to move more rapidly through immigration and security checks (Bowcott 2003). Dealing with the threat posed by air cargo may prove more difficult, since most shipments are not screened, moving through the air transport system unopened from the time they depart the factory or origin warehouse until they arrive at their final destinations. In a more indirect way, airline costs will be raised (or more precisely, not fall as rapidly as they otherwise would have) by the slower adoption of more efficient new aircraft in the financially crippled industry.

The events of September 11 also shook, in a profound way, the predictability of air travel and the confidence that predictability had engendered. Air travel has regained some of its danger and many discretionary travelers have been scared from the sky. After the bombing of Pan Am 103, it took a year for traffic levels to return to normal. That time is likely to be longer after the attacks.

The industry's future course and future geography are intriguingly uncertain. The Airbus A380 and the massive investments by governments worldwide in new and expanded airports suggest the continuation of the industry's long-term trend toward bigger planes and bigger airports.

Conversely, the recent success of the no-frills carriers and the rapid displacement of 200-seat jets by regional jets, and B747s by B767s point in a different direction, one perhaps more consonant with the liberalization that has gained momentum across much of the globe since the 1970s. Liberalization permits more frequent service to more cities by more airlines. That potential proliferation of services may be inherently incompatible with the concentration of services in space and time to the degree that characterized the industry before 2001.

The future course of the industry, whether towards concentration or fragmentation, will have profound implications for the cities and regions the airlines link, for the growing number of firms whose daily operations rely on air cargo, and for the immigrants and executives, tourists, and sojourners whose mobility is dependent on air travel. The breadth of that impact is testament to the far-reaching significance air transport has attained a century on from Kitty Hawk.

REFERENCES

Airports Council International (ACI) (2003). The Worlds Busiest Airports. www.airports.org

Baker, C. and OToole, K. (2002). Negative Growth: Airports Everywhere Suffered from Last Years Crisis Airline Business June: 51.

Bartlett, N. (2002). Worlds Largest Airliner Takes Shape, Global Design News September 1: 19.

Boeing Corporation (2002a). World Air Cargo Forecast: Executive Summary. www.boeing.com/commercial/cargo/exec_summary.html

Boeing Corporation (2002b). World Air Cargo Forecast: World Overview. www.boeing.com/commercial/cargo/world_overview.html

Bowcott, O. (2003). Iris Recognition: A New Game of Eye Spy to Speed the Passengers Journey, Guardian July 29: 10.

Bowen, J.T. (2002). Network Change, Deregulation and Access in the Global Airline Industry, Economic Geography 78: 425-39.

Bowen, J.T. and Leinbach, T.R. (1995). The State and Liberalization: The Airline Industry in the East Asian NICs, Annals of the Association of American Geographers 85: 468-93.

Bowermaster, D. (2002). Boeing to Cut 5,000 More Jobs, Seattle Times November 21: A1.

Button, K. and Stough, R. (2000). Air Transport Networks: Theory and Policy Implications. Northampton: Edward Elgar.

Ciccantell, P.S. and Bunker, S.G. (1998). Introduction: Space, Transport, and World-Systems Theory. In Ciccantell, P.S. and Bunker, S.G. (Eds.) Space and Transport in the World System, 1-18. Westport: Greenwood Press.

Dempsey, P.S. (2000). Airport Planning and Development Handbook: A Global Survey. New York: McGraw Hill.

Done, K. (2003). Efficiency Replaces Speed in Sky Wars, Financial Times June 21: 14.

Doyle, A. (2002). Lure of the East Flight, International October 29: 42.

East, G.W., Sealy, K.R., and. Wooldridge, S.W. (Eds.) (1957). The Geography of Air Transport. London: Hutchinson University Library.

Fallows, J. (2001). Free Flight. New York: Public Affairs.

Fisher, J.W., Cantor, D.J, Harrison, G.J., and Sek, L.M. (1992). Airbus Industrie: An Economic and Trade Perspective. Congressional Research Services, Economics Division. Washington DC: U.S. GPO.

Glassner, I.M. (1993). Political Geography. New York: John Wiley.

Graham, B. (1995). Geography and Air Transport. Chichester: John Wiley and Sons.

Hall, D. and Braithwaite, A. (2001). The Development of Thinking in Supply Chain and Logistics Management. In Brewer, A., Button, K.J., and Hensher, D.A. (Eds.) Handbook of Logistics and Supply Chain Management, 81-98. Amsterdam: Pergamon.

Heppenheimer, T.A. (1995). Turbulent Skies: The History of Commercial Aviation. New York: John Wiley.

Hugill, P. J. (1995). World Trade since 1431: Geography, Technology, and Capitalism. Baltimore: Johns Hopkins University Press.

ICAO (2002). Events of 11 September Had Strong Negative Impact in Airline Financial Results for 2001. Press release dated May 28.

ICAO (2003). Airlines Curb Financial Losses in 2002. Press release dated May 26.

Irving, C. (1993). Wide-Body: The Triumph of the 747. New York: William Morrow and Company.

International Trade Administration (ITA) (2002). Basic Market Analysis Program: U.S. Citizen Air Traffic to Overseas Regions, Canada and Mexico 2001. www.tinet.ita.doc.gov

Iyer, P. (1995). Where Worlds Collide, Harpers Magazine August: 50-57.

James, P.E. and Jones, C.F. (1954). American Geography: Inventory and Prospect. Syracuse: Syracuse University Press.

Kim, W.C. and Mauborgne, R. (1999). How Southwest Airlines Found a Route to Success, Financial Times (London). May 13: 16.

Knox, P., Agnew, J., and McCarthy, L. (2003). The Geography of the World Economy. 4th ed. London: Arnold.

Leonhardt, D. and Markels, A. (2002). Airlines Realize Its Time for a New Flight Plan. New York Times December 8: C1.

McMillan, M. (2003). World Assembly Required? Wichita Eagle March 2: 1.

National Research Council (NRC) (1994). High Stakes Aviation: U.S.-Japan Technology Linkages in Transport Aircraft. Washington, DC: National Academy Press.

Newhouse, J. (1982). The Sporty Game. New York: Alfred A. Knopf.

Odell, M. (2003). An Ignominious Step Backwards: Supersonic Flight Financial Times (Aerospace section) June 16: 2.

Osborne, L. (2002). The Scramjet, New York Times Magazine December 15: 119.

Pfleger, K. (2002). Airbus Pursuing Pentagon Contracts: European Rival Challenging Boeing in Another Arena, Seattle Times October 28: 1A.

Philips, D. (1990). Avianca Crash a Fatal Misunderstanding, Washington Post, June 25: A5.

Place Your Bets on the Future of Flying (2002). Economist March 30: 57.

Prystay, C. (2002). The Skys the Limit: Upstart Carrier AirAsia is Revolutionizing Travel in Malaysia, Far Eastern Economic Review December 5: 34.

Raguraman, K. (1997). Airlines as Instruments for Nation Building and National Identity: Case Study of Malaysia and Singapore, Journal of Transport Geography 5: 239-56.

Rodgers, E. (1996). The Story of Boeing and the Rise of the Jetliner Industry. New York: Atlantic Monthly Press.

Sealy, K. W. (1966). The Geography of Air Transport. London: Hutchinson University Library.

Smith, B. A. (2003). Desert Bird-Watching, Aviation Week and Space Technology March 17: 37.

Solberg, C. (1979). Conquest of the Skies: A History of Commercial Aviation in America. Boston: Little, Brown and Co.

Southwest Airlines Herb Kelleher: Unorthodoxy at Work (1995). Management Review 84: 9-13.

Stein, N. (2000). Regionals Join the Jet Set, Fortune September 4: 287-91.

Sudjic, D. (1993). The 100 Mile City. London: Flamingo.

Taaffe, E. (1952). The Air Passenger Hinterland of Chicago. Chicago: University of Chicago Press.

Taaffe, E. (1956). Air Transportation and the United States Urban Distribution, Geographical Review 45: 219-38.

Taaffe, E. (1959). Trends in Airline Passenger Traffic, Annals of the Association of American Geographers 49: 393-408.

Taaffe, E. (1962). The Urban Hierarchy: An Air Passenger Definition, Economic Geography 38: 1-14.

Taylor, A. (2000). Little Jets Are Huge! Fortune September 4: 274-78.

The Way We Fly Now (2001). Economist July 21: 64.

Thomas, G. (2002). Asias Absent Revolution, Air Transport World 39: 42-48.

Thuermer, K. (2000). International Logistics Hubs, Business Facilities November: 10-12.

Ullman, E.L. (1954). Transportation Geography. In James, P.E. and Jones, C.F. (Eds.) American Geography: Inventory and Prospect, 310-32. Syracuse: Syracuse University Press.

U.S. DOT (U.S. Department of Transportation), Bureau of Transportation Statistics (2002). U.S. International Travel and Transportation Trends. BTS02-03. Washington, DC: U.S. GPO.

U.S. DOT, Bureau of Transportation Statistics (2000). The Changing Face of Transportation. BTS00-007. Washington, DC: U.S. GPO.

U.S. FAA (Federal Aviation Administration) Estimated Airplane Noise Levels in A-Weighted Decibels. Advisory circular: 36-3H.

U.S . House of Representatives (2000). The Future of Aviation Technology: Is the Sky the Limit? Hearing before the Subcommittee on Aviation of the Committee on Transportation and Infrastructure. 106th Congress, 2nd Session. May 16. Washington DC: U.S. GPO.

Wallis, R. (1993). Combating Air Terrorism. Washington DC: Brasseys.

Webster, B. (2001). Pilot Killed as French Refuse to Speak English, The Times (London) July 24: 1.

Yau, W. (2002). Outbound Travel Boom Forecast, South China Morning Post September 3: 4.

Yung, K. (2000). Regional Jets Are Airlines' New Competitive Arena, Seattle Times June 12: 1.

CHAPTER 14

DAVID R. RAIN
SUSAN R. BROOKER-GROSS

A WORLD ON DEMAND: GEOGRAPHY OF THE 24-HOUR GLOBAL TV NEWS

Abstract   The news is in our homes, our workplaces, and the landscape. The views of the world represented by 24-hour global news media have their own spatial dimensions. Given access to new reporting and analytical technologies, geographers are well equipped to study media impacts, especially those of cable, broadcast, and satellite TV news that purport to present all the news while it is happening. This chapter explores the geographies of news events, the history of media technology from the fixtures of the past to tomorrow's personal communication networks, and concludes with a discussion of possible collaboration between geographers and media.

Keywords   news, media, communication, journalism, television, localism

1. WHAT IS THE GEOGRAPHY OF THE NEWS?

Most elementary to geographers is the idea that actions occur in specific places. But seen as a form of geography, the 24-hour global news media can seem disconcertingly placeless: a stream of images and commentary flowing *somewhere* into your living room. Representations of news events exist as virtual replicas of the world—albeit highly distorted ones at times—and yet the news events themselves occur in actual places and are processed, packaged, and delivered over IP (Internet Protocol) networks, cable, fiber, and through satellite and cell-phone transmissions to viewers in specific communities. Geographers have been late to discover formally that the veritable explosion of new ways of delivering content carries rewards, especially for those concerned with how information is consumed by the public. Geographically enabled media studies can illustrate the axiom that each successive wave of new technology carries its own new set of social relations. Media relations in particular are invested with cultural and economic power beamed from and

Stanley D. Brunn, Susan L. Cutter, and J.W. Harrington, Jr. (Eds.), Geography and Technology,
315-337. © 2004 Kluwer Academic Publishers. Printed in the Netherlands.

to specific places. These dimensions should be appealing to those interested in putting a geographical spin on the study of communications media.

Modern Western societies have come to be dominated by the blue glow of television, with 98 percent of households owning one or more TV sets in the U.S., 99 percent in Japan, and 94 percent in the U.K.[1] Twenty-four-hour global news was essentially invented by Cable News Network (CNN) in 1980, but it was events such as the first Gulf War, the contested 2000 U.S. Presidential election, the attacks on New York and Washington on September 11, 2001, and the wars in Afghanistan and Iraq that drew viewers increasingly to their all-news networks at all hours. Yet, despite the ubiquity of television and no shortage of events that draw viewers, the TV news medium as a study subject has been mostly ignored by geographers.[2] Early works (Abler et al. 1975; Jakle et al. 1976; Burgess and Gold 1985; Brunn and Leinbach 1991) drew attention to the need for geographical media study but they did not result in TV becoming a research focus. The media, while an integral part of popular culture, have not been on the cutting edge of geographical inquiry. Why news media have not been seriously studied by geographers is an interesting question. Perhaps our analytical methods do not mesh well with the nature of the subject matter or content. Media, it could be argued, do not seem "real" in the way most subjects that geographers study are, and media are not amenable to the traditional field and survey type research techniques used by geographers. Content analysis of TV news is rather weakly developed for the task with only a few accessible tools to isolate and identify in studying geographical dimensions in the vast amount of material produced. Furthermore, the attitudes of scholars themselves could be considered a limitation. Media and particularly television are subject to bashing by academics, child learning specialists, and environmentalists whose distaste sometimes borders on revulsion.

Television news also can be sensationalized, yet it is often insipid and repetitive. According to recent polls of media use in the U.S., more adults obtain their news from local TV than from any other source, including newspaper and radio,[3] with 57 percent of those polled reporting their local TV station as a source of news. Western society and its media have coevolved to where media flow all around us like air and become fixed in our lived environments, inundating us with images, jingles, and sound bytes in a stream that never subsides. Media critic Todd Gitlin (2001) states that the torrent of information passing through TV screens in a single minute contains more pictures than a prosperous 17th century Dutch household saw in several lifetimes.

What Marshall McLuhan described nearly four decades ago has come to pass: "We have extended our central nervous system itself in a global embrace, abolishing both space and time as far as our planet is concerned" (McLuhan

1964, 3-4). While it may not have abolished space, this central nervous system has expanded into distended landscapes of TV production and consumption, spatially uneven news values, and splintered audience demand. The proverbial parallel universe of TV car chases, shoot-outs, hold-ups, beer advertisements, and stormy weather that superior alien life forms may see when they view us is simply another reflection—yet it is not one without spatial dimensions, or implications for our terrestrial existence.

We are surrounded constantly by commercial and informational messages in our homes, our workplaces, and our landscape, from product placement in movies to outdoor advertising to network news being packaged for long-distance jet passengers. These messages may or may not create permanent or measurable imprints in landscapes, but may represent merely the trappings (ephemera such as crime scenes, police lines, and helicopters) of a news event's significance, so geographers may not find a handhold, something to grasp firmly to guide one's way in research. While it may simply not seem practical to extend traditional geographers' field and survey-type research techniques to examinations of the news media, we might instead need to devise new methods.

Given access to new reporting and analytical technologies, geographers may be well equipped to study this "always on" phenomenon of cable, broadcast, and satellite TV news transmissions that purport to present *all* the news, often while it is happening. Geographers can also make a case based on the common ground that media and geography share: both are visual and thematic, and both at their root are stories about the world. As a social scientist would have no trouble understanding, however, it is not the totality of its media coverage about a given local place that lasts, but rather how this coverage works into the place's geography. This chapter explores the chasm and suggests some bridges. First, we describe some of the geographies conjured up by the relationships among news events, reporters, recording devices, and audiences, by examining the geographical dimensions of the news production and the news consumption environments, the power of place depictions, and the history of media technology from the limited standard fare of the past, to today's cacophony of voices and images, to tomorrow's do-it-yourself personal communications networks. After considering the geographical dimensions of modern television news journalism, we examine the question of news values and audience demand in a globalized marketplace of viewing publics. We then explore some new technologies such as web logging, cell swarming, and digital video recording devices in transforming geographies of the news, and conclude with a discussion of some possible future collaboration between geographers and media.

2. LANDSCAPES OF NEWS PRODUCTION AND CONSUMPTION

Imagine a three-dimensional world map composed of layers of place-based knowledge. Sources for these layers would come from the life experiences of residents and workers and also from tourism, personal contacts, and the news reported and produced there. On such a map the layer representing global news production would appear highly concentrated spatially. Pinnacles and contours representing broadcast studios in Manhattan, pool reporters' offices in major cities, network affiliates in major metropolitan areas such as Chicago and Atlanta, Washington offices of networks and major newspapers, the entertainment industry of Burbank and greater Los Angeles, and the place called "CNN Center," which is actually in metropolitan Atlanta, would draw the eye. Extending around the globe, peaks of activity would appear in London, Paris, Berlin, Brussels, Rome, Tel Aviv, Mumbai, Shanghai, Tokyo, Singapore, Sidney, and other world cities, with vast empty spaces in between that represent smaller markets or the absence of a perceived demand for news.

Largely coastal, but also overwhelmingly metropolitan, the configuration of news production is an information system operating from electronic promontories, focusing attention, eyeballs, notoriety, and historical significance upon some areas and not others. Transmitted through coaxial cable, fiber-optics, and electron beams, the information world consists of intensively scrutinized places interspersed with vast underserved and overlooked (empty) spaces. This uneven distribution of news interest creates a semiotics of places that can have striking consequences for the places themselves. The uneven distribution of knowledge about regions, countries, and cities around the world, and how a country is presented in the media, are hugely influential for international trade and investment, tourism, international diplomacy, and political relations.

Developed as an ad hoc tag or shorthand for certain scrutinized places where news occurs, *metaplaces* act as a coded shorthand in news stories and tend to crop up in policymakers' rhetoric, such as America's radar screen, the nation's schools, "Main Street," our living rooms, our boardrooms, and our bowling alleys. While this kind of shorthand is literally groundless, the punchy messages that are conveyed spare journalists a few precious words and often create a kind of visual lexicon for the correspondent, the West Bank being an example, or the iconic visually referencing backdrop of the White House, which is used by correspondents to lend authority to their reporting. In his book *Democracy and the News* (2003), sociologist Herbert Gans asserts that not only are people disempowered by what he calls "top-down journalism" of official pronouncements and palace intrigue, but so are the reporters and editors that produce it. Telling people what their top elected and appointed officials are doing and saying is important, but it is hardly the only information citizens need to participate in politics. By producing top-down news, journalists often become unintentional publicists for the government.

Ignorance of geography in television news often appears in the way that place names appear as inadequate markers of where news occurs. A volcano erupts on an island "300 miles from Jakarta," when at best only a small portion of the viewing audience can even place Indonesia roughly on a world map. Or a kidnapping occurs in a "California town" in a state with more than 35 million people. It is often very difficult for news media providers to supply necessary context to place a story. Even with the increasing use of maps in broadcasting and in print, the locators are often weakly developed or lacking altogether.

Spatial bias against poorer countries in news coverage has been documented by Wu (2003), who studied determinants of news flows in 44 developed and developing countries. Media from these countries were examined, including both print and broadcast media. He found that the greater the economic power of the country where the story was occurring, the greater the likelihood of coverage back in the viewing country. He also found that the more international news agencies positioned in a country, the more the country generated news. Not surprisingly, he also discovered that "the most covered countries are all world powers" (p. 20) with the U.S. projected as having the brightest spotlight.

The spatial configuration of news production is partly a function of the geography of media ownership, highly concentrated in only a few global cities. Worldwide, there is wholesale consolidation of media, with companies branching out from their traditional domains in search of increased market share. One limiting factor is regulation, but in the U.S., the Federal Communications Commission pursues strategies of deregulation.

Established in 1934 by the Communications Act, the FCC pursued a policy of localism, an early and long-lasting theme in telecommunications regulation (Benjamin et al. 2001, 23). By restricting the number of stations in large markets, the FCC tried to promote television stations in smaller markets and to encourage local ownership. Initially, only three stations nationwide could be owned by the same entity, but the limit was raised over time to twelve, plus a limit of 25 percent of the audience, with UHF viewers counting as half of the actual audience count (Benjamin et al. 2001, 316). The limit on number of stations was eliminated in 1996 with the "reach" limit extended to 35 percent.

In the initial decades of commercial television, the physical geography of broadcasting played a significant role in coverage areas, a role that is lessened by the "wired" and new wireless media. A traditional role of the FCC is to allocate stations so as to minimize interference, placing like channels at some distance from one another, and unlike channels in the same city. The landscape of channel allocation was relatively stable from the end of the 1948-52 "freeze" on new television license applications to the present. The intertwining of channel allocation, station allocation, localism, and corporate ownership is illustrated in the example of WABC in New York City. Established in 1947, ABC's flagship station was instrumental in seeing the network through its financially unsettled beginnings. With the possibility that the FCC would retain only the upper band of TV channels (channels 7-13), after the elimination of the old Channel 1, decisionmakers at ABC chose to establish Channel 7 as the ABC-owned station in five major cities including New York, hoping that Channel 7 would be the new "lowest" and therefore strongest channel (Allen 1997, 5-6).

In recent years, the Federal Communications Commission moved further down the deregulation path by favoring a higher concentration of media ownership. Proponents of deregulation argued that new competition was with broadcast and multiple-cable channels and that broadcast television needed to be able to better compete. Proponents also argued that relaxed regulation would safeguard any regulation in the face of legal challenges. Opponents argued that the resulting media concentration was simply too great, and the ownership mergers would be facilitated by the relaxed rules. These issues were still being debated in summer 2003.

The deregulation question was further complicated by the definitions of media concentration. The Republican FCC Commissioner who supported the change argued that popularity should not be punished. Critics interpreted the same fact differently, viz., that the most popular media are already highly concentrated. Critics also worry that the new rules, by permitting more concentrated ownership, will permit large companies to merge, further decreasing the number of corporate voices in the market.

Thus "localism" as a theme persists in rhetoric more than in reality. With alternate means, cable TV, satellite TV, and the Internet, "local to whom" becomes a question. As early as the 1980s, 30 percent of viewers in cable-equipped households in Bryan-College Station, Texas, watched nonlocal "local" news. Viewers found the nonlocal news more informative, better at reporting weather and sports, and some (12 percent) said that they actively wanted "local" news of the city to which they were tuned (Hill and Dyer 1981). Whether the Internet becomes a countervailing force in providing local news remains to be seen. One study called local television stations' websites the "metropolitan wide web," finding that larger station websites maximize their niche of providing local news, a function more difficult for the national and international news providers to fill (Singer 2001).

Deregulations, and the formidable debts accumulated by companies consolidating their holdings, have created some strange new hybrids masquerading as "local news." In the name of money-saving, the Sinclair Broadcast Group sends out the same signal from suburban Baltimore to Flint, Rochester, Raleigh, and Oklahoma City, and blends it with local stories, creating an amalgam of weather service and traffic reports and video links from far-flung studios that provide the impression of being right there (Farhi 2003; see also Rogers 2003). Such technical sleight of hand has been dubbed "centralcasting" and can produce what is presented as the "same news" with fewer employees.[4]

Radio faces the same kinds of financial pressures and is responding in much the same way, begetting deejays and traffic reporters with multiple personalities beamed to numerous local markets (Fisher 2003). Clear Channel Communications, Inc. grew from 40 stations before the 1996 deregulation to currently more than 1,200 and pipes in a mix of popular formats to different demographic markets in the same cities. When a derailed train in Minot, North Dakota released a cloud of toxic gas in 2002, Minot police hoped to use local radio stations to get the word out about the safety hazard. But all six local stations were owned by Clear Channel and were funneling in music from somewhere else, so nobody was home at the station to pass on the warning to listeners.

Media consolidations are not new and some of the largest date back to the 1980s when media began to involve some of the largest players in the corporate world, including the Walt Disney Company, General Electric, and Rupert Murdoch's News Corporation. These created new entities were called "content providers" (Emery et al. 2000; Downie and Kaiser 2002). Disney combined the ABC network news and local television affiliates with Internet ventures and its entertainment empire of movie studios, cable networks (including ESPN and the Disney Channel), theme parks, and sports teams. Viacom/CBS combined the CBS network and dozens of television and radio stations with movie studios, cable networks, and book publishers. NBC News was absorbed into General Electric, which created a partnership with Microsoft to create MSNBC cable news network on television and MSNBC.com on the Internet. At present, chains own 80 percent of America's newspapers, and 25 corporate owners control nearly half of the commercial television stations in the U.S. (Downie and Kaiser 2002).

Equally important to the geography of news production is the question about where news content is consumed. Americans tend to rely on television news, both locally and nationally broadcast, to obtain their news, usually when they are doing something else, such as cleaning house or making breakfast. The microenvironments of households and the placement of televisions within them, in living rooms, kitchens, bedrooms, and also in workplaces, schools, and transportation centers have what may be a measurable impact on how news is processed by the audience, with survey- and focus group-led programming pushing news agencies toward stories with a more domestic slant, or emphasizing threats to personal safety or celebrity intrigue.

In a society with smaller households and less free time, television serves as more than a home appliance and source of information. The growth in nonfamily and single person households in the U.S. has been a significant change. One-quarter of the American adult population lives alone, outnumbering married couples with children for the first time[5] (U.S. Census Bureau 2000). For those living alone, TV is the companion, the chatty neighbor, and the morning breakfast conversation all in one. A second important demographic reality is the effect of the aging of the U.S. population on TV viewership, and the distribution of wealth among older Americans as compared with other Western countries. Elders have the means to pursue leisure activities without the comparative burdens of fulltime work and family commitments, which creates demographic audience mixes that content providers target.

3. GATHERING THE NEWS: SPATIAL DIMENSIONS AND TECHNOLOGIES

To understand how 24-hour global news media have ascended to their dominant position, we must examine their roots in traditional journalism. However spatially concentrated their production presence is, media content providers have to travel to the locations where the news is actually happening, and these requirements have a set of spatial particulars all their own. Where a story occurs determines in large part whether it becomes a story or not, and if we believe that news is not what actually happens but what someone reports has happened or will happen (Sigal 1987), then accounts are by their nature dependent on someone getting to the scene. Reporters are seldom in a position to witness events firsthand, but have to rely on the accounts of others. Some developments such as socioeconomic trends, swings in public opinion, and shifts in official thinking may not manifest themselves in events. Journalists act as intermediaries and gatekeepers between the event and the interested audience.

Changes in technology of reporting have had significant impacts in the ubiquity of news-gathering. From attempts in the early 1800s of eastern seaboard port city newspapers to send fast boats to meet the incoming transoceanic vessels, news organizations have sought to gather news faster than the competition. Horse expresses (such as the famous Pony Express), the telegraph, and subsequent technologies assisted in moving the news faster. Still, the difficulty of knowing about newsworthy events in the first place remains. "Stringers," newspaper affiliated staff who can be paid for a reported story, send in reports, but the structure of newsgathering itself also matters.

In traditional journalism, other geographies of the news include "the beat," places such as police stations, the courts, schools, or subjects like crime, health, or sports, where newspapers assign their reporters to cover potential stories. The best newspaper reporters cultivate sources on their beat, read related documents, follow people and events over extended periods of time, accumulate expertise in their subject, and try to persuade their sources to divulge privileged information (Downie and Kaiser 2002). The result is that the reporter becomes steeped in expert knowledge about his/her beat and can present the best of the findings to the readership.

Both press releases and press conferences amount to the management of location by the source. Press conferences must be scheduled in advance to allow reporters to travel to the conference location, and in selecting these sites, urban hierarchy still matters. After journalists record the necessary information and ask questions, they return to their newsrooms to continue researching and writing their stories. In traditional newsrooms, editors and reporters are organized by geography—the various local beats, a Washington correspondent, sports, business and entertainment reporters, and local and national editing staffs.

Transmitting the news similarly involves a geographical process of moving the news as quickly as possible across often great distances. In the early days, the postal system was used, and later news was transmitted electronically through wires and then wireless. The flows of information through such media as the telegraph and the telephone were later revolutionized through the advent of computerization, which has made news production far easier through, for example, experiments in pagination in the 1970s and 1980s, the advent of the personal computer in the late 1970s, and still later, desktop publishing (Emery et al. 2000).

The early history of radio was marked by battles between government and the private sector for control of the medium. In the first national radio conference, organized by Secretary of Commerce Herbert Hoover in February and March 1922, order was brought to frequency allocations, with medium waves set aside for commercial use, leading to rapid development of commercial broadcasting (Hugill 1999). The frequencies allocated concerned local radio rather than wide-area radio. Lowest frequencies were reserved for government use. Short-wave was not recognized for long-distance potential until the agreements were in place to use short-wave for amateurs. Subsequent

American governmental and big-business attempts to reclaim the short wavelengths by political action in Congress were defeated by the large number of amateur radio operators who were able to lobby to preserve their interests.

The growing power of mass media thus reflects the potential of its technologies. A turning point in the realization of press power in the conduct of government came in 1950 when presidential news conferences moved from the Oval Office to a larger room in the Executive Office Building (Emery et al. 2000). The move occurred because the number of Washington correspondents grew; this growth formalized the presidential press conference. A dedicated press corps traveling with presidential candidates, and privy to opportunities such as exclusive interviews and TV sound bytes, reflects the importance of the press's positioning with respect to power. It also provides some reporters an all but unlimited vantage point to power and illustrates the growing role of national media in political affairs.

4. ADVENT OF TELEVISION NEWS

In print journalism as well as in radio, the reporter goes to the scene, conducts interviews and other legwork, then reflects on the newsworthiness of the story before composing it. In a sense, he/she is distilling out of the cacophony of data a clear message that will interest the presumed audience. Before TV, newsworthy events occurred in other places, and the public relied on the intermediary and interpretive role of the journalist to bring the news home. TV replaced the intermediaries by going directly to where news was happening and presenting the events themselves with less filtering by the news reporter (Downie and Kaiser 2002).

During TV's ascendancy, there was little expressed awareness of its potential as a conveyor of geographic information, but its growing ubiquity throughout the middle of the 20th century made such reflections unnecessary. By 1952 over one-third of U.S. homes had TV. "Coast-to-coast" broadcasting was made possible with the development of coaxial cable, extended first in 1946 between New York, Philadelphia, and Washington, by 1947 to Boston, and by 1948 to the Midwest. By 1951 AT&T had a microwave relay system to the West Coast. Most network programming still originated in New York (Emery et al. 2000), led by a triumvirate of three broadcasting networks, CBS, ABC, and NBC, which held sway over commercial broadcast frequencies for several decades.

The fantastic profits made from selling advertising during news and other programming made the 1960-70s the "Golden Age" for network broadcasting companies (Downie and Kaiser 2002), particularly for international news coverage through extensive networks of correspondents and foreign press offices. Access to news events such as the Vietnam conflict enhanced the power of TV news personalities such as Walter Cronkite to influence foreign policy.

A challenger arrived in June 1980 when the Cable News Network (CNN) was launched by Ted Turner. CNN's 24-hour service included hourly news summaries, heavy treatment of sports and business, news specials, and lengthy interviews on various subjects throughout the day. The initial experiment met with mixed reviews, as critics complained about poor visual quality and inexperienced reporters. The quick access to news and a direct, business-like approach, however, attracted many viewers. By the mid-1990s, CNN had a large Atlanta staff, a large Washington staff, eight news bureaus in the U.S., and 20 overseas (Emery et al. 2000). Satellite coverage was an important component of CNN success.

Satellites had been used earlier in the news business. The first private domestic satellite was Western Union's Westar, launched in 1974, followed by RCA, Comsat, and Western Union.[6] Time, Inc., through HBO, linked satellite and cable systems in 1975 to show the Ali-Frazier flight, known as the "thrilla in Manila." Satellites were used to offer video versions of press association content. Hubbard Broadcasting's CONUS system was established in 1984 as a consortium of local stations along with Group W's Newsfeed. CONUS joined with the Associated Press to form TV Direct to distribute TV news of Washington events along with AP newsphotos. By the 1990s, *USA Today, the Wall Street Journal, the Christian Science Monitor* and *the New York Times* all used satellites for domestic newspaper delivery. As a commentary on the changing technologies used with both print and broadcast journalism, AT&T closed its last teletype in 1991 (Emery et al. 2000).

The development of domestic satellites led to the quick expansion of the cable television system, with entrepreneurs like Turner starting their own news, sports, and religious networks and competing with the three major networks. By 1998, cable TV served 64 million homes. But before the nation's cable pattern was established, the FCC in 1982 approved roles for direct broadcast satellites. Rival companies began planning for home delivery of high-definition images in competition with cable operators and regular broadcasters.

CNN used satellites to reach viewers throughout the world simultaneously, but its audience did not materialize until the network outplayed the established networks in covering the Gulf War in 1991. This success was followed by expansion of the international cable systems, the development of more original domestic cable programming, and the addition of other companies to their list.

By the mid-1990s, new products of media collaboration were introduced at a rapid rate; they were based on the technological potential offered by digital electronics, which convert information, sound, video, text, and images into a single transmittable code. Participating in these developments were the new communication, computer, electronic, and entertainment conglomerates. Future predictions were made that high-definition television, interactive services, and image processing would be brought to home and office over broadband fiber-optic or cable lines. "Convergence," as it is called, has been billed as a kind of consumer media "one-stop shopping," where a DSL (Digital Subscriber Line) fiber or cable modem line provides access to broadband Internet, and eventually HDTV (high-definition TV) and first-run Hollywood entertainment.

When the effects of corporate consolidation and cost-cutting began to be felt in the broadcast media world, the specifications for what news was reported on TV began to change, affecting notably the number of foreign correspondents. This change in direction was rationalized by the news producers' parent corporations as necessary cutting. But it also reduced the extent to which TV news correspondents could develop "beats" the way their print counterparts could. It also reduced the breadth and depth of international news coverage.

The current atmosphere of cost-cutting, combined with the prevailing perception of the audience's lack of interest in international stories, have meant that foreign correspondents and stringers have been replaced by a kind of "base camp" mentality (Downie and Kaiser 2002). Instead of developing contacts and achieving what might be called a contextual knowledge about a place, correspondents fly out of their base camp just in time to reach the place where news is happening. This capability is made possible through another ubiquitous technology, jet air travel. "Just-in-time reporting" assumes correspondents and camera people have the background and contacts to make sense of events literally on the fly. The prize is good footage and a scoop over other news gatherers, but at the cost of depth and insight that comes from having a beat.

It is unfortunate in the reality of TV news that the traditional geographies of news collection such as the newsroom and the beat have been replaced by reliance on the AP Wire, windowless television studios, and target audience data which drive the selection of program content. The base-camp mentality can produce some surprising omissions, such as for instance the "invisible" war in the Congo where over 3.5 million people were killed during the early years of this decade in interethnic fighting but very little was reported because the area was inaccessible and inhospitable to foreign correspondents (Astill 2002). In fact, there are a startlingly large number of places and regions where little or no news occurs on a daily basis, including most of Africa, South

America, South Asia, and even Canada, unless a conflict, disease outbreak, or natural disaster gets it into the evening's headlines.

5. THE ROVING NATURE OF NEWS INTEREST IN THE GLOBAL ARENA

Since the early 1990s, the advent of global news has taken hold in the form of BBC World Service and CNN International serving as the main providers. Cable and satellite television networks such as Fox and MSNBC now also provide the information that audiences demand and tailor it to shorter attention spans. CNN Headline News runs like a stock ticker or a busy Internet site with moving headlines but little context or insight.

The significance of these content providers as global news institutions should not be lost. These networks cross international boundaries and present self-consciously globalized information to the world. After all, it was CNN that banned the term "foreign" in its broadcasts because CNN has pledged not to represent any specific nation, including the U.S., in its coverage. As a testament to CNN's power, it gets fire from all sides. Its position of power invites charges of bias, such as when Peter Arnett's reporting from Baghdad during the Gulf War was criticized inside the U.S. for being pro-Iraq.

A legitimate question that has been raised is: Precisely how international are the cable networks? or are they simply American formats broadcast to other countries? The controversy over CNN's coverage of the Israel-Palestinian conflict during the past decade concerned how Israeli and Palestinian households were faring under the current security arrangements. In the Middle East, where tensions were already high, CNN found itself having to defend its founder, Ted Turner, after he ignited a firestorm in Israel when he said both the Israelis and the Palestinians were engaged in "terrorism," compounding an already tense situation over the coverage of suicide bombers' families. CNN is often perceived in the Arab world as being biased toward Israel, while viewers within Israel consider CNN's coverage as tilted toward the Palestinian cause.

International stories gained a perverse boost after the events of September 11, 2001, when the viewing public in the U.S. and elsewhere demanded explanation for the conspiracy aimed at tarnishing American symbols of wealth and power. Network and cable TV news viewership grew for about six months after 9/11, but by mid-2002, it was back to pre-9/11 levels.[7] September 11 could be seen as an attack on a node of the central nervous system, in that the New York target was close to Wall Street, media outlets, and telephone and cell phone nodes.

The media also found themselves after 9/11 having to deal with a viewing public in need of consolation, explanation, and an outlet for anger. But the stories were often repetitive, especially the footage of the planes crashing into the World Trade Center. Michael Massing of the Columbia Journalism Review observed a lack of interest in dissent and debate:

> Many talk shows came across as exercises in enforced conformity. All the bad habits that TV talk shows have developed in the last few years–the arrogant shallowness, the false contentiousness, the endless hours of empty analysis and know-nothing opinion–have carried over into the current crisis. (2001, 23)

The spike in international news coverage after 9/11 eventually ended, and news producers were forced back into pondering why they cannot seize the audience's interest in stories from other parts of the world. One way to do this has been to tie coverage back to the lives of the readers, a technique that worked very well during the second Iraq war in March 2003, when approximately 500 journalists received permission from the military commanders to live and travel with the troops and report on their activities in real time, a practice that was termed "embedding." While observing military action, reporters were forbidden from reporting about ongoing missions (unless cleared by on-site commanders) or on specific results of completed missions, breaking embargoes, divulging specific numbers of troops, aircraft, or ships, and traveling in their own vehicles. Included with American troops were reporters from Al-Jazeera, Itar-Tass, and China's Xinhua, about 300 news organizations in all (Bushell and Cunningham 2003).

Hailed by the Pentagon as representing a new era in openness and press-military cooperation, embedding helped personalize the experience of war and thereby appeal to families and supporters of the troops back home. One of the criticisms of embedding was that it also created a "cameo of perception" rather than a total picture of the impact of war on a large scale. Another criticism was that embedding distorted the war because some kinds of action (for example, fire fights) get covered disproportionately because they are considered visually interesting.[8]

An international example of the potential transferability of Western technology in the production of 24-hour global news is the Qatari satellite news network, Al-Jazeera (el-Nawawy and Islander 2002), which gained new prominence in the aftermath of 9/11. With a viewership of approximately 35 million and about 200,000 subscribers in the U.S. and Canada, Al-Jazeera can claim to be a global news network on a par with CNN and the BBC, although it broadcasts exclusively in the Arabic language. Owned and financed by the government of Qatar, its slick, high-production-value broadcasting combined with its controversial coverage of Osama bin Laden and the Afghanistan war earned it millions of new viewers. Currently it employs 350 journalists and 50 foreign correspondents working in 31 countries (mainly in the Arabic-speaking world but also in Europe and the U.S.).

One reason for Al-Jazeera's meteoric rise in the Arab world was its embrace of controversial topics and dedication to presenting differing sides to each story, then and still unusual for Arabic language broadcasting. The network's motto is: "The opinion, and then the other opinion" (el-Nawawy and Islander 2002, 11). Al-Jazeera provided its global Arabic language audience with the first exposure to opposing voices using the medium of television. For instance, Al-Jazeera was the only Arab network to cover the release of a critical Amnesty International report on Saudi Arabia in 1999. Comparisons to CNN are inevitable. Although Al-Jazeera is not a Western media network, its managers and producers have taken a page from the Western media's playbook. Like CNN, Al-Jazeera is frequently criticized for its views, and some Arab regimes have accused it of being a megaphone for dissident voices and a conspirator in antigovernment movements, while others have acknowledged the network's role as the sole voice of journalistic objectivity in a conflict-ridden region. But its similarity to "Western" networks such as CNN raises questions about content and technology. Is the format of the 24-hour news program inherently "Western?" Is technology itself "Western?"

6. CHANGING NEWS MEDIA REFLECT CHANGING LIFESTYLES

The high value placed on images in television news means that reporting of some places is inherently more appealing editorially than others. While there is a need to avoid perceived geographic bias, that is, spreading the originating locations around, this is balanced by a need to reinforce those places that are important to the audience (or the reporters). Trading partners and places of importance to sponsors get more attention, as do places with demographic and cultural similarities, exotic or unusual places, and nearby places. Mostly though, stories can occur anywhere as long as they interest the targeted audiences in the U.S. who are often largely suburban and middle-class consumers, fueling a trend toward "infotainment" rather than hard news. Important subjects in "new news" (Downie and Kaiser 2002) become those that concern families, including health and science issues, generally not world affairs or politics, which is ironically not seen as "real" in people's lives. For local TV news, what viewers are perceived to want based on polls and viewer metering is weather, traffic, health, sports, safety, and shopping.

Yet the process of determining news values and the subjective nature of consumer demand that underlies it is still a crude science, and TV content providers themselves freely admit that there is "no real sense of what the audience wants" (ABC anchor Peter Jennings, as quoted in Downie and Kaiser 2002, 138.) Despite the stampede toward market research, there is a vague sense that the audience is not getting what it wants, since network television viewership continues to decline. This is in spite of the fact that much of the decline can be attributed to the growth of cable TV. Reversing this trend may require satisfying the desire for interactivity, the ability to talk back, which is now only possible through an Internet portal.

Given the extent to which American society is already awash in information created through marketing and demographic studies to suit perceived interests, there are signs of viewer adaptation. And the rebellion on the part of media consumers ironically works to the advantage of media providers. Everyone wants to feel "connected" and will choose from any number of media to do so. This "freedom" will eventually spell doom for traditional methods of transmitting news and advertising messages. Communication can be done through email, cell phones, the rise of compiled web content, web 'zines, and self-published material, all examples of do-it-yourself technology. But broadcasting arrives at the end-user's set prepackaged; it only goes one way.

Certainly this tendency toward interactivity is fueled by the broadband; i.e., the convergence of content and delivery vehicles (cable, Internet, film entertainment). Such a migration will offer more choices than ever, but its impact on the news media is as yet unseen. The emergence of "new media" may provide some clues about the nature of future demand for information and entertainment. The Internet has emerged as a legitimate source of information. Over one-half of U.S. households now use a computer at home, and over 40 percent go online (U.S. Census Bureau 2001). The Internet has opened up countless new horizons, from on-line libraries to chat rooms to live-streaming media to digital compilations of local and international newspapers. The Internet also opens up another set of options for "news grazers" (Pew Center 2002), who are younger and use a varying combination of television, Internet, and radio to obtain their news. Grazers have a lower level of involvement and view only when "something is happening." In short, a smaller place- or event-specific knowledge base is created.

New technology makes "being there" easier, which has certain implications for the geography of news. Internet-based journalism is revising the rules of reporting and redefining the relationship between the observed and the observer. People around the world are using digital cameras for local access TV shows and also producing print or web 'zines featuring their own particular view of the world. Interactive and gloriously seat-of-the pants, this do-it-yourself digital journalism is changing the way audiences will use media in the future (Cox 2002). Three short examples follow:

First, *Washington Post* writer Joel Garreau (2002) recently reported on the phenomenon of "cell swarming," in which ad hoc networks of friends connected by mobile telephones share information in a local network and coordinate their activities instantly and leaderlessly. A locally based form of news is prescient because it portends a future era of ego-centered news. Swarming turns the idea on its head that geography in an Internet age has become irrelevant, because the idea is to get people together in one location for face-to-face contact. In addition, technology can also serve as an enabler of political mobilization, such as when cell phones were used in the upheaval that overthrew Philippine president Joseph Estrada in 2002, and in a similar way that fax machines made the Tiananmen Square uprising in 1989 possible, cassette recordings inspired the Iranian revolution in 1979, and shortwave radios aided the French Resistance. Media technology is life altering, with new kinds of commitment, new views of time and space, and new methods of social and political mobilization.

Second, conceiving the future as a kind of do-it-yourself journalism is the aim of a new set of Internet practitioners called "bloggers," or web loggers. A "blog" is a constantly updated combination of diary and link collection[9] (Shachtman 2002; also see Sullivan 2002). These developments threaten to turn the traditional world of journalism on its head. MSNBC, Fox News, and Slate have all added blogs to their websites, and in 2003, the University of California, Berkeley offered its journalism students training in blogging. *Washington Post* media critic Howard Kurtz (2002a) sees the explosion of media formats as allowing more voices into the grand echo chamber that used to be controlled by a handful of huge media corporations. While it is difficult to imagine this development ever replacing even a small portion of the audience that now tunes into cable or broadcast news, it does encourage ordinary citizens to be empowered by the information they can access on the Internet. And it helps reporters and correspondents get closer to stories and obtain better information. The year 2003 saw the advent of map-based blogs in some coastal cities, compilations of locally based information linked to a common spatial referent.

In September 2002, the search engine firm Google introduced a new service called Google News, a kind of news search that combs daily web sites for important stories and passes them through a common Internet portal, untouched by human hands. To media critic Howard Kurtz, it raised a question:

> Is news in the Net age just a bloodless compilation of electronic connections to global media outlets, allowing users to point-and-click their way to bleary-eyed nirvana? Or does it require sharp editing judgments, seasoned beat reporters, provocative columnists and a small dollop of personality? (2002b)

Either way, the success of Google News suggests that the Internet is and will be into the future the site for innovation in media.

A final new technology worthy of note is the digital video recorder (DVR), also viewed by the industry as a warning sign of TV's impending doom. An example of a DVR, TiVO® is an electronic box that sits on your television set and records television programs digitally to a hard disk and lets the viewer fast forward through the advertisements. Analysts cited by *Newsweek* predict that nearly half of American households will own DVRs by the end of the decade, thereby threatening to destroy the 30-second ad and to open up TV broadcasts to Napster-like file sharing. Industry representatives warn that $60 billion in advertising revenue is imperiled.

Television news broadcasting is profitable and successful, but it is also in a sense doomed: in the future, it will not be able to find the masses of viewers because the masses themselves will be so fragmented. Search engine outputs can be tailored to individual tastes, challenging the likes of CNN and Fox to identify an audience it can pitch its appeal to. One key to success for media companies will be to make the news more relevant for its viewer-consumers, and doing so by necessity, they must bring in a stronger and more nuanced conception of geography. This innovation means more *locally* relevant *global* stories that demonstrate interconnectedness and human interest; more stories that excite viewers' curiosity by providing what geography is good at, the "why behind the where;" more context that allows viewers to set the story better in time and space, and finally more insight on the significance of the news. We need to better understand how news viewing is worked into people's daily lives. News has to go beyond simply providing immediacy to also providing context and perspective. Can we enrich these geographies?

7. MEDIA AS MAPS

The key for news media to capture new viewership is through seizing the audience's interest. Success in this endeavor can currently be measured through targeted surveys, focus groups, and polls. This process works and is informative but will not result in better content until viewership can be interpreted more accurately, which means not just counting eyeballs but understanding responses and putting them into context. We need to yoke more closely together the interests of the audience and the news organizations' ability to locate, interrogate, and present content that is relevant to the audience. The question is, what can geographers contribute?

Digital video recorders present some opportunities. Georeferenced databases of media content, combined with analyses that compare base statistics such as home ownership, demographics, consumer behavior, and crime statistics with media reports to assess impacts on home values, may finally convince news producers that audiences actually live somewhere and seek to make connections between their worlds and the larger ones. We suggest a hybrid approach that could help shift the emphasis from reporting to monitoring and interactivity, and one that is inspired by a more informed notion of viewer impacts, possibly through blending of news with statistics, polling, and surveys as gauges of public sentiment in specific local places. We would call this *geo-journalism*.

The deepest need amid the voluminous amounts of data is insight, that is, the means to discover order in the daily torrents of news. In the splintered and market-segmented world of the future, there will be no single perspective, no Walter Cronkite "to tell it the way it was." The ultimate goal should be an educated public. Are media doing this? How can this be measured? To remedy the fact that those employed in local stations often have a weak notion of geography, we suggest some agenda items. First, examine the spatial distribution of news stories by location. Georeferenced digital databases and archives could be based where stories can be sorted by location or theme (or any other user-specified variable). These would create new datasets for analysis. Digital video recorders would also appear to have great potential for research on media content. Devising research methods that compare media representations with statistics by location could illustrate mismatches. A good place to start will be with crime coverage on local TV news. Finally, time and space allocation studies that look at media use across different demographic groups (wealth, age structure, education level, and household type) could help answer questions about the level of satisfaction with the information provided. This information would help inform an understanding of how people work news viewing into their daily lives and determine the influence on their behaviors, such as voting and consumer preferences.

Going further, we suggest that formats for a more informed and contextualized geography news be taken up by both news organizations and geographers. One well-intentioned and very successful experiment thus far is the new National Geographic Channel, with its redrafting of news about the natural world, culture, and technology. More potential collaborations could be arranged between local media and community groups, local governments, and civic organizations.

Looking ahead, we envision continued growth of cable, satellite, Internet, and wireless news providers. Their success lies in selling the idea that planetary monitoring on a real-time basis is something that audiences want, regardless of location, culture, or income level. But what does 24-hour news on demand tell us about U.S. society, our ignorance of places, U.S. hegemony, and our world view? Difficult to measure but intuitively robust is the notion of our own rooted geographical context growing weaker as we get plunged into the bath of impressions that is 24-hour global news. What does "local" mean now, anyway? Geographers and others have assumed that to be real, a representation has to have geographical dimensions or consequences. We believe that media providers themselves will soon come to face this issue as local programming and perspectives continue to be eroded by deregulated corporate control of the airwaves and the bandwidth. Then we might just have a revolution on our hands. As the technical details of point-to-point distributed network communications become more routine, people will rely increasingly on their own trusted networks for information. They will mistrust official and commercial sources and cultivate independent unbiased voices.

CNN invented the always-on global view of the newsmaking world. Twenty-four-hour global news is an American form, and with the advent of Al-Jazeera, it is growing into an international force. The success of embedding journalists during the 2003 Iraq war suggests that more of this form of newsmaking will soon come our way. Perhaps Hollywood is already thinking of ways to combine "reality TV" with straight journalism to produce even more sensationalized coverage of actual events. We would not put it past them. We would also hope, however, to see the return of more intrepid shoe-leather kind of journalism that exposes the hypocrisies of power. Perhaps the 24-hour eye on the world will make hiding important information from the public harder to get away with. It is journalists' and geographers' jobs to provide better explanations, more perspectives, and more opinions.

What is the future of journalism? Is it a geographical future? We expect to see the rise of Internet newspapers, blending the name brands and comprehensiveness of the old press with the newness and multimedia ubiquity of the Web. Beyond this, we expect to see far more "new media" in the form of do-it-yourself personal communications networks: web logs, PDA- and cellphone-based devices, or map-based real-time venues to information, opinion, and entertainment.

A deeper question related to the meaning of the news is what happens when that news is no longer solely text-based but is almost entirely visual. If people do not "read" the news anymore but "see" what they know or want to know, will this threaten the informational purpose of journalism? Yes and no. Image-based news may be intriguing, but it cannot replace the word, either spoken or written, as a conveyor of information. For media to "go visual" entirely may mean that insight increasingly will be found elsewhere, in face-to-face encounters and small-group formats, as well as in chat sessions online or in web logs. Most of this new content will be of a fleeting nature and far from our standard notion of the enduring nature of things that is the ken of most geographers. This future should not discourage us; we need to bridge the gap.

NOTES

1   Of U.S. households, 98.2 percent own one or more television sets (2000), a higher percentage than that of telephones (94.4 percent). The numbers are quite similar in other countries: Norway (97 percent), Finland (98 percent), Egypt (90 percent), and Algeria (66 percent) (U.S. Census Bureau 2002). All 2001 data are from the UN Statistics Division.

2   In the compendium of U.S.-based geographical research, Geography in America, little formal study in the form of journal articles and book chapters have been devoted to the media in recent years, particularly TV news (Gaile and Willmott 1989; Personal correspondence with Gary Gaile, August 16, 2002).

3   A Pew Center release (2002) "Trend in Regular News Consumption," p. 2, found the following sources cited as news media sources used "yesterday" as of April 2002: local TV news 57 percent, cable TV news 33 percent, nightly network news 32 percent, network TV magazines 24 percent, network morning news 22 percent, radio 41, call-in radio shows 17 percent, National Public Radio 16 percent, newspaper 41 percent, and online news 25 percent.

4   Gannett has a plan to produce a national channel called "America Today" which will be an assortment of "local news" from 22 affiliate stations and aired on a grid schedule so travelers away from home can tune in and watch local news beamed across the country.

5   Married couples with their own children under 18 years: 23.5 percent, as compared with Householder living alone: 25.8 percent.

6   Most such traffic is now carried by fiber-optic cable, due to its lower cost.

7   During the time immediately after Sept. 11, 2001, interest in international news spiked. By the next year, however, interest had returned to its earlier levels (Pew Research Center 2002).

8   Thomas Lippman, *Washington Post Live* Online Chat, April 13, 2003.

9   For a more critical assessment, see Alex Beam (2002), who downplays the significance of blogging.

REFERENCES

Abler, Ronald, Janelle, Donald, Philbrick, Allen, and Sommer, John (Eds.) (1975). Human Geography in a Shrinking World. North Scituate, MA: Duxbury Press.

Allen, Craig (1997). Tackling the TV Titans in Their Own Backyard: WABC-TV, New York City. In Murray, Michael D. and Godfrey, Donald (Eds.) Television in America: Local Station History from Across the Nation, 3-18. Ames, IA: Iowa State University Press.

Astill, James (2002). Congo: An Everyday Story of Horror and Grief: Tentative Talk of Peace Matters Little to Victims Who Have Lost Everything, Guardian Foreign Pages July 24.

Beam, Alex (2002). In the World of Web Logs, Talk is Cheap, Boston Globe April 2.

Bushell, Andrew and Cunningham, Brent (2003). Being There, Columbia Journalism Review 41: 18-21

Benjamin, Stuart Minor, Lichtman, Douglas Gary, and Shelanski, Howard A. (2001). Telecommunications Law and Policy. Durham, NC: Carolina Academic Press.

Brunn, Stanley D. and Leinbach, Thomas R. (Eds.) (1991). Collapsing Space and Time: Geographic Aspects of Communications and Information. Boston: Harper Collins.

Burgess, Jacquelin and John R. Gold (Eds.) (1985). Geography, the Media and Popular Culture. New York: St. Martins Press.

Downie, Leonard, Jr. and Kaiser, Robert G. (2002). The News About the News: American Journalism in Peril. New York: Alfred A. Knopf.

el-Nawawy, Mohammed and Iskander, Abel (2002). Al-Jazeera: How the Free Arab News Network Scooped the World and Changed the Middle East. Cambridge, MA: Perseus Books.

Emery, Michael C., Emery, Edwin, and Roberts, Nancy L. (2000). The Press and America: An Interpretive History. Boston: Allyn and Bacon.

Farhi, Paul (2003) TV News Central: One Source Fits All. Washington Post, May 31.

Fisher, Marc (2003). Sounds Familiar for a Reason, Washington Post, May 18.

Gaile, Gary and Willmott, Cort (1994). Geography in America. Columbus, OH: Merrill Publishing Co. 2nd Ed. forthcoming.

Gans, Herbert J. (2003). Democracy and the News. Oxford: Oxford University Press.

Garreau, Joel (2002). Cell Biology, Washington Post July 30.

Gitlin, Todd (2001). Media Unlimited: How the Torrent of Images and Sounds Overwhelms Our Lives. New York: Henry Holt.

Hill, David B. and Dyer, James A. (1981). Extent of Diversion to Newscasts from Distant Stations by Cable Viewers, Journalism Quarterly 58: 552-55.

Hugill, Peter J. (1999). Global Communications Since 1844: Geopolitics and Technology. Baltimore, MD: Johns Hopkins University Press.

Jakle, John A, Brunn, Stanley D., and Roseman, Curtis C. (1976). Human Spatial Behavior: A Social Geography. North Scituate, MA: Duxbury Press.

Kurtz, Howard (2002a). Washington Post Media Backtalk, March 19. www.washingtonpost.com/wp-srv/liveonline/02/politics/kurtz031902.htm

Kurtz, Howard (2002b). Robotic Journalism: Google Introduces Human-Less News, Washington Post Media Notes, September 30.

Massing, Michael (2001). Talking Heads Go to War, Columbia Journalism Review 11: 4.

McLuhan, Marshall (1964). Understanding Media: The Extensions of Man. Cambridge, MA: MIT Press.

Pew Center for People and the Press (2002). Publics News Habits Little Changed by September 11, June 9.

Rogers, Patrick (2003). Television: More is Less, Columbia Journalism Review 41: 5.

Shachtman, Noah (2002). Blogging Goes Legit, Sort of Wired News, June 6. www.wired.com/news/print/0,1294,52992,00.html

Sigal, Leon V. (1987). Who? Sources Make the News. In Robert Karl Manoff and Michael Schudson (Eds.) Reading the News, 9-37. New York: Pantheon.

Singer, James B. (2001). Metropolitan Wide Web: Changes in Newspapers Gatekeeping Role Online, Journalism and Mass Communications Quarterly 78: 65-80.

Stone, Brad (2002). The War for Your TV, Newsweek July 29.

Sullivan, Andrew (2002). The Blogging Revolution, Wired Magazine May 10. www.wired.com/wired/archive/10.05/mustread.html

U.S. Census Bureau (2002). Table DP-1 Profile of General Demographic Characteristics: 2000. Washington, DC: U.S. Census Bureau.

U.S. Census Bureau (2002). Statistical Abstract of the United States. 122nd ed. Table 1103: Utilization of Selected Media, 1970 to 2000. Washington, DC: U.S. Census Bureau.

U.S. Census Bureau (2002). Statistical Abstract of the United States. 122nd ed. Table 1159: Internet Access.Washington, DC: U.S. Census Bureau.

Utne, July-Aug 02. Craig Cox. Launch Your Own Media Empire.

Wu, H. Denis (2003). Homogeneity Around the World? Comparing the Systemic Determinants of International News Flow Between Developed and Developing Countries. Gazette: The International Journal for Communications Studies 65: 9-24.

CHAPTER 15

JOANNA REGULSKA

DEMOCRACY AND TECHNOLOGY

Abstract   Does access to information and communication technologies (ICTs) reshape the meaning, understanding, and practices of democracy? Will citizens feel less alienated, and will they become more engaged? This chapter focuses on two dimensions of ICTs, cyberdemocracy and e-government, to illustrate how the spread of ICTs in central and east Europe has begun to impact democratic practices and democratic communications. While successful participatory initiatives are few, nevertheless the countries in these region have begun to create mechanisms which allow individuals and groups to enter virtual political space.

Keywords   e-democracy, e-government, cyberdemocracy, information and communication technologies, central and east Europe

Does access to information and communication technology reshape the meaning, understanding, and practices of democracy? Is the Internet different from previous communication media and, if so, will access to the Internet bring new possibilities for democratic communication? Will citizens feel less alienated, and will they become more engaged? Will the gap between those marginalized and powerful vanish, increase, or stay the same? Scholars, citizens, politicians, and policymakers have been raising such questions with increasing frequency as they debate the implications of the spread of information and communication technologies (ICTs) on the ways in which politics and business are conducted, social spaces are constructed, participatory practices are altered, and new attitudes, values, and behaviors are exhibited by citizens (McCaughey and Ayers 2003; Saco 2002; Ebo 2001; Meyer and Hinchman 2001; Norris 2001; Porebski 2001; Rengger 1999; MacGregor Wise 1997). This chapter focuses in particular on one aspect of this debate: will democratic practices be altered as a result of ICTs dissemination, and if so, how?

The rapid spread and great interest and hopes engendered by new ICTs have resulted in the creation and use of a wide range of terms and concepts, and of a plurality of definitions. Thus the terminology use ranges from

Stanley D. Brunn, Susan L. Cutter, and J.W. Harrington, Jr. (Eds.), Geography and Technology,
339-361. © 2004 Kluwer Academic Publishers. Printed in the Netherlands.

teledemocracy (Arterton 1987 in Porebski 2002), cyberdemocracy (Hagen 1997; Ogden 1994; Gibson 1984), virtual democracy (Norris and Jones 1998), digital democracy (Hague and Loader 1999; Norris 2001), electronic democratization and electronic-democracy (e-democracy) (Hagen 1997), to e-governance (Norris 2001). The result, not surprisingly, is that there is still no agreement on terms to be used when talking about a new model of democracy in relation to information and communication technology. As Porebski notes, not only are many different terms used to describe similar phenomena, but proliferation of such a range of terms and concepts, while an important advancement in itself, does not facilitate discussions and debates (Porebski 2002).

The relatively short history of electronic mail, digital and wireless technologies, and computer conferencing reveals that fundamental alterations are occurring, not only in some people's daily practices (at home and at work) and the way they engage in social relations, but also in the multilayered communities that are emerging. It is too early to provide a thorough evaluation, and in fact, there is still a shortage of systematic analysis of what is being done, where, and by whom in the area of electronic democracy and governance. Nevertheless, it is clear that information is available faster and in larger quantities, and that contacts are no longer restricted by place and time: being place-independent and of almost instant 24/7 availability, they cross nation-state borders.

Enthusiasts are also increasingly aware of many challenges and negative implications. How does the fact that not all citizens have and will have access, skills, or resources to become users of ICTs impact whose voices wish to be heard? How is social communication constructed and conducted? How are working relationships structured, and how are politics being accomplished? Does the fact that power relations and, often, the social and cultural markings of individuals are hidden reaffirm or weaken socioeconomic, racial, ethnic or gender divisions? Are governments indeed trying to create new participatory possibilities or do they only attempt to refurbish ways in which things are already done?

In this chapter, I focus on two dimensions of ICTs—cyberdemocracy and e-government—in order to show how the spread of ICTs in one particular region of the world has begun to impact democratic practices and democratic communication. The region of central and east Europe (CEE) will serve as a lens through which this analysis is carried out. This chapter consists of two parts and a conclusion. The first part examines different notions of e-democracy and e-government and looks at the benefits and challenges that their adaptation brings. The second focuses on applications of ICTs and specifically the use of the Internet in CEE countries and the countries of the former Soviet Union

(FSU). I also review recent e-democracy and e-government initiatives. The chapter concludes by arguing that while fully participatory e-democracy initiatives remain few, the countries of the region have embraced its two other functions, viz., the provision and exchange of information. The widespread use of the Internet by nongovernmental organizations, associations, foundations, and formal and informal groups indicates the creation of new mechanisms through which individuals and groups can enter political space. E-government initiatives on the other hand seem to have done little more than provide a new means for delivering a standard and traditional set of services.

1. INFORMATION AND COMMUNICATION TECHNOLOGIES, E-DEMOCRACY, AND A BETTER TOMORROW?

The disagreements about the definitions of the new "e" vocabulary as well as about how different societies will be able to benefit from acquiring new technology are plentiful. In a thoughtful typology, Hagen (1997) distinguishes among teledemocracy, cyberdemocracy, and electronic democratization. He argues that the necessity to recognize and clearly define each term stems from several facts. First, each uses different technologies to advance its aims with teledemocracy (cable TV), cyberdemocracy, and electronic democratization (computer networks). Even so, this latter difference may eventually be eliminated with the shift to a greater use of computers.

Second, Hagen contends that these concepts differ according to the form of democracy (direct or representative) and the mode of political participation (direct participation through voting, activism, information exchange, discussions). For him, the split between the desire to establish direct democracy as articulated by teledemocrats and cyberdemocrats and the desire only to fix the malfunctioning of the representative system, as advocated by electronic-democratization, lies at the core of this difference. Thus, while the first two forms will attempt to foster development of new ways through which citizens would enhance their ability to vote, gain information (teledemocracy), and engage directly in political activities (cyberdemocracy), the supporters of electronic democratization would focus only on the enhancement of the diffusion of information, without changing and reconceptualizing much of what they do. All three will, however, foster greater discussions between and among citizens, groups, and politicians, and in his opinion, teledemocracy and electronic democratization will also facilitate the provision and exchange of information. The omission of cyberdemocracy from this latter category is actually surprising given that cyberdemocracy embraces concrete political actions that cannot actually take place without information being exchanged.

Sakowicz (2003) on the other hand includes e-democracy as one of the dimensions of e-governance, which covers also e-services, e-management, and e-commerce. From his perspective, the preference for e-governance accounts for a shift away from seeing governance as limited to service delivery only (with clients passively accepting services) to a more inclusive concept. He suggests that e-governance is based on "the online engagement of many stakeholders in the process of governing and implementing policies" (Sakowicz 2003, 25). Such a multilayered definition focuses on the participation of different groups, including citizens, nongovernmental organizations, policymakers, politicians, business, and academia as they jointly take part in the decisionmaking processes that directly affect their communities. For Sakowicz, e-democracy is defined as

> the use of ICT as an instrument to help set agendas, establish priorities, make important policies and participate in their implementation in a deliberative way. (2003, 37)

He stresses that the goal is to increase citizens' involvement, through which "newly empowered citizens may emerge in the form of Internet-based alliances responding to issues" (2003, 37).

In the study on integrated e-government conducted by the Bertelsmann Foundation, these distinctions are blurred even further, as in their definition of e-government:

> combines electronic information-based services for citizens (e-administration) with the reinforcement of participatory elements (e-democracy) to achieve the objective of "balanced e-government." (Bertelsmann Stiftung 2002, 4)

Norris, however, rejects this notion by pointing out that while cyberoptimists suggest that "the main potential for digital technologies for government...lies in strengthening policy effectiveness, political accountability," public participation will be affected to lesser degree (2001, 113). Thus, in her opinion, e-governance at best

> holds great promise for the delivery of many types of public services from housing and welfare benefits to community health care and the electronic submission of tax returns, reconnecting official bureaucrats with citizen-customers. (Norris 2001, 113)

The linguistic chaos and commotion over definitions are no doubt further reinforced by contentions over the impacts that ICT has and will have. Supporters, optimists, and/or technoromantics have often seen the Internet as bringing unlimited advantages for the strengthening of democracy. Among those advantages most frequently cited are broadening citizens' access to information and enhancing political participation opportunities, in decisionmaking as well as in actual organizing and mobilizing, and also disseminating and exchanging information (including political information) (Meyer with Hinchman 2002, 119; Norris 2001; Porebski 2001). Norris's (2001) extensive study of 179 countries found that among the key variables affecting the number of Internet users is not only GDP, but also R&D, literacy, secondary education, and democratization. In short, while economic development plays a significant role, other factors, such as the level of democratic consolidation, human capital, and no doubt the cultural context are equally significant. Herron concluded his analysis of government generated and disseminated information in the FSU by indicating that the "democratization, economic development and open provision of information are closely related" (1999, 58). In practice, this means that certain groups of citizens will become further empowered, better informed, and more autonomous as political subjects because of the access and use of ICTs, but also many others will not.

Porebski (2002) argues that as the use of the Internet reduces drastically the costs of political participation and provides for greater flexibility of running political campaigns; the process of political mobilization becomes easier, more accessible and therefore more effective. In particular, he points out that individual organizations could draw extensive benefits as their communication efforts are enriched by better information, and groups are more effective in reaching their target audience, as these efforts are less expensive. O'Lear echoes some of these attributes in her study of grassroots environmental activist groups in Kaliningrad, Russia:

> e-mail is a valuable communication medium that eases interaction among people in disparate places who wish to participate in and contribute to a long-distance, politically active community. (1999, 76)

Indeed, Norris's studies concluded that the biggest beneficiaries would be smaller groups that can more easily adjust to changing circumstances as they are more flexible, have fewer liabilities, and often are quicker to adopt innovations. This development, she argues, is

> particularly important for the process of democratic consolidation, and for the opposition movements seeking to challenge authoritarian rule around the globe. (Norris 2001, 21)

Simultaneously, arguments are made that because of unequal access, ICTs threaten to undermine democratic principles. Low computer literacy, lack of access to technical infrastructure, and/or high costs of use will contribute to further social stratification and will determine who can become a user. The already existing deep divisions along race, class, and ethnicity often parallel the visible alienation of these groups, their growing indifference, and the decline in active political participation; these present challenges to democracy. One of the concerns frequently echoed has to do with worries that the spread of Internet use will actually increase the socioeconomic divide between users and nonusers and create yet another dimension along which societies will be fragmented. Referring to the digital-divide argument (and what Norris refers to as the "social divide," which for her represents one dimension of the digital-divide, the other being global and democratic), Norris (2001, 4) claims that new class divisions are unavoidable as often, for large segments of society (because of their class, ethnic, or gender status), access to the Internet is neither affordable nor available. Thus new divisions could reinforce an already existing divide between younger, skilled, well-educated, and professional populations, who often represent a more independent thinking segment of society, and leave behind those who show a lower degree of engagement, as being "more politically and socially passive" as well as having lower incomes and education (Meyer with Hinchman 2002, 120). The initial required input of time and the high costs involved in obtaining the necessary equipment and learning of computer skills prevents those who already find themselves socially and economically marginalized from having opportunities to reap the benefits from the new technological developments. The still persistent and widespread barriers to political participation, especially among groups such as women, the elderly, and people with different degrees of ability or of a different ethnicity or race, even in the countries where e-democracy has made inroads, reflects the difficulty of changing past exclusionary patterns and building a more inclusive and engaged society.

There is yet another danger that the Internet could foster, which could potentially further the disintegration of civil society. This fragmentation has to do with the creation, by the Internet, of an environment conducive to the emergence of hate groups and reinforcement of the lack of tolerance (Flint 2003; Gallaher 2003). The Internet allows for greater anonymity. Users can easily hide their identity and mislead others about their physical location as well as institutional affiliations, or about the intent of their actions. This inability of others to verify users' identities weakens a sense of being accountable to others. Inasmuch as accountability is one of the basic tenets of democratic practices, its erosion presents a direct threat to democracy (Meyer with Hinchman 2002, 119). Scholars worry about the strengthening of factional divisions, especially in the regions where ethnic strife, the instability of political systems, and a still weak political culture create greater opportunities for groups to be unaccountable for their actions. The reemergence of nationalism and lack of tolerance fostered by past oppressive regimes, such as in CEE and FSU, provide a fertile ground for such groups to emerge (Glassman 2000).

The above discussion reinforces the need to place conversations about electronic democracy and the Internet within the specific political, social, and cultural context of a particular region and/or country. For example, while Hagen (1997) argues that the U.S. crisis in public participation is blamed to a large degree on the media, especially for its lack of accuracy and information manipulation and hence debates on electronic democracy, such arguments could not be advanced in CEE and FSU countries (Hagen). In the region where for decades the media have been controlled ideologically and economically from the top down, where a free press is still struggling to establish itself, and citizens are thirsty for information, the debates about electronic democracy are more concerned with the distribution of information, transparency of governance, and governmental (both national and local) accountability. On the other hand, arguments about limited access and potential fragmentation of societies are quite real. While western scholars, politicians, and/or policymakers attempt to identify how the Internet can assist in engaging the alienated, the new democracies rather uncritically are full of hopes for solving the problems encountered during democratic consolidation. In that sense, there is some similarity as both new democracies and older ones hold out many of the same hopes for the technological magic. The next section takes a closer look at some of the advances and challenges of Internet within the context of central and east Europe.

2. THE ICT AND THE CASE OF CENTRAL AND EAST EUROPE

The regime changes of 1989 brought to CEE and to the countries of the FSU a fundamental restructuring of political, economic, social, and cultural relations between and among individuals, groups, and institutions. The increased ability of citizens to exercise their political and civil rights created new freedoms and the sense of greater possibilities to access, receive, and exchange information, purchase and read books and newspapers, express their disapproval of political institutions located at diverse scales (local, national, international), and lobby for social and political changes. These new freedoms coincided with the rapid development of ICTs in western democracies. The result of this fusion was the emergence of unprecedented circumstances that facilitated, to an extent unmatched by any other region, widespread and rapid diffusion of ICTs across the region (Emery and Bates 2001).

Many reasons can be stated as contributing to the creation of these special circumstances. The elimination of the oppressive political regime that had as its top priority the control of flow and use of information through censorship of media, centrally restricted and ideologically driven monitoring of publishing outlets, and prohibition of gatherings and associations now opened new possibilities for expression and exchange of thoughts. No longer, at least in theory and in most CEE and FSU countries (except in several cases as discussed below), is the national and local state collecting information on its citizens to be used to intimidate, threaten, and sustain a certain level of fear in order to rule. Besides these political legacies, the technological challenges were equally conducive to acceptance of ICTs. It is well known that communist regimes were notorious for allocating inadequate fiscal support for the development of infrastructure (a clearly political decision to keep citizens unconnected to each other and to minimize the possibility of collective actions). The low number of telephone lines built in CEE is only one such example (although there were widespread differences between countries, with Albania's 17 lines per 1,000 people compared to Estonia's 337) (Franda 2002, 107). The lack of support and maintenance services further reinforced the negative impact of low access. But the poor state of infrastructure fostered a rapid embrace of wireless technology, satellite phone lines, and satellite television (Emery and Bates 2001). These, combined with VCRs, fax and copy machines, video cameras, and digital cameras, cellular phones and the Internet, opened new possibilities for communication and information exchange.

The "external" global world of foreign governments, international organizations, commerce, and business was equally eager for CEE/FSU countries to become more technologically advanced and therefore keen to support efforts that would make these countries' markets more open and accessible. In fact, some argue that the pressures for development are not so much generated domestically, but rather represent responses to external forces, such as the desire to join EU by many CEE countries (Regulska 2002). Not unexpectedly, the main foreign governmental assistance, both from the U.S. (through USAID and other federal institutions) and Europe (through the European Union) provided funds, equipment, services, connectivity, and training (Franda 2002; Emery and Bates 2001; Herron 1999). Private foundations were equally supportive. They established their own projects and networks such as those initiated by the Soros Foundation, the Open Society Institute Regional Internet Program and sustained numerous country-specific projects, including the Mellon Foundation initiative in support of higher education (Soros[1] Foundation 2003; Regulska 1999; FSLD 1996). On a smaller scale, individual embassies of western democracies located in CEE/FSU were providing small grants to nongovernmental organizations to purchase equipment, pay for Internet access and training, and pay rent for space or simply for better security of premises where equipment was stored (Regulska 1998a).

As much as external pressures guided the spread of ICTs, its ultimate diffusion to a large degree will be a function of domestic political and economic decisions and of each country's ability to maximize on its specific geopolitical position and role, both past and present (Brunn et al. 1998). For example, Hungary and the Czech Republic's widespread privatization of telecommunication, and the ability to forgo governmental control over development of the basic network permitted rapid and large foreign investment into ICTs. While the long-term implications and consequences of such a strategy remain to be seen (with obvious dangers of increased costs and of denied accessibility for groups socially and economically marginalized), for the moment both countries are enjoying the economic benefits of these decisions. Hungary also made an early strategic decision (both political and economic) to develop its partnership with Germany. This initiative brought heavy investment into the country, as Germany saw Hungary as a "potential 'transit' hub for telecommunication activities in Central and Eastern Europe" (Franda 2002, 133). Slovenia's entry into the world of ICT was aided by Austria, the Vatican, and Germany, whereas Croatia was aided by Germany, Italy, and the large Croatian diaspora in Australia, Argentina, Canada, and the U.S. (Gosar 2003). Bosnia benefited mostly from support from Saudi Arabia and Iran, which competed for influence, and much less support from Italy and Germany (Gosar 2003). Bulgaria's legacies as a never-realized center of computer development, as designated by the Soviet Union in the 1960s and 1970s, left the country with one of the region's highest densities of telephone lines, albeit of poor quality (poor performance, lack of digital connections, etc.) (Franda 2002, 138). Nevertheless, Bulgaria is trying to make a breakthrough by joining various transnational organizations and initiatives including e-Europe+ Action Plans, eSEEurope Action Plan, and the forthcoming Southeast Europe Telecommunication Agency (Yonov 2003). The Caucasus and Central Asian states were aided in their ICT initiatives, under the Virtual Silk Highway Project, through funds in the NATO Science Program and equipment from the multinational electronics firm Cisco Systems, and also DESY (Deutsches Elektronen-SYnhrotron) in Hamburg, which had experience in working on satellite communication in the FSU (www.NATO.int/science and www.silkproject.org).

Baltic countries, on the other hand, despite being under Soviet control that had far deeper and more severe economic consequences than those visible in CEE, were able to benefit from their geographic proximity to the Nordic states. As a result, they received much quicker and greater support for their technology development, which translated into their becoming the most advanced part in the region, in fact, on a level comparable to west European countries (Franda 2002, 141-43). Estonia was assisted especially by Finland, and Lithuania by Germany, with historical ties to Königsberg and Memel. Denmark, Sweden, and the Netherlands also were active in ICT programs in the Baltic states (Löytönen and Helantera 2003). In Estonia, the highest-level political leadership decided to connect the country via a cyberspace network, with one of its top priorities the connection of all secondary schools to the Internet. There is no question that the small size of this country facilitated the projects' implementation, yet without political will and commitment, its realization would have not been possible.

The use of ICTs should also be mentioned in the context of its usefulness to undermine oppressive political regimes and its role in overthrowing a bankrupt leadership. The case of B92 radio in Belgrade and its use of the Internet to provide up-to-the minute information during the anti-Milosevic protests of 1996-97 represent probably the most well known example. Not only were the managers of the B92 station one step ahead of Milosevic's regime in its attempt to jam broadcasts and ultimately to force closure of the radio station, but the fact that foreign radio stations were willing to pick up the information and further distribute it created a far greater and more widespread impact of B92's activities (Emery and Bates 2003; personal communication, Dasa Duhacek). In the final confrontation with the Milosevic regime, B92's Internet and radio, television, and newspaper premises were raided and equipment seized. Although some of these items were regained, B92 continued to struggle to secure appropriate space and transmitting capabilities, even after the new regime of Kostunica took over (Franda 2002). In Estonia, on the other hand, in 1991, it was the state that used the Internet for security purposes during the attempted coup by Moscow (Celichowski 2001). But in wars in Bosnia and later in Kosovo, citizens, journalists, international relief workers, and refugees were aided by gaining access to the Internet in search of information and legal documents/advice, and also having the ability to communicate and search for family members (Franda 2002).

Despite many success stories, ICTs and especially the Internet are not welcome everywhere in CEE/FSU countries. Several governments in the region openly resist implementation and use of ICTs by its citizens, or even worse, they use such technology themselves to exercise 21st century-style oppression. The return to spying, intimidation, censorship, and surveillance has become quite visible during the last few years. While the idea of control has remained the same, the old techniques have become appropriately modernized. It is not surprising that Russia leads the way. In a country where 98 percent of the citizens lack access to the Internet, Putin championed and government institutions embraced the unwritten "information security doctrine" which believes that "government's role in monitoring information flow" should be strengthened (Franda 2002, 112). Despite protests, Putin continues to increase greater control over communication technology and media, and his state institutions, such as Federal Security Service (the old KGB), reinforced its position as a gatekeeper and user of often illegally gained information. Similar tensions emerged in Belarus, where the oppressive regime of Lukashenko continues to govern. In the most visible case, the Belarusian government forced the closure of the Soros Foundation and demanded an extraordinary payoff of $3 million dollars for alleged improprieties in tax payments. A similar fear of ICT and a desire for full governmental control were evident in Ukraine when President Kuchma allowed Ukratelecom to have a monopoly as the Internet provider.

The examples cited above reinforce Norris's point that the Internet offers an opportunity to challenge oppressive regimes and helps those who are flexible and open to new practices as "strengthening the bonds has the capacity to produce sudden disruptions to politics as usual" (Norris 2001, 20). The potential collective power of these new arrangements is in fact also acknowledged by fears and desire for control as expressed by the authoritarian regimes who, when threatened, could "cut off this communication channel by eliminating telephone access" (O'Lear 1999, 177). Yet Herron's conclusions are also correct when he states:

> Although greater use of technology has often been associated with an increase in central governments' ability to monitor the citizenry...the actual development of electronic communication has facilitated the reverse. (1999, 64)

In the remaining part of this chapter, I provide a few specific examples from CEE and the countries of the FSU to show how e-democracy has begun to be used by citizens and political institutions and for service delivery. Following Hagen's typology, I argue that two dimensions of e-democracy are especially visible in the region, cyberdemocracy and electronic democratization, with each being adopted by different groups of users. While cyberdemocracy seems to be engaging at this point predominantly citizens, activists, and organizers (although a few examples indicate that some governments occasionally engage participant mechanisms), the electronic democratization thus far has attracted chiefly governmental institutions, political parties, bureaucrats, and managers. Given that the greatest interest in electronic democratization has been expressed by governments and their units, I focus in my examples on e-government only.

2.1. Cyberdemocracy

Cyberdemocracy, by Hagen's (1997) definition, brings together political activism and discussion; the key forms of political participation. This direct form of democracy, fostered by decentralized and nonhierarchical networking, creates virtual and nonvirtual communities where citizens become empowered and engaged. What seems to be missing in Hagen's definition is the information exchange element. The information-deprived societies of the CEE and FSU are particularly drawn to the Internet and other ICTs, through the promise of fast and easy access. The information exchange in this context (and there is no question that information represents a new form and source of power) is critical for users in helping them to achieve their goals, but also in reinforcing their sense of freedoms and of new position and status.

For activists in CEE and FSU, the increased access to information via the Internet has enhanced their abilities to organize and mobilize individuals and groups (locally and transnationally) in order to foster political action and engagement (O'Lear 1997; Penn 1997; McCaughey 2003). Cyberdemocracy opened possibilities for a wide spectrum of electronic activism. As Internet use has exploded and the number of websites has emerged with unprecedented speed, the frequency of contacts and interactions between individuals, NGOs, and formal and informal groups in virtual communities became almost unlimited. These new linkages permitted coalition building, the creation of new alliances, sharing of the latest information, planning of new mobilizations and actions, and cross-border networking (O'Lear 1999). In just a few years, the Internet created an unprecedented openness for that region in communication among individuals, groups, institutions, and nations. For those involved, it often provided a sense of engagement and feeling of fulfillment of their agenda, even though their contradictory interests became increasingly visible.

Yet as Franda argues, "To suggest...that the Internet is already creating a publicum that embodies meaningful civic discourse on a day-to-day basis seems premature" (2002, 151). These sentiments are echoed by others who underscore that the Internet is not yet part of the daily practices (Iordanova 2000, 110). Some make even stronger arguments that not only are ICTs accessible only to elites, but that despite these promises, "the medium has so far exercised a reverse potential to drown the message" (Horvath 2000, 97). Such pessimism is countered by claims that the issue of access will become of lesser importance as countries develop their e-plans for implementation of ICT. Rather the challenge will lie in "prioritizing that [information] which is most important considering the growing prominence of e-democracy and transparency applications of e-government" (Majcherkiewicz 2003, 42).

Despite the claims that the Internet and ICTs have a limited audience and role in fostering democratic practice, the evidence "on the ground" seems to reinforce the sense of increasingly widespread instances of cyberdemocracy or cyberactivism across the region. While it is impossible to present a comprehensive survey, a few examples of organizations that use Internet to advance democratic practices are indicative of the emerging trends.

We begin with two examples of organizations working at the national level, one in Latvia aiming to increase citizens' political participation, and one in Poland advancing Polish women's rights, but also crossing national borders by specifically focusing on the EU's eastern enlargement from the perspectives of women.

2.1.1. Latvia

In Latvia in July 2001, Soros Foundation-Latvia initiated a public policy website, called Politika.lv. Its mission states:

> The primary objective of the web-site is to contribute towards raising the quality of public policy decisions in Latvia by promoting policy-making based on policy analysis, as well as to promote public participation in the policy process. Quality participation requires resources. Information is one of those resources, therefore we aspire to become a comprehensive source of policy studies and critiques. We wish to develop this website as a meeting place for a virtual community - a public policy community constituted by researchers, analysts, decision-makers, non-governmental organizations, journalists and everyone concerned about Latvia's development. We offer an environment for critical discussions where professionals can debate about the research published, professional standards and methodological issues. The website is also a place where new talent can get a good start, where a researcher can find professional growth opportunities, partners in cooperation and identify financing possibilities for future projects. (http://www.policy.lv/index.php?id=100373&lang=en)

2.1.2. Poland

Politika.lv. site is divided into seven topic areas: policy process, social integration, rule of law, human rights, information society, civil society, education and employment, and foreign affairs. Each area provided numerous policy documents, articles, and other related resources, both in Latvian and English. Although the site is not interactive, the wealth of information and linkages provided is extremely thorough as a user can link with any Latvian governmental office, seat of local government or numerous institutes, and policy-related sites. Politika.lv already has made significant contributions to the policymaking arena, as it has influenced the content of legislations, placed issues on the agenda, and supported NGOs activities (Baumane 2003).

A second site, the Network of East-West Women-Polska (NEWW-Polska), provides a wealth of information of interest to women in Poland as well as the region. Its current projects include Gender and Economic Justice in EU Accession and Integration, Women's Economic Justice Network: Strengthening Feminist Economic Expertise in CEE/NIS, translation and adaptation of the book *Our Bodies, Ourselves*, Pre-election Polish Women's Coalition, and Legal Counseling for Women Victims of Violence. Among the organization's many activities, its biweekly news updates Access News, which is part of the Gender and Economic Justice in EU Accession and Integration has gained widespread recognition. The project is run jointly with the Karat Coalition, a CEE women's coalition, and is supported by UNIFEM (United Nations Development Fund for Women). The unique updates are posted on website http://neww.org.pl and serve as the most comprehensive source of information on European integration and the EU eastern enlargement, the accession process, and women's rights. The website provides more than 80 additional links to women's NGOs in the region, west Europe, and the U.S. as well as with EU institutions and other relevant international organizations and to a variety of information resources. Its book section has more than 450 titles thematically arranged and its discussion section spans a wide range of topics (see UNDP Gender Virtual Library Project at http://gender.undp.sk/ for a similar efforts). Its legal counseling section provides information on how and where such support can be obtained as well, showing samples of complaint letters and other documents that women, especially victims of violence, may need when taking legal action. In another new section, called Forum, users have posted letters that address a variety of health experiences, their visits to doctors, and their struggles with lack of appropriate information. Users have repeatedly rated the NEWW-Polska site as excellent in terms of content, lay out, and navigation.

2.1.3. East-West

Transnational networking and building partnerships across borders and continents are also well represented virtually. These sites combine information dissemination and exchange, foster creation of new alliances and coalitions through listservs, and focus on development of new activities that would empower participants. The two examples presented below are that of the Network of East-West Women (NEWW) and the Information Technology, Transnational Democracy and Gender (ITTDG).

The Network of East-West Women (NEWW) was created in 1991 by a group of feminist scholars and practitioners from CEE and the U.S. As one of its first activities, NEWW took upon itself to establish the NEWW On-Line project (http://www.neww.org). Shana Penn, Executive Director (1994-96), took a leading role in 1994 in creating this first women's electronic communications network in CEE and FSU, and subsequently publishing a first guide for women to the wired world (Penn 1997). This initial virtual community included women and women's groups from more than thirty countries and serves

community of academics, journalists, human rights activists, parliamentarians, lawyers, businesspeople, technology and medical professionals, environmentalists, private foundation and government officers, and United Nations program officers

- all of whom are interested in women's swiftly changing position in societies that are making the transition to democracy (Penn 1997, 2).

By 2003, NEWW On-Line had five listservs in English: academic-resources, neww-rights, opportunities, women-east-west, and women-in-war, as well as two conferences in Russian: glas.sisters and glas.women's rights. The listserv, has well over 1000 subscribers on a regular basis. While the lists are unmoderated, NEWW regularly hears from people that the information shared on these lists is very useful to their work as the lists transmit information that subscribers do not find on other lists.

Like many ICT projects in the region, NEWW On-Line was launched with support from the Soros Foundation. Other NEWW projects include Legal and Economic Fellows, the Economic Justice Project, and a Book and Journal Project (aiming at providing books and journals to women's studies libraries and programs across the region). NEWW also serves as a clearinghouse for information (on women in CEE and FSU) to individuals, organizations, policymakers, state and international organizations, and public and private funders. Currently NEWW has 255 individual members from 48 countries around the world including Afghanistan, Pakistan, UAE, Nigeria, Zaire, Ghana, Kenya, Nepal, India, New Zealand, France, Germany, the U.K., the U.S., Austria, the Netherlands, Portugal, and Turkey, and many countries from CEE and FSU. In addition, NEWW has 61 organizational members from 26 countries. NEWW's membership database is now managed electronically on NEWW-Polska's webpage.

2.1.4. West-East

The Information Technology, Transnational Democracy and Gender Network (ITTDG) was created in 1999. While it is coordinated from Norway, it brings together members from the Nordic countries, Northwest Russia, and the Baltic states. This interdisciplinary group includes technoscientists as well as political and social scientists whose main objective is the articulation of new knowledge, new agencies, and new practices—a matter, in short, of working toward a new configuration of "the political." While the group uses different strategies to achieve these goals, its predominant format has been yearly workshops, seminars, and conferences. The topics addressed so far include: understandings of technologies; information technology, gender and prospects for democratization from below; citizenship and identity in emerging information societies; mainstreaming as a strategy for feminism, infrastructure, and border crossing, as well as virtual workshops on cyberspace and identity.

In its mission, ITTDG states:

The research network...will accept the challenges to contribute to the creation of different meanings of information technology, politics and gender. Our aims are to:

• produce situated (new) knowledge about gender, technology and politics in transformation;

• build critical knowledge and understandings of IT in the contradictions and tensions between equal access and dominating discourses of technology;

• give a contribution to the debates on the democratization of the transnational political institutions from below - on the construction of the civil society and on strengthening of the political citizenship, including its gender, cultural and regional aspects by concentrating on the perspectives of previously excluded and marginalized groups, women, cultural minorities, people living in peripheral areas. (http://www.luth.se/depts/arb/genus_tekn/ITTDG.htm)

The group's activities are conducted on a rotating basis in member's countries. While some events are open to nonmembers, most are for members only.

The above examples represent just a few among thousands of initiatives that have been generated across region or with participation of groups from the region. Many of these initiatives employ innovative ways to provide information, build virtual communities, and engage citizens. While some initiatives are open to any citizen interested in the issue, not necessarily requiring membership in the particular organization, others attempt to retain established links by engaging users in discussion groups, listservs, and chat rooms, or by soliciting their input and opinions. The examples above also raise interesting questions about the notions of engagement and activism. Is reaching and spreading information or joining a listserv a new form of activism? Does such activism need to be repeated on a regular basis to be perceived as sustained engagement? When is such engagement considered as being political? While often it is too early to evaluate the actual political impact of these and other sites, nevertheless their existence seems to make a difference. What still also remains unclear is to what degree these virtual initiatives continue to "engage the engaged" (Norris 2001, 22) or do they indeed create an environment for more inclusive and wider political participation?

2.2. E-Government

The post-1989 position of the state has been drastically redefined. Not only have state institutions begun to be subject to greater scrutiny by citizens who have demanded more transparency and accountability, but also these citizens have pushed for greater openness of bureaucracy and of the development of user-friendly practices. State institutions responded to these demands in a variety of ways. One of these responses is a slow emergence of e-government. Supporters argue that at least two areas will benefit from the adoption of e-government: (1) better, faster, more accessible, and lower-cost service delivery (including the argument about locating services "closer to users"), and (2) an increase in transparency and accountability of governments across geographical scales.

Pessimists stress that as with the hopes for e-democracy and its limited ability to reengage citizens in political activism and policy debates, e-government will not eliminate widespread corruption, inefficiency, and passivity of bureaucracy, as "Technological innovations will not change the mentality of bureaucrats who do not view the citizens as customers of government or participant in decision-making" (Dempsey 2003, 22).

So far, with the exception of a few cases, most CEE state institutions have given priority to the distribution of both national and local-level governmental and parliamentary generated information. These efforts vary in terms of the depth of information provided (e.g., a list of descriptions of activities vs. expectations of programs and projects, a list of legislation enacted vs. description, and an analysis of such legislation with possibilities for citizens' input and queries, information on elected representatives, and how to contact them, etc.), and audiences considered as potential users (e.g., professional, business, citizens). A few selected examples of e-government initiatives are discussed below.

2.2.1. Estonia

Among central and east European countries, Estonia is repeatedly cited as being at the forefront of creating possibilities for citizens' involvement, the exchange of information and enhancement of their political participation. Named as a "leader in the development of an information society" (Celichowski 2001) and cited as undertaking innovative efforts "to turn the digital service and political participation into a permanent part" of the e-government strategy (Bertelsmann Stiftung 2002), Estonia has gained wide recognition for its commitment to the implementation of ICT. While its initial focus had been on public administration and on the use of Internet for internal governmental purposes, more recently the commitment to an information society resulted in the initiation of diverse projects related to e-democracy and e-government (http://www.vm.ee/estonia/kat_175/pea_175/2972.html). For example, the citizens' legislative initiative allows for increased citizens' participation in decisionmaking by engaging them in the legislative process. Citizens can introduce legislative proposals or comment on the draft laws. The only requirement is that they register by providing their name and e-mail address. In cases when citizens want to introduce a new proposal, a consensus among participants needs to be reached and a majority vote is required for the proposal to pass in order for it to be submitted to the government (Celichowski 2001). Citizens can also engage with each other in debates and discussions. Government officials provide the feedback and updates on what is happening with the particular proposal or why it cannot be implemented. Celichowski cites the figure of 300,000 visitors during the three weeks (June 25-July 18, 2001) after the portal, TOM (Today I Decide <tom.riik.ee>), was established, and of 1,300 registered users participating in legislative dialogue. Future Estonian governmental plans call for introduction of electronic voting in 2005. Citizens will be equipped with a digital signature certificate and voting will take place only on advance polling days. Other initiatives include Look@World project aimed at a rapid increase of Internet users through training, with the goal of 100,000 Estonians being trained by 2004.

2.2.2. Bulgaria

Similar efforts to create an interactive portal between the government and citizens have been successfully launched by Bulgaria, where citizens have an

> opportunity to discuss government policies and to send their questions and comments and receive answers from the ministers who oversee the departments fielding the queries. (Georgiev 2003, 11)

2.2.3. Riga City Council

Riga City Council, on its website, under the heading Public Debate, posts questions for the public to comment. Subjects put forward relate to the city's local development issues, such as street redevelopment, historic monuments renovation, or city development. The public is invited to comment on the proposals, but it can also view what others have said on the particular topic or access archives of the past public debates (http://www.rcc/lv). The site is available in three languages (Latvian, Russian, and English) although citizens' interactions, while accessible through English and Russian sites, are conducted only in Latvian (for more explicit discussion on language use, see Herron 1999). Guests and businesses are welcome to browse through separate links connected to the city's administrative units. They can also access a list of Riga's twin or sister cities and read information posted for the press and other notices published by the city council.

2.2.4. Poland

In Poland, a clear distinction can be observed in the way in which the national government tailors its websites to professionals, while local governments and administrations focus on citizens and attempt to anticipate their needs. While national governmental sites are more focused on the provision of information without much possibility for interaction (http:// www.egov.pl/), local authorities are beginning to experiment with interactive models (http://www.zmp.poznan.pl/). For example, the City Council in Szczecin, Poland, with support of the Regional Training Center of the Foundation in Support of Local Democracy in Szczecin, computerized its offices and developed an interactive system whereby citizens can login to locate their application, petition, and/or query (http://www.frdl.szczecin.pl/ centrum/projekty.html). The system provides information on which stage a particular file is at, which office has it, and when it will potentially be resolved. Majcherkowicz points out that while Poland has a national strategy called e-Poland 2001-2006, so far

> the growth of the information society may be largely attributed to bottom-up processes, for example, local government associations promoting the use of the Internet in administration. (2003, 42)

2.2.5. Romania

In Romania, public authorities focus far more on reaching the business community as "they favored business-oriented e-government applications over citizen-centered ones" (Stoica 2003). The increasingly widely available e-procurement and e-tax measures, set up by the Ministry of Communication and Information Technology, however, do little to advance citizens' access and participation, especially when only 13 percent of Romanians access the Internet (Stoica 2003, 17). The overall low level of economic development translates into few computers, few phone lines, and therefore a limited number of Internet users. In addition, the generally not very efficient local public administration structure is still operating along old patterns and does not create an environment conducive to wide acceptance and use of e-government. Indeed, Stoica (2003, 17) argues that "at this point it seems the local authorities realize neither the importance nor complexity of e-government and the advantages it could bring." In cases where local government does have its own websites, they tend to offer very basic information, which is generally not very useful for users. The few information centers for citizens are not interactive and thus do not offer many possibilities for citizens' engagement.

It seems that thus far governmental institutions feel most comfortable with simply distributing information. In some ways, however, this is a big step forward. Considering that only a decade ago, a centralized, top-down, and ideology driven political and decisionmaking system was widespread, providing information to citizens on a regular basis is an achievement reflecting some degree of change. But on the other hand, in order to make further progress, not only do technical issues need to be taken into account and greater access provided, but the role of government needs to be redefined (Ducatel et al. 2000; Regulska 1998b). Sandor argues that a government which "does not see its citizens as an integral part of itself will not be able to 'go electronic'" (2003, 36). Norris has made a similar point when arguing that

> Established political institutions, just like major corporations, can be expected to adapt the Internet to their usual forms of communication, providing information on line, but not reinventing themselves or rethinking their core strategy in the digital world, unless successfully challenged. (2001, 19)

The evidence available thus far does seem to confirm that such challenges may not be forthcoming.

3. CONCLUSIONS

This paper takes a closer look at how the development of information and communication technology has begun to be used in one of the rapidly changing regions of the world, viz., the CEE and the FSU. There is no doubt that the speedy, albeit still limited, spread of ICTs, including the Internet, has had visible impact on many people's lives, their work habits, and daily practices. Groups and individuals have begun to use Internet technologies to advance democratic politics, including issues of sustainable development and environmental quality. The Internet has provided them with space to experiment, not only with new ways of defining and carrying, but also with political organizing and actual political struggles. Citizens are also trying through these new means to make governments more transparent and accountable, and thus to engage two core dimensions of democracy: participation and accountability. Given the political will of many governments in the region and the short time it took them to reach the present level of ICT use, one may speculate that further rapid expansion of users will be taking place.

Despite these positive signs, it remains to be seen how much these initiatives will move beyond the supply and exchange of information and contribute to strengthening of democracy in countries undergoing transformation, such as in CEE and the FSU. Can groups that are marginalized and excluded from the political process, which was denied to them in the past, become engaged? Is technology indeed empowering citizens or is it possible that the advanced technology allows the state to exercise greater control over its citizens? As with any new and emerging technology, the evidence on this front is not only sketchy, but also inconclusive. The abundant barriers of economic, social, and technical nature present at the moment major challenges to the creation of widespread access. Yet as many authors and the selected examples have indicated, even if these barriers were to be eliminated, it is still not clear how much citizens would use the Internet and other ICTs to engage in political organizing for social change.

Finally, what the above discussion suggests is a great plurality and significant degree of ambiguity of how scholars, practitioners, and users conceptualize the virtual reality of what is being done and by whom, and how they envisage what will be happening in the future. This multiplicity of opinions reflects a wide range of disciplinary perspectives and tensions between and among social scientists, communication theorists, practitioners, managers, technocrats, policymakers, government bureaucrats, nongovernmental organizations, and citizens. It indicates great fluidity and uncertainty about the directions of future developments in the area of ICTs and of the societal, political, and economic implications that these advances will bring to the constructions of democratic space.

NOTE

1   Soros Foundations Network was established by Hungarian billionaire George Soros. One of its initiatives included the establishment of the first high-speed local Internet connections in the CEE and FSU through satellite. Its Information Program promotes "the equitable deployment of knowledge and communications resources for civic empowerment and effective democratic governance" (http://www.soros.org/).

REFERENCES

Arterton, C. (1987). Teledemocracy. Can Technology Protect Democracy? London: Sage Publications.

Baumane, K. (2003). The Politika.lv E-Democracy Experience, Local Government Brief. Winter: 18-19.

Bertelsmann Stiftung (2002). Balanced E-Government—Connecting Efficient Administration and Responsive Democracy. Gutersloh, Germany: Bertelsmann Stiftung.

Brunn, S.D., Dahlman, C.T., and Taylor, J.S. (1998). GIS Uses and Constraints on Diffusion in Eastern Europe and the Former USSR, Post-Soviet Geography and Economics 39: 566-87.

Celichowski, J. (2001). Estonia e-Democracy Report, July 4, 2003. http://www.osi.hu/infoprogram/e-government% 20in%20estonia%20proof%20read.htm

Dempsey, J.X. (2003). What E-Government Means for Those of Us Who Cannot Type, Local Government Brief. Winter: 22-24.

Ducatel, K., Webster, J., and Herrman, W. (2000). The Information Society in Europe: Work and Life in an Age of Globalization. Boulder, CO: Rowman and Littlefield.

Ebo, B. (2001). Cyberimperialism? Global Relations in the New Electronic Frontier, Westport CT and London: Praeger.

Emery, M. and Bates, B.J. (2001). Creating New Relations: the Internet in Central and Eastern Europe. In Ebo, B. (Ed.) Cyberimperialism? Global Relations in the New Electronic Frontier, 93-110. Westport, CT and London: Praeger.

Flint, C. (Ed.) (2003). Spaces of Hate: Geographies of Hate and Intolerance in the United States of America. New York: Routledge.

Foundation in Support of Local Democracy (FSLD) (1996). Annual Report. Warsaw, Poland: Foundation in Support of Local Democracy.

Franda, M. (2002). Launching Into Cyberspace: Internet Development and Politics in Five World Regions. Boulder, CO and London: Lynne Rienner Publishers.

Gallaher, C. (2003). On the Fault Line: Race, Class and the American Patriot Movement. Lanham, MD: Rowman and Littlefield.

Glassman, E. (2000). CyberHate: The Discourse of Intolerance in the New Europe. In Lengel, L. (Ed.) Culture and Technology in the New Europe: Civic Discourse in Transformation in Post-Communist Nations, 145-164. Stamford, CT: Ablex.

Georgiev, I. (2003). E-Eastern Europe, Local Government Brief. Winter: 10-13.

Gibson, W. (1984). Neuromancer. New York: Ace Books.

Gosar, A. (2003). Email communication, July 20.

Hagen, M. (1997). A Typology of Electronic Democracy, June 25, 2003. http://www.uni-giessen.de/fb03/vinci/labore/netz/hag_en.htm

Harasim, L. (1993). Global Networks: Computers and International Communication Cambridge, MA and London: The MIT Press.

Hauge, B. and Loader, B. (Eds.) (1999). Digital Democracy. Discourse and Decision Making in the Information Age. New York NY: Routledge.

Herron, E.S. (1999). Democratization and the Development of Information Regimes, Problems of Post-Communism July-August 46: 56-68.

Horvath, J. (2000). Alone in A Crowd: The Politics of Cybernectic Isolation. In Lengel, L. (Ed.) Culture and Technology in the New Europe: Civic Discourse in Transformation in Post-Communist Nations, 77-104. Stamford, CT: Ablex.

Iordanova, D. (2000). Mediated Concerns: The New Europe in HyperText. In Lengel, L. (Ed.) Culture and Technology in the New Europe: Civic Discourse in Transformation in Post-Communist Nations, 107-131 Stamford, CT: Ablex.

Löytönen, M. and Helantera, A. (2003). E-mail communication, July 22.

MacGregor Wise, J. (1997). Exploring Technology and Social Space. Newbury Park, CA: Sage Publications.

Majcherkiewicz, T. (2003). E-Poland and E-Poles, Local Government Brief. Winter: 42-43.

McCaughey, M. and Ayers, M.D. (2003). Cyberactivism: Online Activism in Theory and Practice. New York and London: Routledge.

Meyer, T. with Hinchman, L. (2002). Media Democracy: How the Media Colonize Politics. Cambridge, UK: Polity Press.

Norris, P. (2001). Digital Divide: Civic Engagement, Information Poverty and the Internet Worldwide. Cambridge, UK: Cambridge University Press.

Norris, P. and Jones, D. (1998). Virtual Democracy, Harvard International Journal of Press and Politics 3: 1-4.

Ogden, M. (1994). Politics in the Parallel Universe. Is There a Future for Cyberdemocracy?, Futures 26: 713-29.

O'Lear, S. (1999). Networks of Engagement: Electronic Communication and Grassroots Environmental Activism in Kaliningrad, Geografiska Annaler B 81: 165-78.

O'Lear, S. (1997). Electronic Communication and Environmental Policy in Russia and Estonia, Geographical Review 87: 275-90.

Penn, S. (1997). The Womens' Guide to the Wired World: A User-Friendly Handbook and Resource Directory. New York: Feminist Press.

Porebski, L. (2002). Three Faces of Electronic Democracy. Paper presented at the European Conference on Information Systems, June 6-8, 2002, Gdansk, Poland. http://eta.ktl.mii.lt/~mask/varia/ECIS2002proceedings/leszekporebski.pdf

Porebski, L. (2001). Elektroniczne Oblicze Polityki: Demokracja, Panstwo, Instytucje Polityczne w Okresie Rewolucji Informatycznej (Electronic Dimension of Politics: Democracy, State, Political Institutions in the Period of Information Revolution). Krakow, Poland: Uczelniane Wydawnictwo Naukowo-Dydaktyczne.

Regulska, J. (2002). Der Gleichstellungsdiskurs der Europaischen Union und seine Folgen fur Frauen in Polen (EU Discourse on Women: Will it Matter for Polish Women), Yearbook for European and North American Studies, No 6: 121-51.

Regulska, J. (1998a). The Rise and Fall of Public Administration Reform in Poland: Why Bureaucracy Does Not Want to be Reformed. In Barlow, M., Lengyel, I., and Welch, R. (Eds.) Local Development and Public Administration in Transition. Jozef Attila University: Szeged, Hungary.

Regulska, J (1998b). Building Local Democracy: The Role of Western Assistance in Poland, Voluntas: International Journal of Voluntary and Non-Profit Organizations 9: 1-20.

Rengger, N. (1999). E-Governance: Democracy, Technology and the Public Realm. Edinburgh: The Scottish Council Foundation.

Saco, D. (2002). Cybering Democracy: Public Space and the Internet. Minneapolis, MN: University of Minnesota Press.

Sakowicz, M. (2003). Electronic Promise for Local and Regional Communities. Local Government Brief. Winter: 24-28.

Sandor, S.D. (2003). Thinking Large but Starting Small in Romania, Local Government Brief. Winter: 34-36.

Soros Foundation (2003). www.soros.org

Stoica, O. (2003). Shepherding Electronic Sheep? Romania's New Governance, Local Government Brief. Winter: 17-18.

Yonov, L. (2003). Bulgaria's Big Leap to a Single Click, Local Government Brief. Winter: 14-15.

WEBSITES

Association of Polish Cities. http://www.zmp.poznan.pl/

Estonia. http://www.vm.ee/estonia

Information Technology, Transnational Democracy and Gender. http:www.luth.se/depts./arb/genus_tekn?ITTDG.htm

Foundation in Support of Local Democracy. http://www.frdl.org.pl/

Foundation in Support of Local Democracy—Regional Training Center in Szczecin. http://www.frdl.szczecin.pl/centrum/projekty.html

Network of East-West Women. http://www.neww.org

Network of East-West Women-Polska. http://www.neww.pl

Republic of Poland. http://www.egov.pl/

Politika.lv. http://policy.lv

Riga City Council. http://www.rcc./lv

Soros Foundation. www.soros.org

United Nations Development Program. http://gender.undp.sk/

Virtual Silk Highway Project. www.silkproject.org and www.NATO.int/science

CHAPTER 16

MICHAEL GREENBERG

TECHNOLOGIES APPLIED TO PUBLIC HEALTH

Abstract   Health needs drive technology, but technology also steers public health investments and practice. Technology-health interactions are illustrated by case studies of weapons of mass destruction, environmental cancer, elderly health care, urban redevelopment and sprawl, and the spread of HIV/AIDS and other infectious agents. Technology has speeded up detection, evaluation, forecasts of the ebb and flow of risk, and the development and use of medicines and risk management equipment. Yet technology has created new public health risks and increased the potency of existing low-risk hazards. Effective balancing of the technology-health interactions depends upon economic, social and political health of nations and states, with the residents of the North American, Western European, and a few Asian nations enjoying a much longer life expectancy and higher quality of life than their counterparts in many other nations. Differences between technology-health rich places and technology-health poor ones are likely to increase in the near future.

Keywords   cancer, HIV/AIDS, elderly, urban development, weapons of mass destruction

1. INTRODUCTION

Technology has made it practical to use geographic theories and methods in public health research and practice. Books, some written by geographers, describe theories and methods, and provide hundreds of applications of the disease ecology and medical geography traditions to public health (Albert et al. 2000; Cromley and McLafferty 2002). The availability of this excellent literature allows me the luxury of focusing this essay on five of the most challenging human health hazards and the exciting applications of geographically grounded technologies that these challenges have engendered: (1) the legacy of weapons of mass destruction, (2) environmental cancer, (3) elderly health care, (4) urban redevelopment and sprawl, and (5) spread of HIV/AIDS and other infectious agents.

Stanley D. Brunn, Susan L. Cutter, and J.W. Harrington, Jr. (Eds.), Geography and Technology,
363-381. © 2004 Kluwer Academic Publishers. Printed in the Netherlands.

As you read this chapter, three themes will emerge. First, technologies discussed in this paper have allowed more precise and frequent monitoring and assaying of hazards, sophisticated and rapid statistical modeling of risk associated with these hazards, and mapping, global positioning, and telecommunicating of critical information about hazards and risks. Second, technology combined with population growth and resource exploitation have created new risks and have hastened the diffusion of risks. Third, effective application of risk management technology is only as good as the ability of host political and social systems, that is, some countries can effectively apply technology in some places to minimize risk while others cannot or will not. This third theme is the most important, and I have ended each section with comments on ability of places to apply salutary technology and/or adapt to dangerous technology.

Not all of these applications are in the published literature; some are in the gray literature or have only been presented at conferences. After presenting these applications, I will suggest where societal need is pushing technology and where technology is driving public policy.

2. THE LEGACY OF WEAPONS OF MASS DESTRUCTION

Since September 11, 2001, public attention has been riveted on terrorism and weapons of mass destruction. Long before September 11, 2001, however, the U.S. had major programs to destroy and manage its own chemical and nuclear weapons and environmental legacy. Geographic-based theories and technologies are at the heart of those efforts to manage this unenviable legacy.

The U.S. has about 31,500 metric tons of chemical warfare agents (National Research Council 2001). Only the Russian Federation is thought to have more of a stockpile (approximately 40,000 tons) (Committee on Foreign Relations 1996). Consisting primarily of blistering agents and nerve gases (there are a few other types), the U.S. material is stored at eight sites on the continental U.S. Chemical weapons formerly were stored at Johnston Island in the Pacific Ocean, but these have been destroyed. Chemical agents can deliver lethal human doses within minutes if the material agent is pure, and if the exposed person is not protected. Hence, the U.S. government began a program of destroying its stockpile over a decade ago and signed an international treaty requiring its stockpile to be destroyed by the year 2007.

During the Cold War, the U.S. manufactured tens of thousands of nuclear weapons. Although the U.S. government has dismantled many of these weapons and has agreed to dismantle many more, the reality is that the weapons development, testing, and production has left a massive environmental legacy of radiological and chemical contamination at more than 100 sites. The

Department of Energy spent $60 billion during the 1990s on cleanup, and the additional cost of remediating this legacy has been estimated at between $150 and $350 billion (Top-to-Bottom Review Team 2002). The Savannah River nuclear weapons site in Aiken, South Carolina and the Hanford nuclear weapons sites located near the tri-cities (Pasco, Richland, Kennewick), Washington contain thousands of contaminated tanks, buildings, land, and water bodies.

2.1. The Geography of Risk at Stockpile and Legacy Sites

Individuals who work at the chemical weapons stockpile sites and visitors to these sites are required to carry a gas mask and syringes with them when they are near the stockpile. Obviously, they could be exposed. I focus here, however, on off-site exposure to residents of surrounding communities. Nearly all the sites are located miles away from the nearest towns. Nevertheless, under the worst-case scenario, clouds of lethal chemical warfare agents could escape from the sites and head toward these towns.

Technology grounded in geographic theories and methods is being used to lessen the potential of a catastrophic exposure event. Each morning the commander of the chemical weapons stockpile base meets with staff to plan activities. Once a decision is made, which may, for example, mean processing 50 rounds of rockets with nerve gas, that plan is entered into mathematical air pollution dispersion models. The models use the activity data along with up-to-date weather information to predict what could happen under the worst-case failure circumstances. A GIS map is drawn around the area where the plume would head, and estimates of fatalities and injured people are calculated. This information is transmitted to local health officials in surrounding communities. As the day progresses and conditions change, the information is updated and transmitted to local officials.

The sites also contain up-to-date weather monitoring technology, which informs the commander and operations staff of the approach of hail, thunderstorm, strong winds, or other conditions that might compromise the removal of the weapons from their storage sites, transportation of the weapons to the destruction site, and unloading of the weapons from the vehicles to the processing lines. Operations must cease if these conditions are identified on the radar. These sophisticated air pollution, weather, and communications technologies are early warning signals to the surrounding public and have helped the chemical weapons command to begin to build trust with local officials and some members of the public. In short, modern weather forecasting technologies, air plume models, and land use/demographic information are joined to produce near instantaneous simulations of possible chemical weapons-related accidents and are used to inform the army personnel and local health officials in real time so that they can be prepared to implement their evacuation, shelter in place, and other protective protocols.

At the major U.S. nuclear weapons sites, technology has been used to put together a remarkable database and to use it to simulate potentially hazardous what-if exposure scenarios. A group of university-based researchers has built the most complex geographically structured data file about a single location that I have ever encountered (Balakrishnan et al. 2002, Georgopoulus et al. 2002, Isukapalli and Georgopoulus et al. 2001, Wang et al. 2002). For example, the Savannah River nuclear weapons site, an area of more than 300 square miles, contains hundreds of contaminated sites and about 50 tanks with some of the most hazardous radioactive and chemical waste in the world. Obtaining remote sensing, census, and a great deal of other data from more than two dozen government sources, the researchers created GIS-based overlays about geology, soil, ecology, current land use, contamination, location of population and workers, transportation, wind patterns, and numerous others characteristics that directly bear upon risk. One of their objectives is to use the database and accompanying simulation models to determine optimal geography for sampling the environment for chronic exposures. That is, how many samples should be taken for mercury and cesium in water, and where should ground water, surface water, and air quality monitoring be done to find evidence of contamination? How often should species be trapped and monitored to determine if contamination is being carried?

Their second objective is to run simulations of what hazardous events could occur and thereby help the Department of Energy prepare for natural hazard events and system failures. For example, fires have occurred at the Department of Energy's Los Alamos Lab in New Mexico. Radioactive tumbleweed was dispersed by winds around the site. Using their database, Georgopoulus and colleagues constructed a mathematical model to illustrate the dispersion patterns of fires. In other words, the model figuratively starts a fire and then follows the plume, keeping track of what is dispersed and what is deposited. Their approach has also been applied to recreate the dispersion of the plume and exposures from the collapse of the World Trade Center Towers in New York City on September 11, 2001. In short, the ability to gather formerly disconnected datasets, build computerized overlays of them, and link them instantaneously to high-speed environmental dispersion models makes it possible to quickly estimate exposure to chemical and nuclear weapon-related operations that go awry, and natural hazard events that can trigger a serious hazard event. I cannot overstate the potential importance of this capacity to government agencies and the surrounding public.

2.2. Communicating to the Population at Risk

As the steward of arguably the most dangerous stockpile of weapons and residuals in the world, the managers of the U.S. Departments of Defense and Energy realize the importance of developing innovative ways of gathering and communicating information to the public. Burger and colleagues' (1999, 2001) work at the Savannah River nuclear weapons site is illustrative of a blend of new and old technology. Over the half-century since the "bomb plant" (its local street name) was built along the Savannah River in South Carolina, cesium and metals have seeped into the Savannah River in low concentrations. Mercury in fish is a hazard to those who eat the fish. Burger's group caught fish, filleted them, and using new assaying technologies determined the concentration of mercury in the fish. Then the group walked the streams and talked with fishermen to determine how often they ate fish caught in the river and how those fish were cooked. The net result was that a new quantitative risk assessment was done and fish warnings were written and distributed to fishermen. This activity combined old communication technologies with some of the newest contaminant sampling and assaying technologies.

The case of mercury contamination in Savannah River fish underscores one of the key advances in the geography of exposure assessment: the development of mass spectrometers, personal monitors, and other technologies that allow scientists to detect concentrations in the parts per trillion. Twenty years ago, it was not possible to detect such low concentrations of contaminants, and hence it was not feasible to do accurate fate and transport modeling of contaminants. Today, the combination of sensitive monitors and computer capabilities means that the transportation and ultimate fate of contaminants can be predicted and health-protective policy actions can be taken.

The fish fact sheets prepared by Burger represent a major step forward for an organization that, until the end of the Cold War, deliberately did not inform the public about what was going on at their sites. During the last decade, the Department of Energy (DOE) and Department of Defense (DOD) have tried a variety of communication methods. The DOD has tried speaker's bureaus, located communication centers in nearby towns, made presentations at country fairs, and tried other standard communication technologies. They built and have begun to use innovative Internet technologies that send out information and questions and ask for public responses (National Research Council 2000). The electronic-based technologies have provided access to people who otherwise had not communicated with the DOD.

Without doubt, the greatest communication challenge is about the legacy of contamination left at some of the nuclear weapons sites to future generations who might breach the containments. Some of the radioactive elements have half-lives of thousands of years. DOE risk analyses expect containment of high-level waste to be for 10,000 years! How can communications be maintained for such a long period of time? In science fiction novels and in some government publications, the legacy is communicated with easily recognizable symbols on giant obelisks. But before we can contemplate communicating over thousands of generations, there is a more immediate need to communicate to the local community about the legacy and policy choices. Geographer Christie Drew (2000; also see Drew and Nyerges 2001) has developed a decision-mapping system for the Hanford nuclear weapons site that is a prototype for the DOE and a demonstration of the innovative linking of databases. Noting that decision processes are complex, that values embedded in decisions are not explicit, and that documentation is vast and difficult to understand, she has designed a transparent series of webpages for public use. Users can find data, decisions, and the reasons for the decisions. The author notes that her project is a pilot and that application to the many contaminated sites at the site will require substantial commitment of resources both now and in the future. That caveat noted, however, Drew has clearly built a cornerstone for a data structure that will need to be built and maintained for the indefinite future. This geographic-based database management system allows residents to immediately access key information about the hazards they live with.

I close this section by noting that the U.S., while it has a large fraction of the world stockpile of chemical and nuclear weapons, has the political and social will to apply technology to the destruction and management of weapons of mass destruction. As a member of National Research Council Committees regarding weapons of mass destruction and wastes derived from their production and use, I am much less sanguine about the ability of some other nations, most notably Russia, to apply technology to the management of these weapons.

3. THE GEOGRAPHY OF CANCER

We have known for many decades that some cancer diseases are more prevalent in some countries than in others. For example, Eastern Europeans had higher rates of stomach cancer than Americans, and Americans had higher rates of intestinal cancer than Eastern Europeans; Japanese Americans had low rates of breast cancer, but after a few generations in the U.S., their breast cancer rate resembles those of Caucasian Americans (Greenberg 1983). These so-called "migrant studies" led to provocative ideas about the role of nutrition and hormones in cancer. Yet it was not until the availability of substantial computing capacity in the late 1970s that we could look at county and city scales to see startling differences in the geography of cancer, which riveted attention on environmental exposures and cancer.

A group of epidemiologists at the U.S. National Cancer Institute (NCI) used a 1970s computer to compile a county-by-county file of more than 30 kinds of causes of cancer death in the U.S. for more than 3,000 counties for the period 1950-1969. They calculated age-specific and age-adjusted cancer mortality rates, and then 95-percent confidence limits for those rates (Mason et al. 1975).

The sheer magnitude of the database (millions of cancer death certificates) made it infeasible to develop this database without computer technology. The database was printed in a book about the size of a large telephone book. But it received almost no attention, until Bill McKay, an NCI computer analyst with some cartographic training, took all the data and made maps at the county scale. He used red to signal an extremely high rate and orange to signal a high rate. The result was an atlas that showed extremely high rates of cancer in the area geographer Jean Gottman had called "Megalopolis," and urban areas around the Great Lakes and Louisiana Delta (Mason et al. 1975). This was the first national disease atlas where the units of analysis were local governments. Now the U.S. government has atlases that cover heart attacks, stroke, and other diseases, and other national and state governments have also developed atlases of disease that focus public attention on their disease problems.

Until that atlas was published, little attention was paid to cancer as a statewide issue in New Jersey, or in Louisiana, Rhode Island, and some other areas. Based on the work of these National Cancer Institute researchers, New Jersey was sarcastically called "Cancer Alley." After its publication and the subsequent public outcry for action, cancer became a hot political issue. For example, the State of New Jersey became a leader in environmental studies of cancer, including setting up a cancer registry, allocating funds to monitor industrial emissions, studying toxins in the water supply, and in other ways, taking the lead in investigating and developing policy responses to the strongly believed relationship between cancer and pollution.

Geographers and the tools they used have played a prominent role in these investigations. For example, Greenberg (1983) led a group that studied the changing geography of cancer in the U.S. He found that the gap between cancer in New Jersey and the rest of the U.S. has been decreasing rather than increasing. Furthermore, using other datasets that had become available as a result of computerization of files, he was able to show an association among cancer rates, the spread of a national consumer-oriented culture, including smoking and alcohol consumption, diet, industrialization, and medical care across the U.S. In short, the availability of high-speed computers has allowed the accumulation, analysis, and mapping of huge databases. Many states and the national government have become much more focused on environmental causes of cancer as a result of the public concern that followed the application of these technologies.

Because of their great fear of cancer, the public has not been satisfied with periodic cancer atlases. So-called "cancer clusters" has become a nationally prominent issue; many members of the public perceive that toxins are clustering in their neighborhood, leading to elevated cancer risk, birth defects, and other ailments. Members of the public who perceive elevated cancer rates in their neighborhood initiate most cluster investigations (Greenberg and Wartenberg 1991). The efforts of neighborhood cluster research have been aided by computerized databases put together by the federal government and used by some state governments and not-for-profit organizations such as the Natural Resources Defense Council and the Environmental Defense Fund. Dating back to the initial efforts in New Jersey's industrial carcinogen survey, the widespread availability of toxic release inventory (TRI) data has allowed people to form their own opinions about the geography of cancer in their communities. The pressure to reduce environmental cancer and other environmentally linked disease outcomes will continue to mount as a result of increasing database systems that link sources of contamination and high-speed computing and mapping to the local public's distress about environmental causes of illness and death.

But some cancer cluster investigations have required more than high-speed computing and mapping. For example, one of the most prominent hypotheses of the last decade is that energy transmission lines create electromagnetic fields (EMF), and that these fields interfere with the body's immunological systems, and hence are part of the chain that leads to cancer. In order to determine if electromagnetic fields contribute to elevated cancer rates the analyst (1) needs a detailed cancer registry, (2) needs to be able to create a precise map of the fields around the transmission lines, and (3) needs to identify the residential units that lie within the fields. This is challenging because the magnetic field drops off in a nonlinear rate from the lines. A 10-meter difference in accuracy is significant in measuring the potential hazard of the field. Wartenberg et al. (1993) used global positioning technology (GPS) to accurately define the EMF around the transmission lines, and they are currently linking the EMF data to cancer registry data in New York State.

They also needed the most recent housing data. This was available by an overflight of the transmission line corridor and the air photography that followed. Last, detailed U.S. Census Bureau files helped them estimate the number of people who lived in the residences exposed to the fields. Without the GPS, the ability to have an overflight and rapidly produce air photos, and the ability to match these to geo-coded census files, the project would not have been done.

When I was studying medical geography in the late 1960s, no one could do any of the studies described in this environmental cancer section because there were no computerized databases and only crude devices existed for measuring exposure to carcinogens. I recall talking about a cancer atlas of the U.S. in 1966, but the U.S. National Cancer Institute did not produce one for more than a decade, even though it had been collecting death certificates since 1933. When I wrote *Urbanization and Cancer Mortality* (Greenberg 1983), I had access to computerized files of all the death certificates in the U.S. Only one computer at my university had sufficient capacity to run my programs for sorting over a million cancer deaths and calculating death rates. I went to the computer center at 1 a.m. and ran the calculations until 3 a.m. on many occasions. Today, that same set of calculations is not necessary because the data are already sorted, and take less than a minute to run on my personal computer. In short, the major message of this section is that we have made incredible strides in detecting potential causes of environmental cancers primarily because of technology. But the "we" refers to North America, Western Europe, Japan, Korea, and a few other affluent countries. The technologies I have described are either not available in much of the rest of the world, or if they are present, are not applied to public health.

4. HEALTH CARE SERVICE DELIVERY TO THE ELDERLY AT RISK

The fastest growing segment of the U.S. population is 85 years of age and older. A physical obstacle that was easily avoided at age 50 (e.g., a flight of stairs that was quickly navigated) and nutritional habits that were ignored, become major health risks for most people who are 85. In 1980, 2.2 million Americans were 85 and older (about one percent of the population). By 1990, the number was 3 million, and it reached over 4 million by 2000. The U.S. Bureau of the Census (1999) estimates the 85 and older population will grow to 6.5 million by 2020 and 18.2 million by 2050. These numbers correspond to 2 percent of the population in 2020 and 4.6 percent in 2050. The health care needs of this population have grown and the expense of providing the very elderly with services has grown markedly. For example, in 1985, health care was 4.7 percent of average annual consumer expenditures in the U.S., and it was 5.3 percent in 1997. But it was over 14 percent for those 75 and older (U.S. Bureau of the Census 1999).

Research shows that women, the poor, and elderly are most likely to neglect their health care needs because of the friction of distance, unless there is an emergency (Goodman et al. 1997, Haynes et al. 1999, Young 1999). Without an emergency, appointments are postponed. Geographic-based modeling can help those who do want to make their appointments. For example, maps can be drawn to allow the population to pinpoint the location of essential health services, as well as shopping, and other needs. Bus routes, taxi services, costs, and times can be provided (Morrison et al. 1999) by applying standard geographical mapping procedures.

The big challenge is to help those who do not want to travel or cannot travel because of poverty, fear, and a fragile health, and cannot afford home aid to monitor their condition (Love and Lundquist 1995). We know that unattended chronic diseases worsen, the individual becomes more morbid, the cost of treatment increases exponentially, and it may be too late to help them.

The idea of using telecommunications to monitor isolated elderly people first came to the public's attention during the outer space flights when astronauts were closely monitored. Rural areas of the U.S. have become the focus of this effort. Elsewhere, I wrote about a young physician who benefited enormously from being tied into a telecommunications system that would allow diagnosis and consultation of injuries and illnesses. Located at an Indian Reservation at the extreme northwest tip of the continental U.S. in the state of Washington, his clinic was more than four hours from Seattle and hours away from the nearest hospital (Greenberg 1999).

The exciting new idea is to apply the same technologies to cases of isolated urban people. Isolated urban may seem like an oxymoron. Yet it is a reality for many elderly and disabled urban residents who have limited mobility and have chronic conditions that need to be monitored. The fact that the fastest growing portion of the American population is 85 and older underscores the need to provide monitoring of chronic conditions such as diabetes and heart disease. Sponholz (2002) studied the capacity of New Jersey inner-city hospitals to offer telemedicine. The capacity is built around personnel who can monitor a dozen or more patients from a central station, but it is also dependent upon technology, notably computers linked to phone lines and phone lines attached to monitors that are attached to the patients. She reports that patients who have been linked to telemedicine are more likely to follow instructions and are less frustrated because they are not required to make so many physician visits. While the cost-effectiveness of this technological application needs to be demonstrated and the willingness of urban hospitals to adopt it needs to be proven, it seems clear to me that this combination of technologies is going to be adopted and be in common use because of patient needs as well as to save money. The next step in the evolution will be video monitors in rooms that can be monitored from a central location.

I am excited about the possibilities of reaching underserved populations using new monitoring and communication technologies. Yet I must confess that the U.S. system of health care delivery makes me worry that this opportunity might be lost or at least distorted because of the economics of medical care in the U.S. Specifically, Sponholz's contacts, while excited about the idea, were unwilling to pay for the new technology and were desirous of making the technology profitable as soon as possible. This means that the economically strongest hospitals and clinics will quickly adopt this technology. Many of these are located in the most distressed neighborhoods. But some areas where this technology would be most useful are served by small and economically distressed hospitals and clinics that cannot afford the innovation. Unless funding becomes available and savings are obvious to the health care delivery service industry, some high-risk elderly will be served and other equally needy elderly will not.

5. HEALTHY PEOPLE IN HEALTHY NEIGHBORHOODS

The degraded condition of many inner-city neighborhoods does not compete with weapons of mass destruction, cancer, and the growing needs of the elderly for national headlines. Nevertheless, it represents a hard challenge to society as a whole and to our geographical theories and tools. For over half a century, suburbs have been growing and old industrial and city neighborhoods declining. Cities like St. Louis, Detroit, and Newark have lost one-third to more than half of the population they had in 1940.

Within those cities, some neighborhoods have lost two-thirds of their population and nearly the entire local employment. What remains is abandoned and deteriorated structures, a good deal of it contaminated, and relatively poor people in need of jobs and desperately in need of health services. Wallace and Fullilove (1991) meticulously documented the reality that HIV/AIDS, tuberculosis, acute alcoholism, lead-poisoned children, and other disease rates were closely matched to the geography of distressed neighborhoods in New York City. Hence, it is essential that city planners, for-profit and not-for-profit developers, and local health officers work with the remaining community leadership to develop a neighborhood redevelopment plan that enhances neighborhood and public health.

GIS-based tools that are linked to databases can be valuable to the groups involved in planning. Maps showing vacant property, the location and conditions of infrastructure, schools, bus routes, flood-prone areas, and various other contextual data that are critical to get all the parties working together.

Even more valuable is the ability of the newer GIS technologies to produce simulations of what the neighborhoods will look like under different redevelopment scenarios (Brail and Klosterman 2001). The GIS model is even more valuable if it is tied to a database that can immediately show the users the relationships between design changes and social, economic, and environmental impact. For example, Mayer et al. (2002) have been engaged in a study of small cities impacted by Hurricane Floyd in 1999. The floods physically and economically devastated the towns. They have been working desperately toward a new image for their cities. The GIS models are critical because they show the impact of rebuilding at high and low densities, emphasizing housing versus emphasizing commercial impacts. The models underscore the need for regional planning because the models instantaneously show that the towns will be overburdened by crawling automobile traffic if the redevelopable space is fully committed to commercial activities.

More important for this example, these interactive GIS-based models graphically show the advantages of clustering development. That is, elected officials and community representatives see the advantages of building housing and play areas around new or rebuilt schools, of making the school into a facility that can be used for local health care, job training, and other services.

Audiences immediately begin asking the analysts to try out more options. The ability to provide tables and two-dimensional maps within a minute is incredibly effective at stimulating community participation. Even more exciting is the ability to provide three-dimensional views, in which the participant gets an aerial view of what their neighborhood will look like if the proposed changes are made (Brail and Klosterman 2001). This technology is expensive and very data-demanding, but clearly is the next wave and will become a major tool in future efforts to tie together neighborhood health and personal health.

The need for this communication-enhancing technology is apparent in developed nations. But it is perhaps even more acutely needed in developing nations where neither technology nor professional planning expertise is present. When they are present, the combination is often used without any community involvement and for the benefit of multinational corporations and a privileged few residents of the host nation.

6. SPREAD OF HIV/AIDS AND OTHER INFECTIOUS AGENTS

Some of the most spectacular applications of technology have prevented and treated infectious diseases. Catastrophic disease outbreaks have been prevented by vaccines and drugs and improvements in sanitation and transportation. Drugs and improved nutrition have helped many recover. Yet technology has helped create two worrisome trends. One is the emergence of some infectious agents as serious risks because technology has provided an environment conducive to their growth. The second is the increased potential for rapid spread of infectious agents associated with increased international travel, migration, and trade. The list of technology-infectious agent links is distressingly long. Breiman (1996) collapsed them into five groups, which I have paraphrased and slightly altered: (1) sealed aircraft and poorly ventilated buildings that expose people to viruses, bacteria, and agents; (2) long-distance international shipments and preparations of food and other products that have led to gastrointestinal illnesses, meningitis, spread of mosquitoes, and botulism; (3) widespread adoption of cooling towers and other cooling systems that led to a growth environment for Legionnaire's disease; (4) inadequate water purification and sewage system management leading to exposure to cholera, *Cryptosporidium,* and other infectious agents; and (5) surgery, immunological system-suppressing drugs, transfusions, and transfer of patients that have led to the spread of bacteria resistance, Acquired Immunodeficiency Syndrome (AIDS), hepatitis, malaria, and Severe Acute Respiratory Syndrome (SARS).

While SARS is the most recent and prominent illustration of a technology-infectious disease linkage, I do not think that we understand enough about it at this time to use it as an illustration. The most prominent technology-infectious disease link during the last two decades has been HIV/AIDS. HIV is a sexually, perinatally, and parenterally transmitted infection that usually causes AIDS without antiretroviral therapy. The technology-agent links are multifaceted. Arguably the disease began in Africa. High-speed transportation of infected people across the globe facilitated the spread.

Technology was developed to detect the virus. Drugs can slow down or halt the impact of the virus. The geographical distribution of the disease can be mapped and forecasted (Shannon et al. 1991). Epidemiology has been used to collect data that can identify high-risk people. Moreso than the other examples in this chapter, however, HIV/AIDS is a case where a lack of political and economic capacity hinders the use of the technology, whether the technology is drugs or counseling.

Within the U.S., 850,000 to 950,000 people are currently infected, and 40,000 new people become infected every year, including 300 infants (Umar 2003). The Centers for Disease Control and Prevention (U.S. Department of Health and Human Services 2003) announced in June 2000 an initiative to make HIV testing a routine part of medical care, create new protocols for testing for HIV outside medical settings, work with high-risk people with HIV to prevent new infections, and decrease prenatal spread by working with pregnant women. This program makes sense as a national effort. But it needs to be tailored to high-risk populations. For example, Swartz (2003) reports that HIV infection rates in homeless communities range from 3 to almost 20 percent of the population, compared to less than one percent of the population as a whole. HIV/AIDS is both a cause of homelessness and a result of homelessness. People who are homeless are more likely to be intravenous drug users, have been involved in prostitution and/or been raped, be undernourished, have less access to health care, and be less educated (see for example, Kushel et al. 2002). The U.S. Department of Health and Human Services (2003) has a new initiative aimed at decreasing the number of homeless people with HIV/AIDS. While it is true that it is difficult to track a transient population, this program, if implemented with sufficient resources, can be a positive step in developing a strategy to improve housing and reduce the impact of HIV/AIDS in this high-risk population. Frankly, I am skeptical about the program being implemented with sufficient resources to make a big difference. But the U.S. has at least identified a high-risk population, developed a sensible programmatic response, and is talking about allocating resources to deal with the problem.

Another concern is the resurgence of HIV among men who have sex with men (MSM). Gross (2003) reports that new HIV diagnoses increased 14 percent among U.S. MSM between 1999 and 2001. He points to failures to follow behavioral changes in the MSM community, unproven assumptions about the effectiveness of behavioral change protocols, and to a lack of economic and moral investment (see also Cochran et al. 2002). Gross identifies a failure to conduct clinical trials of social intervention methods, as well as racism and stigmatization as major reasons why rates are rapidly increasing among young men of color. He calls for development and testing of counseling methods, and argues that this second wave will wipe out much of the MSM population unless this is done.

The international picture can only be described as frightening. Umar (2003) reports almost 22 million deaths, more than 36 million currently infected, and disappointing vaccine trials. The highest toll is in Sub-Saharan Africa. Eberstadt (2002) notes that as of late 2001, almost 3 of every 4 HIV carriers in the world lived in Sub-Saharan Africa, almost 10 percent of the population aged 15-49 was a carrier, and that 20 million had already died from it, or about 1/5 of the deaths in the region. Mother-to-child transmission is growing and theoretically can be prevented. John and Kreiss (1996) identified antiviral therapy, topical antiseptics, caesarean section, immunoglobulin and vaccination, avoidance of breastfeeding, and vitamin therapy as possible technological responses. But they point out that these approaches will be difficult, if not infeasible, to implement in poor nations. There is not much local wealth to bring technologies to bear on the HIV/

AIDS problem of Sub-Saharan Africa, nor does much of the rest of the world appear to care very much. Eberstadt (2002, 23) says that "the explanation for this awful dissonance lies in the region's marginal status in global economics and politics."

Eberstadt's grimmest words are for Eurasia. He forecasts massive outbreaks in Russia, India, and China. These nations, Eberstadt argues, have been behaving as if HIV was a minor problem and under control, when data suggest that the highly publicized problems of drug users, homosexuality, prostitution, use of unsafe blood, increasing population mobility, and lack of targeted education have created a volcanic-like eruption of new cases.

Normally, *Foreign Affairs* does not publish public health articles, focusing instead on factors that alter the world balance of power. Eberstadt's analysis extends far beyond public health into economic growth and military power. Like the geographer Mackinder (1942), Eberstadt emphasizes the strategic importance of Eurasia and asserts that the HIV/AIDS epidemic will be so severe that it will undermine the economic prospects of these nations and with it, alter the military balance of the world. In other words, the economic and political failure to face up to and address the HIV/AIDS problem will undermine these nations. The SARS epidemic, if it gets any further out of control, will add to the Eurasian burden.

The bottom-line observation derived from this infectious disease section is that higher income, stable economic growth, and stable political systems increase the capacity to purchase economic goods and services that are the basis for public health, and that public health has clearly become a requirement for achieving economic growth and stability (see Subramanian et al. 2002). These infectious disease data are the major reason that, absent a vaccine for HIV/AIDS and rapid control of SARS, we can expect the 20+ year difference between the two dozen most healthy and two dozen least healthy nations to increase, not decrease.

7. WHERE IS TECHNOLOLGY LEADING AND WHERE IS IT BEING LED?

Technology drives applications and vice versa in public health. I have little doubt about where the technology is heading, but I am much less certain about the public policy response to the increased availability of it. With regard to the disease ecology tradition, equipment will continue to be miniaturized, making it more portable, and it will become cheaper. More environmental data showing low concentrations of contaminants will be gathered more rapidly. These data will be linked to two and three-dimensional mapping packages and to advanced probabilistic-based simulation models that will statistically disperse contaminants to locations where humans live and work and ecosystems function. Furthermore, humans, their homes and workplaces will soon be equipped with miniaturized monitors that can estimate direct cumulative dose to a variety of canary-like contaminants. People will want to know what exposures mean and will want the exposures eliminated.

At the macro and meso geographical scales, the geography of cancer and cancer clusters will be extended to other causes of death, illness, and injury. Not only experts, but the general public will be able to generate their own maps of cancer, birth defects, and other abnormalities, which will surely lead to a public effort to identify and root out clusters of excess diseases. Explanations for excess morbidity and mortality will be demanded; government will need to become proactive in monitoring and surveillance to avoid being embarrassed by citizen-led efforts to identify excess disease and injury.

The challenge to elected officials, ethicists, health scientists, and others is to cope with what certainly will be a growing call for better science and no exposures to toxins. I know from personal experience that it will be difficult to convince skeptical people and build trust using arguments that increased surveillance may be responsible for increased disease identification, that clusters of high disease happen by chance, and that populations must be patient to wait for a causal hypothesis to be tested before spending large amounts of money on research and changing policies. Elected officials and state commissioners of health and environmental protection will feel the political pressure of this technology diffusion.

With regard to the medical geography tradition, technology and policy will be driven by the rapid increase in health care costs. When it is proven that health can be more effectively monitored by miniaturization of personal and home monitors, and that costs can be reduced or at least contained, then the telemedicine experiments that are nascent today will be diffused as widely as VCRs, radios, and televisions. Unlike VCRs and television, which appealed to the public, I suspect the demand for home and personal monitors will originate from a combination of suppliers of the technology, medical professionals, and medical insurers who cannot meet the challenge of tending to all their morbid patients one by one in their office. Overall, there is no escaping the overarching role of the market and political forces in driving the development and application of technology in public health.

Finally, the surreal portrait of the AIDS/HIV infection in Sub-Saharan Africa and forecasts for Eurasia can only be described as sobering. It is clear to this author that many nations cannot realistically cope with aggressive infectious agents because of their economic status, political, and social systems. Frankly, I believe technology quickly developed and widely administered is the only mechanism that can hold off massive deaths and/or draconian political measures taken long after less aggressive ones should have been taken.

ACKNOWLEDGMENTS

I would like to thank Joanna Burger, Christine Drew, Panos Georgopoulos, Dona Schneider, and Jane Sponholz for sharing their research with me. The accurate description of that work and the interpretation drawn from it are solely the responsibility of the author.

REFERENCES

Albert, D., Gesler, W., and Levergood, B. (Eds.) (2000). Spatial Analysis, Gis, and Remote Sensing Applications in the Health Sciences. Chelsea, MI: Ann Arbor Press.

Balakrishnan, S., Roy, A., Ierapetritou, M., Flack, G., and Georgopoulos, P. (2002). Uncertainty Reduction and Characterization of Complex Environmental Fate and Transport Models: An Empirical Bayesian Framework Incorporating Stochastic Response Surface Methods, Environmental Science and Technology, in Review.

Brail, R. and Klosterman, R. (Eds.) (2001). Planning Support Systems: Integrating Geographic Information Systems, Models, and Visualization Tools. New York: ESRI, Inc.

Breiman R. (1996). Impact of Technology in the Emergence of Infectious Diseases, Epidemiologic Reviews 18: 4-9.

Burger, J., Stephens, W., Boring, C., Kuklinski, M., Gibbons, J., and Gochfeld, M. (1999). Ethnicity and Risk: Fishing and Consumption in People Fishing Along the Savannah River, Risk Analysis 19: 427-38.

Burger, J., Gochfeld, M., Powers, C., Waishwell, L., Warren, C., and Goldstein, B. (2001). Science, Policy, Stakeholders and Fish Consumption Advisories: Dealing a Fish Fact Sheet for Savannah River, Environmental Management 27: 501-14.

Cochran, B., Stewart, A., Ginzler, J., and Cauce, A.M. (2002). Challenges Faced by Homeless Sexual Minorities: Comparison of Gay, Lesbian, Bisexual and Transgender Homeless Adolescents with Their Heterosexual Counterparts, American Journal of Public Health 92: 773-77.

Committee on Foreign Relations, U.S. Senate (1996). The Chemical Weapons Convention, September 1, 1996, 104th Congress, 2nd Session, Executive Report 104-33, Jesse Helms, Chair.

Cromley, E. and Mclafferty, S. (2002). GIS and Public Health. New York: Guildford Press.

Drew, C. (2000). Decision Mapping to Promote Transparency of Long-term Environmental Cleanup Decisions, Seattle, WA: Cresp and Department of Geography, University of Washington.

Drew, C. and Nyerges, T. (2002). Decision Transparency for Long-term Stewardship: A Case Study of Soil Cleanup at the Hanford 100 Area, Journal of Risk Research, in Review.

Eberstadt, N. (2002). The Future of Aids, Foreign Affairs 81: 22-45.

Georgopoulos, P., Wang, S., Balakrishnan, S., Chandrasekar, A., Roy, A., and Efstathious, C. (2002). Refinement of Source-To-Dose Models: Forest Fire Dynamics and Emissions, Transport, and Fate. New Brunswick, NJ: Cresp Report.

Goodman, D., Fisher, E., Stukel, T., and Chang, C. (1997). The Distance to Community Medical Care and the Likelihood of Hospitalization: Is Closer Always Better? American Journal of Public Health 87: 1144-50.

Greenberg, M. (1983). Urbanization and Cancer Mortality. New York: Oxford University Press.

Greenberg, M. and Wartenberg, D. (1991). Newspaper Coverage of Cancer Clusters, Health Education Quarterly 18: 363-74.

Greenberg, M. (1999). Restoring Americas Neighborhoods: How Local People Make a Difference. New Brunswick, NJ: Rutgers University Press.

Gross, M. (2003). The Second Wave Will Drown Us, American Journal of Public Health 93: 872-81.

Haynes, R., Bentham, G., Lovett, A., and Gale, S. (1999). Effects of Distances to Hospital and GP Surgery on Hospital Inpatient Episodes, Controlling for Needs and Provision, Social Science and Medicine 49: 425-33.

Isukapalli, S. and Georgopoulos, P. (2002). Characterization and Reduction of Uncertainty in Environmental Fate and Transport Models: Application to Fact (Flow and Contaminant Transport). Groundwater Model at Savannah River Site. New Brunswick, NJ: CRESP.

John, G. and Kreiss, J. (1996). Mother-to-Child Transmission of the Human Immunodeficiency Virus Type I, Epidemiologic Reviews 18: 149-57.

Kushel, M., Perry S., Bangsberg D., Clark R., and Moss, A. (2002). Emergency Department Use Among the Homeless and Marginally Housed: Results from a Community-Based Study, American Journal of Public Health 92: 778-84.

Love, D. and Lundquist, P. (1995). Geographical Accessibility of Hospitals to the Aged: A Geographic Information Systems Analysis within Illinois, Health Services Research 29: 629-51.

Mackinder, H. (1942). Democratic Ideals and Reality. New York: H. Holt and Company.

Mason, T., Mckay, F., Hoover, R., Blot, W., and Fraumeni Jr., J. (1975). The Atlas of Cancer Mortality for U.S. Counties, 1950-1969. Washington, DC: U.S. Department of HEW, NIH 75, 780.

Mayer, H., Danis, C., and Greenberg, M. (2002). Smart Growth in a Small Urban Setting, the Challenges of Building an Acceptable Solution, Local Environment 7: 340-62.

Morrison, D., Alexander, D., Fisk, J., and Mcguire, J. (1999). Improving Delivery of Health and Community Services yo Welfare Recipients, Columbia, SC, Journal of Public Health Management and Practice 5: 49-50.

National Research Council, Committee on Review and Evaluation of the Chemical Stockpile Disposal Program (2001). Occupational Health and Monitoring at Chemical Agent Disposal Facilities. Washington, DC: National Academy Press.

National Research Council, Committee on Review and Evaluation of the Chemical Stockpile Disposal Program (2000). A Review of The Armys Public Affairs Efforts in Support of the Chemical Stockpile Disposal Program, Letter Report. Washington, DC: National Research Council.

Shannon, G., Pyle G., and Bashshur, R. (1991). The Geography of Aids: Origins and Course of an Epidemic. New York: Guildford Press.

Sponholz, J. (2002). Telemedicine as Viable Health Delivery, Ph.D. Thesis, New Brunswick, NJ: Rutgers University.

Subramanian, S., Belli, P., and Kawachi, I. (2002). The Macrodeterminants of Health, Annual Review of Public Health 23: 287-302.

Swartz, A. (2003). Twin Crises: Homelessness and HIV/AIDS, Hiv Impact May/June, 1-2.

Top-To-Bottom Review Team (2002). Review of the Environmental Management Program of The United States Department of Energy. Washington, DC: U.S. DOE.

Umar, K. (2003). What Now? Aids Vaccine Trial Results Disappointing, HIV Impact May/June 13-14.

U.S. Bureau of the Census (1999). Statistical Abstract of the United States, 1999. Washington, DC: U.S. Department of Commerce.

U.S. Department of Health and Human Services (2003). Ending Chronic Homelessness: Strategies for Action. Accessed June 3, 2003 at http://www.ich.gov

U.S. Department of Health and Human Services (2003). Advancing HIV Prevention: New Strategies for a Changing Epidemic–United States, 2003. Accessed June 3, 2003 at http://www.cdc.gov/mmwr/pdf/wk/mm5215.pdf.

Wallace, R. and Fulllilove, M. (1991). AIDS Deaths in the Bronx 1983-1988: Spatiotemporal Analysis from a Sociogeographic Viewpoint, Environment and Planning A 23: 1701-23.

Wang, S., Ouyang, P., and Georgopoulos, P. (2002). Systematic Model Reduction for Efficient Multimedia/Multipathway Exposure and Dose Assessments. Paper presented at International Society of Exposure Analysis Meeting, 2002, Vancouver, Canada, August 11-15.

Wartenberg, D., Greenberg, M., and Lathrop, R. (1993). Identification and Characterization of Populations Living Near High-Voltage Transmission Lines: A Pilot Study, Environmental Health Perspectives 101: 626-32.

Young, R. (1999). Prioritising Family Health Needs: A Time-Space Analysis of Womens Health-Related Behaviours, Social Science and Medicine 48: 797-813.

CHAPTER 17

PAMELA MOSS
MEI-PO KWAN

"REAL" BODIES, "REAL" TECHNOLOGIES

Abstract   With the influential rise of poststructuralist theory in the social sciences, understandings of the notions of women and space are moving away from tightly wound, monolithic packages toward loosely linked, contingent ensembles. Alongside the ontological challenges of these discursive categories has been a shift in the way that body and technology are popularly understood. Drawing on feminist work on bodies and technology, we offer a reading of "real" bodies and "real" technologies that focuses on *both* the social construction *and* the tangible, concrete expressions of the body and technology *at the same time*.

Keywords   body, technology, women, health, discourse, ontology

1. INTRODUCING BODIES AND TECHNOLOGIES

Women and space have long been a concern for feminists in geography (for overviews, see Domosh and Seager 2001; McDowell 1999; Women and Geography Study Group 1984, 1997). An exploration of the links between the category of "woman" and spatial phenomena at various scales has produced rich literatures in many subdisciplines in geography. One of the key areas feeding the burgeoning literature about women and space within and outside geography has been the influence of feminist theory, postmodernism, and the resurgence of radical approaches in the new cultural geography. With a focus on subjectivity, self, and identity, encounters between feminism, poststructural thought, and social constructionist approaches yielded a plethora of studies about the body as both a theoretical field of inquiry and a unit of analysis (e.g., Blunt and Wills 2000; Longhurst 2001; Moss and Dyck 2003a, b; Teather 1999). The most significant contribution of these works is twofold: (1) sociospatial relations play an important role in the construction of women's identities and constitution of their subjectivities and, (2) place matters as women negotiate the cultural inscriptions of their bodies in the context of

Stanley D. Brunn, Susan L. Cutter, and J.W. Harrington, Jr. (Eds.), Geography and Technology,
383-399. © 2004 Kluwer Academic Publishers. Printed in the Netherlands.

their own experiences and in ways that are specific to their immediate environments.

The investigation of space and technology in geography has primarily been located in urban studies about the Internet, cyberspace, and digital landscapes (e.g., Crang et al. 1999; Dodge and Kitchin 2001a, b; Holloway 2002; Valentine and Holloway 2002; Kwan 2002a; Wheeler et al. 2000). This has also been true of women, space, and technology (e.g., Kwan 2003; Light 1995; Rommes 2002; Wakeford 1998). Feminists in geography have extended these analyses topically into areas that tease out the implications of gender beyond individual people, toward the masculine nature of knowledge production (Kwan 2002b; Schuurman and Pratt 2002) and toward the possible uses of spatial technologies for women (Kwan 2002c; Hanson 2002; McLafferty 2002; Pavlovskaya 2002). Outside geography, feminists interested in women, space, and technology have turned to theorizing technology as it relates to women, work, or machine (e.g., Hawthorne and Klein 1999; Haraway 1991; Napier et al. 2000; Prophet 1999). These feminist visions of technology recognize the process of gendering beyond that of sexuality, beyond the presumption that all technology necessarily has to be masculine, and beyond the conflation of the genesis and use of knowledge.

For us, the works cited above pave the way to produce a feminist reading of space *through* bodies and technologies. Our reading presented here differs from many previous feminist geography studies as we direct our attention toward bodily images and the processes through which they are created. There seems to have been a movement away from representing the body as an entity to be accessed through the senses, and toward representing both individual and collective bodies as fragments of data (e.g., Birke 1999b; Mazzio and Hillman 1998; Waldby 2002). The notion of the *body as data* is a way for us to convey the significance of how the shattering of not only identities of individuals, but also their bodies, has drastically altered the way in which bodies are represented (see Doane 1990; Gear 2001). The fragmentation notion of the body as data is important because increasingly, the body, in a wide range of discourses, is being carved up and offered as an assemblage of loosely connected morsels, tailored for particular audiences. These morsels, then, through different analytical methods, are arranged and rearranged to produce different bodies out of a body's data. A concern some feminists have is that this fragmentation shatters the sense of the body as a whole, as connected, as biological and physiological systems working together (Birke 1999b). Attempts to map body as data (as well as technology and culture) end up producing borderless, discontinuous, ruptured, turbulent entities with endless perturbations, kinks, and snarls that seem almost useless in the face of what appears to be a disconnected series of string figures (Haraway 1989; Martin

1996). Thus, at issue is not the age-old debate about "the whole is greater than the sum of its parts"; rather it is the reinforcing of a separation of function for each body part—the heart, the brain, the hand—each of which can be accessed through an idyllic image (or a reasonable facsimile) as its form, without a following through on understanding the consequences of the intricate links among these body parts. For us, these images are interesting in another sense. Both these images of the bodies and the process of creating these images have come not only to represent a particular body, but also, in some cases, to have taken the place of a "real" body—by "real" we mean the concrete, tangible bodies through which individuals exist within and experience the world.

Bodies are often dependent on specific technologies for their representation. What feminist visions of technology have to offer feminists interested in reading space is that the images of the bodies themselves need to be scrutinized. In order to maintain our focus on feminist readings of space, rather than focus on technology as the *object/subject* of inquiry, we choose to use technology as *site* of inquiry, that is, one that assists us in making sense of women's bodies as they exist in and move through spaces. In examining two types of technologies–body imaging and health self-monitoring gadgets–we explore the processes through which knowledges produced by particular technologies come to be masculine as well as come to create ontological dissonance in representing materiality.

In the last decade, feminist researchers and analysts have demonstrated the masculinity of knowledge production in many fields of geography (e.g., Morin and Berg 1999; Cope 2002; McDowell 1991, 1992; Rose 1993; Schuurman and Pratt 2002). Geographic knowledge is not only masculine in the sense of who produces it, but also in how it gets produced, who uses it, and what claims are being made from and because of its production (Berg 2002; Hall et al. 2002; Kobayashi 2002; Monk and Hanson 1982; Rose 1993). The most important points that we take from combining these literatures is that "real" technologies–by "real" we mean the concrete, tangible expressions of technologies themselves–are not ontologically gender-specific; they merely come to be seen that way.

We do not mean to insinuate that conceiving the body as data always has negative consequences. As feminists have been striving to demonstrate that women's identities are fractured, we want to emphasize that the process of representing "real" bodies through "real" technologies is itself frayed because of the ways we come to make sense of bodies and technologies. In this paper, we concentrate on creating a feminist reading of space in the context of "real" bodies and "real" technologies that force us to think through what fragmentation means for bodies and their relationship to technologies. We first set out an approach we find useful in showing how unsettling tight connections among images and reality opens up different interpretations of "real" bodies and "real" technologies. Next, we use the topics of medical imaging and health self-monitoring gadgets to demonstrate how this approach works. Finally, we close by commenting on what types of studies we would like to see that would assist in unraveling the links among bodies and technologies.

2. UNSETTLING CONNECTIONS

In creating our feminist reading of space, we have found it imperative to deal with the unease with which feminism exists alongside masculinist science, bodies alongside their images, and technologies alongside women. Our unease indicates that we need to focus on making soluble the ontological glue holding such notions and concepts together. One way to loosen these tight grips is to recognize contradictory knowledge claims manifesting themselves in our everyday lives (after Nairn 2002). Such a recognition creates room to maneuver through the firm bonds that link certain things or notions together, as if they were "normal" relationships, for example, men, science, and technology, or objects and their representations. Three works that approach topics related to our inquiry were useful in our attempt to unsettle connections among women, space, and technology. All three endeavor to break seemingly stringent ties between two central constructs—for Victoria Lawson, it is numbers and quantitative methods; for Lynda Birke, competing representations of the heart; and for Abigail Bray and Claire Colebrook, self-negation and anorexia. These sets of dual constructs are not binaries in the sense of a Cartesian dualism; rather, they are constructs whose links have been normalized, even naturalized, through multiple discourses and specific social practices.

Lawson (1995) approaches the framing of the choice of methods in research in ontological terms. She argues that thinking about feminist research methods as part of a "quantitative versus qualitative" debate prevents feminists interested in using poststructural theory from realizing the potential that numbers and counting have for research. By uncoupling numerical data and statistical methods from masculinist science, she is able to situate counting as an act congruent within feminist political goals.

Birke (1999a) being both serious and playful, explores a set of binaries associated with the heart—reason/emotion, rationality/irrationality, nature/culture. She wends her way through metaphors in medicine and science overlain with not so sheer veneers of sexism, racism, and colonial conquest. She reads the heart as the location of love while at the same time reading it as a piece of machinery. When read together, these metaphors give rise to texts claiming that although the heart is the center of life emotionally and physiologically, it remains "a simple machine with a sacred mission" (Birke 1999a, 117). These musings lead her to speculate that in order to understand the experiences of people with heart disease, "we need to unravel all these influences in the context of the intertwined strands of cultural meanings about the inner workings of the body" (Birke 1999a, 132).

Bray and Colebrook (1998) offer a reading of anorexia that goes against the dominant readings of the disease as a negation of patriarchal values (cf. Bordo 1993). They suggest that Deleuze's (1994) notion of positive ontology offers feminism "the possibility of a positive, active, and affirmative ethics" as a way "in which bodies become, intersect, and affirm their existence" (Bray and Colebrook 1998, 36). In rethinking anorexia, they claim that by conceiving the sensation of hunger as desire rather than conceiving the act of starvation as denial, then the activities that women diagnosed with anorexia nervosa engage in—counting calories, weighing themselves, measuring body fat— can be recast as self-formative instead of self-destructive. If this were the case, then anorexia would then be considered "a series of practices and comportments; [and] there [would be] no anorexics, only activities of dietetics, measuring, regulation, and calculation" (Bray and Colebrook 1998, 62). Such a numerical organization would continually generate more measuring activities. Thus the discourses that would be identified as shaping the body would be those of regulation, control, and quantification and not slenderness, fitness, and patriarchy. Although Bray and Colebrook make no suggestions about the material consequences of this conceptual shift for the women diagnosed with anorexia, it makes sense that treatments would shift, too. For example, in counseling, rather than focusing on the notion that starvation is bad and one needs to accept the body "as is," counselors might work toward redefining hunger as a different type of activity or, perhaps, substitute what a woman defines as desire. In a clinical sense, however, shifts in treatment are not so obvious; it is also not clear whether acts of forced feeding and constant surveillance of the women would cease as intervention techniques.

We juxtapose these arguments so as to highlight the unease each has with the ontological status of their chosen topic. Lawson's argument rests on the premise that "qualitative methods are no more essentially feminist than quantitative techniques are essentially masculinist" (Lawson 1995, 450). She, like many feminists, refutes essentialist claims as to what is and is not feminist or masculinist, an act that allows feminist uses of numbers. She also suggests that changing the social practices within academia will challenge the intellectual sustainability of implausible dualisms in the choice of research methods, as for example, quantitative/qualitative and masculinist/feminist. Lawson's antiessentialist argument feeds into Birke's reading of metaphors of the heart. Birke demonstrates that representations of the heart in science and medicine, although contrasting, work together to exalt the *form* and *function* of the heart as an organ. She supports this view by drawing attention to the ways popular discourses about science, as well as everyday language, pick up these meanings and deploy them in varying situations, for example, diagnosis and treatment of heart disease and consolation of an emotional ache. Like Birke, Bray and Colebrook seek to work out the relationship between the physical body and the ideas about the physical body. Bray and Colebrook reconceptualize the ontological status of an illness that uses technologies for self-surveillance. Rather than reconfiguring binary constructs, as for example, health/illness, mind/body, and self/other, they reread the representation and construction of anorexic bodies. For them to understand anorexia is not a matter of negating the genesis of an origin (via recognizing *lack* in the self); rather, it is a matter of looking at events, acts, and connections that produce an entity known as an anorexic body.

All these authors allude to the normalization processes that operate within, but not exclusively so, those discourses and social practices associated with the engagement of various research data collection and analytical methods, biomedicine, and the practice of medicine and science. Also because of the relative ease with which individuals are able to shift from one set of representations to another, it makes sense that competing representations arising in different discourses are supported, and even merge, over long periods of time (such as understanding an acceptance of multimethod research or the sacredness of a muscle sustaining life). One way to disrupt these dis/comforting normalization and naturalization processes is to loosen the binds that tie the constructs together, that is, dissolve the ontological glue. By intensifying ontological dissonance between an object and its representation, all authors were able to relax the connections between the constructs they were working with in hopes of uncoupling or unlinking these various representations from their presumed origins. Thus the act of feminists counting with numbers, employing metaphors of the heart, and monitoring weight exemplify positive acts that originate neither in discourse nor in materiality. These acts are productive in and of themselves. Recasting the ontological status of each as positive permits (nonnormalized and nonnaturalized) connections and events to emerge without being predetermined by a particular origin or shaped by an assumed path of development.

In the rest of this chapter, we show how these arguments support our feminist reading of space through "real" bodies and "real" technologies. We use technology as a site to explore some *positive* events and connections among women and space and use the bond between body and health as the topical site for inquiry. In order to disrupt the ontological status of the connections between body and health, we explore two related technological practices.

First is the process of imaging as a way to represent the "real" body as part of literally seeing the body. Second is the increase in health monitoring as part of an enhanced self-surveillance through the use of technology in the form of portable machines.

3. IMAGING "REAL" BODIES THROUGH "REAL" TECHNOLOGIES

Visual images dominate learning the body (Birke 1999b; Martin 1994; Haraway 1997). Visualizing the body has developed as a case of opening up the interior—laying it bare—working from the assumption that seeing has a direct connection to knowing. There is also a premise about seeing that makes things less mysterious, makes unknown things knowable. This perspective was especially true in the early part of the establishment of medicine as a field of study. For example, grave robbing and vivisection were fairly common practices, ones that literally prepared bodies, especially women's bodies, to be opened up for public viewing. This viewing, if not for medical surgery theaters (for both training and public spectacle for the rich), was accomplished through anatomically correct drawings—a direct, translation image of the "real" body. Margrit Shildrick (1997, 17-18) makes the point that the way in which the modern body was ontologically differentiated from the mind permitted a reductive representation of the body as a collection of body parts arranged into systems (akin to Birke's observation of the heart being read as form and function). Medicine has continued to embrace this representation evidenced by the ever-increasing specializations in training and practice. For example, there are numerous fields of studies focusing on body parts and systems—neurology, nephrology, urology, hematology, genetics, rheumatology, and gastroenterology, all of which use body imaging for diagnosis and the basis upon which treatment of a nonhealthy body is designed. It is noteworthy that nonspecialists around the world are paid less and have less status, even though they are supposed to have a wider range of knowledge.

This drive to *see*, and subsequently to *map*, the body and all of its parts has only intensified throughout the development of technologies used to enhance the practice of medicine—microscopy, x-ray, laparoscopy, CAT (computed axial tomography), and MRI (magnetic resonance imaging). Less than fifty years ago, there were x-ray machines in shoestores to determine the exact size and shape of a foot so that the shoe salesclerk could sell the most appropriate shoes, especially for children. Since the 1970s, mammograms have been used routinely as an early detection system for breast cancer. Although increasingly becoming discredited in biomedical fields and something which is being aggressively challenged by women's health advocates, such imaging has produced desexualized breast images, that is, ones that are merely pictures of an internal body part that, again, are supposed to have high correspondence to "truth." And, as the Genome Project continues, the link between the image of the body via its DNA will no doubt solidify the ontological glue maintaining the ties between "real" bodies and "real" technologies. The explicitness (as a normalizing tendency) in both the imaging itself and the connection between genes and bodies makes it even more of a challenge (and indeed more urgent) to unsettle the connection between the "real" body and the "ideal" body, the body and technology, and body and health.

These advanced technological imaging processes that read and map the internal body as a space that (partially) constitutes an individual are setting more highly attenuated norms against which "real" bodies are measured. As Jody Berland (1996) notes:

> Bioengineering plays a special role, here; founded on the remapping of the human body as digital information, it transforms and cements the relationship between human and technological developments, which now replicate one another in more explicit ways. (Berland 1996, 247)

Consider also the Visible Human Project:

> The Visible Human Project® is an outgrowth of the NLM's [National Library of Medicine] 1986 Long-Range Plan. It is the creation of complete, anatomically detailed, three-dimensional representations of the normal male and female human bodies. Acquisition of transverse CT, MR and cryosection images of representative male and female cadavers has been completed. The male was sectioned at one millimeter intervals, the female at one-third of a millimeter intervals.

> The long-term goal of the Visible Human Project® is to produce a system of knowledge structures that will transparently link visual knowledge forms to symbolic knowledge formats such as the names of body parts. (National Library of Medicine 2001)

Although images of individual body parts are not available for sale, there are six sets of tapes full of images organized by body region: head, thorax, abdomen, pelvis, thighs, and feet. Two additional tapes include all the images produced for each body and a sampling of bits of body parts from the male. Each set of tapes costs US$150, or all sets for one body cost US$1,000— prices double for places outside the U.S. (National Library of Medicine 2001). These are cryogenic renditions of two *specific* bodies, cadavers chosen for their lack of any identifying marks physiologically or biologically and a natural death. "Adam" and "Eve" have become the definition of "normal" and "real" for they are closest to the "truth" of the body.

Kember (1999, 30) sees The Visible Human Project® as "a story of autonomous creation and of medicine's attempt to father itself," much like

Mary Shelley's story of *Frankenstein,* written at the emergence of medicine that showed the power of medicine as a discourse. We think that this intensification is linked to the notion of exposing the biomedically structured ontological condition of the body in its ability to produce images that have some truth correspondence to the real, to the material, and to spaces of everyday existence. It is not just the image of the inventory of body parts (or the anatomy of the body per se), but the attempts to document processes in the body that lead one to think that science as creation is viable. There is also now the capability to scan the brain while a person is undertaking ordinary daily tasks to see which part of the brain is being used (via PET, positron emission tomography). Also, with the Genome Project, the visibility of the gene ostensibly makes prediction, prevention, and cure possible, with little discussion of treatment of the conditions produced by the genes.

This discussion of imaging supports our idea about ontological dissonance. What emerges from these imaging activities produces a "normal" body, one naturalized through the practice of truth correspondence. Yet the spatial and place-specific contours of the "real" body are left out, excluded from making existence messy in any way. Imaging, in all its forms, and as a practice that physicians and biomedical healthcare practitioners engage in, reinforces the link between body and health (read "normal") through assuming that "real" bodies can be accessed through "real" technologies. Thus, if any abnormality shows up, then the person is defined as ill, which can be read as freakish, odd, deficient, or deviant.

4. POSITIVE ACTS OF HEALTH

Alongside these "true" images of bodies, icon images of body parts have transpired as a way to represent the "normal" body. For example, a brain on a flow chart represents an idea, a heart on a Valentine's Day card means love, and a hand in a grasping position on a computer screen signifies an option to choose. The use of icons in these ways allows medicine to seep into and appear innocuously in popular discourses. Icons have also emerged about disease that points toward the esteem in which society holds biomedical knowledge—a double helix of DNA to represent life, ribbons to support research in breast cancer and AIDS, and a bum for heightening awareness of colon and rectal cancer. These body parts recur repeatedly and often come to replace the sense of the body as connected, as a whole, as working together and to reinforce a separation of form and function for each body part. This dissonance between truth correspondence and iconic representation remystifies the body, stripping its context and "real" texture, and paves the way for the icons themselves to take the place of the body, body part, or disease process.

For example, the commonsensical and nearly unquestioned way that blood cells have been transformed into numbers is staggering. Although it is easy to take pictures of blood cells, the image is not as important as the context within which the blood cell exists in relation to other blood cells. Images cannot tell this story. The associations and correlations between form and function of each type of blood cell are represented as a set of numbers. These numbers have become the icon for the blood, and by extension, the body, the status of the body, and the health of the person. Slides, smears, and magnified photos of blood are easily taken and quickly interpreted and reported in the form of a printout of numbers alongside what can be considered "normal" readings for women and men.

Together with this iconic transformation has been a vernacularization of biomedical knowledge. It certainly is not uncommon to have a casual conversation with questions like: what was your white blood-cell count? What about your cholesterol? Is that the good fat or the bad fat? This vernacularization has brought with it an explosion of the ways in which individuals can and have come to be expected to monitor their own health by scrutinizing their body.[1] In addition to the already prevalent health promotion activities around nutrition (in the form of the food pyramid) and what constitutes a healthy weight (in the form height/weight charts and body mass indices), there are gadgets that "measure" health. The production of accessible technological machines, such as blood pressure cuffs for daily use in the home, heart-rate monitors for the chest and wrist, and diabetic test strips, have facilitated the enactment of self-surveillance for those who are considered to be ill as well as those considered to be healthy.

Also the extensiveness to which health-monitoring machines have become part of our public and private lives is also significant. Frequently, physicians ask patients to take blood pressure readings at the same time every day, or even twice a day, in public places such as pharmacies in discount stores like Target and Wal-Mart. Each one of these machines produces numbers that are then charted as low, normal, or high, creating a quick method for self-assessment.[2] What comes to matter as a result of these measuring activities is the icon! What becomes problematic, however, is that the reading in the form of an iconic representation may not correspond to the bodily experience of self-surveillance.

Rather than thinking about self-surveillance in terms of empowerment, self-determination, and choice, we want to engage the role technology plays in facilitating a positive production of what constitutes a body. Figuring out what to do with the experiential disjuncture between a lived body (the one we live through/in/with/of) and an idealized one (as mapped out through science and accessed through self-monitoring technologies) is important. What does it mean to scrutinize one's body closely without the security of necessarily knowing that an ill body can be treated, or a healthy body can become more fit? At issue is not so much how *true* the experience of health or of illness is—true in the sense that there is some truth correspondence to a particular claim in reality or, in other words, how out of whack with reality an idealized image is. Rather, the issue is how one comes to *represent* a specific experience of health or illness. It may be prudent to conceive experience not as a tightly bounded end product (as in "I am sick") conveyed through language (in a physician's office or in a hospital emergency room as a description of bodily sensations), but as a set of things (e.g., subjectivities, identities, lives, bodies) coming together at a particular point in time that produces interim entities (nameable identities, such as a patient, a sick person, a woman engaged in counting activities around food and exercise) that are then subject to various movements in and through space and time (one that is ill in a public space, one that uses a diabetic strip, one that is healthy, abled, disabled, fit, or unfit in the workplace, and so on). Implications of moving toward a focus on experience as a process instead of an outcome might influence notions of health, which may influence how we can make sense of self-surveillance publicly and privately. Just as experience could feed into developing personalized treatment regimes designed to deal with specific sets of symptoms, personalized self-surveillance regimes could be designed to deal with specific "real" bodies through "real" technologies, without the ever-present, assumed directive that bodies need regulation to attain health, fitness, and wellness.

In trying to unlink the tight binding between body and health (read as the body being normalized and naturalized into health), we suggest that rethinking the use of these types of medical technologies within a positive ontological framework might open up interpretations outside that which is possible now. Implicit in the active participation in these self-surveillance methods is the idea that these acts will produce a healthy body. But we know this is not the case, even as we participate in them. We can eat the right foods, have a regular heartbeat, normal cholesterol levels, and a normal ratio and distribution of multiple blood cells, and yet still be ill (after all, the "Adam" and "Eve," no matter how anatomically ideal, are dead). The question for us, then, is what are people producing themselves as when they engage in such activities? Illness preventers? Illness monitors? People with health obsessions? Active hypochondriacs? Fitness nuts? Are they casualties of healthcare cutbacks living out the effects of the downloading of healthcare? Are they empowered beings taking control of their own body and health via technology? Are they resisters to hegemonic notions of what counts as healthy, well, or fit? Or is it that they are actively producing selves that are "healthy" in a nonbiomedical sense, "fit" in a nonsport sense, "normal" in a nonnaturalizing kind of way? Clearly, these machines are not supposed to make an ill body well; they simply indicate, via a set of numbers, that at some point in time and at a particular place, something in the body is functioning or existing within what has been determined to be "low," "normal," or "high" by the conventions of medicine and science.

5. FURTHER THOUGHTS

Our chapter is meant to continue the discussion about women, space, and technology. Like many feminists, we are concerned with unsettling categories that appear to be fixed and stable. Unlike many feminists, we are not interested in focusing solely on "woman" as a category to disrupt through "troubling" sex and gender (see Webster 2002, 200). Our intent is to use the analytical tools, approaches, and methods developed and enhanced by feminists to understand how bodies and technologies articulate with/in space.

To further open up discussion about and increase the uncertainty over the ever more convoluted relationships among people and technology, we would like to see more feminist (and nonfeminist readings) offered about "real" bodies and "real" technologies, and not just in the sense of illness and health. In addition to gendered, racialized, colonized, disabled, and classed bodies, explorations of technology vis-à-vis food, sleep, pain, pregnancy, suicidal acts, and workplace are welcomed (e.g., see Evans and Lee 2002; Longhurst 2001; Williams and Bendelow 1998). Also important is tracing the implication of (visual and nonvisual) "real" technologies in scrutinizing bodies in particular places, such as airports, ATMs, grocery stores, shelters, close-circuited public parks, embassies, and beaches on the construction of space (see Dodge and Kitchin 2003; Koskela 2000; Warf 2000).

Our arguments are also highly relevant to investigations of other types of "real" bodies and "real" technologies. Movements in and through the global economy set the stage for "real" bodies to become consumable products via their links with various "real" technologies. For example, trafficking in organs is a practice that brings highly valued foreign currency to poor people across the South and Eastern Europe. Over the past two decades, trafficking in organs emerged as an important human rights issue. Yet it is only now that the issue is being encouraged to be part of more mainstream government policy (see Parliamentary Assembly 2003). Trafficking in women and children for pornography and prostitution, too, deals with "real" bodies being commodified through "real" technologies. Secret (and not so secret) electronic information and extensive distribution networks between the South and richer countries in the North move women and children like simple goods for sale. Our approach, using notions of ontological dissonance, body as data, and positive acts, can assist in figuring out how bodies and technologies are connected by loosening the naturalized links and identifying paths of resistance. These paths of resistance need not be "real," in the sense that they can stop trafficking of organs or trafficking of women and children in their tracks. Rather, as an alternative reading, our perspective can perhaps provide an entry point to changing the way connections between bodies and technologies are envisioned and used to marginalize, discipline, control, oppress, and displace women around the globe.

Our arguments in no way are limited to *women's* "real" bodies, even though our empirical interests (cited in this chapter) remain located with the "real" bodies of women. Integrating feminist visions of technologies into interpretations of space at the microscale level of body parts allows for a closer look at the implications of conceiving the body as data. For no matter how the body is *conceived*, "real" bodies comprise the fodder for our ideas, ones that have been made specific through the spatially grounded sets of social relations of their immediate environments. Going back and forth between the scale of the body part and the body within the context of the technologies we use to see and map the body, as well as monitor and scrutinize it, shows how integral the relationship between human bodies and health technologies is to making sense of the "real" aspects of living. Although the more overtly spatial aspects of the "real" bodies and "real" technologies are not addressed here, the tightness of the connections of concepts linking bodies and health, health and technology, and technologies and bodies provides ample room to explore the subtleties of how space and place may be implicated in our everyday lives.

ACKNOWLEDGMENTS

We presented a version of this paper at the annual conference of the Association of American Geographers in New Orleans in March 2003. Part of the argument was developed through a presentation by Moss at the Human and Social Development Faculty Research Day, University of Victoria, Canada in April 2003, and published in the conference proceedings. We thank Stan Brunn for his patience and confidence in us.

NOTES

1.   Women have been used to monitoring and scrutinizing their bodies, especially around reproductive processes, and have been subject to close scrutiny by physicians because of their anatomy and their capacity to reproduce. Because *seeing* the body is a more valued way of *knowing* the body, information from a physician's surveillance is often times taken more seriously even when contraindicated by women's experiences of their own bodies. It is only recently that health researchers (at least in Canada) are confirming

what some women have known for years: that it is possible for women to ovulate more than once during a menstrual cycle (CIHR/IRSC 2003; http://www.cihr-irsc.gc.ca/news/cihr/2003/menstrual_secret_e.shtml). Implications of women's self-scrutiny are vast and varied. For a discussion of self-surveillance, images, and body, see Traweek (1999). For an example of the extent of possible venues for surveillance of the body and why results of self-surveillance vary, see Ditton (2002).

2.   This increase in opportunity for self-surveillance has mixed effects. Some groups of people have been empowered to "take control" of their lives, especially educated people who can translate biomedical knowledge into everyday-life situations. For others, increased self-surveillance has meant less access to formal healthcare, especially those people who are unable or refuse to interpret the numbers provided to them by various medical technologies. On pap smears, see Howson (1998); hormone replacement therapy, see Harding (1997); and anorexia, see Eckermann (1997).

REFERENCES

Berg, L.D. (2002). Gender Equity as Boundary Object: Or the Same Old Sex and Power in Geography All Over Again?, The Canadian Geographer 46: 248-54.

Berland, J. (1996). Cultural Technologies and the Evolution of Technological Cultures. In Herman, A. and Swiss, T. (Eds.) The World Wide Web and Contemporary Theory, 235-58. New York: Routledge.

Birke, L. (1999a). The Heart—A Broken Metaphor? In Feminism and the Biological Body, 112-34. New York: Routledge.

Birke, L. (1999b). Feminism and the Biological Body. New York: Routledge.

Blunt, A. and Wills, J. (2000). Embodying Geography: Feminist Geographies of Gender. In Dissident Geographies: An Introduction to Radical Ideas and Practice, 90-127. New York: Prentice Hall.

Bray, A. and Colebrook, C. (1998). The Haunted Flesh: Corporeal Feminism and the Politics of (Dis)embodiment, Signs 24: 35-67.

Canadian Institutes of Health Research/Instituts de recherche en santé du Canada (CIHR/IRSC) (2003). New Canadian Menstrual Cycle Research May Alter Medical Texts, press release, July 2003. http://www.cihr-irsc.gc.ca/news/cihr/2003/menstrual_secret_e.shtml

Cope, M. (2002). Feminist Epistemology in Geography. In Moss, P. (Ed.) Feminist Geography in Practice, 43-56. London: Blackwell.

Crang, M., Crang, P., and May, J. (1999). Virtual Geographies: Bodies, Space, and Relations. New York: Routledge.

Deleuze, G. (1994). Difference and Repetition. New York: Columbia University Press.

Ditton, J. (2002). Technical Review. Hair Testing: Just How Accurate Is It?, Surveillance and Society 1: 86-101.

Doane, M.A. (1990). Technophilia: Technology, Representation, and the Feminine. In Jacobs, M., Keller, E. Fox, and Shuttleworth, S. (Eds.) Body/Politics: Women and the Discourses of Science, 163-76. New York: Routledge.

Dodge, M. and Kitchin, R. (2001a). Atlas of Cyberspace. New York: Addison-Wesley.

Dodge, M. and Kitchin, R. (2001b). Mapping Cyberspace. New York: Routledge.

Dodge, M. and Kitchin, R. (2003). Flying Through Code/Space. Paper presented at the Annual Conference of Irish Geographers, Trinity College, Dublin, Ireland, April.

Domosh, M. and Seager, J. (2001). Putting Women in Place: Feminist Geographers Make Sense of the World. New York: Guilford Press.

Eckermann, L. (1997). Foucault, Embodiment and Gendered Subjectivities: The Case of Self-Starvation. In Petersen, A. and Bunton, R. (Eds.) Foucault: Health and Medicine, 151-69. New York: Routledge.

Evans, M. and Lee, E. (Eds.) (2002). Real Bodies: A Sociological Introduction. New York: Palgrave.

Gear, R. (2001). All Those Nasty Womanly Things: Women, Artists, Technology and the Monstrous-Feminine, Womens Studies International Forum 24: 321-33.

Hall, J., Murphy, B.L., and Moss, P. (2002). Focus: Equity for Women in Geography, The Canadian Geographer 4: 235-40.

Hanson, S. (2002). Connections, Gender, Place and Culture 9: 301-03.

Haraway, D.J. (1989). Primate Visions: Gender, Race and Nature in the World of Modern Science. New York: Routledge.

Haraway, D.J. (1991). Simians, Cyborgs, and Women: The Reinvention of Nature. New York: Routledge.

Haraway, D.J. (1997). Modest Witness@Second Millennium.FemaleMan Meets OncoMouse: Feminism and Technoscience. New York: Routledge.

Harding, J. (1997). Bodies at Risk: Sex, Surveillance and Hormone Replacement Therapy. In Petersen, A. and Bunton, R. (Eds.) Foucault: Health and Medicine, 134-50. New York: Routledge.

Hawthorne, S. and Klein, R. (Eds.) (1999). Cyberfeminism: Connectivity, Critique and Creativity. North Melbourne: Spinifex Press.

Holloway, L. (2002). Virtual Vegetables and Adopted Sheep: Ethical Relation, Authenticity and Internet-Mediated Food Production Technologies, Area 34: 70-81.

Howson, A. (1998). Embodied Obligation: The Female Body and Health Surveillance. In Nettleton, S. and Watson, J. (Eds.) The Body in Everyday Lif, 218-40. New York: Routledge.

Kember, S. (1999). NITS and NRTS: Medical sScience and the Frankenstein Factor. In Cutting Edge, The Womens Research Group (Ed.) Desire by Design: Body, Territories and New Technologies, 29-50. London: I.B. Tauris.

Kobayashi, A. (2002). A Generation Later, and Still Two Percent: Changing the Culture of Canadian Geography, The Canadian Geographer 46: 245-48.

Koskela, H. (2000). The Gaze Without Eyes: Video-Surveillance and the Changing Nature of Urban Space, Progress in Human Geography 24: 243-65.

Kwan, M.-P. (2002a). Time, Information Technologies and the Geographies of Everyday Life, Urban Geography 23: 471-82.

Kwan, M.-P. (2002b). Is GIS for Women? Reflections on the Critical Discourse in the 1990s, Gender, Place and Culture 9: 271-79.

Kwan, M.-P. (2002c). Feminist Visualization: Re-envisioning GIS as a Method in Feminist Geographic Research, Annals of the Association of American Geographers 92: 645-61.

Kwan, M.-P. (2003). Gender Troubles in the Internet Era, Feminist Media Studies forthcoming.

Lawson, V. (1995). The Politics of Difference: Examining the Quantitative/Qualitative Dualism in Post-Structuralist Feminist Research, The Professional Geographer 47: 449-57.

Light, J. (1995). The Digital Landscape: New sSpace for Women?, Gender, Place and Culture 2: 133-46.

Longhurst, R. (2001). Bodies: Exploring Fluid Boundaries. New York: Routledge.

Martin, E. (1994). Flexible Bodies: The Role of Immunity in American Culture from the Days of Polio to the Age of AIDS. Boston: Beacon Press.

Martin, E. (1996). Citadels, Rhizomes, and String Figures. In Aronowitz, S., Martinsons, B., and Menser, M. (Eds.) Technoscience and Cyberculture, 97-109. New York: Routledge.

Mazzio, C. and Hillman, D. (1998). The Body in Parts: Fantasies of Corporeality in Early Modern Europe. New York: Routledge.

McDowell, L. (1991). Life Without Father and Ford: The Gender Order of Post-Fordism, Transactions, Institute of British Geographers NS 1: 400-19.

McDowell, L. (1992). Doing Gender: Feminism, Feminists, and Research Methods in Human Geography, Transactions, Institute of British Geographers NS 17: 399-416.

McDowell, L. (1999). Gender, Identity and Place: Understanding Feminist Geographies. Oxford: Polity Press.

McLafferty, S.L. (2002). Mapping Womens Worlds: Knowledge, Power and the Bounds of GIS, Gender, Place and Culture 9: 263-69.

Monk, J. and Hanson, S. (1982). On Not Excluding Half of the Human in Human Geography, The Professional Geographer 34: 11-23.

Morin, K.M. and Berg, L.D. (1999). Emplacing Current Trends in Feminist Historical Geography, Gender, Place and Culture 6: 311-30.

Moss, P. and Dyck, I. (2003a). Embodying Social Geography. In Anderson, K., Domosh, M., Pile, S., and Thrift, N. (Eds.) Handbook of Cultural Geography, 58-73. London: Sage.

Moss, P. and Dyck, I. (2003b). Women, Body, Illness: Space and Identity in the Everyday Lives of Women with Chronic Illness. Lanham, MD: Rowman and Littlefield.

Nairn, K. (2002). Doing Feminist Fieldwork About Geography Fieldwork. In Moss, P. (Ed.) Feminist Geography in Practice, 146-59. London: Blackwell.

Napier, J., Shortt, D., and Smith, E. (2000). Technology with Curves: Women Reshaping the Digital Landscape. Toronto: HarperCollins.

National Library of Medicine (2001). The Visible Human Project. July 13, 2003. http://www.nlm.nih.gov/research/visible_human.html

Parliamentary Assembly (2003). Trafficking in Organs in Europe. Social, Health, and Family Affairs Committee, Adopted Text No. 1611, 25 June. Strasbourg: Council of Europe. http://assembly.coe.int/Main.asp?link=http://assembly.coe.int/Documents/WorkingDocs/Doc03/EDOC9822.htm

Pavlovskaya, M.E. (2002). Mapping Urban Change and Changing GIS: Other Views of Economic Restructuring, Gender, Place and Culture 9: 281-89.

Prophet, J. (1999). Imag(in)ing the Cyborg. In Cutting Edge, The Womens Research Group (Ed.) Desire by Design: Body, Territories and New Technologies, 51-60. London: I.B. Tauris.

Rommes, E. (2002). Creating Places for Women on the Internet: The Design of a Womans Square in a Digital City, The European Journal of Womens Studies 9: 400-29.

Rose, G. (1993). Feminism and Geography: The Limits of Geographical Knowledge. Minneapolis: University of Minnesota Press.

Schuurman, N. and Pratt, G. (2002). Care of the Subject: Feminism and Critiques of GIS, Gender, Place and Culture 9: 291-99.

Shildrick, M. (1997). Leaky Bodies and Boundaries: Feminism, Postmodernism and (Bio)Ethics. New York: Routledge.

Teather, E.K. (Ed.) (1999). Embodied Geographies: Spaces, Bodies and Rites of Passage. New York: Routledge.

Traweek, S. (1999). Warning Signs: Acting on Images. In Clarke, A.E. and Olesen, V.L. (Eds.) Revisioning Women, Health, and Healing: Feminist, Cultural and Technoscience perspectives, 187-201. New York: Routledge.

Valentine, G. and Holloway, S.L. (2002). Cyberkids? Exploring Childrens' Identities and Social Networks in On-Line and Off-Line Worlds, Annals of the Association of American Geographers 92: 302-19.

Wakeford, N. (1998). Urban Culture for Virtual Bodies: Comments on Lesbian Identity and Community in San Francisco Bay Area Cyberspace. In Ainley, R. (Ed.) New Frontiers of Space, Bodies and Gender, 176-90. New York: Routledge.

Waldby, C. (2002). Biomedicine, Tissue Transfer and Intercorporeality, Feminist Theory 3: 229-54.

Warf, B. (2000). Compromising Positions: The Body in Cyberspace. In Wheeler, J.O., Aoyama, Y., and Warf, B. (Eds.) Cities in the Telecommunications Age: The Fracturing of Geographies, 54-68. New York: Routledge.

Webster, F. (2002). Do Bodies Matter? Sex, Gender and Politics, Australian Feminist Studies 17: 191-205.

Wheeler, J.O., Aoyama, Y., and B. Warf (Eds.) (2000). Cities in the Telecommunications Age: The Fracturing of Geographies. New York: Routledge

Williams, S.J and Bendelow, G.A. (1998). The Lived Body: Sociological Themes, Embodied Issues. New York: Routledge.

Women and Geography Study Group (1984). Geography and Gender. London: Hutchinson.

Women and Geography Study Group (1997). Feminist Geographies: Explorations in Diversity and Difference. Harlow: Longman.

CHAPTER 18

MARK W. CORSON
EUGENE J. PALKA

GEOTECHNOLOGY, THE U.S. MILITARY, AND WAR

Nothing during the past fifty years has exerted so great an influence on geographic cartography as has the occurrence of two world wars. (Arthur H. Robinson, 1954)

Abstract   Geotechnologies have always been an integral part of military training and war. Maps as keys to effective strategies have been and remain essential in planning and combat. They have been supplemented by aerial photography, remotely sensed images, and recent advances in GIS and GPS. This chapter focuses on the uses of major geotechniques by the U.S. Army and Air Force during major wars of the past century. Many of these innovations in military geotechnology can also be used in peacetime and nonwar arenas. The demand for geographic information and spatial analysis continues with new technologies to map, represent, and analyze spatial and environmental data.

Keywords   cartography, remote sensing, satellite images, digital technologies, military geography, war, geoinformation

1. INTRODUCTION

Since prehistoric times, humans have used technology to overcome their physical limitations. Violent action, whether hunting large predators for food or engaging in warfare with neighboring tribes, spurred rapid technological advancement. The human propensity to engage in armed conflict accelerated this leveraging of technology to produce more efficient warriors and armies.

Geographic technology made significant contributions to military effectiveness, while war and preparation for war provided an impetus for the rapid development of geographic technologies. The purpose of this chapter is to detail how U.S. military activities and warfare have both affected and been affected by the evolution of geographic technology. Geographic technologies are defined as the geographic information processing techniques, which most often include cartography, remote sensing, geographic information systems (GIS), and global positioning systems (GPS). Other technological

Stanley D. Brunn, Susan L. Cutter, and J.W. Harrington, Jr. (Eds.), Geography and Technology,
401-427.

innovations important in subdisciplines, such as meteorology, are mentioned where appropriate.

2. PRE-WORLD WAR I

In hunting and gathering societies, knowledge of the landscape was a matter of survival. Knowing the spatial distribution and attributes of hunting grounds, water resources, and sheltered places for encampments was an absolute necessity. Knowing how to get from one place to another was equally important. Much of this information was passed down from generation to generation through oral tradition, experience, and rudimentary drawings on cave walls (Roberts 1993). Knowledge of the landscape and the location and attributes of both friendly and enemy forces has also been essential to warriors since the beginning of armed conflict. Graphic representations of terrain, hydrographic features, settlements, and friendly and enemy forces were a critical tool for commanders. Early unscaled sketches were eventually replaced by scaled graphical representations known as maps (James and Martin 1993). Cartography (the art and science of making maps) was one of the earliest contributions of geographic technology to warfare, and the science of cartography was advanced by the patronage of leaders engaged in military activities.

Maps serve multiple purposes for the warrior. As the U.S. Army map reading manual states,

> No one knows who drew, molded, laced together, or scratched out in the dirt the first map. But a study of history reveals the most pressing demands for accuracy and detail in mapping have come as the result of military needs. Today, the complexities of tactical operations and deployment of troops is such that it is essential for all soldiers to be able to read and interpret their maps in order to move quickly and effectively on the battlefield. (Department of the Army 1993, 2-21)

Soldiers use maps to understand the nature and location of the natural and human-engineered features on the landscape on which they operate. Topographic maps, originally developed to enhance artillery effectiveness, show both landforms and elevation through the use of contour lines. They enable soldiers to visualize the terrain they occupy, and perhaps more importantly, to visualize and plan operations on enemy-held terrain they intend to attack. Thus, maps can be used for terrain visualization and understanding; terrain analysis to identify cover and concealment and routes of movement, navigation, and command and control to communicate maneuver instructions to friendly forces and to target enemy ones. The map in support of these functions has evolved since the first recorded Sumerian and Babylonian mapping efforts, and it remains the premier instrument of military intelligence, decisionmaking, and command and control (O'Sullivan 1991).

Observation of the actual terrain has been nearly as important as mapping it. The original method to increase observation was to seize the high ground. As early as the 18th century, military forces experimented with sending observers aloft in tethered balloons to increase the range of their observation. This method was largely unsatisfactory, however, until the development of photography in 1826 made possible the recording of ground activity for later analysis by intelligence experts. Balloon observation was used during the American Civil War but the fluidity of the battle precluded its use with cameras, and it provided little usable information. The U.S. Army also experimented with cameras mounted on large kites; some of these were reportedly used in the Spanish-American war. In some cases, surprisingly good results were obtained, but the difficulties of sufficient wind from the needed direction made this an unwieldy technology that was soon rendered unnecessary by the Wright brothers and their airplane. Military establishments in Europe and the U.S. were quick to recognize the potential of the airplane as an observation platform. The first aerial photograph was probably taken near Le Mans, France in 1908. In the fall of 1911, the U.S. Army Signal Corp established a flight training school at College Park, Maryland and the American Army began experiments with aerial reconnaissance. The U.S. Army successfully used visual and photoreconnaissance between 1913 and 1915 in the Philippines and along the U.S.-Mexican border (Stanley 1981).

3. WORLD WAR I

World War I prompted a number of advances in cartography and especially the nascent field of photoreconnaissance, which would evolve into the modern field of remote sensing. As William Burrows (1986, 32) said, "If the camera and airplane were the mother and father of photoreconnaissance, then World War I was its midwife." The stalemate in the trenches coupled with the generals' fear of what the other side was preparing forced both sides to look for ways to see beyond the front. While the collection component of aircraft, camera, and film technology evolved rapidly, the analysis component of image interpretation also developed and proved its worth. Comparative coverage or change detection was a cornerstone of image analysis and developed early-on. Interpreters were taught to spot points of interest and to "exploit" what they saw to draw valid conclusions about enemy intentions. Stereoviewing to see the battlefield in three dimensions, target graphics, strip coverage, and photomosaics that showed large areas of the battlefield were other innovations from World War I (Stanley 1981).

While cartography had been important before World War I, the expertise of "geographic cartographers" coupled with other geographers proved essential once again. Topographic mapping provided tools for intelligence, planning, movement, logistics resupply, artillery bombardment, and command and control. Many geomorphologists and geologists were consulted concerning the effect of soils, hydrography, and the underlying bedrock, which proved to be critical knowledge in the era of trench warfare (Russell et al. 1954). Geographers were also involved with the peace conference which followed the war (James and Martin 1993; Palka 1995).

4. WORLD WAR II

The U.S. emerged from the interwar isolationist period badly unprepared for the coming Second World War. Geographers were quickly pressed into service for a multitude of tasks including mapping, area studies, photo interpretation, meteorology, and geomorphology. Many of these geographers were in Army intelligence while others were scattered throughout the War Department and other services (Palka 2002). In addition to mapping technologies, other innovations emerged that would later prove to be of enormous value, such as radar, sonar, and ballistic missiles.

World War II produced more cartographic activity than had been seen in more than a decade. As Robinson (1954, 558) notes, "Probably more maps were made and printed during the five years from 1941 to 1946 than had been produced in the aggregate up to that time." During World War II, the use of aerial photography and photogrammetry to provide base data for maps was perfected, thus contributing to a significant improvement in geodetic control and the massive increase in world topographic coverage. Wartime cartography progressed along four lines: (1) compilation of map information, and a program to publish maps, (2) map intelligence, (3) place-name intelligence, and (4) terrain modeling. For the first time in the U.S., standardization of place names became critical because the best maps of certain theaters of the war were lettered in alphabets other than Latin (Russell et al. 1954). In addition, World War II prompted great strides in mass production techniques, the development of aerial charts for aerial navigation, special purpose or interpretative (thematic) maps, and techniques for producing terrain models (relief maps) (Robinson 1954).

While many elements of aerial photography and image interpretation were developed in World War I, large strides in both areas and especially in the airphoto coverage of the world were made during World War II (Kline 1954). By 1940 the interpretation of airphotos had grown from a tool of tactical operations specialists and cartographers to a full-fledged discipline. Aerial photoreconnaissance and interpretation also took on a strategic role with the advent of long-range strategic bombing of the enemy's industrial heartland. Long-range strategic reconnaissance with the associated aerial photographs was essential for target analysis, selection, and subsequent bomb damage assessment (Stanley 1998). Technical improvements in aircraft, cameras, and film supported both the tactical and strategic efforts. The widespread use of triple lens cameras provided both vertical and two oblique photos producing horizon-to-horizon coverage. Auto-compensating long focal-length lenses of up to 240 inch (610 cm) focal lengths enabled aircraft to take crystal-clear pictures from as high as 40,000 feet (12,192 meters) thus avoiding anti-aircraft fire and enemy fighter planes. New film came into use, such as normal color film and infrared film, which proved to be very useful at finding a camouflaged enemy. Late in the war, the British Royal Air Force pioneered the use of "radar reconnaissance cameras" that could penetrate clouds and darkness. Photo interpreters, many of them women, specialized in particular geographic areas, weapons systems, or engineering types to the point where interpreting changes in their area became intuitive. One of Britain's most celebrated photointerpreters, Constance Babington-Smith, confirmed the existence of the German V-1 "Buzz Bomb" vengeance weapons, thus providing early warning of the terror attacks that were soon to befall Britain (Burrows 1986).

Geographers made important contributions to the war effort above and beyond the application of geotechnologies (Palka 2002). For example, regional geographers made substantive contributions to the war effort. The American military, on the eve of World War II, was not prepared to fight a global war in environments ranging from the desert to the Arctic. In 1941 the Army Quartermaster had only three standard issues of uniform and equipment— temperate, torrid, and frigid—with boundaries based on lines of latitude. Most of the geographers' effort was focused on the preparation of area intelligence studies that involved the military, economic, and administrative aspects of potential areas of operations. Reports for operations planning emphasized descriptions of topography, soils, vegetation, drainage, and the human elements of urban areas and transportation. Probably the finest examples of wartime reports were the Joint Army and Navy Intelligence Studies that were produced by a team of experts in various fields but were directed and coordinated by professional geographers (Russell et al. 1954, Palka 1995, Palka and Galgano 2000). Geomorphologists were instrumental in providing analysis of soils' trafficability for armored movement and beach suitability for amphibious assaults. All sides in the war relied on accurate weather information for ground and especially air operations (Bates and Fuller 1986). Military forces on both sides established weather stations throughout their areas of operation and even in meteorologically strategic places such as Greenland so as to be able to provide early warning of storms and accurate forecasts.

Perhaps the most critical decision of the war based on a weather forecast was General Dwight D. Eisenhower's D-Day decision. The Allied amphibious assault into Normandy was a momentous undertaking. The dawn landing required a combination of environmental conditions, including a low tide to reveal beach obstacles, three miles of visibility for naval gunfire support, clear skies for air support, a full moon to enhance a large-scale night-time airborne assault, calm seas for the landing craft, and a light wind to disperse smoke. These conditions needed to last at least 36 hours if not for several days.

Group Captain James M. Stagg of the Royal Air Force was appointed chief meteorological officer for Eisenhower's headquarters. His task was to provide accurate five-day forecasts, given this was the time needed for embarkation and transit of the massive force. His analysis, based on averaging weather data from previous years, showed that the necessary conditions were likely in only a few short windows in April, May, or June. The invasion was originally scheduled for May 1944, but due to force changes, it was postponed until June 4, 5, or 6. If the invasion did not occur during this time, the next window was not until June 19. On June 1, the weather appeared to be bad for an invasion with the possibility of an extratropical cyclone sweeping over the invasion area. Early on June 4, Eisenhower postponed the landing for a day. As the day progressed, Group Captain Stagg and his meteorologists recognized that the storm systems were turning north and that a following high-pressure cell would provide a 48-hour window. As Winters et al. describe in *Battling the Elements* (1998, 28),

> At 2130 on 4 June Group-Captain Stagg presented this information to General Eisenhower. Fifteen minutes later, about 30 hours before the first wave of troops would land on the beaches, Eisenhower ordered the landings to take place on 6 June. This decision involved tremendous responsibility and, no matter how the operation turned out, it would affect history in a most profound way.

Of course the invasion was a success, and on June 19 (the next invasion window), Normandy experienced the largest spring storm of 1944 (Winters et al. 1998).

A number of technologies that were not specifically geographic, but would prove invaluable to geographers at a later time, emerged during the Second World War. These included radar, sonar, and ballistic missiles. Gustav Herz discovered the existence of electromagnetic waves in 1888, but he attempted no practical application. In the years before World War I, there was some civil experimentation, but surprisingly there was no military research effort in World War I. The U.S., France, Germany, and the Soviet Union experimented with radar in the interwar years and made some rudimentary progress. Britain, however, was the first country to develop and field a practical air defense system based on radar. The "Chain Home" radar system coupled with the eight-gun fighter would ultimately assure Britain's survival in the Battle of Britain (Latham and Stobbs 1996, 1998). Radar technology developed rapidly during the war, and by 1945, Royal Air Force planes were using a rudimentary radar "camera" to obtain images at night and through clouds. This technology was the precursor to later technology that would prove invaluable to the remote sensing community. Another technology developed during the war was sonar, which used sound waves underwater rather than the reflected radio waves of radar. Sonar was used to detect vessels at sea, especially submarines. It is now a standard remote sensing technique in oceanography. The Germans developed the first ballistic missile as a terror weapon to wreak havoc on London in the Second Blitz. The V-2 was the precursor to both the Inter-Continental Ballistic Missiles (ICBM) with their nuclear payloads that would terrorize a generation and the rocket boosters that would later lift remote sensing satellites into space (Hartcup 2000).

5. THE COLD WAR

It did not take long after World War II for the former allies of the Soviet Union and those of the U.S. to set their sights on one another in a forty-year conflict that came to be known as the Cold War. A significant factor in keeping this Cold War predominantly cold was the geographic technology of overhead imagery intelligence or remote sensing. While vast strides in aerial photography, photogrammetry, and photo interpretation had been made in World War II, even greater technological achievements and military applications were to come.

By the early 1950s, the Cold War was well under way and the threat of nuclear war between the Soviet Union and the U.S. was a serious possibility. The U.S. Air Force, lacking good intelligence, posited that the Soviet Union had built a massive fleet of strategic bombers and thus a so-called "bomber gap" existed. They used this argument to circumvent President Dwight D. Eisenhower so as to convince the U.S. Congress of massive funding needs for a large increase in American nuclear bombers. Both the Air Force and the Central Intelligence Agency wanted new reconnaissance aircraft that could fly so high and far that they could spy on the Soviet Union with impunity. Eisenhower, not trusting the Air Force, gave this mission to the CIA, which with the help of the famous aeronautical engineer Kelly Johnson and his "Skunkworks," built the U2 spy plane.

The U2 was a marvel of engineering for its time. It could fly at 70,000 feet (21,212 meters) and it had a very long range. The U2's main camera set the standard for future reconnaissance cameras and was a quantum leap ahead of its World War II predecessors. A sophisticated system of precise image-movement compensation overcame the problem of vibration that had plagued all aerial cameras since World War I. This system took into account the motion of the plane, the vibration of its engine, and the movement of the new Kodak fast, high-sensitivity film. This compensation was essential to acquire clear pictures from such a high altitude. The system had 60 lines of resolution per millimeter and could distinguish objects the size of a basketball from 13 miles (21 km) high. The aircraft could also carry other sensor packages to include a radar imaging system. On July 4, 1956, a U2 flew a mission from Germany and passed over Moscow, Leningrad, and the Baltic coast. Over the next four years, the U2s would fly 20 deep-penetration missions over the Soviet Union and prove that there was no bomber or missile gap. In 1960 the Soviets finally succeeded in shooting down a U2 piloted by Francis Gary Powers. Eisenhower was forced to cancel the overflights, but a new technology emerged that made such dangerous flights unnecessary (Burrows 1985).

The U.S. military and intelligence communities knew that the U2 was vulnerable, and that one would eventually be shot down. Thus in conjunction with the Rand Corporation and other elements of the aeronautical and scientific communities, they began to study the feasibility of using the new ballistic missile technology to launch a surveillance device into orbit. The advent of the Soviet *Sputnik* program gave them added incentive. As usual, the Air Force and CIA had competing visions and fought a bureaucratic battle over technology and control. Eventually the first U.S. imagery intelligence satellites would emerge under the codename CORONA. This very expensive project sought to overcome substantial technical difficulties and would prove to be the basis for the subsequent civilian remote sensing systems such as LANDSAT that would follow some 13 years later. The CORONA Program was revolutionary but fraught with problems. In fact, the program suffered 12 mission failures from February 1959 until the first successful mission in August 1960 (Day et al. 1999).

The CORONA Program was comprised of six satellite models with three different intelligence objectives. In 1962 these satellites were given the codename "Keyhole" or KH for short. Keyhole systems referred to orbital platforms such as CORONA, while the code word "Talent" referred to suborbital systems such as the U2. The KH-1 through KH-6 satellites comprised the CORONA family. As Day and his colleagues note (1999, 7),

> CORONA achieved a number of notable firsts: first photoreconnaissance satellite;
> first recovery of an object from space (and first mid-air recovery of an object from

space); first mapping of the Earth from space; first stereo-optical data from space; and first program to fly more than 100 missions in space.

Most of the CORONA satellites were "bucket dumpers," meaning they returned their film to earth in capsules that were retrieved in mid-air by specially equipped Air Force cargo planes. The film was then developed on the ground and distributed to the various agencies. While the image quality was great (up to 12-inch spatial resolution), the time lag in delivery of the photos was long. The KH-5 LANYARD was equipped with a low-resolution videocon camera and radio-link transmitter that could beam the images back to a receiving station. This was a technology before its time as the images were of such poor quality as to be nearly worthless. The CORONA satellites were considered the first and second generation of imagery intelligence satellites (Burrows 1986; Day et al. 1999).

In August 1966, the third generation KH-7 and KH-8 "Gambit" systems came into service. The KH-7 and KH-8 worked in conjunction. The KH-7 was a low-resolution surveillance craft equipped with a Multi-Spectral Scanner (MSS) and a radio-link transmitter. Later KH-7s carried a Thematic Mapper device that had three times the spatial resolution of the MSS and operated in seven bands (blue, green, red, near infrared or IR, first mid-IR, second mid-IR, and far IR or thermal infrared—TIR). The single-band panchromatic images could be overlaid in various combinations to produce color images. The KH-8 was a close-look system used to examine targets of interest identified by the KH-7. KH-8s carried a long focal-length camera with a spatial resolution of six inches as well as a low-quality thermal imaging system, MSS, and TM. The KH-8s flew at very low altitudes (as low as 69 miles (112 km) which gave the camera a resolution of three to four inches) but had very short lifespans due to atmospheric drag. They carried two to four "buckets" for film return; some 752 were launched between 1966 and 1985 when they were retired. The second-generation systems were a significant improvement with their MSS and TM systems, but they were still primarily bucket dumpers and were limited to daylight imaging (Burrows 1986, FAS 1997).

In 1971, the fourth generation KH-9 Hexagon (known lovingly as the "Big Bird" because it was as large as a Greyhound bus) was placed into service. The Big Bird had a "folded" 20-foot focal-length mirror that took excellent pictures with a one-foot or better resolution. It also carried secondary TIR, MSS, and TM systems with a television download capability, and it could produce three-dimensional images. Most importantly, it carried a photomultiplier that intensified the available light, thus making this the first night-capable system. The Big Bird carried both low-resolution, large-area surveillance systems, and a high-resolution camera with several buckets, thereby merging the functions of the KH-7 and KH-8 into one platform

(Burrows 1986, FAS 1997). The Big Bird, however, was still a bucket dumper, which limited its life span, substantially increased its cost, and failed to provide the Holy Grail of "real time" intelligence.

The military and intelligence communities had come to rely on their eyes in the sky, but the systems were not adequately responsive in providing high-quality imagery in a timely fashion. What was needed was photo-quality imagery that could be telemetered back to earth. The answer came from the Bell Telephone Laboratories in 1970 with the invention of the Charged-Coupled Device or CCD. The CCD is a light-sensitive electronic device that senses and records the amount and wavelength of photons striking it. These values are recorded in digital format and then reassembled by a computer from a digital image. This digital data can be telemetered via radio back to an earth station where it is reassembled into an image (Burrows 1985). These images are manipulated by computer so as to be sharpened, error corrected, rectified to map coordinates, and classified. These techniques would become known as Digital Image Processing or DIP and would become a mainstay in the geographic techniques (Jensen 1996).

The first KH-11 "Kennan" or "Crystal" was launched in 1978. The platform carried new CCD-based digital sensors that included TM, MSS, TIR, and a photomultiplier light-intensification system. The system could also produce three-dimensional images, but the spatial resolution of the sensors remained a closely guarded secret. Various sources suggest the KH-11 could clearly image a license plate from space, giving it a one-inch or less resolution. The most important KH-11 feature was its ability to telemeter its data to the ground, thus precluding the need for buckets and substantially increasing the lifespan of the satellite (an important consideration given a KH-11 cost about $1 billion). The KH-11s carried a large amount of hydrazine maneuvering fuel, giving them a service life of about three years. Depending on their maneuvering, some lasted much longer—up to a decade (Burrows 1985; FAS 1997; Lindgren 2000).

The significance of imagery intelligence in preventing World War III should not be underestimated. From the earliest contentions of a "bomber gap" in the 1950s and the latter "missile gap" in the 1960s, the spy planes and satellites gave leaders on both sides the confidence that they were not facing a "nuclear Pearl Harbor." Indeed the capability to see and verify the number, type, and location of enemy missiles, planes, tanks, and other weapons made the concepts of arms control and arms reduction possible, thus stopping the nuclear arms race, reducing nuclear arsenals, and diminishing the threat of global thermonuclear war (Peebles 1997; Day et al. 1999; Lindgren 2000).

The advancement in remote sensing brought about by the CORONA systems were matched by CORONA's contribution to both American military and civilian mapmaking. CORONA's images from space forced the military Mapping, Charting, and Geodesy (MC&G) community to develop entirely new systems and methods to deal with this fantastic data source. The CORONA data also necessitated an entirely new geodetic control system, and prompted a number of geography departments across the country (starting with Ohio State University, which imported an entire geodesy faculty from Europe) to offer courses in the new technologies, thus expanding the realm of the geotechnics. As the U.S. Geological Survey (USGS) became involved with CORONA, it was able to revise its maps of the U.S. in a rapid and cost-effective manner while improving the organization's technical skills.

The Cold War brought about a number of other innovations that would have a great impact on civil geographic techniques, including thermal imaging, laser range finding, the Internet, digital mapping, and the Global Positioning System. Thermal imaging devices were developed early-on for airborne platforms, but became important tools for ground combat. By the early 1980s, main battle tanks in western armies were outfitted with sophisticated thermal imaging systems, which when slaved to the tank cannon, became thermal sights. Thermals are used as the primary fire control mode because they can image targets through darkness, smoke, and dust. Thermal imaging systems found later use in many civil applications (see Hodgson and Jensen's chapter in this volume).

Laser range finders were another military technology installed on tanks in the early 1980s. A laser beam is bounced off a target and reflects back to the sensor. A computer then determines the range based on the elapsed time. While laser range finders have uses in engineering, extensions of this technology such as LIDAR (Light Detection and Ranging—basically radar with light instead of microwaves) and Laser Induced Fluorescence (LIF) are remote sensing systems used in the geotechniques (Lillesand and Kiefer 1994).

The Internet, which has arguably revolutionized how we communicate and work, began as the ARPANET, a project of the Defense Advanced Research Projects Agency (DARPA). ARPANET was a Cold War effort to interlink defense, industry, and academic research computers in such a way that the system would automatically reroute data around any damaged elements or the network. The idea was for the network to be robust enough to survive any level of attack up to a nuclear strike. The net eventually evolved far beyond what its developers had intended into the Internet and World Wide Web of today. The Internet and Web have had a major impact on how geographers conduct research, and how we acquire, store, analyze, and distribute spatial data products from the geotechniques. Military planners, operators, and staff sections regularly tap into a whole range of databases within the public domain in order to develop and support operational decisions. This practice has become increasingly widespread, effective, and efficient, thanks to the Internet. Examples include USGS, NOAA (National Oceanic and Atmospheric Administration), ESRI, NIMA (National Image and Mapping Agency), UN, CIA, PRB (Population Reference Bureau), CDC (Center for Disease Control), U.S. Census Bureau, climate sites, etc.

The advent of precision-guided munitions ranging from ICBM to cruise missiles was predicated on the possession of extremely accurate earth data. The need for such data spurred the early CORONA earth-mapping missions. Early cruise missiles were a boon to the military because of their accuracy and the fact that human pilots were not at risk. These early-version cruise missiles used a terrain-following radar that matches the missiles' position with a digital map. Thus digital mapping of most of the world became an important task for the military, as these data were essential for new generations of precision weapons (Larson and Pelletiere 1989). The Defense Mapping Agency (which later merged into NIMA) put great emphasis on digital mapping of the world, and made much of these data and techniques available to the USGS and the geotechnical community (Larson and Pelletiere 1989).

The predecessor to the NAVSTAR GPS was the U.S. Navy's TRANSIT navigation system designed to accurately locate ballistic missile submarines and surface vessels. TRANSIT consisted of four satellites, was slow and prone to error, but it did open the era of satellite navigation systems. The Air Force began work on a multisatellite navigation system in 1963 and after years of testing and modifications, launched the first NAVSTAR satellite in 1978. The GPS operated today by the Air Force has 24 operational satellites that provide precise around-the-clock, all weather, three-dimensional navigation information. While the system is owned and maintained by the military, civilian use of GPS has blossomed (see Hodgson and Jensen's chapter).

6. POST-COLD WAR ERA

Technologically, the post-Cold War Era has been an extension of the Cold War. Advances in satellite reconnaissance systems to include radar satellites and the weaponization of GPS are two major trends. The leveraging of information technology to include GIS, geographic visualization, and use of the Internet/WWW as an information delivery system also are trends worth noting. There has been a significant change, however, in the relationship between civil developments of geographic technology and military applications.

Prior to the end of the Cold War, military requirements, research, and application drove much of the cutting-edge research in the geotechniques. The impacts of the World Wars and the Cold War on cartography, aerial photography, photo interpretation, and satellite imaging are well-documented. In the post-Cold War Era, however, the relationship has been reversed. While each of the military services has its own research and development components (R & D), none is in the business of fully developing technologies, equipment, ordnance, etc. The services articulate a "needed capability," and private corporations attempt to satisfy the needs of the potentially lucrative customer. It is the private sector that has made rapid advances in information processing technologies for civilian use that the U.S. military looks to for developing cost-effective, off-the-shelf solutions to suit military needs. We have evolved to the point where the military relies more than ever on outsourcing. Meanwhile, "beltway bandits," defense contractors, and megacorporations around the country have filled their ranks with former military personnel in order to gain ties to the services and acquire insights regarding "projected capabilities." As such, the private corporations are usually ahead of the military's R&D community in developing technologies, systems, and equipment. The former "shop around" their products early-on in the development process with hopes of acquiring interest and feedback, and landing a multimillion (or billion) dollar contact, or better yet, a succession of contracts. The corporation may in turn subcontract services and components related to the main product. Thus military procurement processes have blossomed dramatically in recent years and have major cultural/social/political implications in the U.S.

Computers and computer software are a major case in point. Years of development by the various armed services have produced a plethora of information systems that are obsolete, expensive, and that will not interface with each other. The solution has been to adopt industry-standard off-the-shelf products such as personal computers and the Microsoft Windows operating systems. Fueling this trend has been increasing reliance on contracted services. While the military has developed specialized systems in the post-Cold War Era, they are often based on civil and/or commercial systems.

In terms of post-Cold War satellite reconnaissance systems, the military-intelligence community advanced two major systems. The KH-12 or Improved Crystal is a military version of the Hubble Space Telescope. It is highly classified, but the Federation of American Scientists (FAS) believes it has improved electrooptical sensor systems; some reports speculate it can image a postage stamp. A major feature of this platform is that it was designed for deployment by the space shuttle, and it can be refueled in space by the shuttle, thus giving it a very long operational life. It is also highly maneuverable since fuel consumption is no longer the limiting factor, and it can operate at multiple altitudes, indicating it is a very versatile system.

All of the camera or electrooptical imaging satellites, however, share a major drawback. They cannot image through clouds, smoke, or dust. While side-looking airborne radar (SLAR) has been an option in aircraft, the U.S. sent up the first dedicated radar imaging satellite in 1988. The Lacrosse or Onyx satellites launched in 1988, 1991, and 1997 can see through darkness, clouds, smoke, and dust using synthetic aperture radar (SAR, a type also used on the space shuttle). FAS estimates that Lacrosse has a maximum spatial resolution of about one meter, but that the satellite can vary its resolution to cover large areas or to zoom in (FAS 2002).

The first post-Cold War conflict to confront the American military was the Persian Gulf War of 1990-91. Geotechnology played a very important role in this war especially in terms of remote sensing, precision-guided weaponry, early use of GIS, and widespread use of GPS down to small-unit level. Satellite imagery played a critical role in the Gulf War, especially in the defensive buildup phase of Desert Shield when aircraft could not overfly Iraq. The U.S. had KH-11 and Lacrosse satellites available. Of interest is that the Coalition purchased large amounts of LANDSAT and SPOT commercial imagery to supplement their own systems. They also purchased all the commercial imagery so the Iraqis could not buy it. The satellites provided thousands of images that supported the threat estimation and targeting effort and the preparation of the land, air, and sea war plans. The U.S. also deployed two E-8A Joint Surveillance and Target Attack Radar Systems (JSTARS) aircraft. The JSTARS was still in development but was rushed to the war. JSTARS uses a multimode SLAR with a very high resolution. It flies 50-70 miles (80-113 km) behind the front and can provide information on vehicle location, number, and movement over an area of 30,000 sq. mi. (77,700 sq. km.). Another interesting remote-sensing innovation was the first widespread use of Unmanned Aerial Vehicles (UAVs) equipped with video cameras that beamed back images of enemy areas for tactical intelligence and targeting purposes. A major lesson from the Gulf War was that satellite imagery was critical, but that it was not timely enough and adequately available to help battlefield commanders plan and make decisions. After the war, the call went out for smaller, less expensive and more versatile satellites that could meet the needs of the military (Cordesman and Wagner 1996).

The Gulf War was one of the first conflicts to make extensive use of precision-guided munitions launched from aircraft, surface ships, and even submarines. Aircraft used laser and television-guided bombs and missiles, while conventionally armed cruise missiles launched hundreds of miles from their targets used terrain-following radar and digital earth data to find their targets. Three-dimensional terrain visualization came of age in 1995 during the NATO air campaign over Bosnia. The U.S. Air Force introduced a computer system called PowerScene that modeled the terrain of Bosnia including target areas and enemy air defense sites. Aircrews could virtually fly through their missions and see the terrain before ever sitting in a cockpit. PowerScene helped to improve the accuracy of bombing raids (thus reducing civilian casualties) and reduced the risk to the aircrews, as they were familiar with the hazardous terrain and location of enemy defenses. PowerScene was later used to great effect in the Dayton Peace Accord negotiations where leaders from Serbia, Croatia, and Bosnia were able to visualize the impact of their boundary negotiations by "flying" the border in PowerScene (Corson and Minghi 1996).

During the U.S. war in Afghanistan, there were a number of geotechnology uses. One involved the use of UAV technology, which was enhanced by robust communications links that enabled one to identify a target in an otherwise remote or inaccessible location, follow it or conduct surveillance, decide when and where to interdict it, attack it (with either ordnance carried by the UAV or another platform), and acquire battle-damage assessment, all in real time and on a wide-screen display within a command and control facility. A much simpler form of technology, also used during the Afghanistan war, was the use of the chat room concept to report information from lower units to higher headquarters. This technique enabled rapid dissemination of information as opposed to the traditional "stove-piping" of radio communication up and down the chain of command in an inefficient (and sometimes ineffective) fashion. A third area involved the use of virtual 3-D maps, as opposed to traditional paper maps, during the planning process for a tactical mission. Despite advancements in mapmaking, 2-dimensional maps still require the user to develop a 3-D mental image of the terrain, which may require extensive training and/or years of experience. The 3-D renditions, however, make it possible for all users to identify and focus on relief variability, vantage points, natural routes and corridors, etc. Programs such as FALCON VIEW were especially beneficial to ground forces and staffs, and simplified the detailed planning of flight routes, air assaults, tactical operations, communications, logistics, and supporting fires. Finally, munitions and robotics were designed and used in concert with geological and geographical information to target and exploit specific types of caves that were used by Al Qaeda. The nature of the cave, especially composition (that is, granite vs. limestone or schist), depth, and the extent of the underground network posed challenges that were addressed by these emerging technologies.

Some early use of GIS for battle management and control also made their debut in the Gulf War and were fully operational in the Afghanistan and Iraq missions. The systems would eventually become coupled with GPS and communications systems to provide digital command and control systems.

The role of GIS in "battlespace management" is covered later in the discussion of the revolution in military affairs.

Global Positioning Systems proved extremely effective and popular during this period. It is very difficult to navigate in the desert (especially at night) due to a lack of landmarks to orient a map. The U.S. Army used a number of navigational systems including the maritime LORAN system to orient their formations. Those units fortunate enough to have GPS receivers found them invaluable for navigation and for locating the enemy in order to call for artillery or air strikes. Troops were so enamored with the technology that they wrote home asking family members to purchase civilian receivers to send to them in the desert, as there were not nearly enough military issue units to go around. The military procured many commercial receivers under an emergency procurement action, but these commercial receivers were not "crypto-capable" or rugged enough for military use (DOD 1992).

By the late 1990s, GPS was adapted as a guidance system for precision-guided munitions (Air Force New Service 1998). Second-generation cruise missiles were fitted with a GPS receiver, in addition to terrain following radar, thus increasing their accuracy. Precision-guided bombs that use laser or television guidance requiring terminal guidance to the target have degraded capability in bad weather. The introduction of the inexpensive Joint Direct Attack Munition (JDAM) tailkit enabled the U.S. military to transform its large inventory of 1000 and 2000-pound general purpose "dumb bombs" into GPS guided precision weapons. The JDAM is a kit for conventional bombs consisting of an inertial navigation system/GPS guidance kit and steering "strakes" that enable the weapon to be launched from up to 15 miles (24 km) from the target in all weather. The bomb is a "fire and forget" weapon in that once launched, it needs no further input from the aircraft as it will follow its GPS coordinates (if the GPS loses signal the inertial navigation system takes over) to hit within 13 meters (43 feet) of the target. The weaponization of GPS is significant to the American military, because the large stock of Vietnam-era dumb bombs can be turned into precision-guided munitions. This fact means that far fewer aircraft can effectively attack far more targets in all weather, day or night, thus reducing the risk to aircrews, and the risk of unintended civilian casualties (and/or collateral damage) (FAS 2002).

In the decade since the Gulf War, geographic technologies have continued to play an important role in military activities, but the trend of the military adopting technology produced by the civilian sector is even more pronounced. This is especially true in the area of GIS, which is gaining increasing military acceptance across the spectrum from peacetime garrison operations to conventional warfare.

The military spends most of its time in garrison or in training areas. The U.S. military has adopted GIS for a range of tasks to include facilities management on its many bases and the environmental protection of its limited training areas (Chang 2002). After early attempts to develop in-house GIS solutions (e.g., the U.S. Army Corps of Engineers GRASS GIS), the U.S. military increasingly uses commercial off-the-shelf GIS solutions to manage its hundreds of installations and millions of acres of land.

The U.S. Army Training and Doctrine Command (TRADOC) is responsible for training the American Army on 16 installations housing 150,000 people on two million acres of land. TRADOC, supported by a contractor, developed a GIS-based decision support system known as the BASOPS Corporate Database. This system connects all 16 installations and interfaces with other DOD systems; it contains more than 450 map and aerial photographic data layers plus data from other federal, state, and local agencies. The system enables analysts to collaborate across the network to develop recommendations to help decisionmakers solve problems (ESRI 2001a, 2001b, 2001c).

Computer simulations are also used in training as well as during "real world" operations. One of the major benefits that is often overlooked is their contribution to conserving resources (ammunition, vehicle or aircraft miles, fuel, wear and tear on equipment, training lands, and perhaps most of all, time). Simulations can enhance decisionmaking at all echelons and in a variety of situations. They also can provide training opportunities for individuals as well as units, and they can facilitate resource management.

Peacetime military training causes extensive damage to the military's limited training lands. The Army developed the Integrated Training Area Management (ITAM) system to overcome the apparent conflict between force readiness and environmental stewardship. The ITAM system consists of four components: Land Condition Trend Analysis (LCTA), Training Requirements Integration (TRI), Land Rehabilitation and Maintenance (LRAM), and Environmental Awareness (EA). The LCTA is a land-use decision support GIS that tracks the use of training lands and identifies when lands require rest or restoration. The other systems interface with the LCTA to provide training requirements and land restoration methods (the EA module is a means to educate users on their environmental stewardship responsibilities). The ITAM system is the major method for ensuring that training and environmental stewardship are properly balanced (DA 2002).

GIS is finding increasing use in humanitarian and peacekeeping operations. Early efforts to use GIS in Bosnia were expanded for the Kosovo mission. The Kosovo Force (KFOR) headquarters in Pristina had a multinational staff of military cartographers and GIS analysts available to provide custom maps and GIS products to the multinational force. An example application was the use of GIS to analyze the distribution and cleanup of landmines in the province. Mine Action Centers in each of the Multi-National Brigade Headquarters tracked the location of minefields and unexploded ordnance, logged the detection of new hazards, and tracked the nature and progress of demining operations. Another GIS application (used by nongovernmental institutions in support of the International Criminal Tribunal for the Former Yugoslavia) in Kosovo was the War Crimes Documentation Database developed by the Illinois Institute of Technology's Inter-Professional Studies Program. This project consists of a traditional database documenting war crimes linked to locational information. The information on war crimes is compared with information on troop movements to both validate the integrity of the evidence and assist prosecutors in identifying suspected perpetrators (Atkins 2001).

Digital maps are increasingly used in the U.S. military. The Force XXI experiment involved equipping combat vehicles and headquarters units with digital map displays, which when coupled with GPS and secure data and voice communications, enabled commanders and vehicle crews to see the same battlefield picture of friendly and enemy locations. The actual use of GIS analytical capabilities is in the early stages of experimentation and implementation.

The military logistics community is also leveraging information technology and GIS to improve their operations. This development is especially true in the transportation community, which plays a key role in deploying forces from bases in the continental U.S. to hotspots around the world. As an example, the Military Traffic Management Command's Transportation Engineering Agency (TEA) maintains GIS databases on strategic seaports, military installations, the National Highway Planning Network, National Bridge Inventory, National Railway Network, and strategic highway and railway networks. The TEA uses sophisticated transportation analytical models to determine transportation infrastructure capabilities and requirements. The TEA is working on a system to make this information available to users over the World Wide Web (Corbley 2000).

The military transportation community also uses a number of systems to track cargo through the transportation system. In the ocean transportation system, equipment is loaded on ships using the Improved Computerized Deployment System or ICODES, which is essentially a GIS that has the layout and dimensions for each deck and hold of strategic sealift ships. The system also contains digital templates of all military vehicles and cargo that might be transported. The number and types of vehicles to be transported are collected at the origin and electronically forwarded to ICODES, which then creates an optimum stow plan for each hold of each ship. This capability dramatically reduces the time it takes to load a ship. The World Wide Port System (WPS) tracks the specific location of each piece of cargo and reports this information to the Global Transportation Network (GTN). This system provides in-transit visibility and accountability of all cargo. The U.S. Air Force has a similar system for the optimal loading of transport planes. These systems are critical in ensuring optimum utilization of very limited transport assets and the accountability of billions of dollars of equipment and supplies (Corson 2000).

7. THE REVOLUTION IN MILITARY AFFAIRS

The experience of the Persian Gulf War of 1991 and the recent military operations in Afghanistan and Iraq led a number of commentators to suggests that we were or soon would be experiencing a "Revolution in Military Affairs" (RMA). Such revolutions have occurred in the past with the advent of gunpowder, railroad transportation, and the aircraft carrier. This one, however, is predicated on the idea that a rapid pace of technological innovation is altering the nature of modern warfare and the basic foundations of security (Martel 2001). Michael O'Hanlon (2002, 83) of the Brookings Institution explains:

> Due to the excellent performance of American high-technology weapons in the 1991 Persian Gulf War, as well as the phenomenal pace of innovation in the modern computer industry, many defense analysts have posited that a revolution in military affairs (RMA) is either imminent or already under way. The RMA thesis holds that further advances in precision munitions, real-time data dissemination, and other modern technologies, together with associated changes in war-fighting organizations and doctrines, can help transform the nature of future war and with it the size and structure of the U.S. military. RMA proponents believe that military technology, and the resultant potential for radically new types of war-fighting tactics and strategies is advancing at a rate unrivaled since the 1920s through 1940s when blitzkrieg, aircraft carriers, large scale amphibious and airborne assault, ballistic missiles, strategic bombing, and nuclear weapons were developed.

Barry Schneider (1995, 43) of the Air War College defines RMA as "a fundamental change, or discontinuity, in the way military strategy and operations have been planned and conducted." He suggests that RMAs are driven by technological innovations such as nuclear weapons at the end of World War II, operational innovations such as the German blitzkrieg, societal changes such as Napoleon's conscripted national army, or a combination of developments. Many defense analysts argue that we have been in the midst of an integrated RMA since the start of the Gulf War and that it is accelerating and becoming a mature system that integrates logistical, organizational, and technological capabilities across all the operational mediums of sea, land, and air. New warfare applications areas are emerging including long-range precision strikes, information warfare, dominating maneuver, and space warfare (Schneider 1995). A review of these new warfare areas shows that geography and geotechnology are integral and essential to all of them with perhaps the exception of information warfare.

Long-range precision strike is the ability to locate high-value enemy targets and destroy them quickly while causing minimal collateral damage. The U.S. has been a leader in this area since early efforts with laser-guided bombs in the Vietnam War. The Gulf War was characterized by substantial use of precision-guided munitions (PGM), and while the PGMs got most of the press coverage, the majority of the bombs dropped were the unguided "dumb" bombs of previous conflicts. In contrast, precision-guided munitions were extensively used in Afghanistan and Iraq to target terrorist threats, while reducing the collateral damage and casualties among non-combatants.

The NATO air campaign against Serbia in 1999 (Kosovo Crisis) was the first true application of PGMs as the dominant weapon. The essence of the precision strike is to detect the enemy deep in their rear areas, recognize their concept and strategic plan, and select and prioritize the critical targets to attack. These attacks must be synchronized in time and space to deal a devastating blow from which the enemy will not easily recover. Since Operation Desert Storm, U.S. commanders have had continuous wide-area surveillance and target acquisition systems with capabilities that continue to evolve (McKitrick et al. 1995). These surveillance and target acquisition systems are based on remote-sensing technologies discussed earlier such as orbital and aerial imagery, data from JSTARS aircraft, and data from unmanned aerial vehicles. The precision-strike munitions are also based on geographic technologies to an ever-greater extent. Cruise missiles use terrain-following radar that compares the return against a digital map, and later models have a GPS guidance system as well. The JDAM mentioned earlier makes large stocks of conventional bombs available as PGMs and overcomes the limitations of laser and other guidance systems from weather or limited visibility situations.

Maneuverability has always been a critical element of warfare. The ability to reposition forces globally or locally, on a much compressed time scale, and with highly lethal but greatly reduced forces, is the potential of this new warfare area. Dominating maneuver is defined as the positioning of forces—in coordination with the other three warfare areas—to attack decisive points, destroy the enemy's center of gravity, and attain the war objectives (McKitrick et al. 1995). Dominating maneuver is predicated on identifying the enemy's center of gravity and understanding where to position forces to render the enemy position untenable. Remote sensing is an essential tool in identifying enemy dispositions and thus their center of gravity. Cartographic data of the world, GPS, and GIS analytical tools are all essential in repositioning the forces. The transportation automation tools described earlier that facilitate rapid strategic mobility are a critical element in this new warfare area.

Space warfare is the exploitation of the space environment to conduct full-spectrum, near-real-time, global military operations. The U.S. military relies to a great extent on space-based systems during its daily operations. Space operations currently support earth-bound forces with satellites that enable communications, remote sensing, timing, and navigation. Future antisatellite weapons, an orbiting missile defense system, space strike systems that can hit targets on earth, and transatmospheric transports have the potential to expand the realm of space operations as did the fighters, bombers, and transport planes of the Second World War. Because orbital dynamics require operating speeds of about 17,000 miles (27,350 km) per hour, space operations will occur an order of magnitude faster than traditional air operations. Once again these envisioned technologies would rely on the geotechnologies of remote sensing and GPS at a minimum. There may ultimately be a role for GIS and cartography as well (imagine mapping and spatial analysis on the Moon or Mars).

A major area of the RMA that is maturing today is the use of information technology in automating command, control, and communications systems. The ultimate goal has been characterized as information dominance (Libicki 1997) or dominant battlespace awareness (DBA) (McKitrick et al. 1995). Both of these concepts are fundamentally based on geography and geographic technologies.

It should be borne in mind that one of the ultimate goals for geotechniques within the military is to provide total situational awareness for unit commanders at all echelons. Various systems, as noted above, already contribute to this goal to a limited degree. The objective is to have real-time knowledge of the enemy, terrain, weather, and the location and disposition of all friendly forces in order to command and control operations, to facilitate coordination and changes to the plan when the unexpected occurs, to avoid fratricide, and to minimize collateral damage. Research in this area will continue for the foreseeable future because of the major communications and compatibility changes that exist between systems. Information capacities and capabilities are related to the geotechniques being developed and adopted.

The concept of information dominance is defined as superiority in the generation, manipulation, and use of information sufficient to afford its possessors *military* dominance. It can be analyzed in terms of command and control, intelligence, and information warfare. Much of the U.S. military's investment in information technologies is focused on improving knowledge of *where* and *when*. The U.S. Army Force XXI project envisions outfitting every fighting vehicle with information systems that have digital maps. GPS linked to radios keep the maps updated with the current location of all friendly forces. Remote-sensing systems such as Predator UAVs along with reports from scouts and combat units identify the location and activity of enemy units, which are automatically updated on the digital map display in every vehicle. Staff officers can use various analytical tools (to include spatial analytical tools such as terrain analysis software) to develop plans. Operations orders and the accompanying graphics can be sent directly to the digital map displays in every combat vehicle. This concept was tested by a brigade of the 4th Mechanized Infantry Division at the National Training Center in the Mojave Desert of California in 1997. Observers reported that operations could be planned and carried out in half the time. The U.S. Navy and Air Force are working on similar concepts (Libicki 1997).

Dominant Battlespace Awareness (DBA) envisions total information dominance over an enemy. A military force would know where the enemy is, what it is doing, what it intends to do, and where its critical points are. It would deny the enemy this information about itself. The implications of DBA are significant. The ability to target the enemy with long-range precision-strike weapons might enable the bulk of forces to stand off and thus minimize casualties. The DBA also implies the power to do more with less. If the DBA can reveal where the main enemy attack is coming from, it may be possible to defeat them with a smaller force that will not have to cover a large front. Much lighter forces such as air assault or amphibious forces might also be used. These maneuvers would be less risky because planners could select landing sites they know to be free of enemy forces. Finally, a smaller DBA-enabled force might launch a successful counterattack sooner because they would know the path of least resistance and could focus fire support and reserves at the critical time and place (McKitrick et al. 1995).

The DBA and information dominance ideas are certainly not yet mature, but they have demonstrated potential. Consider the lessons of using special operations forces (SOF) along with remote-sensing systems to identify key targets, then having the SOF call in air strikes that used PGMs to destroy enemy forces with minimum firepower and collateral damage. These concepts have been used extensively in Afghanistan and Iraq, but they continue to evolve and may prove to be even more and more important in future conflicts.

It bears mention that not all technological innovations have to be cost-prohibitive to have an impact on performance. In attempting to improve the cultural awareness of U.S. troops destined for Afghanistan, Iraq, and North Korea (recognizing that approximately 35,000 troops have been postured south of the DMZ for more than fifty years), the geography faculty at the U.S. Military Academy prepared three separate handbooks with accompanying CD ROMs (Palka 2001; Malinowski 2002; Palka and Galgano 2003). Each of these are nothing more than focused regional geographies, made possible by information-gathering technologies, GIS, and mapmaking and publishing software. The CD ROMs provide an e-book, along with digital maps, and a program that enables one to select a route between two points and experience an actual fly-through. The idea was to enhance unit and staff preparation and mission effectiveness by providing deploying units with increased cultural awareness about the people of the country (friendly or otherwise) and a preview of the terrain and climate prior to their arrival. The idea for these three publications stemmed from the knowledge that other geographers had made during wartime, such as contribution to the JANIS books or the Admiralty Handbooks in the U.K. (see Palka 1995, 2002). In this case, however, technology facilitated the production of user-friendly references in an extremely short amount of time and at a minimal cost.

The key issue for the geographer in these ongoing discussions is the recognition of the role of geographic technologies. The command and control systems associated with these concepts have at their very core digital terrain data that are cartographic representations. Identifying the position of friendly forces is based on GPS technology. Detection and reporting on enemy forces, especially at long range, relies more than ever on the use of remote-sensing systems, whether they be satellites, aircraft, UAVs, or ground sensors. Spatial analysis as embodied in GIS has great potential but has seen limited use to date.

The potential to leverage spatial analysis for battlefield use at various scales is enormous. Military intelligence officers could use GIS spatial analysis tools for what the U.S. Army calls intelligence preparation of the battlefield (IPB). This includes determining avenues of approach (mobility corridors), templating enemy forces based on their doctrine (until the remote sensing systems determine their actual location), and developing likely enemy courses of action based on terrain, weather, and enemy capabilities and intent. Operations planners could use spatial analysis to develop and war game their plans, conduct intervisibility studies, determine time lines based on distance and terrain, and for many other tasks. Logisticians could use site selection tools to locate their logistics bases, and route optimization tools to maximize the use of their limited transportation resources. As all of these data are digital, they could be transmitted to the digital map displays in subordinate headquarters and combat vehicles in far less time than traditional methods. The U.S. military is a leader in this area, but has only conducted some rudimentary experiments. The potential of the combination of GIS spatial analysis, remote sensing, and GPS overlaid on digital maps is very significant to the RMA.

Most of the components of the RMA rely to some extent on geographic technologies, thus geography and geotechnologies, whether done by professional geographers or someone else, will play a key role. Presumably, professional geographers in government, industry, and the military will make significant contributions. It remains to be seen if academic geographers will play any role in the development of these concepts and technologies and thus have any voice or influence on how the RMA develops.

8. CONCLUSION

This chapter has described how military activities and warfare have both affected and been affected by the evolution of geographic technology. Military requirements have had a major impact on the development of the geographic technologies of cartography, remote sensing, GPS, and GIS. Cartography was employed early-on to provide military maps, and many of the advances in topographic mapping were in response to the needs of improved artillery. Cartography continued to serve and advance with the two World Wars and was forced to develop entirely new approaches with the advent of CORONA satellite imagery. These great strides were translated into gains for the civilian population of the U.S. through maps produced by the USGS. These maps were significantly improved by the contributions of both aerial photography and the CORONA satellite images.

Aerial photography, airphoto analysis, and photogrammetry were born with the advent of the airplane and came of age in the two World Wars when military necessity prompted tremendous technological innovation. Generations of photo interpreters would go on to apply their skills in civilian endeavors after both wars. The techniques pioneered in the military form the basis of modern academic programs in aerial photo interpretation and photogrammetry.

Satellite remote sensing was born because of the Cold War need to prevent a "nuclear Pearl Harbor." The sensors that would eventually become standard on LANDSAT and SPOT satellites were invented, tested, and perfected for a decade in the Keyhole series of U.S. imaging satellites. CORONA and its successors not only helped prevent World War III, but they provided the basis for civil satellite programs, and recently declassified imagery is a treasure trove for scientists utilizing change detection procedures. In addition, the need for trained geodesists in the CORONA program prompted American academic geography programs to begin formal training programs, which formed the basis of for much of our automated geographic technology training programs.

The Global Positioning System is a military initiative that has become a critical technology in civil life. GPS has revolutionized many fields to include surveying and mapping, aerial and marine navigation, and even fishing.

In the late Cold War and post-Cold War eras, however, we have seen a shift. The military has pursued lower cost, off-the-shelf information technologies. GIS in particular has come into widespread military use, but now it is the commercial providers that are showing the military how to use existing commercial products for military applications. The military has embraced GIS for a broad range of functions ranging from environmental management of military training lands to battlefield analysis (Chang et al. 2002)

Geography and military activities always have been intimately linked. Every soldier is a geographer at heart and must understand and appreciate terrain, weather, climate, and the human environments in which they operate. Geographic technology has always been critical to soldiers in that maps tell them where they are, where the enemy is, and where they must go. Other geographic technologies are an extension of these functions. Geography and geotechnology will play a critical role in any so-called revolution in military affairs. If history is any guide, then military systems and approaches that are derived from future revolutions in military affairs may spur the geotechniques into new directions and facilitate the development of new ones.

Finally, applications of geographic information, tools, and techniques are arguably just as important during peacetime and military operations other than war (MOOTW) as during wartime. Technologies used to support missions within the peacetime and MOOTW arenas include humanitarian assistance, land use management, protection of endangered species, management of water resources, responding to natural disasters, and fighting forest fires. These are causes that we consider socially responsible uses of geotechniques developed to solve military problems (Palka 1995, 2000, 2002).

REFERENCES

Air Force News Service (1998). Global Positioning System Marks 20th Anniversary. http://www.fas.org/spp/military/program/nav/n19980224_980226.html

Atkins, A. (2001). Project Kosovo. GEO: connexion. http://www.geoconnexion.com/magazine/article.asp?ID=123

Barnes, S. (2002). Salt Lake Hosts Spatial Olympics. GeoSpatial Solutions 12: 26-33.

Bates, C.C. and J.F. Fuller (1986). Weather Warriors 1814-1985. College Station: Texas A&M University Press.

Burrows, W.E. (1986). Deep Black: Space Espionage and National Security. New York: Random House.

Chang, K. (2002). Introduction to Geographic Information Systems. New York: McGraw-Hill.

Corbley, C.P. (2000). GIS and Transportation: Web-Accessible GIS Helps Military Move Em Out! GeoWorld. http://www.geoplace.com/gw/2000/0100/0100web.asp

Cordesman, A.H. and Wagner, A.R. (1996). The Lessons of Modern War vol. IV: The Gulf War. Boulder: Westview Press.

Corson, M.W. and Minghi, J.F. (1996). Powerscene: Application of New Geographic Technology to Revolutionize Boundary Making? International Boundaries Research Unit Boundary and Security Bulletin 4: 102-05.

Corson, M.W. (2000). Strategic Mobility in the 21st Century: Projecting National Power in a MOOTW Environment. In Palka, Eugene. J. and Galgano, F.A. (Eds.) The Scope of Military Geography: Across the Spectrum from Peacetime to War, 233-62. New York: McGraw-Hill Primus Custom Publishing.

Curtis, G.B., Stoddard, R., Kim, D.Y., and Devasundaram, J.K. (2002). Anthrax: GIS Helps Investigators Hunt for the Deadly Spores, GeoWorld 15: 34-37.

Day, D.A., Logsdon, J.M., and Latell, B. (Eds.) (1999). Eye in the Sky: The Story of the Corona Spy Satellites. Washington, DC: Smithsonian Institution Press.

Department of the Army (DA) (1993). FM 21-26 Map Reading and Land Navigation. Washington, DC: Headquarters, Department of the Army.

Department of the Army (DA) (2002). Integrated Training Area Management. http://www.army-itam.com/home.jsp

Department of Defense (DOD) (1992). Conduct of the Persian Gulf War. Washington, DC: Department of Defense.

ESRI (200a). Military Installations: Taking Command of Your Base with GIS. Redlands, CA: ESRI.

ESRI (2001b). ArcGIS Military Analysts: GIS Tools for the Defense and Intelligence Communities. Redland, CA: ESRI.

ESRI (2001c). GIS for Defense. Redlands, CA: ESRI.

ESRI (2002). Homeland Security: GIS for Community Safety. Redlands, CA: ESRI.

Federation of American Scientists (FAS) (1997). FAS Space Policy Project Military Space Programs: IMINT Overview. http:www.fas.org.spp/military/program/imint/overview.htm

Federation of American Scientists (2002). Joint Direct Attack Munition (JDAM). GBU-29, GBU-30, GBU-31, GBU-32. FAS Military Analysis Network. http://www.fas.org/man/dod-101/sys/smart/jdam.htm

Hartcup, G. (2000). The Effect of Science on the Second World War. New York: St. Martins Press.

James, P.E. and Martin, G.J. (1993). All Possible Worlds: A History of Geographical Ideas, 3rd ed. New York: John Wiley and Sons.

Jensen, J. R. (1996). Introductory Digital Image Processing: A Remote Sensing Perspective. 2nd ed. Upper Saddle River, NJ: Prentice Hall.

Kline, H.B.D. Jr. (1954). The Interpretation of Air Photographs. In James, Preston E. and Jones, Clarence F. (Eds.) American Geography: Inventory and Prospect, 530-46. Syracuse: Syracuse University Press.

Larson, J.L. and Pelletiere, G.A. (1989). Earth Data and New Weapons. Washington, DC: National Defense University Press.

Latham, C. and Stobbs, A. (1996). Radar: A Wartime Miracle. London: Sutton Publishing Ltd.

Latham, C. and Stobbs, A. (1999). Pioneers of Radar. London: Sutton Publishing Ltd.

Libicki, M.C. (1997). Information Dominance; National Defense University Strategic Forum Number 132. http://www.ndu.edu/inss/strfourum/forum132.html

Lillesand, T.M. and Kiefer, R.W. (1994). Remote Sensing and Image Interpretation. 3rd ed. New York: Wiley and Sons.

Lindgren, D.T. (2000). Trust But Verify: Imagery Analysis in the Cold War. Annapolis: Naval Institute Press.

Malinowski, Jon C. (2002). Iraq: A Geography. West Point, NY: U.S. Military Academy Press.

Martel, W.C. (2001). Introduction. In The Technological Arsenal: Emerging Defense Capabilities, xi-xix. Washington, DC: Smithsonian Institution Press.

McKitrick, J., Blackwell, J., Littlepage, F., Kraus, G., Blanchfield, R., and Hill, D. (1995). The Revolution in Military Affairs. In Schneider, B.R. and Grinter, L.E. (Eds.) Battlefield of the Future: 21st Century Warfare Issues, 65-98. Maxwell Air Force Base: Air University Press.

OHanlon, M.E. (2001). Defense Policy Choices: For the Bush Administration 2001-05. Washington, DC: Brookings Institution Press.

OSullivan, P. (1991). Terrain and Tactics. New York: Greenwood Press.

Palka, Eugene J. (1995). The U.S. Army in Operations Other Than War: A Time to Revive Military Geography. GeoJournal 37: 201-08.

Palka, Eugene J. (2001). Afghanistan: A Regional Geography. West Point, NY: U.S. Military Academy Press.

Palka, Eugene J. (2002). Perspectives on Military Geography. The Geographical Bulletin 44: 5-9.

Palka, Eugene J. and Galgano, F.A. (Eds.) (2000). The Scope of Military Geography: Across the Spectrum from Peacetime to War. New York: McGraw-Hill Primus Custom Publishing.

Palka, Eugene J. and Galgano, F.A. (2003). North Korea: A Geographic Analysis. West Point, NY: U.S. Military Academy Press.

Peebles, C. (1997). The CORONA Project: Americas First Spy Satellites. Annapolis: Naval Institute Press.

Roberts, J.M. (1993). History of the World. New York: Oxford University Press.

Robinson, A.H. (1954). Geographic Cartography. In James, P.E. and Jones, C.F. (Eds.) American Geography: Inventory and Prospect, 553-77. Syracuse: Syracuse University Press.

Russell, J.A., Booth, A.W., and Poole, S.P. (1954). Military Geography. In James, Preston E. and Jones, Clarence F. (Eds.) American Geography: Inventory and Prospect, 484-95. Syracuse: Syracuse University Press.

Schneider, B.R. (1995). Overview: New Era Warfare? A Revolution in Military Affairs? In Schneider, B.R. and Grinter, L.E. (Eds.) Battlefield of the Future: 21st Century Warfare Issues, 43-45. Maxwell Air Force Base: Air University Press.

Stanley, R.M. (1981). World War II Photo Intelligence. New York: Scribner and Sons.

Stanley, R.M. (1998). To Fool A Glass Eye: Camouflage Versus Photoreconnaissance in World War II. Washington, DC: Smithsonian Institution Press.

U.S. Geological Survey (2002a). The National Map: Topographic Maps for the 21st Century. USGS Fact Sheet 018-02. Washington, DC: Department of the Interior.

U.S. Geological Survey (2002b). Homeland Security and The National Map. USGS Fact Sheet 061-02. Washington, DC: Department of the Interior.

Winters, H.A. et al. (1998). Battling the Elements: Weather and Terrain in the Conduct of War. Baltimore: Johns Hopkins University Press.

PART IV

THE ENVIRONMENT AND TECHNOLOGY

CHAPTER 19

DOUGLAS J. SHERMAN
ANDREAS C.W. BAAS

EARTH PULSES IN DIRECT CURRENT

We feel the long pulsation, ebb and flow of endless motion,
The tones of unseen mystery...
And this is ocean's poem. (Walt Whitman, *Leaves of Grass*)

Abstract   The Earth's pulse is evident in a variety of geomorphic processes that shape its surface. This chapter describes how electronic instrumentation has dramatically increased our capacity to investigate sediment transport by wind and water and to relate processes to morphological change. A number of instruments and techniques for measuring fluid flow and sediment transport in fluvial, coastal, and aeolian environments are discussed, including current meters, current profilers, pressure transducers, optical backscatter sensors, anemometers, hot-film probes, photoelectric erosion pins, sediment traps, and saltation impact responders. The deployment of such instruments is placed in the context of scale, methodology, limitations, and interpretation of spatiotemporal records of measured processes.

Keywords   geomorphology, geomorphic processes, sediment transport, fluid flow, instrumentation, measurement, methodology, time series.

1. INTRODUCTION

Wind and water flow across the surface of the earth, sculpting and reshaping aeolian, coastal, and fluvial environments particularly. The study of these flows (processes) and their interactions with associated sediment systems (responses) is a central theme of modern geomorphology. The measurement and recording of these processes and responses relies substantially on instrumentation systems capable of high-speed data acquisition and storage. The geomorphological expressions of technoearth are manifest in digital files stored in a spectrum of media, reconstituted in graphical and analytical representations, and recreated in computer simulations. The emphasis of this chapter is on the description of the instrumentation systems that provide a foundation for geomorphology's technoearth.

Stanley D. Brunn, Susan L. Cutter, and J.W. Harrington, Jr. (Eds.), Geography and Technology,
431-460. © 2004 Kluwer Academic Publishers. Printed in the Netherlands.

The title of this chapter was inspired by years of observations of waves, formally and informally, for business and for pleasure. The rhythmic breaking of ocean waves is strongly suggestive of heartbeat, as recognized by Whitman (quoted above). Our measurements of that pulse are often expressed as a DC voltage signal that subsequently is recorded digitally. For example, Figure 19-1 depicts a two-minute time series of water surface measurements from Galveston, Texas, sampled 50 times per second (50 Hz). In this record, we can see what prompted Sylvia Plath to write, in Whitsun, "The waves pulse like hearts." In this case, at about ten beats per minute.

The possession of a digital time series such as that shown in Figure 19-1 allows us the opportunity to analyze the record statistically, and to decompose the record numerically in order to detect the presence and magnitude of secondary motions. For example, at the beginning of the 20th century, Cornish (1934) measured characteristics of individual waves using his understanding of linear wave theory and the geometry of ships at sea. This method, however, did not allow rapid and accurate measurement of series of waves. One result of this limitation was the development of the concept of significant wave height, an observational convention that relates the distribution of heights in a wave field to the heights most apparent to the human eye, equivalent to the highest 1/3 of the waves present (e.g., Aagard and Masselink 1999).

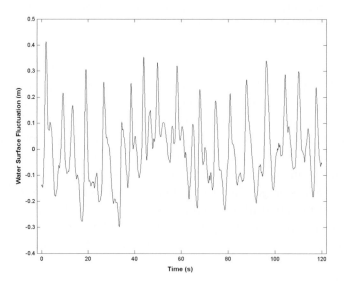

Figure 19-1. Two-minute time series of water surface fluctuations measured with a pressure transducer at Galveston, Texas, October 2002. Sample rate is 50 Hz, data were smoothed with a 1-second moving average. Source: authors.

Observations of this type limited the description of the wave field to the motions that were visually obvious. This limitation was congruent with general practices in geomorphology where visual observations of most natural phenomena were the norm until the late 1900s. For some applications, this level of detail was often sufficient. For example, visual observations of the waves depicted in Figure 19-1 would probably produce wave characteristics similar to those derived statistically from the digital records. There are suites of processes, however, that are not readily apparent to the eye, but that are often revealed in the instrumented record. The detailed datasets obtained using fast-response instrumentation systems provide us with closer linkages between process and form, and thus greater opportunities for geomorphological interpretation.

2. BACKGROUND

Geomorphology is a discipline that once relied mainly on the mental tools of observation, simple measurement, deduction, and intuition, in its quest to explain the surface features of the earth as exemplified by the works of Davis (1909), Gilbert (1886), or Sauer (1930), for example. These tools, perhaps supplemented by photography, grain size analysis, field mapping, or map interpretation, comprised the methodological components of traditional geomorphological research. The common parlance of geomorphology revolved substantially around concepts such as where, how long, how big, what geometry, what geology. Spatial data, often manifest in maps, sometimes only crudely qualitative, reigned in the era before technoearth. The geomorphologist worked in Lagrangian modes, studying landforms by traverse and perambulation; obtaining a spatial perspective largely through motion.

In recent decades, there has been a pronounced effort to design and utilize instrumentation systems capable of producing digital representations of geomorphic environments and processes. This is the technoearth concept realized for geomorphology. Experimental design usually involves deployment of the systems at a location, and then monitoring processes—earth pulses—at that location. The utilization of the time series of measurements produced by these systems, even with the use of three-dimensional arrays, makes process geomorphology an inherently Eulerian discipline. Many of the sensors (e.g., pitot tubes or anemometers) used by such systems have a long history of development and application, but their linkage to modern data acquisition systems has revolutionized their utility in geomorphological research. Other sensors (e.g., electromagnetic current meters or Acoustic Doppler Current Profilers) are relatively recent innovations, and only realize their optimal performance when coupled with digital recording systems. These new tools provide a theory-driven, empirical basis for understanding the agents of landform change and the rates at which they work. Many such instruments allow high-resolution field studies of complex flow environments or sediment responses. Sets of different instruments allow us to measure detailed behavior of geomorphic systems over relatively small spatial and temporal scales. This inherently reductionist approach will not lead immediately to answers about complex landform and landscape evolution, but it will strengthen the scientific foundation for those answers.

The purpose of this chapter is to describe several modern instrumentation systems, to present examples of their applications and operations, to examine the methodological implications of deploying instrument arrays in field experiments, and to discuss how the resulting measurements contribute to our understanding of geomorphic environments. For this discussion, we are defining an instrument system to include at least a sensor capable of producing an electronic signal and a digital data recorder. For the most part, therefore, we emphasize applications in aeolian, fluvial, and coastal sedimentary systems because they tend to display a close coupling between fluid processes, sediment response, and small-scale landform change. They are also the systems with which we (the authors) are most familiar. We will discuss methods of electronic flow measurement in air and water and the examples of the instruments designed to measure the geomorphic responses. The instruments discussed include anemometers and wind vanes, current meters, current profilers, pressure transducers, optical and acoustic back-scatterance sensors, photoelectric erosion pins, saltation impact sensors, and fast-response sediment traps. Several examples of the performance of such systems, especially in the measurement of aeolian systems, will be presented. All of these instruments respond as a result of direct contact with the geomorphic environment being sensed. We consider this characteristic to be a fundamental distinction from what are typically considered to be remote-sensing technologies, which are not discussed here.

3. HISTORICAL PERSPECTIVES

Many of the sensors associated with modern instrumentation systems have a long history of geomorphological applications. The cup-anemometer, with mechanically recorded output, was invented by Robinson in the mid-19th century. Numerous versions of mechanical tide gauges also began to be deployed at about the same time, and these quickly evolved into the first wave gauges. Gilbert (1914) experimented (unsuccessfully) with a pitot tube in his classic flume research. Bagnold used a vertical array of pitot tubes and manometers to measure wind profiles in the laboratory (Bagnold 1936) and the field (Bagnold 1938). Through the first decades of the 20th century, such applications became increasingly valuable in the study of landform development. This was especially the case for studies of (relatively) fast-response sedimentary systems, where the results allowed fluid behavior to be linked closely to sediment transport, or where systems had characteristics that could not be discerned through simple signal acquisition. The latter case is exemplified by the development of the Vibrotron by Munk et al. (1963) for the measurement of swell and the determination of angle of wave approach.

There are many other examples of instrumentation innovation throughout the 20th century, including thermal anemometry, sonar-based depth detection, pressure sensors, and electromagnetic current meters. Until about 1980, output from instruments such as these, especially during field deployments, was recorded on paper strip charts, audio or video tape, was read from visual displays and recorded manually, or required links to unwieldy digital acquisition and recording systems (e.g., Greenwood and Sherman 1984). The latter typically required substantial custom programming by the scientist, and seldom had a monitor. In situ troubleshooting and preliminary evaluation of data in the field was often impractical. The advent of the personal computer, low-cost and robust data acquisition systems, and packaged software programs added the ingredients needed to complement the instruments themselves. These allowed geomorphologists to begin to take the laboratory to the field with greater mobility and enhanced versatility. Color monitors (especially) allowed real-time instrument monitoring to become commonplace. Troubleshooting was simplified. Quality assessment and control tasks in the field could be routinely conducted. The relatively low cost of such instrument systems made them readily accessible to the community of geomorphologists. We became able to readily take the pulse of technoearth. This development represents a fundamental revolution in our ability to conduct scientific investigations into how the earth's surface is reshaped. We discuss next a few examples that focus on the instruments used and the records obtained rather than the data acquisition and recording components of the systems.

4. MONITORING GEOMORPHIC SYSTEMS CONTROLLED BY HYDRODYNAMIC PROCESSES

Many instruments have been designed for measuring morphological and sedimentological responses to hydrodynamic processes. We consider here those used for flow measurement, water depth, sediment detection, and bank or bed elevation changes.

4.1. Flow Measurement

Impeller-type (or cup) current meters have long been a staple for field experiments, especially in fluvial systems (e.g., Hjulström 1935) and the design remains a U.S. Geological Survey standard for stream flow measurement (Hubbard et al. 2001). These current meters perform in a manner similar to cup anemometers. The propeller (or the cups) rotates at a speed that is proportional to the flow speed. An electronic contact is tripped by each revolution, and this provides a signal that is converted to a speed measurement, and that can be read directly or recorded digitally. Modified versions of these current meters have been used to measure currents in marine and lacustrine systems. Other versions have been adapted for surf zone applications, for example, the ducted, bidirectional current meter (Nielsen and Cowell 1981; Brander and Short 2000).

In the last 25 years, there has been increased use of Acoustic Doppler (AD) and electromagnetic flow sensors. There are several types of AD sensors, sometimes referred to as ultrasonic sensors, but all operate on the basic principle that there is a shift in acoustic frequency when a sound source and receiver are in motion relative to one another. AD sensors emit a focused acoustic beam that bounces off particles and bubbles in the water. The sensors are also receivers for that part of the beam that is reflected directly back to the source. Velocity is calculated as:

$F_d = -2F_s (V/C)$
where F_d is the frequency of the Doppler (returned) signal
F_s is the original source frequency
V is the velocity difference between the instrument and the flow, and
C is the speed of sound.

By focusing a set of sound beams at a specific distance from the instrument, velocity measurements at a specific location may be obtained. This principle is incorporated into the design of several commercially available instruments. Perhaps the simplest is a flow meter where the focus distance of the acoustic beam is fixed. This provides flow measurement at a point. A more complex version of this design is the AD velocimeter where the instrument is designed to measure flow through a very small volume and with high-frequency sampling rates. These measurements can be used to estimate turbulence characteristics, among other properties of the flow.

The most versatile instrument in this family is the Acoustic Doppler Current Profilers (ADCP), which is able to measure the reflected signal at different time intervals, corresponding to distance traveled from and to the sensor. It is thus possible to obtain velocity measurements from a large number of locations, or sample bins (sometimes exceeding 100) along a linear profile, making the ADCP the instrument of choice for many field experiments (e.g., Simpson and Oltman 1993), especially in environments with physical characteristics (e.g., large depth) that make conventional flow measurement, with the sensor mechanically raised and lowered through the flow, difficult. With linked GPS hardware and software, current profiling can be done from a mobile platform, allowing three-dimensional representations to be obtained or the measurement of multiple stream cross-sections at relatively short time increments. Garcia-Goritz et al. (2003) used such a system to obtain detailed measurements of currents in the Mediterranean Sea. Their data allowed decomposition of the flows into the subinertial, inertial, and tidal contributions to the net motion. With downward-looking ADCPs, errors may arise in the computation of discharge characteristics under conditions of substantial sediment transport, because the instruments may not accurately detect the motionless bed through the moving sediments. This situation has been demonstrated to be a critical factor during repeated measurements made of the Amazon River (Callede et al. 2000).

With integrated power supplies and internal memory boards, ADCPs can be deployed autonomously for extended periods of monitoring, especially installed at the bed, in an upward-looking mode, or attached to a mooring or structure (e.g., a pier or bridge) in either upward, downward, or occasionally, sideways-looking modes. The systems can be programmed to sample periodically (e.g., hourly) at designated burst lengths (e.g., ten minutes) and sample frequencies (e.g., 2 Hz). For example, Lopez and Garcia (2003) used an upward-looking ADCP moored at a depth of 280 m over a period of eight months to sample currents at depths between 263 and 33 m, at 10 m intervals, finding anomalously fast currents near the bottom. There is no reasonable alternate methodology for the acquisition of data with this degree of detail over time spans of months or longer.

There are two drawbacks associated with ADCP measurements that limit their utility for some applications. First, they measure velocities through sample volumes that may be large relative to the range of water depths spanned by the profile, so that changes in velocity over small distances may not be discerned. Second, because acoustic signals are reflected strongly from boundaries such as a river's bed and surface, data cannot be collected close to those boundaries, causing "blanking" areas. The inability to measure close to the bed, in particular, limits the utility of ADCPs for some sediment transport applications. These shortcomings can be addressed with AD velocimeters (ADV).

ADVs are designed to obtain high-frequency measurements of three components of flow through very small volumes. ADVs can be used to obtain details of flow structure, including turbulence measurements, in the immediate vicinity of a fluid-sediment interface; also applications in laboratory experiments are common (Laser Doppler Velocimeters (described briefly in Clifford and French 1993) are also common in lab studies, but seldom used in the field). Field applications of ADVs offer great potential for linking the behavior of fluid and sediment systems, or for the elucidation of characteristic flow structures, including coherent flow structures. Frothingham and Rhoads (2003) used an ADV to obtain detailed measurements of the structure of river flow through a meander bend. They sampled the three components of flow across 14 cross-sections, using 40-50 locations per cross-section, by sequentially relocating the ADV over a number of days. They sampled flow at a rate of 25 Hz, for 60-90 seconds at each location, with a precision of 0.1 mm s^{-1}.

Electromagnetic current meters (ECM) have been a mainstay of coastal research since the 1970s, although they have also seen substantial service in fluvial environments (e.g., Best and Roy 1991). Most models allow simultaneous measurement of two orthogonal components of flow at sample frequencies greater than 1 Hz, a substantial asset for the acquisition of data describing the complex flow environments associated with wave-current interactions in the nearshore. The lack of moving parts, and generally robust design also reduces the potential for instrument fouling and failure in extended deployments. Faraday's Law of electromagnetic induction is the principle governing the design of ECMs. Faraday's Law states that a voltage is induced when a conducting fluid (such as water) flows through an electromagnetic field. ECMs generate a controlled electromagnetic field, and measure the flow with a set of four electrodes around a discoidal or spherical sensor head. The technical aspects of ECM performance have been discussed extensively (e.g., Aubrey and Trowbridge 1985; Lane et al. 1993) and are not described further here.

ECMs have seen their most extensive use in studies of nearshore processes. Sherman et al. (1993) used a single ECM to measure flow conditions associated with bedform migration in a rip channel. Nordstrom et al. (2003) linked ECM data with the movement of sediment tracers to assess alongshore sediment transport. Osborne and Greenwood (1993) used a set of ECMs to explore the vertical structure of flow in the nearshore in the context of describing controls on sediment suspension. Deployments of extensive arrays of ECMs have been commonplace in the lengthy series of experiments conducted at the U.S. Army Corps of Engineers Field Research Facility (FRF) at Duck Island, North Carolina. Fedderson and Guza (2003) used data from a two-dimensional array of 25 ECMs deployed at the FRF for almost four months to obtain measurements of along and acrossshore flows to describe nearshore circulation. Experiments of this length require that care be given to avoid biological fouling of the sensor's electrodes, as fouling will significantly degrade the signal.

4.2. Water Depth

Water surface elevations are measured with surface piercing probes, floats in stilling wells, pressure transducers, and other instruments. Surface piercing probes (e.g., resistance or capacitance wave wires) are used mainly in lower energy environments or, laid on their sides, in swash zones (Holland et al. 1995). Stilling wells are used for environments where rates of surface change are relatively slow; for river or tide stage measurements, for example. Pressure transducers are the instrument of choice for precision measurement of small differences in surface elevation and for the measurement of waves. Pressure transducers measure water depths above the sensor according to the hydrostatic pressure equation. As the crest of a wave passes over the instrument, for example, the local water depth is increased temporarily, with a corresponding increase in pressure. A mechanical diaphragm, or a piezoelectric wafer, in the transducer, is distorted as the pressure fluctuates and the degree of distortion produces a proportional change in output voltage.

Meirovich et al. (1998) used pressure transducers in stilling wells to measure the slope of river surfaces over short-reach lengths (less than 100 m). Horn et al. (1998) used high-precision pressure transducers to measure pressure gradients within a beach as part of a study to elucidate water table processes. Arrays of pressure transducers are deployed in large-scale nearshore studies to measure alongshore and acrossshore differences in mean water depth and wave forcing (Fedderson and Guza 2003; Ruessink 2000).

As depicted in Figure 19-1, pressure transducers are capable of reproducing water surface fluctuations in great detail. Because the computer memory requirements to store individual measurements are not great relative to the capacity of desktop or even laptop computers, time series may be acquired simultaneously from large numbers of instruments or for smaller numbers for longer time periods. Thus records of gravity and infra-gravity waves and tidal fluctuations can be determined from the same dataset. For example, Figure 19-2 depicts the entire water surface record from which the Figure 19-1 data are extracted. From this record, we can discern temporal clusters of larger and smaller waves, termed wave groups, caused by interference between different wave trains or other long-wave generation mechanisms. Interpretation of wave groups requires records that are long compared to the periodicity of the groups.

There are several analytical methods available for the interpretation of such records and the estimation of periodicities in surface fluctuations. In Figure 19-3, we see an energy spectrum for this record that indicates periodic

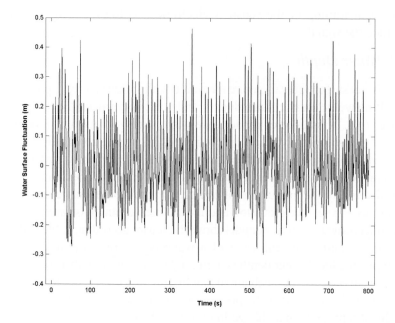

Figure 19-2. Thirteen-minute time series of water surface fluctuations measured with a pressure transducer at Galveston, Texas, October 2002. Sample rate is 50 Hz; data were smoothed using a 1-second moving average. Source: authors.

motions at several temporal scales, most importantly at periods of 7.9 s and 51.3 s.

The application of a low-band pass filter to the original data confirms the presence of a longer period motion, as shown in Figure 19-4. Here the original time-series is smoothed with a 1-minute moving average that removes the signal of the incident waves. The apparent longer period motion would be very difficult or impossible to quantify, or perhaps even discern, through visual observation because of its small amplitude relative to the incident waves.

Individual pressure transducer records are capable of indicating the presence of wave motions at periods longer than that of the incident wave field, but they cannot determine the nature of the motion. For example, fluctuations with periods twice that of incident waves may be caused by standing, reflected waves, or by standing edge waves. Either signal could manifest an identical appearance in a spectral record but demonstrate substantially different impacts on nearshore bathymetry. Records of surface

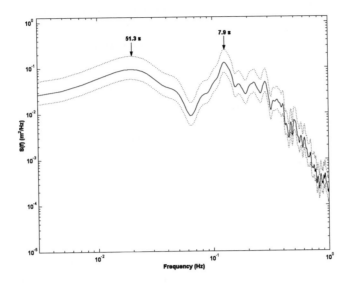

Figure 19-3. Spectral analysis of the Figure 19-2 time series, showing statistically significant peak frequencies at 0.127 Hz (7.9 s) and 0.0195 Hz (51.3 s). Source: authors.

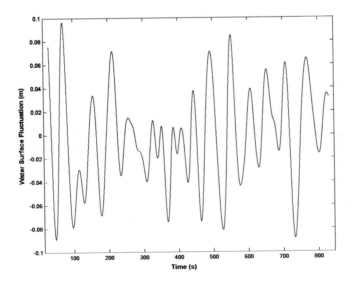

Figure 19-4. Thirteen-minute time series of water surface fluctuations measured with a pressure transducer at Galveston, Texas, October 2002. Sample rate is 50 Hz; data were smoothed using a 1-minute moving average. Source: authors.

Table 19-1. The distinctive hydrodynamic signatures of types of wave common in the surf zone, based upon the relationships between periodic acrossshore (u) and alongshore (v) velocities, and water surface fluctuations (ç).

Wave Type	u - v	u - η	v - η
Progressive gravity	in phase	in phase	in phase
Standing gravity	quadrature	quadrature	in phase
Progressive edge	quadrature	quadrature	in phase
Standing edge	in phase	quadrature	quadrature

fluctuations can be paired with bidirectional current measurements to distinguish the nature of these (and other) important motions because they have distinct signatures (Table 19-1 based on Holman and Bowen 1979). The terminology "in phase" indicates that the maximum fluctuations from the mean occur simultaneously in the paired time series. For example, with a progressive gravity wave (the waves you see on beaches), the maximum acrossshore (u) and alongshore (v) velocities, and water surface fluctuations (ç), occur at the same time. With quadrature, the maximum fluctuations are offset by a period equal to one half of the wave period. For a more detailed explanation, see Komar (1998).

4.3. Sediment Detection

Most of the devices to detect and quantify the presence of sediments in water rely on the reflection or attenuation of optical or acoustic signals. A source and receiver are required for the operation of these instruments. Optical (OBS) and acoustic (ABS) backscatter sensors combine both signal source and receiver. These sensors emit a signal of known intensity. Material in the water, such as sediments or air bubbles or even zoo-plankton (Kringel et al. 2003), will scatter the signal, reflecting part of it back toward the source. An alternate technology uses spatially separated source and receiver and measures the attenuation of the signal between them, as it is scattered by the sediments (e.g., Waddel 1976). The relative intensity of the received signal can be calibrated against known sediment concentrations to translate the voltage signals into sediment equivalents. Further, by measuring the strength of the return at different beam travel times, concentrations at multiple distances from the instrument can be obtained. It is thus possible to gather concentration profiles through a water column. With appropriate calibration, the signal can also be used to estimate particle size (e.g., Law et al. 1997; Schat 1997). Ridd et al. (2001) have used an OBS, installed vertically with the sensor flush with a sediment bed, to measure rates of sediment deposition in a mangrove swamp, a lake, and a riverbank. Data obtained from OBS records are often used with measurements or estimates of flow parameters to assess sediment transport rates in the nearshore (Beach and Sternberg 1988; Osborne and Greenwood 1993; Ruessink 2000), estuaries (Kraus and Ohm 1984) or with estuarine sediments (Kineke and Sternberg 1992), rivers (Lewis 1996) and on the continental shelf (Green et al. 1995). Ridd and Larcombe (1994) designed an automated cleaning wiper for deployments in environments where sensor fouling is a potential problem. Bauer et al. (2002) used a coupled OBS and ECM array to measure sediment suspension (hence erosion) and transport caused by boat wakes impinging on a levee bank. In that study, OBS measurements were converted (via calibration) to sediment concentrations, and the portion of the suspended sediment load that could be attributed to a boat-wake event was converted to a bank loss equivalent. An example of a boat-wake-induced suspension event is illustrated in Figure 19-5. Note that

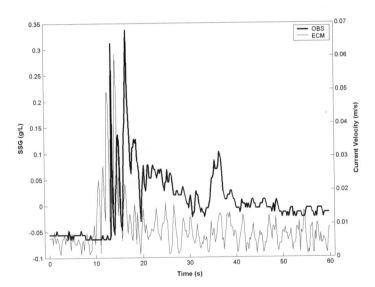

Figure 19-5. Sediment suspension (dark trace) from a levee bank in the Sacramento River Delta (August 2000), caused by a boat wake (light trace). Samples were taken at 5 Hz. Note that there is a distinctive time lag between velocity peaks and concentration peaks and that sediments remain in suspension long after the forcing impact has diminished. Source: authors.

in this environment (described in Ellis et al. 2002), suspension does not occur until the magnitude of the wake velocities exceeds a threshold and the slow settling velocities of the fine sediments allows them to be advected from the site, absent control structures.

4.4. Bed Elevation Changes

The ability to measure bed elevation changes, concurrent with the sediment transport processes that drive such changes, provides an extremely valuable tool for the interpretation of landform changes. Traditional survey methods are well-suited to quantify surface changes over time periods of hours or longer. But the methods integrate change over the interval between surveys. There are several instruments that are capable of monitoring bed level changes at the scale of seconds and for relatively long time periods. These instruments, however, provide point data that then must be extrapolated to areal data. The two types of instruments discussed here are the photo-electronic erosion pin and the sonar altimeter.

The photoelectronic erosion pin (PEEP) was designed originally for monitoring bed and bank elevation changes in fluvial systems (Lawler 1991). The principles of operation are straightforward. The instrument comprises a linearly arranged set of photovoltaic cells, wired in series and housed in a water-tight, transparent tube. The cells generate a signal (VDC) proportional to the quantity of light received by the PEEP. The signal will change as the length of PEEP exposed to sunlight changes or as ambient light intensity changes. Reference cells at the ends of the PEEP sensor monitor the maximum and minimum light intensities at a given time, and these values are used to calibrate the output from the linear array of cells. Voltages can be recorded with autonomously deployed data loggers, for example, to obtain long time series of bank change. A robust version of the PEEP was designed for deployment in the intertidal zone (Lawler et al. 2001). McDermott (2001) deployed a series of PEEP sensors, co-located with pressure transducers, across the foreshore of a beach to measure small-scale sediment level changes associated with groupiness in the incident wave field. PEEP sensors should be well suited to deployment in a range of environments.

Sonar altimeters are precision versions of the common echo sounder used for bathymetric surveys (or fishing!). Their widest applications to geomorphological investigations have been in studies of nearshore environments where the altimeter is positioned at a known elevation above the bed. Thus changes in depth below the instrument are a result of bed elevation changes. An early adoption of this technology was by Dingler et al. (1977), for the measurement of nearshore bed forms, and Greenwood et al. (1985) modified that design to allow remote operation of the bed profiler. In these applications, the sonar altimeter was moved across the bed forms to measure their geometry. Gallagher et al. (1998) used a stationary altimeter to measure the migration of large bed forms below the instrument. Gallagher et al. (2003) attached seven altimeters to a mobile instrument platform—the Coastal Research Amphibious Buggy or CRAB—to measure linear series of bed forms across the surf zone of the Duck Island FRF. Fedderson and Guza (2003), as part of their large current meter array, also had co-located sonar altimeters so that changing nearshore morphology could be measured in real time.

This summary of instrumentation systems to take the pulses of hydrodynamically controlled systems is, of course, incomplete. There are many other types of instruments, some of which have the potential to become fieldwork standards (including some of the laser applications). But there is a reason that medical doctors still carry stethoscopes.

5. MONITORING GEOMORPHIC SYSTEMS CONTROLLED BY AEOLIAN PROCESSES

Motions of air in the atmosphere are another expression of Earth's pulse, manifesting itself over a range of scales. Atmospheric circulation moves air like ocean currents around the globe, wind systems migrate like tides over the land, and gusts bathe the Earth's surface. From storms that buffet a continent, turbulence trickles down in an eddy cascade to wafts that ruffle a blade of grass. Where the winds touch upon the surface, they transport sediments and so change the shape of the landscape, producing ripples, dunes, and a great variety of aeolian landforms. Research in aeolian geomorphology is then concerned with measuring airflow, sediment transport and morphology, and the interactions between these components.

Taking the Earth's pulse with instruments in air is different compared to taking measurements in water. While electronic instrumentation deployed in water needs to be watertight and capable of withstanding significant forces, the dramatically lower density and viscosity of air result in much smaller forces exerted on instruments. On the other hand, airflow and sediment processes occur on much smaller spatial and temporal scales requiring a higher resolution of measurements. Furthermore, aeolian processes take place within a fully three-dimensional setting that is directly analogous to the three-dimensional framework common to hydrodynamic systems. The selection of sampling locations and the temporal resolution of measurements is directly related to the scale of investigation and governs the requirements for the instrumentation that is deployed. Research in aeolian geomorphology covers a vast range of scales. Measurements of detailed transport processes, such as saltation dynamics and turbulence, require instrumentation operating on scales of centimeters and fractions of seconds, while investigations of dune-field evolution and wind climates call for instrumentation measuring on scales of meters to kilometers and days to decades. The current technological challenge lies mainly in instrumentation capable of measuring the small scales involved in airflow turbulence and saltation dynamics.

The monitoring of changes in morphology is an area of highly developed instrumentation. While small changes of sand surface elevation due to erosion or deposition are often still measured with erosion pins—a decidedly low-level approach—the evolution of ripple patterns and changes in dune morphology are now monitored with remote-sensing techniques, such as stereophotography, aerial photography, and satellite imaging. The remainder of this chapter focuses on measuring airflow and sediment transport in aeolian environment. For developments in remote-sensing techniques for monitoring morphology changes, the reader is referred to other chapters in this volume.

5.1. Airflow Measurements

The airflow is the primary forcing agent in aeolian systems responsible for sediment transport and subsequent morphological development. The three-dimensional flow field is usually characterized by a limited set of time-series of wind speed and directions at select locations within a three-dimensional framework. In relation to sediment transport, airflow measurements are primarily intended to ascertain shear stresses acting on the bed surface that drive transport processes. Shear stresses can be derived from wind measurements in two ways. First, a set of horizontal wind speed measurements along a vertical profile can be used to derive the shear stress in a boundary layer from fitting a logarithmic velocity profile, the so-called "law-of-the-wall." Second, high-frequency time-series of the horizontal and vertical components of the wind vector can be split into time-averaged speeds (u, w) and instantaneous fluctuating parts ($u'w'$), according to Reynolds decomposition. The product of the horizontal and vertical fluctuating parts, multiplied with the density of the air ($\rho\, u'w'$), is then equal to the instantaneous shear stress in the airflow.

The primary instrument for measuring wind speed is the anemometer, which can be based on either of three basic working principles: mechanical, sonic, and thermal anemometry. Mechanical anemometers consist of a rotating shaft driven by cups or a propeller in the airflow, where the rate of rotation is proportional to the wind speed. Important instrument specifications for these anemometers are their threshold speed and distance constant. Mechanical anemometers are not able to adjust instantaneously to changing wind speeds because of the inertia of the rotating assembly. Instead, a certain amount of airflow is required to pass the anemometer before it is able to adjust its rate of rotation to a new wind speed. This effect is quantified in terms of a distance constant, which is defined as the length of airflow required to pass in order to adjust the sensor speed to 63 percent of a newly imposed (equilibrium) wind speed (Schubauer and Adams 1954; Camp et al. 1970; Stout and Zobeck 1997). The distance constant can be converted to a dynamic response time by dividing the distance constant by the wind speed the sensor encounters, and is therefore a function of wind speed. The dynamic response of mechanical anemometers is often not identical for increasing versus decreasing speeds, but exhibits a hysteresis effect, where response to increasing speeds is faster than response to decreasing speeds (Kaganov and Yaglom 1976).

The most widely used cup-anemometer is the Gill-type three-cup anemometer, which has a distance constant on the order of 2 m and a threshold around 0.5 m s^{-1}. Because it generates a DC signal internally, there is no need for an external power supply, and it is therefore easy to deploy in the field. Mechanical anemometers, however, are subject to fouling of the bearings of the rotating shaft by dust collection when deployed in field conditions with active sediment transport. The resulting additional friction affects both the rate of rotation as related to wind speed and the dynamic response of the sensor. Other disadvantages are the relatively large size of the cup- or propeller assemblies (on the order of tens of centimeters) and the restrictions this imposes to deployment near the surface or spaced close together with other instruments. Distance constants on the order of 2 m s^{-1} result in a dynamic response time on the order of 0.2 to 0.5 seconds under wind speeds ranging from 4 to 10 m s^{-1}, typically encountered in the field. This means that wind-speed fluctuations can only be resolved up to a frequency limit of around 5 Hz under high wind speeds. Because of the dynamic response time, mechanical anemometers cannot be used to derive instantaneous shear stresses through a Reynolds decomposition. Instead, shear stresses are derived from a time-averaged vertical profile of horizontal wind speed, using the law-of-the-wall. Mechanical anemometers are generally robust and inexpensive and can be left in the field for long-term deployment, collecting wind-speed measurements on large time-scales.

Wind direction is usually measured separately with vanes. The response of wind vanes to changes in wind direction is mainly characterized by two parameters: the damping ratio and the vane delay distance, both of which are independent of wind speed (Finkelstein 1981). The damping ratio is a (inverse) measure of the amount of overshoot in the shifting of a vane to a newly imposed wind direction. High damping ratios indicate that overshoot in the adjustment to a new position is low. The delay distance is the distance of airflow past a vane required to shift its position 50 percent towards the final newly imposed wind direction from its previous offset. The delay distance can be converted to a delay time by dividing by the encountered wind speed. In addition to the above response characteristics, other important specifications are the starting threshold, the wind speed above which the vane responds to changing wind directions and the accuracy of direction measurement in degrees. Finkelstein (1981) provides extensive test results for 11 wind vane designs from 7 different manufacturers, many of which are used in contemporary research. Damping ratios generally range from 0.2 to 0.5, while delay distances vary between 0.6 and 2 meters. In the case of a delay distance of 1.0 meter, for example, delay times range between 0.25 and 0.1 seconds under wind speeds of 4 to 10 m s^{-1}, respectively. This indicates that wind-direction fluctuations can be measured up to a frequency limit of around 5 to 10 Hz under high wind conditions.

Sonic anemometers work on the principle that the speed of sound through air is dependent on the wind speed component parallel with the direction of travel of the sound waves. The basic element of a sonic anemometer, therefore, consists of an ultrasonic sound emitter and receiver pair at opposite ends of a sampling volume to measure the speed of sound. Path lengths between emitter and receiver are usually on the order of 0.1 m. Three pairs of emitters and receivers are combined in a 3D mounting framework, measuring the speed of sound along three different axes, to resolve the complete wind vector. Two basic arrangements of emitter-receiver pairs are in use. First, the pairs can be mounted independently, measuring along individual path lengths, often in a mutually orthogonal framework. Second, the mounting of pairs can be arranged so that they measure along paths crossing through a single sampling volume. The advantage of the latter design is that the 3D wind vector is obtained from a single, relatively small sampling volume. Sonic anemometers are able to measure in the 20 to 40 Hz frequency range, with a wind-speed resolution on the order of 0.01 m s^{-1}, and a wind-direction resolution of less than one degree. Their precision and measuring frequency make these anemometers more suitable for detailed turbulence measurements and determining shear stresses from Reynolds decomposition. Furthermore, the sensor does not contain any moving parts and is, therefore, not prone to fouling by airborne dust or sediment. Sonic anemometers are, however, considerably more expensive and cannot be deployed close to the surface because of the size of the mounting framework.

Thermal anemometry deploys a thin, electrically heated, metal wire or film that is cooled by passing airflow. Hot-film or hot-wire probes are usually kept at a certain temperature (around 250 °C) and the amount of power necessary to keep the probe at constant temperature is consequently a measure of the speed of the airflow past the sensor. Hot-wire probes usually consist of a tungsten or platinum wire with a diameter on the order of 5 (m, while hot-film probes are composed of a thin metal film deposited on a core of quartz fiber. In both cases, the sensitive element is usually only 1 to 2 mm long. An overview of various types of hot-wire and hot-film probes used in aeolian research is provided by Butterfield (1999a). Because of their small size, these probes can be deployed on very small scales. Furthermore, thermal anemometry allows for measuring wind speed at high frequencies, ranging from 50 Hz to 100 kHz. Pairs of hot-films or wires can be mounted in mutually orthogonal directions (hence termed "cross-probes") to measure wind speeds in two dimensions. These cross-probes allow for the direct determination of shear stresses according to a Reynolds decomposition.

Regular hot-wire or hot-film probes are, however, very fragile and are easily broken by the impact of saltating grains upon the sensitive element; they cannot be deployed inside an active saltation layer. For this purpose, metal-clad or armored hot-film probes have been developed, where the sensitive hot-film is wrapped with a protective coating of nickel or stainless steel. The resulting probe has a greatly increased diameter, on the order of 0.5 mm, and due to the considerable thermal inertia, the frequency response of these sensors is greatly suppressed, down to a frequency on the order of 10 to 15 Hz.

Thermal anemometry is considerably more expensive than sonic anemometry; it requires additional signal-processing equipment and necessitates a significant power supply. Since the probes are very fragile with respect to handling, thermal anemometry is rarely used in fieldwork and is mostly limited to research in wind tunnels. Research with hot-film probes also imposes some serious calibration challenges. The response of the probes to wind speed is not linear and, therefore, requires careful calibration in a wind tunnel. Hot-film probes also often exhibit significant calibration-drift, so that calibrations must be repeated before and after deployment. When thermal anemometry is deployed successfully in atmospheric environments, however, it generates a wealth of information on small time-scales otherwise inaccessible with cup-anemometry. Figure 19-6 presents two time-series of wind speed obtained simultaneously from a cup-anemometer (top) measuring at 5 Hz, and a co-located armored hot-film probe (bottom) measuring at 20 Hz. The graphs show how the cup-anemometer measurements suffer from lags and suppressed fluctuations due to a slow dynamic response, while the hot-film probe is able to capture gustiness in greater detail as well as the more extreme fluctuations in wind speed over time. Clearly, such information is required for relating wind to sediment transport on small spatiotemporal scales (see Figure 19-7 below).

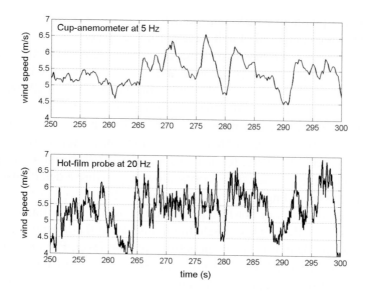

Figure 19-6. Synchronous time-series of wind speed obtained from a cup-anemometer measuring at 5 Hz (top), and a co-located hot-film probe measuring at 20 Hz (bottom). Data collected at Windy Point, California, August 2001. Source: authors.

5.2. Sand Transport Measurements

The desire to assess aeolian sand transport rates in the field has yielded a great variety of instruments. These instruments can be classified according to three working principles: sand capturing devices (traps), optical or remote-sensing devices, and impact responders. Sand traps are traditionally used for measuring sand transport, and these include both manually operated traps and capturing devices with electronic measuring systems. Some examples of manually operated traps are the sand box (Owens 1927; Bagnold 1941), the Leatherman trap (Leatherman 1978), the BSNE sampler (Fryrear 1986), vertical saucer tray stacks (Arens and Van der Lee 1995), the wedge-shaped trap (Nickling and McKenna Neuman 1997), a variety of vertical array traps (Rasmussen and Mikkelsen 1998), and the horizontal water trap (Wang and Kraus 1999). Some of these traps contain vertical arrays of compartments that enable an investigation of the vertical distribution of sand transport. Although manual traps are relatively cheap and easy to produce, their utility is hampered by three major limitations. First and foremost is the lack of information concerning their trapping efficiency. In many cases, traps are manufactured by individual researchers according to various—and subtly differing—designs, and they are often deployed on a per-experiment basis. Although relative transport rates can be determined within a single experiment, the possibility for cross-comparison with other experiments using different traps under different conditions is severely limited. Many traps pose a considerable obstruction to the wind flow and some are not able to adjust to changing wind directions. Aerodynamic designs, porous screens, and wind vanes attached to vertical traps partly alleviate these problems, as in the case of the Leatherman trap and the wedge-shaped trap. Recently, more effort has been directed toward determining trapping efficiencies of several trap designs in wind tunnels (Nickling and McKenna Neuman 1997; Rasmussen and Mikkelsen 1998; Goossens et al. 2000). The second limitation of sand traps is that many (especially vertical traps) do not function properly under wet conditions because trap openings are clogged with moist sand, rendering the trap inoperable. The third limitation is posed by the considerable manual labor involved with operating the traps and the need to capture sizeable amounts of sand for proper weighing analysis. As a result, manual traps must be operated on time-scales of minutes to hours and measurements of spatiotemporal transport variability on smaller scales is severely restricted.

Over the past decades, a number of electronic devices have been developed that address some of the above limitations. The use of load-cells for weighing trapped sand on a continuous basis was initiated by Fryberger et al. (1979) and Lee (1987). Load-cells are able to weigh sand with a resolution of 0.1 to 1 grams at frequencies on the order of 5 Hz. Both Jackson (1996) and Bauer and Namikas (1998) refined the use of load-cells by installing tipping-bucket assemblies that periodically empty the weighing container connected to the load-cell to avoid overloading the system, so that the device can be operated over prolonged periods of time. Namikas (2002) also used load-cells in combination with a horizontally compartmentalized trap and a vertical array trap for high-resolution transport measurements on both temporal and spatial scales. While these devices overcome the limitation of manual traps with regard to the scale of the measurements and ease of operation, they are still subject to issues of unknown trapping efficiency, moist sand conditions, and changing wind directions. Worth mentioning in the context of load-cells is the development of stress plates for directly measuring shear stresses acting on the bed surface (Wyatt and Nickling 1997; Gillies et al. 2000). These devices consist of one or two plates installed flush with the bed that are connected to load-cells that measure the amount of horizontal strain exerted on their surfaces. Such independent assessments of surface shear stress provide complementary data that can be compared with wind velocity profiles or Reynolds stresses.

More recently, instrumentation has been developed that does not require the capture of sand with traps. Noninterfering optical or remote-sensing techniques have been applied to the measurement of sand transport rates. Butterfield (1999a, b) pioneered the use of laser in an optical mass flux sensor that is capable of determining transport rates on a scale of milliseconds with an accuracy close to the mass of individual sand grains. The system consists of a sheet of laser light that is transmitted between two prism posts over a height of 2 cm above the bed, perpendicular to the sand transport direction, onto a detector. This detector subsequently measures the degree of interference or obstruction of the laser sheet caused by the movement of sand grains through it. Because the system is nonintrusive, there are no issues with trapping efficiency, although it must be carefully calibrated with the appropriate sediment. Most optical or remote-sensing techniques, however, have been employed to investigate details of saltation dynamics rather than to obtain direct transport rates. The use of film and still-photography for determining saltation trajectories and grain impact and ejection processes has yielded a great wealth of insight into these small-scale dynamics (Willetts and Rice 1986; Rice et al. 1995; Foucaut and Stanislas 1997; Zou et al. 2001). The velocity of grains in moving sand clouds has also been measured with particle dynamic analysis or phase Doppler "anemometry" (Dong et al. 2002) at frequencies of over 40 Hz. Although the above nonintrusive techniques are able to provide high-resolution measurements, the equipment is relatively expensive, fragile, and elaborate so that their use has been limited to wind tunnel studies.

The third principal technique for measuring sand transport rates is based on devices that respond with electronic pulses to the impact of saltating grains on a small sensitive element, hence called impact responders. In the case of the Saltiphone (Spaan and Van den Abele 1991) the sensitive element consists of a microphone that is able to register discreet impacts of grains larger than 50 çm at frequencies of up to 1 kHz. The physical design of the instrument includes a horizontal cylinder positioned in front of the microphone to guide saltating grains and a wind vane to rotate the microphone. This design introduces some issues, with clogging of moist sand, directional response time to changing wind directions and fouling of bearings. Piezoelectric crystals are used as sensitive elements in two other types of impact responders, the Sensit and the Safire. The Sensit was developed by Stockton and Gillette (1990) and consists of a piezoelectric sensing ring mounted around a vertical post that can be positioned in the saltation layer. This omnidirectional sensor can detect grain impacts at frequencies of up to 55 kHz and registers grains larger than roughly 150 çm. The exposure of the piezoelectric crystal to abrasion by saltating sand may lead to deterioration of the signal, however, and the considerable size of the crystal can lead to internal reverberations of grain impacts. As a result, the Sensit is at present primarily used to detect the onset of saltation. The Safire circumvents these problems by using a damped metal ring around a vertical post from which impact vibrations are transmitted by leads to an internally protected, smaller piezoelectric element. The Safire can measure grain impacts with frequencies of up to 12.5 kHz, filters the signal for vibrations stemming from rain drops or jolts, and registers grains larger than approximately 200 çm (Baas, in press). These types of sensors are fully weatherproof, omnidirectional and easy to deploy. Impact responders suffer from one fundamental limitation, however, because the detection of grain impacts is based on the momentum, rather than the mass, of the saltating grains. Their detection limit is, therefore, a function of both grain-mass and velocity, and they measure discreet grain impacts above this threshold irrespective of grain-size or flight velocity. A translation of such measurements to traditional transport rates requires either co-located sand traps (with related problems) (Gillette et al. 1997) or specific assumptions concerning vertical distributions of grain-size and flight velocity in the saltation layer together with careful sensitivity determinations (Baas, in press).

Despite such calibration issues, the deployment of impact responders and co-located hot-film probes allows for detailed investigations of sediment transport response to changes in wind speed on an unprecedented small scale. Figure 19-7 presents two synchronous time-series from a hot-film probe (top) and a co-located Safire (bottom), both measuring at 20 Hz at 4 cm above a sand surface. The spatial scale of these measurements is on the order of a few square centimeters in frontal area to the flow. The figure shows how individual transport events can be related to high-frequency fluctuations in wind speed to a level not attainable with traditional sand traps and cup-anemometers. Such instrumentation has been crucial for the understanding of the formation and behavior of aeolian streamers (Baas 2003) and is bound to further our fundamental understanding of aeolian transport processes.

6. PERSPECTIVES

It should be apparent, at this point, that instrumentation systems represent one methodological future for geomorphologists. Many themes that had been amenable to examination only in the restricted environment of the laboratory may now be explored in the prototype. Many of the complexities that cannot be reproduced within the scaling limits of the lab may now be explored directly. Three-dimensional processes with large length scales may only be approachable through field experiments. Nevertheless, the scale limitations of instrumented, process-oriented, field studies still imply that there remains

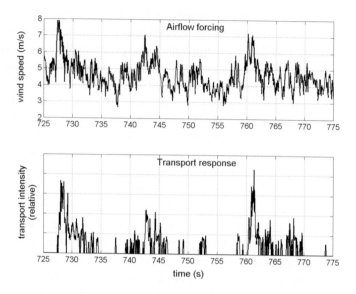

Figure 19-7. Synchronous time-series of wind speed obtained from a hot-film probe (top) and a co-located Safire (bottom), both measuring at 20 Hz at 4 cm above a sand surface. Data from the Safire are in terms of relative transport intensity. Data collected at Windy Point, California, August 2001. Source: authors.

a central need for complementary approaches. In the examples presented here, we commonly described spatial scales of the order of 100 m to less than 10^{-2} m. Time scales range from periods of months to much less than a second. In a consideration of the range of scales included in modern geomorphological analysis, this perspective is analogous to approaching biology at the cellular level. Certainly many important processes or changes in process regimes, and morphological responses, occur at scales much larger than these—a whole-body analogy. And, to risk dog-earring the analogy, sediment transport is the circulation system whose pulse we strive to record.

In this paper, we have paid little attention to two important issues associated with the use of instrumentation systems. First, we have not described considerations for the positioning of single instruments or for instrument sets. In the field, these decisions must be driven by the characteristics of the site, and the rationale motivating the deployment. This task remains part of the art of successful science. Similarly, we have said little about designing temporal sampling protocols, i.e., how fast, how long, and how often? Again, optimal sampling considerations should be driven by the nature of the system to be monitored and the information desired. There are also technical or physical limitations that will influence these decisions. For example, the capacity of digital storage media will limit the total number of samples that can be obtained. The technology in the sensors themselves, or in the data acquisitions systems, will also place limits on maximum sampling speeds. These issues, especially tradeoffs between sampling speed and length of record, also must be confronted by the scientist in the field, although the literature can offer strong guidance.

For most geomorphologists, landform change is a manifestation of net sediment transport. Many of the reductionist approaches to geomorphology have converged on sediment transport as fundamental to developing robust conceptual, quantitative, or physical models. The integration of the reductionist pieces, and, especially, the extrapolation of our understandings to landscape scales, remains a daunting prospect at any level other than the quite general. Our improving ability to move instruments and data acquisition systems into the field, however, provides the data necessary to continue the quest. The data allow us to "see" pulses that are otherwise invisible. The data drive our models, they confuse our diagnoses (Schumm 1991), and tease our imaginations. They provide the solutions and the problems that will motivate the next generations of geomorphologists. The fluids pulse continuously, and we are learning how to listen.

REFERENCES

Aagard, T. and Masselink, G. (1999). The Surf Zone. In Short, A.D. (Ed.) Handbook of Beach and Shoreface Morphodynamics, 72-118. Chichester: Wiley and Sons.

Arens, S.M. and Van der Lee, G.E.M. (1995). Saltation Sand Traps for the Measurement of Aeolian Transport into the Foredunes, Soil Technology 8: 61-74.

Aubrey, D.G. and Trowbridge, J.H. (1985). Kinematic and Dynamic Estimates from Electromagnetic Current Meter Data, Journal of Geophysical Research 90: 9137-46.

Baas, A.C.W. (2003). The Formation and Behavior of Aeolian Streamers. Los Angeles, University of Southern California, Department of Geography, Ph.D. dissertation.

Baas, A.C.W. (in press). Evaluation of Saltation Flux Impact Responders (Safires) for Measuring Instantaneous Aeolian Sand Transport Intensity, Geomorphology.

Bagnold, R.A. (1936). The Movement of Desert Sand, Proceedings, Royal Society of London, Series A 157: 594-620.

Bagnold, R.A. (1938). The Measurement of Sand Storms, Proceedings, Royal Society of London, Series A 167: 282-91.

Bagnold, R.A. (1941). The Physics of Blown Sand and Desert Dunes. London: Chapman and Hall.

Bauer, B.O. and Namikas, S.L. (1998). Design and Field Test of a Continuously Weighing, Tipping-Bucket Assembly for Aeolian Sand Traps, Earth Surface Processes and Landforms 23: 1171-83.

Bauer, B.O., Lorang, M.S., and Sherman, D.J. (2002). Estimating Boat-Wake-Induced Levee Erosion Using Suspended Sediment Measurements, Journal of Waterway, Port, Coastal and Ocean Engineering 128: 152-62.

Beach, R.A., and Sternberg, R.W. (1988). Suspended Sediment in the Surf Zone: Response to Cross-Shore Infragravity Motion, Marine Geology 80: 61-79.

Best, J.L. and Roy, A.G. (1991). Mixing-Layer Distortion at the Confluence of Unequal Depth Channels, Nature 350: 411-13.

Brander, R.W. and Short, A.D. (2000). Morphodynamics of a Large-Scale Rip Current System at Muriwai Beach, New Zealand, Marine Geology 165: 27-39.

Butterfield, G.R. (1999a). Application of Thermal Anemometry and High-Frequency Measurement of Mass Flux to Aeolian Sediment Transport Research, Geomorphology 29: 31-58.

Butterfield, G.R. (1999b). Near-Bed Mass Flux Profiles in Aeolian Sand Transport: High-Resolution Measurements in a Wind Tunnel, Earth Surface Processes and Landforms 24: 393-412.

Callede, J., Kosuth, P., Guyot, J-L., and Guimarães, V.S. (2000). Discharge Determination by Acoustic Doppler Current Profilers (ADCP): A Moving Bottom Error Correction Method and Its Application on the River Amazon at Óbidos, Hydrological Sciences Journal 45: 911-24.

Camp, D.W., Turner, R.E., and Gilchrist, L.P. (1970). Response Tests of Cup, Vane, and Propeller Wind Sensors, Journal of Geophysical Research 75: 5265-70.

Clifford, N.J. and French, J.R. (1993). Monitoring and Modeling Turbulent Flow: Historical and Contemporary Perspectives. In Clifford, N.J., French, J.R., and Hardisty, J. (Eds.) Turbulence: Perspectives on Flow and Sediment Transport, 1-34. Chichester, John Wiley.

Cornish, V. (1934). Ocean Waves and Kindred Phenomena. Cambridge: Cambridge University Press.

Davis, W. M. (1909). Geographical Essays. Johnson, D.W. (Ed.), Boston, Ginn and Co.

Dingler, J.R., Boylls, J.C., and Lowe, R.L. (1977). A High-Frequency Sonar for Profiling Small-Scale Subaqueous Bedforms, Marine Geology 24: 279-88.

Dong, Z., Wang, H., Liu, X., and Zhao, A. (2002). Velocity Profile of a Sand Cloud Blowing Over a Gravel Surface, Geomorphology 45: 277-89.

Ellis, J.T., Sherman, D.J., Bauer, B.O., and Hart, J. (2002). Assessing the Impact of an Organic Restoration Structure on Boat wWake Energy, Journal of Coastal Research SI36: 256-65.

Fedderson, F. and Guza, R.T. (2003). Observations of Nearshore Circulation: Alongshore Uniformity, Journal of Geophysical Research 108: 1-10.

Finkelstein, P.L. (1981). Measuring the Dynamic Performance of Wind Vanes, Journal of Applied Meteorology 20: 588-94.

Foucaut, J.-M. and Stanislas, M. (1997). Experimental Study of Saltating Particle Trajectories, Experiments in Fluids 22: 321-26.

Frothingham, K.M. and Rhoads, B.L. (2003). Three-Dimensional Flow Structure and Channel Change in an Asymmetrical Compound Meander Loop, Embarras River, Illinois, Earth Surface Processes and Landforms 28: 625-44.

Fryberger, S.G., Ahlbrandt, T.A., and Andrews, S. (1979). Origin, Sedimentary Features, and Significance of Low-Angle Eolian Sand Sheet Deposits, Great Sand Dunes National Monument and Vicinity, Colorado, Journal of Sedimentary Petrology 49: 733-46.

Fryrear, D.W. (1986). A Field Dust Sampler, Journal of Soil and Water Conservation 41: 117-20.

Gallagher, E.L., Thornton, E.B., and Stanton, T.P. (2003). Sand Bed Roughness in the Nearshore, Journal of Geophysical Research 108: 1-8.

Gallagher, E.L., Elgar, S., and Thornton, E.B. (1998). Megaripple Migration in a Natural Surf Zone, Nature 394: 165-68.

Garcia-Gorritz, E., Candela, J., and Font, J. (2003). Near-Inertial and Tidal Currents Detected with a Vessel-Mounted Acoustic Doppler Current Profiler in the Western Mediterranean Sea, Journal of Geophysical Research 108: 1- 21.

Gilbert, G.K. (1886). The Inculcation of the Scientific Method by Example, with an Illustration Drawn from the Quaternary Geology of Utah, American Journal of Science 3: 1-13.

Gilbert, G.K. (1914). The Transportation of Débris by Running Water. U.S. Geological Survey Professional Paper 86, Washington, DC: U.S. GPO.

Gillette, D.A., Fryrear, D.W., Xiao, J.B., Stockton, P.H., Ono, D., Helm, P.J., Gill, T.E., and Ley, T. (1997). Large-Scale Variability of Wind Erosion Mass Flux Rates at Owens Lake; 1. Vertical Profiles of Horizontal Mass Fluxes of Wind-Eroded Particles with Diameter Greater than 50 mm, Journal of Geophysical Research 102: 25977-87.

Gillies, J.A., Lancaster, N., Nickling, W.G., and Crawley, D.M. (2000). Field Determination of Drag Forces and Shear Stress Partitioning Effects for a Desert Shrub (Sarcobatus Vermiculatus, Greasewood), Journal of Geophysical Research 105: 24871-80.

Goossens, D., Offer, Z., and London, G. (2000). Wind Tunnel and Field Calibration of Five Aeolian Sand Traps, Geomorphology 35: 233-52.

Green, M.O., Vincent, C.E., McCave, I.N., Dickson, R.R., Rees, J.M., and Pearson, N.D. (1995). Storm Sediment Transport: Observations from the British North Sea Shelf, Continental Shelf Research 15: 889-912.

Greenwood, B., Dingler, J.R., Sherman, D.J., Anima, R.J., and Bauer, B.O. (1985). Monitoring Bedforms Under Waves using High Resolution Remote Tracking Sonar (HRRTS), Proceedings of the Canadian Coastal Conference, Ottawa, Canada, 143-58.

Greenwood, B. and Sherman, D.J. (1984). Waves, Currents, Sediment Flux and Morphological Response in a Barred Nearshore System, Marine Geology. 60: 31-61.

Hjulström, F. (1935). Studies of the Morphological Activities of Rivers Illustrated by the River Fyris, Uppsala University Geological Institute Bulletin 25: 221-527.

Holland, K.T., Raubenheimer, B., Guza, R.T., and Holman, R.A. (1995). Runup Kinematics on a Natural Beach, Journal of Geophysical Research 100: 4985-93.

Holman, R.A. and Bowen, A.J. (1979). Edge Waves on Complex Beach Profiles, Journal of Geophysical Research 84: 6339-46.

Horn, D.P., Baldock, T., Baird, A.J., and Mason, T. (1998). Field Measurements of Swash Induced Pressure Gradients within a Sandy Beach, Proceedings, 26th International Conference on Coastal Engineering (ASCE) 2812-25.

Hubbard, E.F., Scwartz, G.E., Thibodeaux, K.G., and Turcios, L.M. (2001). Price Current-Meter Standard Rating Development by the U.S. Geological Survey, Journal of Hydraulic Engineering 127: 250-57.

Jackson, D.W.T. (1996). A New, Instantaneous Aeolian Sand Trap Design for Field Use, Sedimentology 43: 791-96.

Kaganov, E.I. and Yaglom, A.M. (1976). Errors in Wind Speed Measurements by Rotation Anemometers, Boundary Layer Meteorology 10: 229-44.

Kineke, G. and Sternberg, R.W. (1992). Measurements of High Concentration Suspended Sediment Using the Optical Backscatterance Sensor, Marine Geology 108: 253-58.

Komar, P.D. (1998). Beach Processes and Sedimentation. 2nd ed. Upper Saddle River, NJ: Prentice-Hall.

Kraus, G. and Ohm, K. (1984). A Method to Measure Suspended Load Transport in Estuaries, Estuarine, Coastal, and Shelf Science 19: 611-18.

Kringel, K., Jumars, P.A., and Holliday, D.V. (2003). A Shallow Scattering Layer: High-Resolution Acoustic Analysis of Nocturnal Vertical Migration from the Seabed, Limnology and Oceanography 48: 1223-34.

Lane, S.N., Richards, K.S., and Warburton, J. (1993). Comparison Between High Frequency Velocity Records Obtained with Spherical and Discoidal Electromagnetic Current Meters. In Clifford, N.J., French, J.R., and Hardisty, J. (Eds.) Turbulence: Perspectives on Flow and Sediment Transport, 121-63. Chichester, John Wiley.

Law, D.J., Bale, A.J., and Jones, S.E. (1997). Adaption of Focused Beam Reflectance Measurement to in Situ Particle Sizing in Estuaries and Coastal Waters, Marine Geology 140: 47-59.

Lawler, D.M. (1991). A New Technique for the Automatic Monitoring of Erosion and Deposition rates, Water Resources Research 27: 2125-28.

Lawler, D.M., West, J.R., Couperthwaite, J.S., and Mitchell, S.B. (2001). Application of a Novel Automatic Erosion and Deposition Monitoring System at a Channel Bank Site on the Tidal River, Trent, UK, Estuarine, Coastal and Shelf Science 53: 237-47.

Leatherman, S.P. (1978). A New Aeolian Sand Trap Design, Sedimentology 25: 303-06.

Lee, J.A. (1987). A Field Experiment on the Role of Small Scale Wind Gustiness in Aeolian sand Transport, Earth Surface Processes and Landforms 12: 331-35.

Lewis, J. (1996). Turbidity-Controlled Suspended Sediment Sampling for Runoff-Event Estimation, Water Resources Research 32: 2299-310.

Lopez, M. and Garcia, J. (2003). Moored Observations in the Northern Gulf of California: A Strong Bottom Current, Journal of Geophysical Research 108: 1-18.

McDermott, J.P. (2001). Sediment-Level Oscillations in the Swash Zone. Los Angeles, University of Southern California, Department of Geography. Unpublished MS thesis.

Meirovich, L., Laronne, J.B., and Reid, I. (1998). The Variation of Water-Surface Slope and its Significance for Bedload Transport During Floods in Gravel-Bed Streams,. Journal of Hydraulic Research 36: 147-57.

Munk, W.H., Miller, G.R., Snodgrass, F.E., and Barber, N.F. (1963). Directional Recording of Swell from Distant Storms, Philosophical Transactions of the Royal Society A 255: 505-84.

Namikas, S.L. (2002). Field Evaluation of Two Traps for High-Resolution Aeolian Transport Measurements, Journal of Coastal Research 18: 136-48.

Nickling, W.G. and McKenna Neuman, C. (1997). Wind Tunnel Evaluation of a Wedge-Shaped Aeolian Sediment Trap, Geomorphology 18: 333-45.

Nielsen, P. and Cowell, P.J. (1981). Calibration and Data Correction Procedures for Flow Meters and Pressure Transducers Commonly Used by the Coastal Studies Unit. Coastal Studies Unit, Department of Geography, University of Sydney, Australia, Technical Report 81/1.

Nordstrom, K.F., Jackson, N.L., Allen, J.R., and Sherman, D.J. (2003). Longshore Sediment Transport Rates on a Microtidal Estuarine Beach, Journal of Waterway, Port, Coastal and Ocean Engineering 129: 1-4.

Osborne, P.D. and Greenwood, B. (1993). Sediment Suspension Under Waves and Currents: Time Scales and Vertical Structure, Sedimentology 40: 599-622.

Owens, J.S. (1927). The Movement of Sand by Wind, Engineer 143: 377.

Rasmussen, K.R. and Mikkelsen, H.E. (1998). On the Efficiency of Vertical Array Aeolian Field Traps, Sedimentology 45: 789-800.

Rice, M.A., Willetts, B.B., and McEwan, I.K. (1995). An Experimental Study of Multiple Grain-Size Ejecta Produced by Collisions of Saltating Grains with a Flat Bed, Sedimentology 42: 695-706.

Ridd, P.V. and Larcombe, P. (1994). Biofouling Control for Optical Backscatter Suspended Sediment Sensors, Marine Geology 116: 255-58.

Ridd, P.V., Day, G., Thomas, S., Harradance, J., Fox, D., Bunt, J., Renagi, O., and Jago, C. (2001). Measurement of Sediment Deposition Rates Using an Optical Backscatter Sensor, Estuarine, Coastal and Shelf Science 52: 155-63.

Ruessink, B.G. (2000). An Empirical Energetics-Based Formulation for the Cross-Shore Suspended Sediment Transport by Bound Infragravity Waves, Journal of Coastal Research 16: 482-93.

Sauer, C.O. (1930). Basin and Range Forms in the Chiricahua Area, University of California Publications in Geography 3: 339-414.

Schat, J. (1997). Multifrequency Acoustic Measurement of Concentration and Grain Size of Suspended Sand in Water, Journal of the Acoustic Society of America 101: 209-17.

Schubauer, G.B. and Adams, G.H. (1954). Lag of Anemometers. Washington DC, NBS (National Bureau of Standards) Report 3245.

Schumm, S.A. (1991). To Interpret the Earth (Ten Ways to be Wrong). Cambridge: Cambridge University Press.

Sherman, D.J., Short, A.D., and Takeda, I (1993). Sediment Mixing Depth and Megaripple Migration in Rip Channels, Journal of Coastal Research SI15: 39-48.

Simpson, M.R. and Oltman, R.N. (1993). Discharge Measurement Using an Acoustic Doppler Current Profiler. U.S. Geological Survey Water-Supply Paper 2395.

Spaan, W.P. and Van den Abele, G.D. (1991). Wind Borne Particle Measurements with Acoustic Sensors, Soil Technology 4: 51-63.

Stockton, P.H. and Gillette, D.A. (1990). Field Measurement of the Sheltering Effect of Vegetation on Erodible Land Surfaces, Land Degradation and Rehabilitation 2: 77-85.

Stout, J.E. and Zobeck, T.M. (1997). Intermittent Saltation, Sedimentology 44: 959-70.

Waddel, E. (1976). Swash-Groundwater-Beach Interactions. In Davis, R.A. and Etherington, R.L. (Eds.) Beach and Nearshore Sedimentation, 115-25. Society of Economic Paleontologists and Mineralogists Special Publication 24.

Wang, P. and Kraus, N.C. (1999). Horizontal Water Trap for Measurement of Aeolian Sand Transport, Earth Surface Processes and Landforms 24: 65-70.

Willetts, B.B. and Rice, M.A. (1986). Collision in Aeolian Transport: The Saltation/Creep Link. In Nickling, W.G. (Ed.) Aeolian Geomorphology. London: Allen and Unwin.

Wyatt, V.E. and Nickling, W.G. (1997). Drag and Shear Stress Partitioning in Sparse Desert Creosote Communities, Canadian Journal of Earth Sciences 34: 1486-98.

Zou, X.-Y., Wang, Z.-L., Hao, Q.-Z., Zhang, C.-L., Liu, Y.-Z., and Dong, G.-R. (2001). The Distribution of Velocity and Energy of Saltating Sand Grains in a Wind Tunnel, Geomorphology 36: 155-65.

CHAPTER 20

JULIE A. WINKLER

THE IMPACT OF TECHNOLOGY UPON IN SITU ATMOSPHERIC OBSERVATIONS AND CLIMATE SCIENCE

Abstract   Over the past 100 years geographers have extensively employed in situ observations of the atmosphere in their research endeavors. Technological improvements in instrumentation, communication, and data storage media have played fundamental roles in the expansion of observational networks and in the improved accuracy and precision of measurements. Yet, technological changes, along with technological limitations, such as instrument drift and failure, have introduced unwanted inhomogeneity into the time series of surface and upper-air observations. An additional constraint is that most, if not all, observational networks for the atmosphere were designed for short-range weather prediction rather than for climate monitoring. This essay reviews, for the non-climatologist, the primary surface and upper-air datasets available for climatological research in the U.S. and for global-scale analyses. Particular focus is placed on the known sources of inhomogeneities in these data series. Geographers need to become more aware of the value of in situ surface and upper-air observations and the current limitations and fragility of these networks.

Keywords   in situ atmospheric observations, upper-air observations, surface observations, inhomogeneity, climate monitoring

> All this you surely **will** see, and much more, if you are prepared to see it,—if you **look** for it. Otherwise, regular and universal as this phenomenon is, whether you stand on the hilltop or in the hollow, you will think for threescore years and ten that all the world is, at this season, sere and brown. Objects are concealed from our view, not so much because they are out of the course of our visual ray as because we do not bring our minds and eyes to bear on them; for there is no power to see in the eye itself, any more than in any other jelly. We do not realize how far and widely, or how near and narrowly, we are to look. The greater part of the phenomena of Nature are for this reason concealed from us all our lives. ("Autumnal Tints," *Atlantic Monthly,* Henry David Thoreau, 1862)

Stanley D. Brunn, Susan L. Cutter, and J.W. Harrington, Jr. (Eds.), Geography and Technology,
461-490. © *2004 Kluwer Academic Publishers. Printed in the Netherlands.*

1. INTRODUCTION

The 100 years that have marked the existence of the Association of American Geographers are also a period of remarkable advances in our understanding of the earth's weather. The springboard for this improved understanding has been the development and expansion over the past century of observational networks for describing and monitoring the atmosphere. These systems have, borrowing liberally from Henry David Thoreau, allowed meteorologists to "look for" and to "see" both the "regular and the universal" in regard to the atmosphere. Over the past century, improved observations have brought into focus atmospheric phenomena previously "concealed from view," and they have aided in the development of physical laws to describe atmospheric motion and of numerical models to aid in its prediction. At the same time, observations have revealed atmospheric circulations and phenomena that cannot yet be explained, which in turn has brought about further observational campaigns that have used different sensors or taken observations at different spatial and/or temporal scales. Observations, in other words, have not only assisted us in answering questions but they have also helped us know when and where to look for answers to the many challenging unanswered questions regarding the earth's atmosphere. Not surprisingly, advances in atmospheric observational systems have been intimately tied with advances in technology, particularly with the development of new sensors and to improvements in communications services.

Observations of the atmosphere are also essential for climate monitoring and research. An inescapable obstacle, however, is that most atmospheric sensors and sensor networks were designed for short-range weather prediction and real-time monitoring rather than for climate applications. One reason for this is that many of these networks were initially established at a time when climate was believed to be relatively invariant (Karl et al. 1993). Consequently, there was little or no foresight as to how these observations could eventually be used to study climate variability and trends. The continuing emphasis of the atmospheric networks on short-range weather prediction is in large part a reflection of the primary missions and customer bases of the agencies responsible for maintaining the networks. The upshot is that atmospheric observations frequently are used to study phenomena and address research questions, such as those related to the nature and attribution of climate change and variability, that may not be compatible with the purposes for which the data were originally collected, processed, and summarized (Guttman and Baker 1996).

Users of atmospheric observations must also bear in mind that technological advances function as both "friend" and "foe" when it comes to the quality of observational networks for climatological analyses. Networks designed for short-range weather forecasting strive for accuracy and precision of measurements, and over time new and/or improved sensors have frequently been introduced in order to increase accuracy and precision. On the other hand, a desired characteristic of climatological networks is spatial and temporal consistency of observations. Technological improvements to increase accuracy or precision often introduce unwanted changes in consistency and increase the complexities of using these observations for climate research. Indeed, these inhomogeneities[1] in the time series of atmospheric observations have opened climatological analyses to criticism (Karl et al. 1993), both warranted and unwarranted.

This essay provides an overview of the primary observational datasets available for climate monitoring and research. The essay not only highlights the importance of technology in making these observations possible, but also describes the unwanted inhomogeneities introduced into the time series by technology innovation and by technological limitations and failure. A motivation for providing this overview is the importance of climate observations in many aspects of geographic research, not only for research conducted by geographer-climatologists but also by many others within the discipline. This usage is likely to increase as geographers become increasingly involved in applied environmental research and in global change analyses. Information about available climatological datasets is scattered among a variety of sources, many of which are not very accessible for nonclimatologists. Thus, this essay attempts to fill a perceived gap in the literature for a relatively comprehensive overview of climatological time series directed to the nonclimatologist. The networks summarized below are surface and balloon-borne upper-air observations of the troposphere and lower stratosphere. These in situ observational types were selected based on their past, present, and potential future applicability for geographic research. In situ observations have a considerably longer record compared to remotely sensed observations, and geographers have extensively used in situ observations, particularly surface observations, since the establishment of the AAG. Also, as pointed out by Trenberth et al. (2002), in situ surface and upper-air measurements are invaluable resources that currently operate under serious difficulties. Geographers need to become more cognizant of the value of these observations, their limitations, and the current stresses on the networks. The discussion below focuses on observational datasets available specifically for the U.S. and on datasets that are global in scale. The essay concludes with a summary of the current conversations within the climate community regarding the improvement, organization, and development of networks more suited for climate studies.

2. SURFACE OBSERVATIONS

2.1. National (U.S.) Networks and Databases

Two primary observational networks, the Cooperative Observer Program (COOP) and the Automated Surface Observing System (ASOS), measure surface (or more accurately "near surface") weather and climate phenomena in the U.S. Of the two, the mission of the COOP network is most directed toward climatological research. Established in 1890, the COOP network is the oldest and largest official monitoring network in the U.S. The early mission was to provide weather data for agricultural purposes (Horvitz 2002). This mission has been updated and expanded to include the provision of observational meteorological data for defining and monitoring the climate of the U.S. (Horvitz 2003). Currently, more than 11,000 volunteers use instruments provided by the National Weather Service (NWS) to record daily measurements of maximum and minimum temperature, liquid equivalent of precipitation, and snowfall. Many observers also provide additional hydrometeorological information such as snow depth and evaporation, along with other special phenomena including the number of days with thunder or hail (Horvitz 2002). Observations from the COOP network currently are not available in real time. Instead, each month the volunteer observers send their observations to the National Climatic Data Center (NCDC), where the daily observations are quality controlled and made available to the public.

In contrast to the COOP network, the mission of ASOS is to provide frequent real-time data for weather forecasting and aviation needs. ASOS observations include temperature, relative humidity, pressure, wind speed and direction, rainfall, visibility, cloud ceiling, and type of precipitation (i.e., rain, snow, or freezing rain) (FAA 2002). Measurements are taken automatically minute-by-minute, although hourly or daily values are most frequently used in meteorological applications and climate research. A subset of the ASOS stations is also part of an international synoptic network reporting and transmitting observations at least once every six hours. Currently, the ASOS network is comprised of 569 sites sponsored by the Federal Aviation Agency (FAA) and 313 sites sponsored by the National Weather Service (NWS) (FAA 2002). The majority of these sites are located at airports. The ASOS network was established as part of the recent modernization effort of the NWS, and the first ASOS sites were installed in the early 1990s with the majority of stations commissioned after 1996. ASOS replaced the conventional, manual observations taken at what is usually referred to as "first-order weather stations" (primarily weather service offices and larger airports). Records for many of the first-order stations extend back to the 1930s. Although the ASOS network has a real-time forecasting mission, its importance, along with that of the earlier first-order stations, to climatological research should not be undervalued, as these observations are the primary source of measurements of important climatic variables including pressure and wind. As with the COOP network, NCDC is charged with archiving the ASOS observations.

Both the ASOS and COOP networks suffer from lack of consistency. Beginning with the ASOS network, differences in instrumentation and station location between the automated ASOS and the earlier conventional observations have introduced inhomogeneities at first-order stations. Perhaps surprisingly, changes in station location, even changes on the order of only a few hundred meters, may have introduced considerably more bias compared to instrument changes. A recent study for a one-year overlap period of conventional and automated temperature observations at 10 first-order stations found that the switch in hygrothermometer type with ASOS introduced a modest negative bias in observed temperature on the order of a few tenths of a degree Celsius but siting differences led to bias of more than a degree (Guttman and Baker 1996). Another important change that occurred with the installation of the ASOS stations was the shift from a Universal Rain Gauge to a Heated Tipping Bucket Rain Gauge, which appears to undermeasure liquid precipitation (Butler 1998). A positive aspect of the automation, however, is that it permitted a more than doubling of the number of full-time stations reporting hourly weather observations across the U.S. (NWS OM 1999). It should also be noted that even before the switch to automated observations, significant inhomogeneities existed in the time series of the first-order stations. For example, beginning in the 1940s some, but not all, first-order stations equipped standard rain gauges with wind shields in an attempt to reduce precipitation undercatch (Groisman and Legates 1994). Also, thermometers and sling psychrometers used to measure temperature and humidity were replaced in the early 1960s with lithium chloride hygrothermometers, which in turn were replaced in the mid-1980s with what is known as the HO-83 hygrometer composed of a bead thermometer and a chilled mirror (Gaffen and Ross 1999). The switch to the HO-83 instruments appears to have introduced a bias of around 0.5°C into the maximum temperature series at first-order stations (Karl et al. 1995). Another important source of inhomogeneity occurred from the 1930s into the 1950s, when most first-order stations were relocated from downtown locations to suburban airports.

Instrument changes have also influenced the time series at COOP stations. One change that has received considerable attention is the replacement beginning in the mid-1980s of liquid-in-glass maximum and minimum thermometers in wooden Cotton Region Shelters (CRSs) with thermistor-based Maximum-Minimum Temperature Systems (MMTS) housed in small plastic shelters (Quayle et al. 1991). Although the instrument change appears to have improved the absolute accuracy of the temperature measurements, biases on the order of +0.3°C for mean daily minimum temperature, -0.4°C for mean daily maximum temperature, and –0.7°C for daily temperature range were introduced into the record (Quayle et al. 1991). These biases are of the same order of magnitude as natural and anthropogenic trends in climate records. Reasons for these biases include earlier erroneous observations caused by column separations of the liquid-in-glass thermometers, greater heating of the CRS during the day compared to the MMTS shelter, and greater radiation loss at night from the slatted bottom of a CRS compared to the MMTS shelter (Quayle et al. 1991). Another potential contribution to the bias is that the MMTS units were often installed closer to an observer's home or outbuildings compared to the CRS in order to reduce the length of cable needed to reach the sensor (Quayle et al. 1991). Yet another change in the COOP network is a general increase in the mean angle between rain gauges and nearby obstacles that appears to have caused significant positive trends in liquid and frozen precipitation (Groisman et al. 1996; Peterson et al. 1998). This bias is thought to have come about because of increased awareness by observers to place their rain gauges in more protected areas in order to reduce undercatch due to strong winds (Peterson et al. 1998). Further changes are forthcoming with an on-going two-phase initiative to modernize the COOP network. The first phase, started in 2001, will replace obsolete equipment and evaluate optimum network size and distribution, and the second phase, slated for 2003, will mark a transition to a more automated network (Horvitz 2002). The target goals for the modernization include an approximately 32 x 32 kilometer (20 x 20 mile) monitoring grid, hourly measurements of temperature and precipitation, electronic data communication, and measurements of additional variables such as soil temperature.

The recognition of the potentially deleterious impact of inhomogeneities on climatological analyses, especially those analyses involving temporal trends, has led to considerable efforts to develop methods for adjusting for these inhomogeneities (see Peterson et al. 1998 for an excellent summary of homogeneity adjustments). Homogeneity adjustments range from subjective judgments to a range of relatively sophisticated objective techniques. A widely used method is the Alexandersson's standard normal homogeneity test which compares the ratios (for precipitation) or differences (for temperature) between the time series at a candidate station and a reference time series (Alexandersson 1986). The convention is to adjust the time series to the most recent homogeneous period of record. Many of the adjustment methods require metadata (i.e., data about data) to identify change points, but newer methods have been developed to detect discontinuities when metadata are not available or when change points were not documented (e.g., Vincent 1998; DeGaetano 1999, Lund and Reeves 2002). Also, most adjustments are best applied to monthly or annual means, but more recently methods have been proposed to homogenize nonclimatic discontinuities in daily data and in the time series of extreme temperature exceedences (e.g., Allen and DeGaetano 2000). A still intractable problem is the development of homogeneity adjustments for areal averages, as adjustments are station specific and should not be spatially scaled to larger areas (Quayle et al. 1991).

Concern about data homogeneity led to the development by NCDC of the U.S. Historical Climatology Network (USHCN) (Karl et al. 1990). This network is comprised of a subset of approximately 1200 COOP stations selected on the basis of long periods of record, a small percentage of missing data, and a modest number of changes in station location, instrumentation, and observing time. Available climate variables from the USHCN are monthly averaged maximum, minimum, and mean temperature, and total monthly precipitation. Temperature data, but not precipitation data, from these "high quality" stations have been adjusted for known and unknown inhomogeneities. The sequential adjustment process includes adjustments for time-of-observation bias, the introduction of the MMTS temperature sensors at about half the stations in the network, station moves, and urban warming (Peterson et al. 1998). Adjustments are generally not applied to precipitation observations for the USHCN stations because of the difficulty of computing appropriate adjustment factors. As explained by Easterling et al. (1996), almost all rain gauges significantly undercatch precipitation by around 5 percent for liquid precipitation and by 50 percent or more for frozen precipitation, which complicates any adjustment procedure. Also, the spatial autocorrelation for precipitation falls off rapidly with distance between stations, making the construction of a homogeneous reference series difficult. Although the USHCN is currently considered the network of greatest quality in terms of climate monitoring in the U.S., the UNHCN time series nevertheless should be used cautiously. For example, Balling and Idso (2002) found that the various adjustments made to the USHCN temperature data produce a significantly more positive and perhaps spurious temporal trend compared to the trends evident in high-quality but unadjusted temperature time series. Easterling et al. (1996) advocate that care should be taken when using adjusted time series for individual stations, and that the best use of the USHCN time series is for regional analyses of climate patterns and trends, since a regional climate signal is introduced into the individual station series through the homogeneity adjustment processes. Proposed upgrades to the USHCN should improve its future applicability for regional-scale climate analyses. These upgrades include the selection of currently active stations to continue the record for closed sites, the addition of "recovered data" from the first half of the 20th century, quality control tests for snow observations, improved inhomogeneity detection techniques, and an evaluation technique that verifies or negates proposed adjustments to a station's time series (Williams et al. 2003).

2.2. Global Networks and Databases

Obviously, many climatological applications and analyses require climate observations for a larger area than only the U.S. International surface observations have been much more difficult to acquire compared to national observations, although during the past few years considerable effort has been expended to compile quality international datasets and to provide these data in a user-friendly form. Two technological advances have been particularly significant for the development of these datasets. One is simply improved communications. This is particularly true for the development of global databases of hourly and daily observations. These observations, which are originally taken to support real-time forecasting, are usually transmitted in an internationally approved code and distributed over what is known as the Global Telecommunication System (GTS) maintained by the World Meteorological Organization (WMO). The second technological advance is improved computing power and high-capacity electronic storage devices necessary for the quality control and archival of these voluminous datasets.

Historically, time series of monthly mean values at the global scale have been more readily available compared to hourly or daily observations. The oldest available global climatic dataset is the *World Weather Records*, which began in 1923 and continues to provide decadal statistics of temperature, precipitation, and pressure (NCDC 2002c). Another database that has been available for some time is the *Monthly Climatic Data for the World* published by NCDC in cooperation with the WMO. This database, as its name suggests, contains monthly mean values of key surface parameters including temperature, pressure, and vapor pressure along with monthly totals of precipitation and sunshine duration. The database includes approximately 1200 stations that the WMO has designated as climatology (or CLIMAT) stations (NCDC 1997). These stations transmit observations using a special code (the CLIMAT code) over the GTS. It should be noted that the observations contained in *World Weather Records* and in the *Monthly Climatic Data for the World* have not been assessed for homogeneity.

Global datasets of mean monthly parameters (typically temperature, pressure, and precipitation) have also been compiled by individual researchers. The best known of these was developed by Phil Jones and his colleagues at the University of East Anglia and first published in 1982 (Jones et al. 1982) with several updates since then (Jones et al. 1986; Jones 1994). This dataset includes global monthly mean temperatures for approximately 3000 stations worldwide that were obtained from *World Weather Records* and *Monthly Climatic Data for the World*. The majority of the stations were subjected to visual inspections for homogeneity, and adjustments were applied to time series with distinct breaks. From this adjusted dataset, Jones and colleagues computed their well-known time series of Northern Hemisphere and global mean temperature. These time series have frequently been used by both the popular press and the research community to illustrate potential global warming. A similar, but larger, dataset is the Global Historical Climatology Network (GHCN) which includes approximately twice as many stations as the Jones dataset (Peterson and Vose 1997; Vose et al. 1998; NCDC 2002c). The philosophy that guided the development of the GHCN is comparable to that of the USHCN. That is, time series of surface temperature and precipitation observations are known to contain significant inhomogeneities, and adjustments can be applied to a candidate time series by comparing it to a reference time series developed from nearby observations. Like the USHCN, data from the GHCN are more appropriate for climate analyses at the regional or larger spatial scales rather than at the local scale. The GHCN, however, does not have as extensive metadata as the USHCN. Metadata are limited to latitude, longitude, elevation, and the current population and vegetation type of the region surrounding the observing station (Peterson and Vose 1997). Users of the GHCN may be surprised to find multiple time series for the same station. This interesting characteristic of the GHCN illustrates the difficulties encountered when compiling datasets of this nature and for which there still are no obvious solutions in spite of technological advances. The multiple series do not represent different observational data for a location, but rather the varied ways that different organizations (either national or international) calculate monthly mean temperature from the original measurements. Thus, time series of mean temperature for the same location, but provided by different agencies, can have temperature differences that are greater than the differences between neighboring stations. This makes elimination of station duplicates difficult, and the dataset developers have opted to assign the same station identifier to the duplicate series but not to merge the series (Vose et al. 1998).

Currently, several datasets are available for global climatological analyses that require surface observations at the hourly or daily temporal scale. The U.S. Air Force Combat Climatology Center (AFCCC) surface observations database serves as the foundation or backbone for most of these datasets. The AFCCC database consists of synoptic reports taken at least every six hours, but most often hourly, and transmitted via GTS (Lott et al. 2001). The observations, which are archived at NCDC, include temperature, dewpoint, sea level pressure, wind, cloud characteristics, and precipitation for close to

10,000 stations worldwide (Lott et al. 2001). In addition to the hourly values, daily mean values of temperature, dewpoint, sea-level pressure, and wind speed, along with maximum and minimum temperature, precipitation amount and several other surface variables are available for a subset of approximately 8000 stations. The daily means are archived as the *Global Summary of the Day* dataset which is also available from NCDC (Lott 2003). Observations for both the AFCCC and *Global Summary of the Day* datasets are generally available from 1982 onward, with less complete data extending to 1973 and for some stations back to 1930 (NCDC 1998). Both datasets are invaluable resources for climate studies, but both must be used cautiously as observational practices vary by country, metadata are often missing or inaccurate, and only gross error checking for random errors has been conducted. Also data for some countries are not available, contributing to uneven spatial coverage.

A recent major undertaking at NCDC was to combine hourly observations from multiple sources into a single database referred to as the *Federal Climate Complex Integrated Surface Hourly Data Base* (Lott et al. 2001; Lott and Baldwin 2002; NCDC 2003c). This compilation involved the establishment of a common archiving format, quality control to ensure that data were actually from the same location, conversion of all observation times to Universal Coordinated Time (UTC), and standardization to the same measurement units (Lott and Baldwin 2002). In addition, checks were conducted for validity, extreme values, and internal consistency, although no checks were made for spatial continuity or for inhomogeneities in the time series of individual stations (Lott and Baldwin 2002). Users of the integrated dataset should be aware that the observations are more complete at major locations, and the completeness of the observations is driven by the meteorological practices of the individual countries (Tom Ross, personal communication). Anticipated future improvements for this data base include adding station history and metadata, including information on instrumentation (Lott and Baldwin 2002). An additional dataset available to climatologists requiring observations at fine time scales is the Global Daily Climatology Network (GDCN) which includes daily maximum and minimum temperature and precipitation amount (NCDC 2002b). Here again the observations primarily are obtained from the AFCCC database, although additional observations for the U.S. come from the ASOS network (and the first-order stations that preceded ASOS). Unlike the *Global Summary of the Day*, however, the GDCN observations have been checked for spatial quality control by comparing monthly average values to the gridded monthly mean fields computed by Legates and Willmott (1990a, b) (NCDC 2002b).

Gridded global climatologies of surface variables are another recent development that is in part technology-driven, in this case by the modeling community's need for spatially complete datasets for validating the output from General Circulation Models (GCMs) (New et al. 1999). Improved objective techniques for interpolating irregularly spaced observations to a uniform grid also have contributed to the development of these datasets. The early versions of these gridded climatologies had coarse spatial resolutions. For example, the spatial resolution of the temperature and precipitation climatologies developed by Legates and Willmott (1990a, b) was 5° latitude by 5° longitude. More recent versions have much finer scales. For example, the gridded dataset developed at the Climatic Research Unit (New et al. 1999, 2000) has a 0.5 degree latitude/longitude spatial resolution. This dataset is particularly interesting in that a mean monthly climatology for each variable was calculated over a standard normal period and then used in conjunction with the gridded fields for individual months to compute a time series of gridded anomaly fields extending back to 1901. Data for the gridded fields were obtained from multiple sources including the Jones dataset, GHCN, and through direct contact with national meteorological agencies, personal contacts, and other published sources (New et al. 1999).

3. UPPER-AIR OBSERVATIONS

3.1. Instruments and Coded Transmission Message

As with surface observations, the upper-air observational network in the U.S. and worldwide was designed for operational weather forecasting, and the quality of the these observations for climate research has been questioned by a number of researchers (Gaffen et al. 2000; NRC 2000). Nonetheless, measurements of the lower atmosphere (i.e., troposphere and lower stratosphere) from balloon-borne instrument packages are the only viable historical source of upper-air data currently available (Wallis 1998), and are an essential component of climate change detection and attribution studies (Gaffen et al. 2000). In addition, these observations are used as ground truth for model verification and the calibration of satellite information (Wallis 1998). In the U.S., the upper-air network was established in 1937 (OFCM 1997) and was facilitated by technological developments in instrumentation and radio telemetry. Currently, close to 70 stations in the U.S. and 900 radiosonde stations worldwide report upper-air observations at least twice a day.

Upper-air observations (often called RAOBs) are taken using an instrument package referred to as radiosonde. A radiosonde consists of three basis parts: the meteorological sensors, the data encoding electronics, and the telemetry transmitter (OFCM 1997). For most radiosondes in use today, the instrument package includes an aneroid barometer for measuring pressure, a thermistor for measuring temperature, and a carbon hygristor for measuring relative humidity. The sensors are linked to a battery-operated miniature radio transmitter that generates an FM radio signal which is relayed to a tracking station near the launch site (NCDC 2003b). Currently, data are transmitted at least every 6 seconds throughout the flight which corresponds to about a 10-50 meter vertical resolution, although past observations had a considerably coarser vertical resolution. The radiosonde is carried aloft by a spherically shaped balloon that expands from approximately 2 meters in diameter at the surface to between 8-10 meters before its bursts at the low pressures (10 hPa or lower) higher in the atmosphere (OFCM 1997). The balloon's ascension rate is approximately 275 meters/minute (OFCM 1997), and a typical radiosonde flight lasts approximately 90 minutes and reaches an altitude of approximately 27-37 km. Strong winds blow the instrument downwind of its launch site, and it is not unusual for a rawinsonde to drift 200 km from the release point (OFCM 1997; NCDC 2003b). Wind data are obtained by tracking the azimuth and elevation of the radiosonde using a parabolic dish antenna, which locks onto the radio signal emitted from the radiosonde[2] (OFCM 1997). In addition, geopotential height and dewpoint temperature are calculated from the pressure, temperature, and relative humidity measurements.

Sounding data are exchanged internationally in the form of coded messages referred to as TEMP messages. The coded messages contain only a subset of the actual observation. The current version of the WMO-approved coded message includes what are referred as "standard," "mandatory significant," and "additional" levels.[3] The standard levels include 12 predefined pressure levels below the 100 hPa level and 5 levels above the 100 hPa level (i.e., 1000, 925, 850, 700, 500, 400, 300, 250, 200, 150, 100, 70, 50, 30, 20, 10 hPa). Geopotential height, temperature, dewpoint, wind direction and wind speed are interpolated to each standard level. Mandatory significant levels include the surface, the flight termination level, the tropopause, and the bases and tops of temperature inversions, isothermal layers, or layers with sharp changes in the relative humidity lapse rate. Additional levels are defined with respect to the temperature and humidity profiles and are based on the greatest departure from linearity on a logarithmic pressure scale of at least $\pm 1.0°C$ for temperature and ± 10 percent for relative humidity (OFCM 1997). In addition, coded messages for radiosondes include wind speed and direction for fixed levels (approximately 305-meter intervals in the U.S.) and for significant levels. In the U.S., the raw sounding data are translated automatically using an on-site data processor known as the MicroArt Sounding System, although prior to the early 1980s the message coding was done by hand.[4]

3.2. Sources of Archived RAOBs

The three centers in the U.S. that archive the twice-daily upper-air observations are NCDC, AFCCC, and the U.S. Navy Fleet Numerical Meteorology and Oceanography Detachment. The TEMP coded messages transmitted over the GTS provide the basis for these archives, although since 1995 NCDC also stores high-resolution 6-second sounding data from NWS-operated stations. The 6-second data, however, have not been quality controlled (NCDC, 2003b). Until recently, upper-air data were expensive, as there was no single data base that included all operationally collected radiosonde data for either the U.S. or the world. In addition, data were only available in time series format (i.e., all observations through time for one station) and not in synoptic format (i.e., all stations for a particular time period) (Schwartz and Doswell 1991). Beginning in the late 1980s, NCDC and the NWS Forecast Systems Laboratory (FSL) jointly produced a comprehensive radiosonde archive for North America that can be accessed either by location or by time. The archive extends back to 1946, and was compiled from soundings collected at NCDC that had been interpolated to 50 hPa intervals and from GTS messages collected at FSL (Schwartz and Doswell 1991). The archived data were checked for gross errors and for hydrostatic consistency, although one should not assume that all bad observations have been eliminated by the quality assurance practices (Elliott and Gaffen 1991). The vertical extent of the archive varies with time. Prior to 1994, the highest pressure level included for any sounding was 100 hPa which in effect limits the use of these data to tropospheric analyses. More recent soundings extend to 10 hPa, which permits analysis of the lower stratosphere. An important component of the archive is a station history file that includes station identifiers, changes in station locations, and dates of station moves (Govett 2001; Schwartz and Doswell 1991). More recently, NCDC collaborated with the All-Union Research Institute of HydroMeteorological Information in Russia to produce a global dataset of daily upper-air observations known as the Comprehensive Aerological Research Dataset (CARDS) (Eskridge et al. 1995). At the present time, the CARDS archive begins in 1948 and includes over 27 million quality-controlled radiosonde observations (Durre 2003). Many of the records in CARDS are short or incomplete, however, and spatial coverage is not uniform (Wallis 1998). "Bare-bones" station histories that include WMO number, station name, geographic coordinates (latitude, longitude, and elevation), and period of record are available for over 2400 stations in the CARDS archive (Eskridge et al. 1995).

Upper-air observations summarized by month are also available. An important source of monthly mean values is the CLIMAT TEMP reports,

archived at the Hadley Centre for Climate Prediction and Research in the
United Kingdom (Parker et al. 1997; Gaffen et al. 2000). This data archive
begins in 1959 and includes monthly means and the number of days missing
in each month for the surface and 11 standard pressure levels (Gaffen et al.
2000). The CLIMAT TEMP reports are also the basis for the monthly mean
values of upper-air temperature, dewpoint depression and wind velocities
reported for approximately 500 stations in the *Monthly Climatic Data for the
World* series prepared at NCDC. In addition to these formal publications,
individual researchers have compiled subsets of stations meeting preselected
criteria for data quality or spatial separation. Perhaps the best known is the
network of 63 stations used by Angell and Korshover (1977) to calculate
global trends in upper tropospheric temperature. A more recent monthly dataset
is referred to as MONADS and consists of the monthly mean data for each
station in the CARDS archive (Durre 2003). Attempts have been made to
develop gridded datasets of upper-air observations (e.g., Parker et al. 1997)
although the large spatial separation of upper-air stations and the discontinuous
spatial coverage limit the usefulness of gridded fields.

3.3. Inhomogeneities in RAOB Time Series

Users of archived upper-air observations need to be cognizant of a mind
boggling array of potential inhomogeneities in the data series. A compounding
factor is that international standards have never been established for radiosonde
observations (Schwartz and Doswell 1991). The sources of systematic biases
in the radiosonde archive can be summarized as inhomogeneities due to
changes and/or errors in (1) sensor type, (2) signal processing algorithms and
instrument calibration procedures, (3) ground systems, (4) boundary layer
measurements, (5) reporting and coding practices, and (6) archiving
procedures. The time series of radiosonde observations for the U.S. are used
below to highlight the types of biases that can exist.

In general, a greater variety of sensors have been used for upper-air
observations compared to surface measurements (Elliott et al. 2002). This is
certainly the case in the U.S., where the upper-air network has used sondes
from three different manufacturers. Prior to 1988, the NWS purchased sondes
solely from the VIZ Manufacturing Company. In 1988, however, the NWS
introduced sondes manufactured by the Space Data Division (SDD) of the
Orbital Science Corporation at a subset of 17 stations in the western U.S.
(Elliott et al. 1998), ostensibly to reduce reliance on one manufacturer (Elliott
and Gaffen 1991). The SDD sonde was phased out in 1995 (Luers and Eskridge
1998), while around the same time the NWS introduced sondes made by the
Vaisala Company of Finland at 25 stations (Elliott et al. 1998; Luers and
Eskridge 1998). By 1998, a total of 60 sites had switched to Vaisala sondes

(Elliott et al. 2002; NWS OPS2 2003). Accompanying the changes in sonde manufacturer were changes in instrumentation. Instrument changes also occurred when sondes from the same manufacturer were updated with new technology. Starting with temperature, the VIZ sondes were furnished with ducted thermistors prior to 1960. These instruments were replaced in the 1960s with an unducted thermistor that had a white lead carbonate coating to reflect solar radiation. Like the VIZ sondes, the SDD sonde used a chip thermistor; however the protective coating differed. Several changes in humidity instrumentation have also occurred. First, the hair hygrometer was replaced by a lithium chloride hygristor in 1943 (Elliott and Gaffen 1991). Then between 1963-66, the NWS switched from a lithium chloride to a carbon hygristor (Elliott et al. 1998). In 1973 the housing that shielded the carbon hygristor was changed (Elliott and Gaffen 1991; Zhai and Eskridge 1996). The earlier housing, which also had been used for the lithium chloride strip, was defective in that it allowed sunlight to enter the housing which warmed the inside of the housing and lowered the humidity that the element experienced (Elliott and Gaffen, 1991). A further change occurred in 1988 with the introduction of the SDD sondes at a limited number of stations. Even though the VIZ and SDD sondes employed the same hygristor, the paint on the hygristor duct for SDD sondes was hygroscopic (Schwartz and Doswell 1991). More recently, the switch to Vaisala sondes at some stations has introduced a substantial inhomogeneity in the humidity record due to chemical contamination of the humidity sensor from the sonde packaging material (Guichard et al. 2000; Wang et al. 2002). Previous authors have demonstrated that the changes in humidity instrumentation over time have caused (1) a distinct break in the humidity record around 1963-66 with lower daytime relative humidity data values after the switch to a carbon hygristor (Zhai and Eskridge 1996), (2) an underestimation of relative humidity, especially during the day, by as much as 50 percent for the period 1963-72 due to the inadequate instrument housing (Elliott and Gaffen 1991; Schwartz and Doswell 1991), (3) unrealistic boundary-layer humidity profiles for 1988-1995 at the sites that used SDD sondes during this period (Schwartz and Doswell 1991), and (4) a dry bias for those stations recently equipped with Vaisala sondes. These changes have shown up as statistically significant discontinuities at many stations (Zhai and Eskridge 1996; Elliott et al. 2002).

Temperature and humidity sensors measure electrical resistance, which must be translated into air temperature and relative humidity (Garand et al. 1992). These processing algorithms have changed with time, introducing another source of inhomogeneity. Again beginning with temperature, prior to 1960, the data processing algorithm for VIZ sondes included a correction factor that was applied to measurements between 400 hPa and 100 hPa to account for solar heating of the instrument (Mahesh et al. 1997). The NWS discontinued use of this correction factor in the early 1960s when a reflective coating was applied to the thermistors on VIZ sondes. Other countries, however, continued to use solar energy correction factors, and, in particular, correction factors were used for Vaisala sondes. The current use in the U.S. of sondes from two different manufacturers means that some temperature measurements (those from Vaisala sondes) have received radiation adjustments whereas other measurements (those from VIZ sondes) have not (Elliott et al. 2002). Furthermore, neither the corrective coating nor the radiation adjustment fully accounts for the heating of the thermistors. For example, large differences (<1°C in the troposphere and 2-3°C in the stratosphere) have been observed for the VIZ sondes between daytime and nighttime measurements at the same altitude (Mahesh et al. 1997). Users of U.S. upper-air temperature observations also need to be aware that the NWS has never adopted the practice used in some other countries to correct radiosonde temperature observations for thermistor response time lag (Mahesh et al. 1997). Thermal lag is especially large in the stratosphere because of the low air density at these altitudes (Mahesh et al. 1997; OFCM 1997).

Changes in signal processing algorithms have had an even greater impact on humidity measurements. In the early period of record, the data reduction algorithm used by the NWS resulted in a high bias at relative humidities >90 percent. A revised algorithm was introduced in 1980, but this algorithm underestimated relative humidity at high values (Garand et al. 1992; Elliott et al. 1998). Another discontinuity occurred around 1988-89 when the NWS switched from a VIZ Type A to VIZ Type B sonde. Even though the humidity sensors on the two sondes used different resistors, the resistance factor was not changed in the data reduction software (Elliott et al. 1998). This error, which affected humidity values until a change in software in 1993, caused measurements at high relative humidities to be biased toward lower values (Elliott et al. 1998). Another change involving the VIZ sondes occurred in 1993, when a different calibration curve was introduced for relative humidity measurements below 20 percent (Elliott et al. 1998). This resulted in a high bias in low-end humidity from 1993 until 1997, when the practice of using separate calibration curves was discontinued (Elliott et al. 1998). An error in the data reduction software for the SDD sondes is another source of inhomogeneity. Hygristor resistance depends on both the ambient temperature and humidity, and a correction factor for temperature is required when converting the resistance values to humdity. Software installed at the SDD sites in 1990 incorrectly divided temperatures by 100, which meant that all humidity calculations effectively used a temperature of 0°C (Elliott et al. 1998). This caused either an over- or underestimation of humidity, depending on the temperature. Removal of the influence of temperature is also an issue for the Vaisala sondes. The software developed for these sondes assumes that there is a linear relationship between temperature and hygristor resistance, whereas the actual sensor dependence to temperature appears to be nonlinear (Wang et al. 2002). This misspecification has introduced a dry bias into the humidity time series (Wang et al. 2002). Changes or errors in data reduction software are not limited to temperature and humidity. For example, prior to 1993 it was the practice in the U.S. to use a different value for the earth's gravitational constant in geopotential height calculations compared to all other nations (Eskridge et al. 1995). A further problem with the archived data is that the interpolation scheme used during 1970-79 to compute wind at significant levels contained errors resulting in questionable wind values for that period (Schwartz and Doswell 1991).

The switch from manual to automated data reduction procedures also introduced inhomogeneity. In the U.S., the computer-based automated data processing system installed in 1986, referred to as the Automatic Radiotheodolite (ART) System, has been linked to an increase in gross (i.e., random) errors and to an increase in missing observations (Schwartz 1990; Schwartz and Doswell 1991). The increase in gross errors has been attributed to poorer on-site quality control either because observers are relying too heavily on the automated procedures or because intervention on the part of the observer is difficult (Schwartz and Doswell 1991). Large blocks of missing observations appear in the records for a number of stations soon after the ART system was installed. One cause of the missing observations was the sensitivity of the original ART system to nearby lightning strikes, which resulted in system failures; a number of stations experienced several-month outages due to an inadequate supply of spare parts (Bosart 1990). A replacement system (known as Micro-Art) was installed in 1989 and is considerably less sensitive to lightning strikes, although blocks of missing data are still evident at some stations.

Spurious vertical profiles close to the surface also introduce error and inhomogeneity into the radiosonde network. One cause for the spurious profiles is that the radiosonde is sometimes calibrated in a heated or air-conditioned room and released before the sonde equilibrates to the outside air temperature or humidity (Schwartz and Doswell 1991; Mahesh et al. 1997; OFCM 1997). This error is more frequent for stations located in very cold or very warm climates (Mahesh et al. 1997). A more frequent cause of spurious near-surface profiles is that the pressure, temperature, humidity and wind values for the surface that are included in the GTS coded message are not taken with the radiosonde but rather are measured by the surface observing equipment at the station. Not only do the instruments differ, but the locations of the surface observing system and the radiosonde release often are not coincident. A frequent manifestation of differences in the surface and radiosonde measurements is a superadiabatic lapse rate next to the surface (Slonaker et al. 1996).

Reporting practices have also changed over the history of the upper-air network introducing further systematic biases in the historical record. One example is change in observation time. The current standard observation times of 0000 and 1200 UTC were not established until 1957. Prior to that time, upper-air observations in the U.S. were made at 0300 to 1500 UTC (Zhai and Eskridge 1996). This switch has been shown to have introduced a distinct change point in the time series of temperature and humidity at several upper-air stations (Zhai and Eskridge 1996; Lanzante et al. 2003a). Not as well documented is the impact of the switch in 1955 from a 16-point scale to reporting wind direction in tens of degrees (NCDC 2002a). Changes in the reporting practices for humidity have been frequent and complex, especially in the U.S. Early carbon hygristors could not accurately register humidities below 50 percent at very low temperatures, which led to the practice of not reporting humidity values at temperatures <-40°C (Elliott and Gaffen 1991). Even though sensor accuracy and precision improved as technology advanced, this practice of censoring relative humidity at cold temperatures continued until 1993 when, at the request of users, the policy was dropped (Seidel and Durre 2003). The early hygristors were also not able to accurately measure relative humidity below 20 percent, which led to the censoring of low relative humidities regardless of the ambient temperature for portions of the historical record. Prior to 1965, NWS reported all humidities less than 20 percent as missing (Elliott and Gaffen 1991). This practice was modified in 1965 and humidities between 10-20 percent were reported until 1973. At this time, further concerns about the unreliability of low relative humidity values persuaded the NWS to revert to its earlier practice of not reporting relative humidities <20 percent (Schwartz and Doswell 1991; Garand et al., 1992). Concurrently, the convention was established of reporting the dewpoint depression as 30°C in the GTS coded message whenever the relative humidity was <20 percent (Schwartz and Doswell 1991; Elliott et al. 1998). Beginning in 1993, NWS once again reported humidities below 20 percent (Elliott et al. 1998), although the practice of reporting 30°C dewpoint depressions in the GTS message continues. The differing reporting practices for relative humidity have introduced a moist bias into the record for those periods when the censoring practice was in effect. The censoring practices in the U.S. also have introduced spurious spatial gradients in the global radiosonde archive (Elliott and Gaffen 1991; Garand et al. 1992), as other countries, including neighboring Canada, have not followed the same practices. These temporal and spatial inhomogeneities have lead Elliott and Gaffen (1991) to conclude that humidity data above 500 hPa, and even at 500 hPa for high-latitude locations, are not reliable enough to draw conclusions about temporal changes in upper-level humidity. For example, they found that the abrupt decrease in 500 hPa relative humidity evident at most U.S. stations in 1965 is almost entirely due to the inclusion of drier humidity values and not to climate variability or change (Elliott and Gaffen 1991).

Another potential source of inhomogeneity is the increase with time in the number of archived levels per observation. A typical archived sounding early in the record has approximately 20 vertical levels (Schwartz and Doswell 1991; NRC 2000). The number of archived levels increases substantially around 1960 when significant levels were included along with the standard and pseudo-standard levels (Chernykh et al. 2003), and again in 1983 when wind-only levels were also archived. Recent soundings in the Northern Hemisphere Radiosonde Archive often have more than 100 vertical levels. The higher-resolution coded message and increase in number of archived levels was facilitated by automated rather than manual data processing (Chernykh et al. 2003; Seidel and Durre 2003). Another factor was the improvement of balloon technology and tracking, such that data are now regularly obtained up to 10 hPa or even higher in the atmosphere.

In contrast to surface observations, much less effort has been extended to develop homogeneity adjustment procedures for upper-air data, in part because of the difficulty of deriving such adjustments. For example, the radiative effects on upper-air temperature observations are difficult to adjust for due to multiple sources and sinks of longwave fluxes (i.e., space, ground, clouds, atmosphere, radiosondes, balloon). Also, the sign of these fluxes can differ in the troposphere compared to the stratosphere owing to the reversal of the thermal lapse rate (OFCM 1997). Another factor complicating the derivation of adjustment schemes is that changes in upper-air observational practices, instruments, and data reduction software often take place at the same time at multiple stations, making it difficult to identify inhomogeneities using reference series defined from surrounding stations (Zhai and Eskridge 1996). The large station separation also compromises inhomogeneity evaluations based on "neighboring" stations. A further limitation is that station histories are often missing or, if available, are incomplete, especially for the global upper-air network. This limitation makes identifying possible change points in the record difficult. Results of previous attempts to adjust upper-air observations are mixed, but point to the importance of continued work in this area. Parker et al. (1997) were able to successfully use time series of satellite (i.e., Microwave Sounding Unit) observations to adjust radiosonde temperature series for stations in Australia and New Zealand. Widespread implementation of their adjustment technique is limited, however, by its dependence on detailed station histories which are not available for most countries. Attempts by Elliott et al. (2002) to adjust U.S. temperature observations for the switch from VIZ to Vaisala sondes were not as successful, and they concluded that a networkwide adjustment factor was not possible because the differences between the two sondes depend on a number of factors including altitude, time of day, region and season. The impact of adjustments on the detection of climatic trends is also not clear. Gaffen et al. (2000) found that the magnitude of detected temperature change was very sensitive to the choice of adjustment scheme, although, in general, temporal trends were removed in adjusted series compared to unadjusted series. On the other hand, Lanzante et al. (2003b) found that radiosonde temperature series adjusted for instrument changes displayed an increased tropospheric warming and decreased stratospheric cooling compared to the unadjusted series.

Most efforts to quality control archived RAOBs have focused on gross (i.e., random) error detection. Gross errors are primarily caused by equipment malfunctions, observer mistakes, and transmission errors. Recent estimates suggest that from 5 percent to 20 percent of all upper-air observations contain gross errors (Eskridge et al. 1995), even though observers are required to perform an on-site quality control (OFCM 1997). Gross error checks typically include criteria for identifying unrealistic values of geopotential height, temperature, humidity, wind speed, and wind direction along with internal checks for consistency. For example, the hydrostatic approximation has been used to check both the North American Rawinsonde Archive and CARDS for unrealistic superadiabatic lapse rates, and the geostrophic and thermal wind relationships have been used to evaluate the quality of wind observations outside of the Tropics (Eskridge et al. 1995). Users should not assume that all gross errors have been removed from archived observations, and additional checks for gross errors need to be part of any research design that employs upper-air observations.

Further changes in the U.S. upper-air network are imminent. A replacement system for the antiquated Micro-Art system is currently under development. This system, referred to as the Radiosonde Replacement System (RRS), will employ a Global Positioning System (GPS) to track the sonde, and will provide considerably higher resolution wind observations compared to the current Micro-Art system (NWS OPS2 2003). Another improvement is that the surface weather conditions will be measured directly at the balloon release site. Also, an objective of the proposed system is to provide a high-resolution data archive for users with the goal of 1- second (5-meter) resolution through the atmosphere (NWS OPS2 2003). The planned system will make use of advanced automation technology and is designed to minimize operator interaction and maintenance. Whether this automation will increase or decrease the rate of gross errors in radiosonde observations will need to be closely monitored.

4. CLIMATOLOGICAL OBSERVATIONS FOR THE FUTURE

> How much more, then, it requires different intentions of the eye and of the mind to attend to different departments of knowledge! ("Autumnal Tints," *Atlantic Monthly,* Henry David Thoreau, 1862)

As illustrated above, the current in situ climate-observing network is a patchwork of national and regional networks that were not initially designed for climate monitoring. How, then, can these in situ networks be improved and what efforts are currently underway to enhance the global climate database? An obvious suggestion is to design national and global networks specifically for monitoring climate variability and change, but the cost of establishing and maintaining multiple networks serving different purposes is prohibitive, not only for developing countries but also for developed countries. At least for the near future, observational networks will need to serve multiple functions. Borrowing from Thoreau, the networks will need to "attend to different departments of knowledge," and analysis of the network observations will require "different intentions of the eye and of the mind" depending on whether the observations are being used for weather forecasting or for climate monitoring.

From the viewpoint of climate science, the challenge facing network designers, sensor developers, observers, and data managers is to provide a homogeneous, continuous series of atmospheric observations for long time periods with sufficient accuracy to measure small temporal changes in climate parameters. At the same time, the flexibility to adapt networks to new technologies (Trenberth et al. 2002) and to withstand common discontinuities (e.g., relocations, observer and environmental changes) (Tuomenvirta 2001) must be maintained. Numerous possibilities, some rather straightforward, exist for improving climatological observations, although their implementation depends on financial and personnel resources, international collaboration including the open exchange of data and metadata, and a commitment on the part of international and national agencies, the research community, and even individuals. Some of the most frequently discussed, and the most needed, improvements include (1) the standardization of national and international observing practices including practices for reporting essential metadata (Karl et al. 1993), (2) concurrent observations from old and new sensors for a sufficient period of time (e.g., multiple years) to calculate accurate adjustment factors, (3) real-time monitoring of the quality of observations and the performance of the observing systems so that developing biases are detected and quickly corrected (Trenberth et al. 2002), and (4) archiving both "raw" and processed observations so that reprocessing is possible when improved reduction algorithms are available or if an error in an earlier processing algorithm is detected (Karl et al. 1993). Data archaeology is also essential and includes inferring station histories from secondary sources, converting observations and station histories from hard copy to electronic format for improved access, retrieving data stored on outdated recording media, and processing past observations for gross and systematic errors (GCOS-SC 2003). Other proposals include reintroducing old sensors to test the differences with new instruments and/or the establishment of a museum of old instruments (Karl et al. 1993).

The principal internationally coordinated effort currently underway to improve the quality of climate observations is the Global Climate Observing System (GCOS) established in 1992 by four international organizations: the WMO, the Intergovernmental Oceanographic Commission, the U.N. Environment Programme, and the International Council of Science (Rosner 2002: GCOS Secretariat, no date). GCOS does not take observations directly, but rather provides a framework for coordinating, integrating, and enhancing the observational systems of participating countries and organizations (GCOS Secretariat, no date). GCOS activities are guided by ten climate monitoring principles (Table 20-1) and a five-point observation strategy is envisioned that includes (1) comprehensive global networks of both in situ and remotely sensed observations at fine space and time scales, (2) baseline global observation networks for a smaller number of variables and locations, (3) reference networks at a few locations for calibration purposes, (4) research networks, and (5) ecosystem networks (GCOS-SC 2003).

The GCOS baseline and reference networks are particularly relevant in regard to in situ climate observations and highlight, on one hand, the difficulties of improving the global climate database and, on the other hand, the potential for innovation. An initial priority of GCOS was to identify baseline observing networks, and two networks, the GCOS Surface Network (GSN) and the GCOS Upper-Air Network (GUAN) were recently formulated for this purpose. The approximately 900 GSN stations and 150 GUAN stations were chosen on the basis of record length, past performance, accurate and timely transmission of reports, availability of metadata, and global representation (Peterson et al. 1997; GCOS-SC 2003). Deutscher Wetterdienst in Germany and the Japan Meteorological Agency serve as the GSN Monitoring Centers for real time distribution, and NCDC serves as the GSN Archive and the Analysis Center responsible for higher-level quality control (GOSIC 2003). The U.K. Met Office and NCDC serve jointly as the data analysis centers for GUAN. Of

Table 20-1. Global Climate Observing System (CGOS) Climate Monitoring Principles.

Effective monitoring systems for climate should adhere to the following principles:

1. The impact of new systems or changes to existing systems should be assessed prior to implementation.

2. A suitable period of overlap for new and old observing systems is required.

3. The details and history of local conditions, instruments, operating procedures, data processing algorithms, and other factors pertinent to interpreting data (i.e., metadata) should be documented and treated with the same care as the data themselves.

4. The quality and homogeneity of data should be regularly assessed as a part of routine operations.

5. Consideration of the needs for environmental and climate-monitoring products and assessments, such as the IPCC assessments, should be integrated into national, regional and global observing priorities.

6. Operation of historically uninterrupted stations and observing systems should be maintained.

7. High priority for additional observations should be focused on data-poor regions, poorly-observed parameters, regions sensitive to change, and key measurements with inadequate temporal resolution.

8. Long-term requirements, including appropriate sampling frequencies, should be specified to network designers, operators, and instrument engineers at the outset of a system design and implementation.

9. The conversion of research observing systems to long-term operations in a carefully planned manner should be promoted.

10. Data management systems that facilitate access, use, and interpretation of data and products should be included as essential elements of climate monitoring systems.

Source: GCOS-SC 2003.

considerable concern is the disappointing performance so far of GSN and GUAN networks. Only 50-70 percent of the GSN and GUAN stations routinely report observations to the monitoring centers (Trenberth et al. 2002), and only about 30 percent of the stations have supplied the archive and analysis centers with historical data (GCOS-SC 2003). Also, few of the soundings from the GUAN stations reach the required altitude of 5 hPa, and instrumental and procedural changes have been made without due regard to the GCOS climate monitoring principles (GCOS-SC 2003). The limitations of these networks point to the difficulty of establishing and maintaining baseline networks without a huge input of financial resources and better international exchange of data and metadata.

The philosophy underlying the GCOS reference networks departs considerably from earlier approaches to network design, and these networks have the potential to improve, even revolutionize, the way in which observations are taken and quality controlled. One of the first reference networks is being developed in the U.S. If fully implemented, the U.S. Climate Reference Network (USCRN) will eventually be composed of close to 250 stations; at the present time, prototype stations have been installed at 12 locations for operational testing and evaluation (Franklin 2003b). USCRN stations will be spatially distributed such that the major features of the climate of the contiguous U.S. are represented (NCDC 2003a; Redmond et al. 2003). Preferred sites are locations, such as rural areas or national parks, where little or no human modification is anticipated in the foreseeable future (Redmond et al. 2003). An intriguing aspect of the network is the use of paired stations, separated by 25 km or less, that provides a contingency in case one of the stations is later abandoned (Redmond et al. 2003). Temperature and precipitation are the primary climate variables to be measured by the USCRN stations, although secondary measurements such as solar radiation and wind speed will also be taken (Redmond et al. 2003). The automated instrument packages at each site will be equipped with a standard set of core sensors, and observations will be transmitted hourly in near real time (NCDC 2003a). Each site will have duplicate temperature sensors in order to continuously monitor the temperature measurements for bias introduced by sensor drift and other malfunctions, and to quickly correct for these errors so that time-dependent biases are not introduced into the observational record (GCOS-SC 2003; NCDC 2003a). The reference network is also important for calibration, and transfer functions will be developed to relate observations from the USCRN to nearby observations from conventional networks (e.g., the COOP network and ASOS). This network will allow data adjustment methodologies for historical data to be determined and help relate the USCRN data to the past record (Franklin 2000). The full implementation of the USCRN depends on future availability of funding (Franklin 2003a).

5. CONCLUSIONS

When you come to observe faithfully ... ("Autumnal Tints," *Atlantic Monthly,*
Henry David Thoreau, 1862)

The literally millions of archived atmospheric observations available today for climate monitoring, research, and applications were made possible by technological advances in instrumentation, communication systems, and data storage media. This essay, which is more utilitarian than philosophical in character, attempted to provide a relatively comprehensive overview of the major in situ observational datasets available for climatological research specific to the U.S. and for studies that are global in scale. A key objective was to summarize the many potential sources of inhomogeneities in these datasets in a manner accessible to potential users who are not meteorologists or climatologists. Those of us who use atmospheric observations, whether for applied purposes or for research, must keep in mind that observations need be interpreted in the context in which they were taken and with an understanding of their limitations and potential biases. Otherwise, observations may obscure rather than illuminate. Technology has made it possible to "see" atmospheric and climatic phenomena, but geographers and other users of atmospheric observations must clearly perceive and understand the limitations of the data in order to ensure that what is "seen" is, in fact, true. It is particularly appropriate to situate this essay in the context of geography and geographic knowledge, as not only are geographers frequent users of climate observations, but geographer-climatologists have been intimately involved in the development of observational networks and in the analysis of data inhomogeneities.

In spite of technological advances, the current climate observing network is under considerable stress. Resources in many countries are inadequate for taking even routine observations needed for real-time weather forecasting, let alone climate monitoring. Stations with long-term records continue to close. Observing equipment, even in so-called developed nations, is often in poor condition. Data centers cannot keep up with the influx of new observations or with ongoing changes in storage media. Technological advances in instrumentation and communication have put even greater stress on observation networks, as large monetary and personnel investments are often required to keep sophisticated equipment working. In sum, the condition of the global climate observing systems is of great concern, and those of us who are frequent users of climate observations need to be cognizant of the fragility of the observing networks and become advocates for their improvement. Turning one last time to Thoreau: "observe faithfully." This directive can be expanded to "observe carefully" and "interpret cautiously."

NOTES

1. The convention in climatology is to refer to discontinuities, change points, and biases in time series of climatological variables as "inhomogeneities" rather than "heterogeneities."
2. Historically the term "rawinsonde" was used to refer to upper-air soundings for which wind data were determined, and the term "radiosonde" was reserved for soundings without wind information. The terms "radiosonde" and "rawinsonde" are increasingly used interchangeably, as most upper-air soundings now include wind information.
3. The U.S. terminology differs somewhat with "mandatory" corresponding to the WMO usage of "standard," and "significant" referring to the WMO's "mandatory significant" and "additional" levels.
4. A detailed description of the current version of the TEMP upper-air code can be found in Appendix E-II of OFCM (1997). The code consists of several parts. The TTAA portion of the code includes the geopotential height, temperature, humidity, and wind speed and direction for the standard levels up to and including 100 hPa. The TTBB includes the mandatory significant levels with respect to temperature and humidity, for the surface up to and including 100 hPa. The TTCC and TTDD portions of the code are similar, except that the information is for the standard and mandatory significant levels, respectively, for the portion of the observation above 100 hPa. Finally, the PPBB portion of the code reports wind speed and direction for the wind-only levels.

REFERENCES

Alexandersson, H. (1986). A Homogeneity Test Applied to Precipitation Data, Journal of Climatology 6: 661-75.

Allen, R.J. and DeGaetano, A.T. (2000). A Method to Adjust Long-Term Temperature Extreme Series for Nonclimatic Inhomogeneities, Journal of Climate 13: 3680-95.

Angell, J.K. and Korshover, J. (1977). Estimate of Global Change in Temperature, Surface to 100 mb, Between 1958 and 1975, Monthly Weather Review 105: 375-85.

Balling, R.C. and Idso, C.D. (2002). Analysis of Adjustments to the U.S. Historical Climatology Network (USHCN) Temperature Database, Geophysical Research Letters 29: 1387.

Bosart, L.F. (1990). Degradation of the North American Radiosonde Network, Weather and Forecasting 5: 527-28.

Butler, R.D. (1998). ASOS Heating Tipping Performance Assessment and Impact on Precipitation Climate Continuity [Report can be ordered from Storming Media, 529 14th St. NE, Washington, DC 20002].

Chernykh, I.V., Alduchov, O.A., and Eskridge, R.E. (2003). Reply, Bulletin of the American Meteorological Society 84: 241-47.

DeGaetano, A.T. (1999). A Method to Infer Observation Time Based on Day-to-Day Temperature Variations, Journal of Climate 12: 3443-56.

Durre, I. (2003). Comprehensive Aerological Reference Dataset Global Radiosonde Data and Station History Information. http://lwf.ncdc.noaa.gov/oa/climate/cards/

Easterling, D.R., Peterson, T.C., and Karl, T.R. (1996). On the Development and Use of Homogenized Climate Datasets, Journal of Climate 9: 1429-34.

Elliott, W.P. and Gaffen, D.J. (1991). On the Utility of Radiosonde Humidity Archives for Climate Studies, Bulletin of the American Meteorological Society 72: 1507-20.

Elliott, W.P., Ross, R.J., and Schwartz, B. (1998). Effects on Climate Records of Changes in National Weather Service Humidity Processing Procedures, Journal of Climate 11: 2424-36.

Elliott, W.P., Ross, R.J., and Blackmore, W.H. (2002). Recent Changes in NWS Upper-Air Observations with Emphasis on Changes from VIZ to Vaisala Radiosondes, Bulletin of the American Meteorological Society 83: 1003-17.

Eskridge, R.E., Alduchov, A.O., Chernykh, I.V., Panmao, Z., Polansky, A.C., and Doty, S.R. (1995). A Comprehensive Aerological Reference Dataset (CARDS): Rough and Systematic Errors, Bulletin of the American Meteorological Society 76: 1759-75.

Federal Aviation Administration (FAA) (2002). Automated Surface Observing System. http://www2.faa.gob/asos/asosinfo.htm

Franklin, D. (2000). U.S. Climate Reference Network: Use of the Data. http://www.ncdc.noaa.gov/oa/climate/research/crn/crnuseofdata.html/

Franklin, D. (2003a). U.S. Climate Reference Network Program Overview. http://www.ncdc.noaa.gov/oa/climate/uscrn/programoverview.html

Franklin, D. (2003b). U.S. Climate Reference Network Progress and Milestones. http://www.ncdc.noaa.gov/oa/climate/uscrn/progress.html

Gaffen, D.J. and Ross, R.J. (1999). Climatology and Trends of U.S. Surface Humidity and Temperature, Journal of Climate 12: 811-28.

Gaffen, D.J., Sargent, M.A., Habermann, R.E., and Lanzante, J.R. (2000). Sensitivity of Tropospheric and Stratospheric Temperature Trends to Radiosonde Data Quality, Journal of Climate 13: 1776-96.

Garand, L., Grassotti, C., Halle, J., and Klein, G.L. (1992). On Differences in Radiosonde Humidity Practices and Their Implications for Numerical Weather Prediction and Remote Sensing, Bulletin of the American Meteorological Society 73: 1417-23.

Global Climate Observing System (GCOS) Secretariat (no date). Global Climate Observing System. http://www.wmo.ch/web/gcos/whatisgcos.htm/

Global Climate Observing System Steering Committee (GCOS-SC) (2003). The Second Report on the Adequacy of the Global Observing System for Climate in Support of the UNFCCC. Report GCOS-82 (WMO/TD No. 1143). Published by the World Meteorological Organization, Geneva, Switzerland. http://193.135.216.2/web/gcos/Second_Adequacy_Report.pdf

Global Observing Systems Information Center (GOSIC) (2003). The Operational Observing Systems for the GCOS Surface Network (GSN). http://www.gos.udel.edu/gcos/GSN_flow.htm

Govett, M. (2001). FSL/NCDC Radiosonde Data Archive. http://raob.fsl.noaa.gov/Raob_Software.html

Groisman, P.Ya, Easterling, D.R., Quayle, R.G., Golubev, V.S., Krenke, A.N., and Mikhailov, A.Yu (1996). Reducing Biases in Estimates of Precipitation in the U.S.: Phase 3 Adjustments, Journal of Geophysical Research 101: 7185-95.

Groisman, P.Ya. and D.R. Legates, 1994: The Accuracy of U.S. Precipitation Data, Bulletin of the American Meteorological Society 74: 215-227.

Guichard, F., Parsons, D., and Miller, E. (2000). Thermodynamic and Radiative Impact of the Correction of Sounding Humidity Bias in the Tropics, Journal of Climate 13: 3611-24.

Guttman, N.B. and Baker, C.B. (1996). Exploratory Analysis of the Difference Between Temperature Observations Recorded by ASOS and Conventional Methods, Bulletin of the American Meteorological Society 77: 2865-73.

Horvitz, A. (2002). Coop Modernization. http://www.nws.naa.gov/om/coop/coopmod.htm

Horvitz, A. (2003). What is the Coop Program? http://www.nws.noaa.gov/om/coop/what-is-coop.html

Jones, P.D. (1994). Hemispheric Surface Air Temperature Variations: A Reanalysis and an Update to 1993, Journal of Climate 7: 1794-1802.

Jones, P.D., Wigley, T.M.L., and Kelly, P.M. (1982). Variations in Surface Air Temperatures: Part 1. Northern Hemisphere, 1881-1980, Monthly Weather Review 110: 59-70.

Jones, P.D., Raper, S.C.B., Bradley, R.S., Diaz, H.F., Kelly, P.M., and Wigley, T.M.L. (1986). Northern Hemisphere Surface Air Temperature Variations: 1851-1984, Journal of Climate and Applied Meteorology 25: 161-79.

Karl, T.R., Williams, C.N., Jr., Quinlan, F.T., and Boden, T.A. (1990). U.S. Historical Climatology Network (HCN). Serial Temperature and Precipitation Data. Environmental Science Division, Publication No. 3404, Carbon Dioxide Information and Analysis Center, Oak Ridge National Laboratory, Oak Ridge, TN.

Karl, T.R., Quayle, R.G, and Groisman, P.Y. (1993). Detecting Climate Variations and Change: New Challenges for Observing and Data Management Systems, Journal of Climate 6: 1481-94.

Karl, T.R., Derr, V.E., Easterling, D.R., Folland, C.K., Hofmann, D.J., Levitus, S., Nicholl, N., Parker, D.E., and Withee, G.W. (1995). Critical Issues for Long-Term Climate Monitoring, Climatic Change 3: 185-221.

Lanzante, J.R., Klein, S.A., and Seidel, D.J. (2003a). Temporal Homogenization of Monthly Radiosonde Temperature Data. Part I: Methodology, Journal of Climate 16: 224-40.

Lanzante, J.R., Klein, S.A., and Seidel,D.J. (2003b). Temporal Homogenization of Monthly Radiosonde Temperature Data. Part II: Trends, Sensitivities, and MSU Comparison, Journal of Climate 16: 241-62.

Legates, D.R. and Willmott, C.J. (1990a). Mean Seasonal and Spatial Variability in Gauge-Corrected, Global Precipitation, International Journal of Climatology 10: 111-27.

Legates, D.R. and Willmott, C.J. (1990b). Mean Seasonal and Spatial Variability in Global Surface Air Temperature, Theoretical and Applied Climatology 41: 11-21.

Lott, N. (2003). Federal Climate Complex Global Surface Summary of Day Data Version 6 (Over 8000 Worldwide Stations). National Climatic Data Center. ftp: //ftp.ncdc.noaa.gov/pub/data/globalsod/readme.txt

Lott, J. Neal and Baldwin, R. (2002). The FCC Integrated Surface Hourly Database, A New Resource of Global Climate Data. Preprints, 13th Symposium on Global Change and Climate Variations, Paper 6.2. American Meteorological Society, Boston, MA.

Lott, N., Baldwin, R., and Jones, P. (2001). The FCC Integrated Surface Hourly Database, A New Resource of Global Climate Data. National Climatic Data Center Technical Report No. 2001-01. U.S. Department of Commerce, National Oceanic and Atmospheric Administration, National Environmental Satellite Data and Information Service, National Climatic Data Center, Asheville, NC.

Luers, J.K. and Eskrdige, R.E. (1998). Use of Radiosonde Temperature Data in Climate Studies, Journal of Climate 11: 1002-19.

Lund, R. and Reeves, J. (2002). Detection of Undocumented Changepoints: A Revision of the Two-Phase Regression Model, Journal of Climate 15: 2547-54.

Mahesh, A., Walden, V.P., and Warren, S.G. (1997). Radiosonde Temperature Measurements in Strong Inversions: Correction for Thermal Lag Based on an Experiment at the South Pole, Journal of Atmospheric and Ocean Technology 14: 45-53.

National Climatic Data Center (NCDC) (1997). Data Documentation for Monthly Climatic Data of the World TD 3500. National Climatic Data Center, Asheville, NC. http://www4.ncdc.noaa.gov/ol/documentlibrary/datasets.html

National Climatic Data Center (NCDC) (1998). Global Surface Summary of Day. http://rabbit.eng.miami.edu/info/weather/documentation.html

National Climatic Data Center (NCDC) (2002a). Data Documentation for Dataset 6200 (DSI-6200) NCDC Upper-air Digital Files. Asheville, NC. http://www4.ncdc.noaa.gov/ol/documentlibrary/datasets.html

National Climatic Data Center (NCDC) (2002b). Data Documentation for Dataset 9101 (DSI-9101) Global Daily Climatology Network, V1.0. Asheville, NC. http://www4.ncdc.noaa.gov/ol/documentlibrary/datasets.html

National Climatic Data Center (NCDC) (2002c). Data Documentation for Dataset 9644 (DSI-9644).World Weather Records. Ashville, NC. http://www4.ncdc.noaa.gov/ol/documentlibrary/datasets.html

National Climatic Data Center (NCDC) (2003a). Data Documentation for Dataset 3286 (DSI-3286) Climate Reference Network. Asheville, NC. http://www4.ncdc.noaa.gov/ol/documentlibrary/datasets.html

National Climatic Data Center (NCDC) (2003b). Data Documentation for Dataset 9948 (DSI-9948) Six Second Upper-Air Data. Asheville, NC. http://www4.ncdc.noaa.gov/ol/documentlibrary/datasets.html

National Climatic Data Center (NCDC) (2003c). Integrated Surface Hourly Data. National Climatic Data Center, Asheville, NC. http://www4.ncdc.noaa.gov/ol/documentlibrary/datasets.html

National Research Council (NRC) (2000). Reconciling Observations of Global Temperature Change. Washington: National Academy Press. http://www.nap.edu/books/0309068916/html/

National Weather Service Office of Meteorology (NWS OM) (1999). Automated Surface Observing System (ASOS). http://www.nws.noaa.gov/ost/asostech.html

National Weather Service (NWS) Office of Operational Systems (OPS2) (2003). Upper-Air Observations Program. http://www.ua.nws.noaa.gov/

New, M., Hulme, M., and Jones, P. (1999). Representing Twentieth Century Space-Time Climate Variability. Part I: Development of a 1961-90 Mean Monthly Terrestrial Climatology, Journal of Climate 12: 829-56.

New, M., Hulme, M., and Jones, P. (2000). Representing Twentieth-Century Space-Time Climate Variability. Part II: Development of a 1901-96 Monthly Grids of Terrestrial Surface Climate, Journal of Climate 13: 2217-38.

Office of the Federal Coordinator for Meteorology (OFCM) (1997). Federal Meteorological Handbook No. 3: Rawinsonde and Pibal Observations. FCM-H3-1997. Washington, DC. http://www.ofcm.gov/fmh3/text/default.htm

Parker, D.E., Gordon, M., Cullum, D.P.N., Sexton, D.M.H., Foland, C.K., and Rayner, N. (1997). A New Global Gridded Radiosonde Temperature Data Base and Recent Temperature Trends, Geophysical Research Letters 24: 1499-1502.

Peterson, T., Daan, H., and Jones, P. (1997). Initial Selection of a GCOS Surface Network, Bulletin of the American Meteorological Society 12: 2145-52.

Peterson, T.C., Easterling, D.R., Karl, T.R., Groisman, P., Nicholls, N., Plummer, N., Torok, S., Auer, I., Boehm, R., Gullett, D., Vincent, L., Heino, R., Tuomenvirta, H., Mestre, O., Szentimrey, T., Salinger, J., Forland, E.J., Hassen-Bauer, I., Alexandersson, H., Jones, P., and Parker, D. (1998). Homogeneity Adjustments of In Situ Atmospheric Climate Data: A Review, International Journal of Climatology 18: 1493-1517.

Peterson, T.C. and Vose, R.S. (1997). An Overview of the Global Historical Climatology Network Temperature Database, Bulletin of the American Meteorological Society 78: 2837-49.

Quayle, R.G., Easterling, D.R., Karl, T.R., and Hughes, P.Y. (1991). Effects of Recent Thermometer Changes in the Cooperative Network, Bulletin of the American Meteorological Society 72: 1718-23.

Redmond, K.T., Janis, M.J., and Hubbard, K.G. (2003). Climate Reference Network Site Reconnaissance: Lessons Learned and Relearned. Preprints, 12th Symposium on Meteorological Observations and Instrumentation, Paper #6.4. AMS Meteorological Society, Boston, MA.

Rosner, S. (2002). GCOS Surface Network Monitoring Centre. http://www.gsnmc.dwd.de/Background/background.htm/

Schwartz, B.E. (1990). Regarding the Automation of Rawinsonde Observations, Weather and Forecasting 5: 167-71.

Schwartz, B.E. and Doswell III, C.A. (1991). North American Rawinsonde Observations: Problems, Concerns and a Call to Action, Bulletin of the American Meteorological Society 72: 1885-96.

Seidel, D.J. and Durre, I. (2003). Comments on Trends in Low and High Cloud Boundaries and Errors in Height Determination of Cloud Boundaries, Bulletin of the American Meteorological Society 84: 237-40.

Slonaker, R.L., Schwartz, B.E., and Emery, W.J. (1996). Occurrence of Nonsurface Superadiabatic Lapse Rates Within RAOB Data, Weather and Forecasting 11: 350-59.

Trenberth, K.E., Karl, T.R., and Spence, T.W. (2002). The Need for a Systems Approach to Climate Observations, Bulletin of the American Meteorological Society 83: 1593-1602.

Tuomenvirta, H. (2001). Homogeneity Adjustments of Temperature and Precipitation Series– Finnish and Nordic data, International Journal of Climatology 21: 495-506.

Vincent, L.A. (1998). A Technique for the Identification of Inhomogeneities in Canadian Temperature Series, Journal of Climate 11: 1094-1104.

Vose, R.S., Schmoyer, R.L., Steurer, P.M., Peterson, T.C., Heim, R., Karl, T.R., and Eischeid, J.K. (1998). Global Historical Climatology Network, 1753-1990, Dataset. Available at http://ww.daac.ornl.gov from Oak Ridge National Laboratory Distributed Active Archive Center, Oak Ridge, TN. Previously published as The Global Historical Climatology Network: Long-Term Monthly Temperature, Precipitation, Sea Level Pressure and Station Pressure Data, ORNL/CDIAC-53, CDIAC NDP-041, Carbon Dioxide Information Analysis Center, Oak Ridge National Laboratory, Oak Ridge, TN, 1992.

Wallis, T.W.R. (1998). A Subset of Core Stations from the Comprehensive Aerological Reference Dataset (CARDS), Journal of Climate 11: 272-82.

Wang, J.H., Cole, H.L., Carlson, D.J., Milller, E.R., Beierle, K., Paukkunen, A., and Laine, T.K. (2002). Corrections of Humidity Measurement Errors from the Vaisala RS80 Radiosonde—Application to TOGA COARE Data, Journal of Atmospheric and Oceanic Technology 19: 981-1002.

Williams, C., Vose, R., and Easterling, D.R. (2003). The 2002 U.S. Historical Climate Network Upgrade. Presentation at the Annual Meeting of the Association of American Geographers, March 5, 2003, New Orleans, LA.

Zhai, P. and Eskridge, R.E. (1996). Analyses of Inhomogeneities in Radiosonde Temperature and Humidity Time Series, Journal of Climate 9: 884-94.

CHAPTER 21

STEPHEN J. WALSH
TOM P. EVANS
BILLIE L. TURNER II

POPULATION-ENVIRONMENT INTERACTIONS WITH AN EMPHASIS ON LAND-USE/LAND-COVER DYNAMICS AND THE ROLE OF TECHNOLOGY

Abstract   Technology has played a fundamental role in mapping, monitoring, and modeling land-use/land-cover (LULC) dynamics across a range of spatial and temporal scales and local, regional, and global extents. Spatial technologies, including remote sensing, geographic information systems (GIS), global positioning systems (GPS), data visualizations, spatial and statistical analyses, and models have combined to position people, place, and the environment within a spatially and temporally explicit context. These technologies help characterize the rate, pattern, and composition of LULC dynamics so that associated drivers of land-use change can be related to socioeconomic, demographic, geographic, and environmental dynamics. Special challenges exist because of inherent differences in how people and the environment are characterized in both space and time. Theories and practices from the social, natural, and spatial sciences are integrated to study LULC dynamics within the context of human-environment interactions. The goal has been to characterize the composition and spatial organization of LULC through its structure, function, and change and to relate the drivers of change to observed or simulated LULC patterns at different scales of analysis. Here we emphasize the use of technology for characterizing LULC dynamics, collecting and linking data from households, communities, regions, and nations with spatially explicit data collected, managed, and integrated within a Geographic Information Science (GISc) perspective. We discuss how technology aids in (1) mapping, monitoring, and modeling LULC dynamics by considering remote-sensing systems for LULC mapping, (2) image change-detection approaches for monitoring land-cover dynamics, (3) socioeconomic and demographic surveys linked to place through GPS technology and other approaches for characterizing the human dimension, (4) GIS for deriving and integrating disparate data, and (5) land-cover models for creating multilevel and spatial simulation of LULC dynamics. We describe how technology is being used to consider human behavior and agency in conjunction with a wide variety of processes associated with land-use/land-cover change.

Keywords   land-use and land-cover, land science, GIScience, spatial technologies

Stanley D. Brunn, Susan L. Cutter, and J.W. Harrington, Jr. (Eds.), Geography and Technology,
491-519. © 2004 Kluwer Academic Publishers. Printed in the Netherlands.

1. INTRODUCTION

It is now well recognized, that at local, regional, and global scales, land-use changes are significantly altering land cover, perhaps at an accelerating pace. This transformation of the Earth's surface, particularly through deforestation and urbanization, is linked to a variety of scientific and policy issues affecting the Earth system that includes land degradation, human migration, global climate change, and the loss of species habitat, to name a few (Sala et al. 2000; Lambin et al. 2001; Homewood et al. 2001). Theoretical and practical issues are involved in characterizing land-use/land-cover (LULC) dynamics, monitoring the change in land states or conditions over time and space, creating models of landscape dynamics, and linking observations and spatial simulations of change to policy questions and land-development scenarios.

Spatial digital technologies–such as satellite and aircraft remote sensing, geographic information systems (GIS), global positioning systems (GPS), data visualizations, and spatial analyses and modeling–are a central part of land-use and land-cover characterization and analysis (Tucker et al. 1986; Pastor and Broschart 1990; Phillips 1999a; Millington et al. 2001; Walsh and Crews-Meyer 2002; Fox et al. 2003). Used for characterizing the nature of biophysical, socioeconomic, demographic, and geographical landscapes and assessing their scale-pattern-process interrelationships, geographers are applying conventional approaches, such as map and air photo interpretation, as well as newly emergent technologies and approaches to study human-environment interactions and LULC dynamics (Brown and Walsh 1991, 1992; Millington et al. 2001; Messina and Walsh 2001). Empowered by our growing understanding of biophysical and socioeconomic processes and systems, the availability of spatial digital data, and new spatial digital technologies, geographers are among an emerging community of scholars who examine the landscape (a) at specified space-time scales that extend from fine-grained to coarse-grained resolutions, (b) from local, to regional, to global extents, (c) through endogenous and exogenous factors (Allen and Walsh 1993; Lambin 1996; Walsh et al. 1999), and (d) using interdisciplinary approaches and cross-cutting technologies (Sengupta and Bennett 2002; Li and Yeh 2002; Read et al. 2003).

Spatial patterns of LULC may be mapped for a discrete time through field measurements and/or the use of single-date air photos or satellite images; for "snap-shots" in time through an assembled image time-series; or continuously in time through an assortment of electronic devices such as quantum sensors linked to data loggers or "automated stations" transmitting data to orbiting satellites for downloading to ground receiving stations. GIS are often used to integrate data characterizing a host of space-time elements of variables and system parameters so that process relationships can be examined and complex issues such as feedback mechanisms, switches, and thresholds between system components can be assessed (Malanson et al. 1990, 1992; Phillips 1995a, b; Malanson 2003). Image animations, movie loops, and other forms of scientific data visualizations are used to consider the spatial co-occurrence of variables, spatial-temporal trends and variations in the pattern of variables, and the nature of data assembled from disparate sources and multiple resolutions fused through a variety of data transformations. Finally, geographers are using, among other techniques, (a) multilevel models to integrate effects seen at multiple scales; (b) cellular automata to simulate LULC patterns through rules of behavior, initial conditions, and neighborhood relationships; and (c) agent-based models to consider the effects of nonlinearity, emergence, and evolutionary behavior of systems on LULC patterns. All in all, spatial digital technologies are affecting geography in a number of fundamental ways: they enhance and value-add to the richness of our knowledge, increase our understanding of how our environment is organized and connected in space and time, and graphically represent how variables are organized and how systems function in a host of ecological and geographical settings.

The goal of this chapter is to describe important elements of a selected group of spatial digital technologies that offer geographers immense potential, as well as proven capacities, for studying LULC dynamics and assessing the multi-thematic and scale-dependent drivers of change. We will consider how technology is being applied to the study of LULC dynamics by emphasizing landscape characterization involving human-environment domains; how data can be value-added through the spatial digital technologies; and how different modeling environments can be used to integrate human and environment variables and space-time effects.

2. THE CONTEXT OF LAND-USE/LAND-COVER CHANGE

Grand Challenges in Environmental Sciences (National Research Council 2001) identified an environmental research agenda for the next decade that has important implications for science, technology, and policy. Central to the recommendations of the National Research Council was the interaction of people, place, and environment and the space-time linkages to land-use and land-cover dynamics. Also cited was the importance of the continued development of a suite of spatial digital technologies for integrating scientific theory and information systems. This "challenge" was built upon several international- and U.S.-based initiatives that defined the role of land change, first within global environmental change science (International Geosphere-Biosphere Program and the International Human Dimensions Program (IGBP-IHDP) 1995, 1999), including NASA's Land Cover/Land Use Change (LCLUC) research program, and subsequently in regard to biodiversity and ecosystem initiatives, including the emergence of sustainability science and its links to policy (National Research Council 1999; Kates et al. 2001). Indeed, the National Research Council's (2002) Down to Earth: Geographic Information for Sustainable Development in Africa was delivered by Secretary of State Colin Powell to the World Summit on Sustainable Development held in Johannesburg, South Africa in August, 2002. And as this chapter goes to press, biophysical, remote sensing/GIS, and social scientists of the IGBP and IHDP are creating a new "Land Project" that elevates the understanding to the couple human-environment system, much of which examines the synergies between land-cover and land-use change.[1]

Crews-Meyer (2002a) points out that increasingly policymakers and researchers are confronted with problems that are manifested and function at multiple spatial and temporal scales or extents. Global climate change and land-use/land-cover dynamics are but two of them. The exchanges between people and the environment are often played out through land-development and land-transformation scenarios (Dunning et al. 1995; Janssen 2000). Yet a lack of adequate tools and theoretical understandings across the social, natural, and spatial sciences has traditionally led researchers to focus on relatively coarse grains of analysis where aggregate data are available. But it is at the finer social, biophysical, or spatial scales where spatially explicit information may be more appropriately collected, derived, and applied, because decision-making about the use of the land is often local (Rindfuss et al. 2003; Turner and Geoghegan 2003). Furthermore, data analyses have tended to focus on either people or the environment, without suitable integration across thematic domains, and often occurring without a spatially and temporally explicit context. A number of researchers have linked human and environmental data at the finer household and/or community levels, and have integrated spatial digital technologies as they examine LULC dynamics (see Fox et al. 2003 for a collection of papers on the subject).

Land-use and land-cover change is one cornerstone of the science of global environmental change, sustainability, and increasingly, environment-and-development (Turner 2002). It is so because human use and changes of land surface have a spatial reach and magnitude that affects the structure and function of the biosphere, with assorted impacts on humankind (National Research Council 1999; Steffen et al. 2002; Turner 2002). Several examples illustrate. Human uses of the Earth usurp significant NPP (Net Primary Production) (Vitousek et al. 1997; DeFries at al. 1999). Biomass burning strongly affects nutrient cycling globally (Crutzen and Andreae 1990). Land-base emissions of C02 account for about 33 percent of the atmospheric increase in this "greenhouse gas" atmosphere since the Industrial Revolution (Watson et al. 2000), and land-surface conditions may account for the missing carbon sink (Pacala et al. 2001). Watersheds have been fragmented by the impoundments of 35,300 dams constructed since 1950 (Johnson et al. 2001), one part of the loss of ecosystem services worldwide (Clark et al. 1999; Dailey et al. 2000), including the lost biota (Myers 1997). These and other changes increase the vulnerability of people, places, and coupled ecosystems to hazards of all kinds (e.g., Chamedies et al. 1994; Johnson and Lewis 1995; Kasperson et al. 1995; Matson et al. 1997) and increasingly elevates what Turner (2002) labels, "integrated land-change science."

3. LAND-USE/LAND-COVER THEORY

Land-use and land-cover changes are highly varied in kind, and follow from multiple and complex processes. For this reason, among others, no "theory" of land change per se exists; no general explanation accounts for all types of changes in land-use and land-cover, and no comprehensive explanation treats the synergy of different land changes as they occur in place (Turner 1994). Indeed, much of the understanding and modeling of land change is "empirical" in kind: observation and monitoring generate evidence and data relative to the spatiotemporal patterns of land changes; and narrative or quantitative approaches induce causal connections of these changes (e.g., Geist and Lambin 1996; Lambin 1999; Lambin et al. 2001). Theories applicable to land change tend to be sectoral in kind, such as von Thünen's classic "concentric zonation" of decreasing land-use intensities away from a central market owing to land-bid rents, which also explains urban land-use expansion into other land uses. Here "theory" refers to the explanation of some kind of land change, not the modeling form used to express the explanation (see Irwin and Geoghegan 2001).

Among the sectors receiving significant attention over the past several decades is agriculture, especially mixed subsistence-market cultivation so prevalent throughout third-tier economies in which significant land-cover change has been taking place. Here a number of peasant economic theories explain the expansion and intensification of cultivation linked to changes in demand, either endogenously through household labor-consumer relations or exogenously through market relations (e.g., Turner and Ali 1996). To these can be added political economy explanations focused on structural changes that provide or deny access to land, or that empower or not individuals, groups, and corporations relative to land uses. Most of these theories do not signal "where" land will change, but "why" it changed.

Perhaps no kind of land change has received as much attention as tropical deforestation and woodland loss, including numerous efforts to model the change in various locales around the world. One comparison of case studies (n=180) reveals that multiple causes, up to three and four sets of major "cause clusters" (e.g., market or policy change) are needed to explain most cases of deforestation (Geist and Lambin 2001), although major regional "scenarios" may be applicable (Lambin et al. 2001). Another comparison of tropical deforestation models concludes that no simple set of explanatory factors work across all cases (Angelsen and Kaimowitz 1999). At best, these reviews support the simple and obvious observation that tropical deforestation takes place when and where it is profitable to someone or by the ability of some group to influence policy (Angelsen and Kaimowitz 1999).

This last truism embeds a reality that has hindered the development of a comprehensive land-change theory relevant to human agency and structure. Numerous interactive factors influence the decision to change land uses and covers. These factors involve complex linkages between directed (i.e., policy, planned) and autonomous action, including cascading or downstream consequences that were not intended, but equally intentional (i.e., policy-led timber extraction followed by autonomous settlement of logged areas). Such circumstances lend support to the recognition that significant differences in understanding land change are generated by the locale and scale of the unit examined.

To this complexity must be added that of the biophysical processes affecting land-use and land-cover. These processes are well understood for many coupled human-environment systems, such as the amplification of land-cover taken to cultivation owing to soil degradation on cropped land. Less well understood are those couplings in biophysical systems with multiple equilibria and in which the state condition of land cover is continually in flux. These themes follow largely from work in landscape ecology, much of which does not attempt to explain land change per se, but examines ecosystem consequences of different spatiotemporal patterns of change. Landscape ecology points to the interactions between landscape composition, spatial organization, and time as factors affecting human behavior (Walsh et al. 1999; Walsh et al. 2001; Read and Lam 2002).

Relatively recently, complexity theory has been used to address the rates and patterns of LULC dynamics and the possible nonlinear feedbacks between the processes of change and existing patterns (Manson 2001). Changes in LULC may depend partly on the existing patterns of LULC, which may involve critical points where a small amount of change significantly alters feedback processes and leads to a new pattern or equilibrium (Epstein 1999). Luhman (1985) states that complex systems are those that contain more possibilities than can be actualized. The goal of complexity theory is to understand how simple, fundamental processes can combine to produce complex holistic systems (Gell-Mann 1994). Nonequilibrium systems with feedbacks can lead to nonlinearity and may evolve into systems that exhibit criticality. Agent-based models and cellular automata approaches are being used to examine LULC dynamics and the associated factors of change (Parker et al. 2002). Human and biophysical complexity in land change and the multisectoral or comprehensive understanding of this change in places (locales, regions, and so on) does not equate to partial understanding and does not preclude modeling, as we shall see.

4. REMOTE SENSING FOR LULC CHARACTERIZATION

4.1. Satellite Remote Sensing

Satellite technology is fundamental to the characterization of LULC for single points in time and across multiple time points. An assembled satellite image time-series is commonly derived particularly for remote, inaccessible, and large geographic regions. The spatial, spectral, temporal, and radiometric resolutions of remote sensing systems serve as selection criteria in the choice of reconnaissance platforms and sensor systems. Relating these resolutions to the mapping requirements is basic to remote sensing. The vantage point of Earth observation, large areal extents, computer compatibility, and the orbital characteristics for repeat remote sensing are among the advantages of satellite systems for characterizing state and condition variables as part of LULC mapping (DeFries et al. 1999; Veldkamp and Lambin 2001).

The Landsat systems have gained wide prominence in LULC remote sensing, particularly at landscape levels of analysis, because of its temporal depth, fine to moderate spatial resolutions, large image extents, spectral characteristics extending from the visible to the near-, middle-, and thermal infrared wavelengths, and the 8-bit radiometric properties of the sensors.[2]

Common in remote-sensing studies of LULC is the tracking of LULC dynamics at the pixel level, land management unit, or at some other landscape strata. Remote-sensing analysts frequently integrate multiresolution systems in the study of LULC. For example, NOAA's (National Oceanic and Atmospheric Administration) Advance Very High Resolution Radiometer (AVHRR) data, with a local area coverage of approximately 1 x 1 km cell size and high temporal periodicity, may be used to develop a regional perspective by tracking weekly or bimonthly plant biomass variations over relatively large geographic areas. This regional context may be supplemented by a local perspective afforded by 1 x 1 m panchromatic and/or 4 x 4 m multispectral IKONOS satellite images for specified subareas, whereas Landsat Thematic Mapper might be used for acquiring "wall-to-wall" spectral characteristics of the study area as the primary mapping system. There are many deviations from this somewhat typical scenario. Aerial photography is often used for detailed mapping for control areas, because IKONOS (and QUICKBIRD) data are a relatively new product. MODIS (Moderate Resolution Imaging Spectrometer), too, is supporting local to regional studies, for example, through its 250 x 250 m and 500 x 500 m pixel resolutions (Zhan et al. 2000). Ground-based, aircraft-based, and helicopter-based videography and digital or analog still photography often completes the multisystem remote-sensing design specifications. But the scope of work and the goals of the study define the remote-sensing design, and the image preprocessing, enhancement, and information extraction protocols.

Digital elevation models (DEMs) are routinely used in conjunction with remotely sensed digital data. Often used to correct for illumination bias imposed by topographic orientation or used to topographically stratify a LULC classification, DEMs are an important element of the landscape characterization process. While 30 x 30 m DEMs are the standard product in the U.S., studies placed within a number of international settings are far less fortunate. Often, contour lines from topographic base maps are digitized and DEMs constructed to meet project goals, but they are constrained by the inherent specifications of the base map, particularly the contour interval and the areal extent of the terrain information. New technologies including SAR (Synthetic Aperture Radar) interferometry, LIDAR (Light Detection and Ranging) imaging lasers, and terrain-emphasizing sensors (Shuttle Imaging Radar) on-board the U.S. Space Shuttle offer alternative approaches for the construction of high-resolution DEMs to support LULC studies in remote and distant landscapes (Flood and Gutelius 1997; Jensen 2000).

4.2. Digital Aircraft Data

In addition to satellite technologies for mapping LULC patterns, aerial photography has a rich history of LULC mapping achieved historically through analog approaches, and more recently, through digital technologies. Traditionally, 9 x 9 inch, hardcopy prints have been acquired by placing cameras and film in an aircraft, in captive balloons, and/or positioned on topographic promontories. Panchromatic, natural color, and color-infrared aerial photography are most common. Standard overlap between successive photos and between photo flight-lines insure stereoscopic views of acquired photo-frames. While standard photogrammetric equipment is used for image measurement and interpretation, digital scanners and image analysis software are also used to construct seamless digital mosaics that can be analyzed through standard image-processing techniques. Overlays generated through the interpretation of aerial photographs can also be scan-digitized for subsequent digital analysis. Archived aerial photography is a valuable source of information about historical LULC types and patterns, and as a source product for performing validation of satellite land-cover classifications, change-detections, and enhancements. Aerial photography also has been of considerable value in defining landscape factors that may impact LULC patterns, including the expansion of road networks, changes in settlement patterns, and evidence of land degradation.

With the advance of image processing capabilities and the need for high spatial-resolution remotely sensed data, digital aerial photography has been acquired to meet a host of mission requirements. One such system is the ADAR-5500. The ADAR system operates in four spectral channels extending from the visible into the near-infrared wavelengths. A spatial resolution of 1 x 1-m and image frames measuring 1500 x 1000 pixels in extent are standard specifications. Using an on-board GPS and a network of geodetic control points collected through ground-based rover units, image frames can be acquired and seamless mosaics developed through image-processing approaches. Differentially corrected through the use of a nearby base-station, location-based coordinates of landscape reference points are used to rectify the images through geometric alignment to standard base maps, often contained within a GIS.

The ADAR-5500 system is a second-generation, charge-couple device frame camera system. Typically, the imagery is used for high spatial-resolution studies, performing calibration and/or validation tests for subregions, and for describing landscape variables and systems for singular points in time, possibly as the base period or initial conditions. Following the application of standard geometric and radiometric corrections to the ADAR (similar to the preprocessing of satellite imagery) data to reduce the effects of geometric errors, terrain-imposed illumination variance, and atmospheric inequalities (Walsh et al. 1998; Jensen 2000), the ADAR imagery can be digitally enhanced through a number of approaches. Walsh et al. (2003a) used Principal Components Analysis (PCA) and Wavelet Transform (WLT) to study the alpine treeline ecotone.

In brief, PCA is a data compression and enhancement technique that generates independent, orthogonal axes using the spectral channels as input. The PCA is often calculated using a covariance matrix. WLT is used to enhance both "trend" and "detail" of image properties (e.g., shape, tone, texture, and pattern) expressed as numeric indices (e.g., mean, standard deviation, and number of edge pixels), applied as multilevel vertical and horizontal transforms (Bian 2003). Wavelet transforms were developed as a framework for decomposing images into their component parts at different resolutions (Mallat 1989a, b; Li and Shao 1994; Yocky 1996; Csillag and Kabos 1996; Csillag 1997; Mohanty 1997).

4.3. Image Change Detections and Multiresolution Integration

To characterize LULC dynamics, it is customary to assemble an aircraft or, more commonly, a satellite image time-series and to assess the nature of landscape composition and pattern through selected single image views, pair-wise comparisons, or dynamic views involving multiple images and visualization and/or modeling techniques (Tucker and Townshend 2000). LULC types and plant biomass levels are normally considered, and, more recently, assessed through pattern metrics that define the composition and spatial organization of cover-types and state variables (e.g., plant biomass levels) at a range of ecological scales, typically at the landscape-, class-, and patch-levels (Allen and Walsh 1996; Millington et al. 2001; Read and Lam 2002).

LULC composition and spatial organization at defined time periods are critical indicators of landscape form and function. But so too is the nature of change characterized between selected image dates, where the spatial-temporal and compositional-organizational elements of change can be assessed. Strategically, these change images can be selected to correspond in space and/or time to survey data collected as cross-sectional or longitudinal data. A number of image change-detection approaches can be used to characterize landscape dynamics (Green et al. 1994; Kaufmann et al. 2001; Haan et al. 2002; Walsh et al. 2003b). Among these, change trajectories and change vector analysis have shown potential in defining LULC patterns. Considered analogous to panel data analysis in which survey respondents are mapped over time through multiple observations, change trajectories track "pixel histories" by using LULC classifications to define the nature of change over time and space as part of the "life history" of a pixel across an assembled time-series (Crews-Meyer 2002b).

Two primary benefits of the panel approach are that the current landscape can be understood through the dominant trajectories of landscape change, and that periodic (interannual and otherwise) patterns of LULCC can be distinguished. Similarly, the introduction of both interannual and intraannual images into the panel analysis allows for the assessment of different drivers of LULC flux, such as deforestation (i.e., steady, interannual patterns of change) versus seasonal flux caused by solar, hydrographic, or phenological pulses (i.e., sporadic but cyclical intraannual patterns of change). Connecting these spatially explicit patterns of LULCC to socioeconomic and demographic survey data, and the policy record, enables a more complete assessment of the drivers of LULCC (Wood and Skole 1998; Liu et al. 1999; Mertens et al. 2000; Seto et al. 2002).

In addition, change vector analysis is used to examine the change in LULC for multiple spectral channels between selected dates using Euclidean geometry to derive change magnitude and direction vectors (Allen and Kupfer 2001). The vectors correspond to Landsat TM spectral channels and Tasseled Cap indices—brightness, greenness, and wetness. Spherical statistics are used to assess the relative change among LULC classes as defined through the multitemporal, multispectral change detection.

4.4. Assessing Landscape Structure

Pattern metrics are groups of algorithms that are used to assess the spatial structure, organization, or pattern of landscape state and condition variables. Often applied to a LULC classification, represented as a continuous coverage in a raster format, pattern metrics are calculated at the landscape, class, and patch levels by setting the grain and extent of the study and by defining the nature of the LULC classification scheme (McGarigal and Marks 1993). Landscape is a heterogeneous land area composed of a cluster of interacting ecosystems that is repeated in similar form across space and time. For applying the pattern metrics, the landscape is bounded, for example, by ownership boundaries, watershed extents, and/or by some type of ecological strata or zonation so that the spatial and compositional structure of mapped LULC types can be assessed. The output is an aspatial statistic that describes, for instance, the juxtaposition of cover types, the level of fragmentation, diversity information, and contagion descriptions. Class is the pattern, for example, of forest occurring within the bounded area, whereas patch is a single homogeneous and contiguous cover type occurring within the defined landscape. Landscape patterns are the result of a web of interacting and complex processes that function over a range of spatial and temporal scales (Turner 1990a, b; Ritters et al. 1995). The grain is the unit of observation, whereas the extent is the area of consideration.

Applied to a single image classification, pattern metrics quantitatively describe the spatial and compositional structure of the landscape, defined through the LULC classification (O'Neill et al. 1988; Frohn et al. 1996). Using a satellite time-series, pattern metrics are used to consider the changing nature of landscape composition and pattern over time and space as possible indicators of LULC sustainability, resilience, or dynamism (Allen and Walsh 1996; Walsh et al. 2001; Read and Lam 2002; Walsh et al. 2003).

Pan et al. (in press) modeled the spatial pattern of LULC dynamics of household farms in Ecuador, characterized through pattern metrics and applied to a satellite classification. Key factors predicting landscape pattern were population size and composition, plot fragmentation through land subdivision, expansion of the road and electrical networks, age of the plot for initial clearing stages, and topography. Remote sensing, GIS, a longitudinal socioeconomic and demographic survey, and pattern metrics were integrated to support the Generalized Linear Mixed Models (GLMMs) used to examine 1990 and 1999 land compositions and spatial patterns.

5. LINKING PEOPLE AND THE ENVIRONMENT

Thus far, we have described how remote-sensing technologies are used to characterize the environment by classifying LULC types, assessing plant biomass and greenness levels, performing change detections for selected time-periods, and assessing the composition and spatial organization of the mapped LULC types using pattern metrics. While these methodologies are extensively used to assess LULC types and patterns as a consequence of scale relationships, the associated processes may represent a host of thematic domains, for instance, socioeconomic, demographic, biophysical, and geographical. As a result, studies have considered the theoretical and practical approaches for linking people and the environment, particularly where the spatial digital technologies, including remote sensing and GIS, are used to characterize the environment, and socioeconomic and demographic surveys, among other approaches, are used to characterize the human dimension (Sui and Zeng 2001; Evans and Moran 2002). GPS technologies are used, for example, to characterize the spatial location of LULC types as part of field-validation studies and for spatially describing the location of survey respondents, such as the location of the dwelling units, land parcels used or owned by a household, and the position of important institutions and facilities distributed across the landscape.

Rindfuss et al. (2003) described procedures for linking people and the environment in Thailand where people live in nuclear-village settlement patterns and use land parcels that are areally distributed in discontinuous patterns, a circumstance also addressed in Cameroon by Mertens and colleagues (1997), the Sahel of western Africa by Turner (1999), Mali by Laris (2002), and in the Yucatán by Turner, Geoghegan, and Foster (2003). Walsh et al. (2003b) described research in Ecuador where people live on the land that they own, but where land subdivision through sale and kinship ties has altered land tenure and the corresponding pattern of land use, while McCracken et al. (1999) illustrated changing land decisions by household characteristics in Amazonia. Again, the issue of land-people complexity was addressed by Laney (2002) for Madagascar where considerable borrowing of lands by different households takes place. BurnSilver et al. (2003) described approaches for linking people to the pastoral land that their animal herds graze in Kenya, which has regional LULC implications, as does Robbins (1998) for Rajasthan, India, and Archer (in press) for the Karoo of South Africa.

What is important is to describe whether the linking process will start from the land and connect to people, or from people and connect to the land. There is no one-to-one match between them, and they produce very different selectivities. Land is stationary, continuous, and many of the characteristics of land cover can be assessed through remote-sensing technologies. The spatial linkage of LULC at a georeferenced, pixel location is relatively straightforward, although the characterization of LULC types is less so. People are mobile, discrete, and not easily detected directly by remote-sensing technologies. Land ownership boundaries provide one mechanism to connect a spatial landscape partition within which land-cover changes can be observed and linked to social survey data collected for land managers. This linking process is complicated, however, by the fact that landownership boundaries change over time as land holdings are fragmented by parcelization. Although human signatures or impacts on the landscape can be measured, they range over large geographic areas, may be disconnected in space and time, and can reflect a one-to-one, one-to-many, many-to-one, and many-to-many relationships of people to the land that they affect (Evans and Moran 2002). Linking people to the land is not a straightforward proposition.

6. GIS AND DATA INTEGRATION

GIS technology offers an analytical framework for data synthesis that combines a system capable of data capture, storage, management, retrieval, analysis, and display. From a functionality perspective (Figure 21-1), GIS techniques can (a) examine spatial and nonspatial relationships through analytical tools and techniques, (b) represent an array of landscape perspectives through the integration of geographically registered spatial coverages, (c) efficiently display such information through a variety of data visualization approaches for spatial and temporal pattern analysis, (d) examine the co-occurrence of spatial and nonspatial data through database manipulations, (e) display singular thematic coverages or composited coverages through cartographic and/or statistical approaches, and (f) describe the location and behavior of phenomena through interfaces to models. The power of the GIS is based on its interactivity, integration of information, data visualization tools,

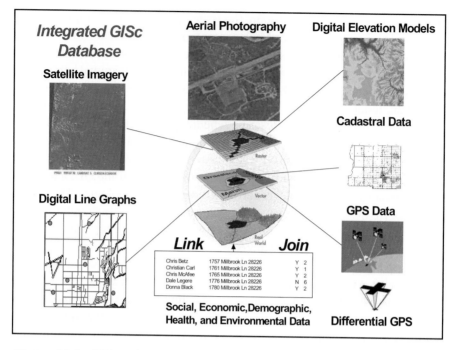

Figure 21-1. GIS and the integration of data and technologies for linking people, place, and environment.

analytical approaches, and spatial analysis tools and techniques offered to the analyst.

6.1. Data Visualizations

Like the old adage, "a picture is worth a thousand words," data visualization seeks to communicate by using maps and graphics that add value to the data and insight to the user. Relying upon both art and science, data visualization increases our understanding of data distributions and relationships by adding meaning to spatially explicit data at various stages of analysis. Whether mapping single variables to learn about spatial patterns of occurrence, mapping multiple variables to consider spatial co-associations, or designing graphics that tell the story of temporal lags or multilevel relationships, data visualization seeks to transform data into information through the use of computer graphics, image animation, image drapes, perspective views, and more. For studies of land-use and land-cover dynamics, data visualizations are used to consider population-environment interactions and the interplay, for example, between deforestation, agricultural extensification, and population migration. These visualization tools have been used within landscape architecture and are now being applied to studies of LULC change research through an integration of visualization, modeling, GIS and remote sensing techniques (Pitt 1992; Batty and Howes 1996; Simpson 2001). Maps and graphics have been developed to show the change in LULC over time and space by deriving, for example, simulations of land-use and land-cover dynamics used to characterize landscape conditions for future time periods, given hypothesized drivers of change, initial conditions, and neighborhood associations. Data are also being fused, for example, by merging high spectral-resolution remote sensing data with high spatial-resolution imagery through the Intensity-Hue-Saturation color model. Also, discrete data such as household dwelling units or community centroids are being transformed to continuous representations to provide cartographic compatibility with representations of the environment for subsequent analyses.

In short, data visualizations effectively render the social and biophysical landscapes through a host of data transformations and derived graphics that communicate through patterns, relationships, and spatial associations (Simpson 2001). Examining the linkages between people and place offer fresh insights into pressing academic and societal issues, and provide a clarity of meaning through visualizations involving maps and graphics that "speak to the user" through colors, symbols, and cartographic representations.

6.2. The Internet and Data Accessibility

Equally vital to the generation of well-crafted data visualizations is the use of the Internet for sharing such data views that effectively communicate with local as well as remote researchers, educators, and policymakers elsewhere in the U.S. and abroad. Meta-search tools are emerging that allow researchers to find publicly available data for domestic and international research (e.g., the Geospatial Data Clearinghouse). Substantial computer power, sophisticated peripherals, complex software, and a well trained staff need *not* be a prerequisite for interacting with data visualizations. Internet technologies merged with data transfer and viewing protocols can be used to "see what we see" and to become fully engaged in the interpretation of data and the extraction of meaning and understanding from information portrayed through maps and graphics. "Data without borders" is the operating paradigm, data visualization is the shorthand for conveying meaning, and web-based technologies are the mechanisms for the engagement of international participants into scientific inquiry and discourse.

With graphics in-hand, conversations between scholars and policymakers can be accommodated using e-mail listservs, discussion forums, and interactive messaging. In such procedures, interactivity is provided to researchers and/or policy specialists who can integrate their ideas through dialogue among all project participants regardless of location. For instance, computer simulations of land-use and land-cover dynamics could be made available to all project participants and interactive dialogue engaged to extract meaning and insights. Issues related to local constraints on the model, policy implications, error and uncertainty on model performance, exogenous factors and their temporal lags, and how pattern and human behavior are interrelated could be readily addressed. Data transfer standards, including metadata standards, and concepts of "truth in labeling" and "fitness for use" set the responsibilities of the producer and user of data respectively. Reporting errors and uncertainty, describing the data-processing lineage, and using data-transfer standards are important elements of data sharing and using archival data for LULC characterizations.

In short, the visualization technology, dissemination approaches, and the local/remote interactivity technology exists where investigators, educators, and policy specialists in developing countries, or in remote locations, can engage in scientific, management, and policy discussions. Today, we have the critical components to inform the disenfranchised, connect the isolated, and integrate voices and perspectives for greater understanding of interactions between people, place, and the environment. Community and local empowerment can be achieved through this information sharing. In the northeastern Ecuadorian Amazon, for instance, Bilsborrow and Walsh (Walsh et al. 2003b) are conducting presentations and workshops on the use of collected and processed socioeconomic and demographic survey data, an assembled remote-sensing image time-series, and GIS coverages for assessing LULC dynamics, urban expansion, road development, and drivers of change within the region. Information is being shared with government and nongovernment organizations for planning and assessment purposes. Together, these organizations are providing a host of vital services within the region associated with health care delivery, road improvements, and land-use planning. Data access through the Internet and through GISc technologies is giving a voice to local people and places often dismissed in the face of petroleum exploration and other national concerns. In Thailand too, Walsh, Entwisle, and Rindfuss (Rindfuss et al. 2003) are sharing the results of spatial simulations of LULC dynamics with local government and nongovernment organizations as well as public schools in the U.S. and Thailand, so that factors influencing deforestation and other forms of land conversion can be visualized and integrated into the planning and assessment process by local constituencies including households who are living and working within the study district.

7. MODELING APPROACHES: SPATIAL AND TIME INTEGRATION

7.1. Multilevel Models and Spatial Simulations

Our ability to link spatially and temporally across the socioeconomic, demographic, biophysical, and geographical domains and across the space-time dimensions are critical to assess the coupling of human and environmental systems (Grove et al. 2002). Computational simulations are a technological tool used by planners to understand the behavior of rural and urban land-cover change processes. A key aspect of these models or simulations is their ability to incorporate spatial and temporal complexity (see Agarwal et al. [2000] for a review). Models which allow the user to explore dynamics at multiple spatial and temporal scales of analysis are important, because of the scale dependence in complex social-biophysical systems (Veldkamp and Fresco 1996; Walsh et al. 1999; Stephenne and Lambin 2001). Multilevel models, for instance, can be used to examine variations in household decisions that represent multiple space and time considerations through, for example, the specification of effects at the individual/household and community-levels, combined with the presence of hierarchical effects and spatial autocorrelation (Laird and Ware 1982). To do this, multilevel models are used to assess the covariance structure of variables, account for within-area heterogeneity (e.g., farms related to households and households related to communities), hierarchical effects (e.g., arising from small areas being grouped into larger areas), and spatial effects (e.g., between household farms and communities).

A key aspect of multilevel models is the estimation and modeling of the impacts of a wide range of factors on land use occurring at different spatial (and/or temporal) scales. Prices of major agricultural commodities produced in a region have well-known effects on what farmers grow and how much land they clear (Bilsborrow 1994). The growth of local towns may also increase the opportunities for off-farm employment and therefore also affect labor allocated to farm versus off-farm use. This is particularly the case near major towns. Of course, prices of certain commodities (e.g., petroleum, timber, coffee) are affected by world market prices.

To integrate these different levels of information, a conceptual multilevel model can be used to estimate multiscale effects in a statistical context. Figure 21-2 shows how a multilevel model might be used to accommodate variables operating at different scales. Two- and three-level hierarchical models with

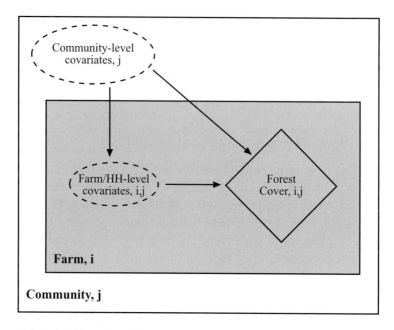

Figure 21-2. Linking the effects of households and communities through a multi-level model.

multiple dependent variables can be utilized to estimate LULC and to distinguish the effects of individual/household and community-level factors.

7.2. Spatial Simulations and Landcover Change Modeling

A variety of different modeling methods have been used to develop complex simulations of land-cover change in urban and rural systems (Fischer and Sun 2001). Dynamic simulation models are effective tools to explore feedbacks and interrelationships in systems (Evans et al. 2001; Stephenne and Lambin 2001). With a few exceptions (Costanza and Ruth 1998), however, these dynamic simulation models are not effective tools for exploring spatial patterns and spatial interactions. Spatially explicit modeling methods such as cellular automata (CA) and agent-based models (ABM) are highly suited to the exploration of spatial factors in land-use systems. A cellular automaton system consists of a regular grid of cells, each of which can be in one of a finite number of k possible states, updated synchronously in discrete time steps according to a local interaction rule (Messina and Walsh 2001). The state of a cell is determined by the previous states of a surrounding neighborhood of cells (Wolfram 1984). The rule contained in each cell is usually specified in the form of a transition function or growth rule that addresses every possible neighborhood configuration of states. The ability of a system to grow and then alter its rate of growth and possibly reverse itself or "die" is a fundamental attribute in biological or human system CA modeling. The systems modeled by Clarke et al. (1996, 1997) and the modeling scheme proposed by Messina and Walsh (2001) attempt to follow biological patterns. CA has the capability to infuse concepts of thresholds, feedbacks, hierarchy, and complexity in the simulation of landscapes (Malanson 1999), particularly those associated with examining the change patterns of LULC as a consequence of scale-dependent forces and processes. Emergent phenomena, persistent patterns occurring in generated systems, are a general goal of CA approaches.

Messina and Walsh (2001) used organic growth rules to spread LULC conversion in the northeastern Ecuadorian Amazon outward from existing developed, deforested, or agricultural areas, representing the tendency of cities to expand through spatial diffusion. Both organic and spontaneous growth rules were modeled by Messina and Walsh (2001). Diffusive growth promoted the random dispersed development of urban centers regardless of proximity functions (Clarke et al. 1996). Spontaneous growth occurred when a randomly chosen cell was located nearby an already urbanized cell, simulating the influence that urban areas have on their surroundings (Clarke et al. 1997). Road-influence growth encouraged cells to develop along the transportation network, replicating the effects of increased accessibility on LULCC, a central issue in the Ecuadorian Amazon, because of the comparative advantage of farmers connected to market towns through the limited and unevenly improved road network. Physical elements such as hydrology, soils, and slope, contained with a GIS, can also be incorporated into the CA model. Soares-Filho et al. (2002) also used a CA model to simulate land-use dynamics, in a frontier region of Brazil. Land fragmentation was considered as an outcome of different colonization projections and dynamic phases of landscape change. The model incorporated a stochastic simulation engine, transition probabilities of LULC change from a satellite time-series, and spatial feedbacks.

CA can be interpreted to represent the decision making of a variety of stakeholders. These decisions are made in the context of an existing pattern of land use. Transition probabilities are used to denote the likelihood of LULC dynamics depending upon endogenous factors and neighborhood associations. Household surveys are an important source of data for determining the transition probabilities. Model runs produce a variety of LULC patterns. Each pattern generated can be compared to the patterns of LULC change observed in the remote-sensing image time-series (Verburg et al. 1999). One test of the model is the degree to which it reproduces a pattern similar to that observed. System behaviors and LULC patterns can also be interpreted within a policy-relevant context by comparing simulated LULC scenarios to targeted land-management outcomes. Multiple LULC change scenarios can be developed around defined policy goals. Model convergence and variable sensitivities can be examined relative to the LULC patterns, model variables, and policy goals and expectations (White and Engelen 1997; Wu 1998).

An emerging modeling technique in land-cover-change analysis is the use of agent-based models (ABM), also called multiagent systems (MAS) (Parker et al. 2002, 2003). Agent-based models were formerly more theoretical than applied, but recently an increasing number of researchers have been applying ABM techniques to field sites supported by rich empirical datasets (Hoffman et al. 2002). Agent-based models are defined by a series of agent-interaction rules that allow individual-level behaviors to be scaled to aggregate or macrolevel outcomes. ABM techniques are useful tools for exploring LULCC, because of the importance of agent-interactions in land-cover-change outcomes. For example, an ABM may be used to explore how the participation of a single landholder in a policy program providing tax incentives for reforestation can result in the dissemination of information about public programs through communication among proximal land-owners. As more landholders become aware of these tax incentive programs, aggregate-level land-cover changes can be observed, such as the decrease in patchiness of landscape.

7.3. Field Methods: Calibration and Validation

Numerous field methods have been developed to support remote sensing studies of LULC characterization and dynamics, as well as for calibrating and validating image interpretations and model runs including spatial simulations (Messina and Walsh 2001; Marcano-Vega et al. 2002). GPS technology is commonly used to define geographic location either by navigating to preselected points for sampling (or performing a census) people and/or the environment, or by randomly or systematically selecting points across the landscape to characterize the composition of LULC or some other important landscape feature. Here our intent is to emphasize technology and not necessarily sampling designs per se.

Among the more useful technologies are spectroradiometers—relatively small, portable devices used for spectral sampling at landscape points through electronic means. Often times replicating spectral regions sensed by satellites and/or aircraft systems, spectroradiometers are used to collect spectral control data for LULC types and water characteristics at specified locations and for documented environmental conditions (Goodin et al. 1993; Ullah et al. 2000). Associated electronic devices include plant canopy analyzers that are used to compute vegetation indices such as the Leaf Area Index (LAI). Normally, the

LAI is computed in near real-time at discrete point locations and subsequently transformed to continuous surfaces by developing correlations between the LAI and satellite-based measures of the Normalized Difference Vegetation Index (NDVI), a measure of plant greenness linked to plant biomass. Finally, quantum sensors, connected to data loggers and possibly vertically positioned along a sensor mast, are used to continuously measure light attenuation through the plant canopy as a function of plant architecture, species type, and/or disturbance characteristics.

In general, the electronic devices cited above are used for basic remote sensing, creating control observations for scale-up to aircraft and satellite remote sensing systems, and as devices for calibrating and validating remote sensing interpretations, as well as model results (Rundquist 2001). But advances have also occurred in data-collection technologies that now include GPS units integrated with hand-held GIS devices for in situ data collection and information display as part of social surveys (Walsh et al. 2003). Particularly useful for navigating through difficult settings and offering on-screen information as part of the data collection, such devices further extend information technology to field settings as thematic and spatial context to the interview process.

8. CONCLUSIONS

Technology and the characterization of LULC patterns and dynamics are intertwined in very fundamental ways. Used for data collection, analysis, and display, technology offers insights into the how, when, where, and why of LULC change. GIScience recognizes the interplay of spatial digital technologies in the study of Land Science by providing the tools and techniques for gathering and integrating multithematic and space-time dependent variables in the study of LULC patterns, change trajectories, and the associated drivers of change. Of particular relevance is the need to characterize landscapes at multiple spatial and temporal resolutions thereby requiring the integration of new and varied remote sensing systems as part of the data collection process. New functionalities in GIS software systems, advances in integrated GIS and 3-D visualizations, improvements in data access and distribution protocols, and new approaches for spatial modeling such as multiagent simulations that incorporate evolutionary computations (i.e., genetic algorithms and genetic programming) offer landscape views and scientific insights not previously available to the land scientists. Advances in technology will continue to expand our understanding of LULC patterns and processes. Equipping the land scientist with suitable theories and practices to guide the best and most appropriate use of technology will continue to be a challenge, but the benefits to be realized are considerable.

NOTES

1   See the following websites for newsletters and additional information about the IGBP [http://www.igbp.kva.se], IHDP [http://www.ihdp.uni-bonn.de], and the NASA LCLUC Program [http://lcluc.gsfc.nasa.gov].

2   See Jensen (2000) for a review of the Landsat systems (and other useful remote sensing systems for LULC mapping), their sensor specifications, and image-processing approaches.

REFERENCES

Agarwal, C., Green, G., Grove, L.M., Evans, T.P., and Schweik, C. (2000). A Review and Assessment of Land-Use Change Models: Dynamics of Space, Time, and Human Choice, Fourth International Conference on Integrating GIS and Environmental Modeling (GIS/EM4), Banff, Canada.

Allen, T.R. and Kupfer, J.A. (2001). Spectral Response and Spatial Pattern of Fraser Fir Mortality and Regeneration, Great Smoky Mountains, USA, Plant Ecology 156: 59-74.

Allen, T.R. and Walsh, S.J. (1993). Characterizing Multitemporal Alpine Snowmelt Patterns for Ecological Inferences, Photogrammetric Engineering and Remote Sensing 59: 1521-29.

Allen, T.R. and Walsh, S.J. (1996). Spatial and Compositional Structure of the Alpine Treeline Ecotone, Photogrammetric Engineering and Remote Sensing 62: 1261-68.

Angelsen, A. and Kaimowitz, D. (1999). Rethinking the Causes of Deforestation: Lessons from Economic Models, World Bank Research Observer 14: 73-98.

Archer, E.R.M. (in press). Beyond the Climate Versus Grazing Impasse: Using Remote Sensing to Investigate the Effects of Grazing System Choice on Vegetation Cover in Eastern Karoo, Journal of Arid Environments.

Batty, M. and Howes, D. (1996). Exploring Urban Development Dynamics Through Visualization and Animation. In Parker, D. (Ed.) Innovations in GIS, 149-60. London: Taylor and Francis.

Bian, L. (2003). Retrieving Urban Objects Using a Wavelet Transform Approach, Photogrammetric Engineering and Remote Sensing 69: 133-42.

Bilsborrow, R.E. (1994). The Roles of Population and Development in Deforestation in Developing Countries, UN Expert Group Meeting on Population, Environment, and Development, New York.

Brown, D.G. and Walsh, S.J. (1991). Compatibility of Non-synchronous in-situ Water Quality Data and Remotely-Sensed Spectral Information for Assessing Lake Turbidity Levels in Complex and Inaccessible Terrain, GeoCarto International 6: 5-11.

BurnSilver, S.B., Boone, R.B., and Galvin, K.A. (2003). Linking Pastoralists to a Heterogeneous Landscape: The Case of Four Maasai Group Ranches in Kajiado District, Kenya. In Fox, J., Rindfuss, R.R., Walsh, S.J., and Mishra, V. (Eds.) People and the Environment: Approaches for Linking Household and Community Surveys to Remote Sensing and GIS, 173-99. Boston: Kluwer Academic Publishers.

Chameides, W.L., Kasibhatla, P.S., Yienger, J., and Levy II, H. (1994). Growth in Continental-Scale Metro-Agroplexes: Regional Ozone Pollution and World Food Production, Science 264: 74-77.

Clark, J.S., Carpenter, S.R., Barber, M., Collins, S., Dobson, A., Foley, J.A., Lodge, D.M., Pascual, M., Pielke Jr., R., Pizer, W., Pringle, C., Reid, W.V., Rose, K.A., Sala, O., Schlesinger,W.H., Wall, D.H., and Wear, D. (1999). Ecological Forecasts: An Emerging Imperative, Science 286: 685-86.

Clarke, K.C., Gaydos, L., and Hoppen, S. (1997). A Self-modifying Cellular Automaton Model of Historical Urbanization in the San Francisco Bay area, Environment and Planning B 23: 247-61.

Clarke, K.C., Hoppen, S., and Gaydos, L. (1996). Methods and Techniques for Rigorous Calibration of a Cellular Automaton Model of Urban Growth, Third International Conference/Workshop on Integrating GIS and Environmental Modeling. Santa Barbara: National Center for Geographic Information and Analysis.

Costanza, R., and Ruth, M. (1998). Using Dynamic modelling to Scope Environmental Problems and Build Consensus, Environmental Management 22: 183-95.

Crews-Meyer, K.A. (2002a). Challenges for GIScience: Assessment of Policy Relevant Human-Environment Interactions. In Walsh, S.J. and Crews-Meyer, K.A. (Eds.) Linking People, Place, and Policy: A GIScience Approach, 1-5. Boston: Kluwer Academic Publishers.

Crews-Meyer, K.A. (2002b). Characterizing Landscape Dynamism Using Paneled-Pattern Metrics, Photogrammetric Engineering and Remote Sensing 68: 1031-40.

Crutzen, P.J. and Andreae, M.O. (1990). Biomass Burning in the Tropics: Impact on Atmospheric Chemistry and Biogeochemical Cycles, Science 250: 1669-78.

Csillag, F. (1997). Quadtrees: Hierarchical Multiresolution Data Structures for Analysis of Digital Images. In Quattrochi, D.A. and Goodchild, M.F. (Eds.) Scale in Remote Sensing and GIS, 247-72. Boca Raton, FL: Lewis Publishers.

Csillag, F. and Kabos, S. (1996). Hierarchical Decomposition of Variance with Applications in Environmental Mapping Based on Satellite Images, Mathematical Geology 28: 385-405.

Daily, G.C., Söderqvist, T., Aniyar, S., Arrow, K., Dasgupta, P., Ehrlich, P. R., Folke, C., Hansson, A., Jansson, B-O., Kautsky, N., Levin, S., Lubchenco, J., Mäler, K-G., Simpson, D., Starlet, D., Tillman, D., and Walker, B. (2000). The Value of Nature and Nature of Value, Science 289: 395-96.

DeFries, R., Field, C., Fung, I., Collatz, G., and Bounoua, L. (1999). Combining Satellite Data and Biogeochemical Models to Estimate Global Effects of Human-Induced Land Cover Change on Carbon Emissions and Primary Productivity, Global Biogeochemical Cycles 13: 803-15.

DeFries, R., Hansen, M., Townshend, J.R.G., Janetos, A., and Loveland, T. (2000). A New Global Dataset of PercentTree Cover Derived from Remote Sensing, Global Change Biology 6: 247-54.

Dunning, J.B., Stewart, D.J., Danielson, B.J., Noon, B.R., Root, T.L., Lamberson, R.H., and Stevens, E.E. (1995). Spatially Explicit Population Models: Current Forms and Future Uses, Ecological Applications 5: 3-11.

Epstein, J.M. (1999). Agent-Based Models and Generative Social Science, Complexity 4: 41-60.

Evans, T.P., Manire, A., de Castro, F., Brondizio, E., and McCracken, S. (2001). A Dynamic Model of Household Decision-making and Parcel Level Landcover Change in the Eastern Amazon, Ecological Modelling 143: 95-113.

Evans, T.P. and Moran, E.F. (2002). Spatial Integration of Social and Biophysical Factors Related to Landcover Change. In Lutz, W., Prskawetz, A., and Sanderson, W.C. (Eds.) Population and Environment: Methods of Analysis. New York: Population Council.

Fischer, G. and Sun, L. (2001). Model Based Analysis of Future Land-Use Development in China, Agriculture, Ecosystems, and Environment 85: 163-76.

Flood, M. and Gutelius, B. (1997). Commercial Implications of Topographic Terrain Mapping Using Scanning Airborne Laser Radar, Photogrammetric Engineering and Remote Sensing 63: 327.

Fox, J., Rindfuss, R.R., Walsh, S.J., and Mishra, V. (Eds.) (2003). People and the Environment: Approaches for Linking Household and Community Surveys to Remote Sensing and GIS. Boston: Kluwer Academic Publishers.

Frohn, R.C., McGwire, K.C., Dale, V.H., and Estes, J.E. (1996). Using Satellite Remote Sensing Analysis to Evaluate a Socioeconomic and Ecological Model of Deforestation in Rondonia, Brazil, International Journal of Remote Sensing 17: 3233-55.

Geist, H. and Lambin, E. (2001). What Drives Tropical Deforestation? A Meta-Analysis of Proximate and Underlying Causes of Deforestation Based on Subnational Case Study Evidence. Luovain-la-Neuve, BE., LUCC International Project Office.

Gell-Mann, M. (1994). The Quark and the Jaguar. New York: Freeman.

Goodin, D., Han, L., Fraser, R., Rundquist, D., and Stebbins, W. (1993). Analysis of Suspended Solids in Water Using Remotely Sensed High Resolution Derivative Spectra, Photogrammetric Engineering and Remote Sensing 59: 505-10.

Green, K., Kempka, D., and Lackey, L. (1994). Using Remote Sensing to Detect and Monitor Land-Cover and Land-Use Change, Photogrammetric Engineering and Remote Sensing 60: 331-37.

Grove, M., Schweik, C., Evans, T.P., and Green, G. (2002). Modeling Human-Environment Systems. In Clarke, K.C., Parks, B.O., and Crane, M.P (Eds.) Geographic Information Systems and Environmental Modeling, 160-88. Upper Saddle River, NJ: Prentice Hall.

Haan, N., Gumbo, D., Eastman, J.R., Toledano, J., and Snel, M. (2002). Linking Geomatics and Participatory Social Analysis for Environmental Monitoring: Case Studies from Malawi, Cartographica 37: 21-32.

Hoffman, M., Kelley, H., and Evans, T. (2002). Simulating Land Cover Change in Indiana: An Agent-Based Model of Deforestation. In Janssen, M. (Ed.) Complexity and Ecosystem Management: The Theory and Practice of Multi-Agent Approaches. Cheltenham UK, Northampton: Edward Elgar Publishers.

Homewood, K., Lambin, E.F., Coast, E., Kariuki, A., Kikula, I., Kivelia, J., Said, M., Serneels, S., and Thompson, M. (2001). Long-term Changes in Serengeti-Mara Wildebeest and Land Cover: Pastoralism, Population, or Policies? Proceedings of the National Academy of Sciences 98: 12544-49.

IGBP-IHDP (1995). Land-Use and Land-Cover Change Science/Research Plan. IGBP Report No. 35 and IHDP Report No. 7. Stockholm: IGBP and Geneva: IHDP.

IGBP-IHDP (1999). Land-Use and Land-Cover Change (LUCC). Implementation Strategy. IGBP Report No. 48 and IHDP Report No. 10. Stockholm: IGBP and Bonn: IHDP.

Irwin, E.G. and Geoghegan, J. (2001). Theory, Data, Methods: Developing Spatially-Explicit Economic Models of Land Use Change, Agriculture, Ecosystems and Environment 84: 7-24.

Janssen, M.A. (2000). The Human Factor in Ecological Models, Ecological Economics 35: 307-10.

Jensen, J.R. (2000). Remote Sensing of the Environment: An Earth Resource Perspective. Upper Saddle River, NJ: Prentice Hall.

Johnson, D.L. and Lewis, L.A. (1995). Land Degradation: Creation and Destruction. Oxford, UK: Blackwell.

Johnson, N., Revenga, C., and Echeverria, J. (2001). Managing Water for People and Nature, Science 292: 1071-72.

Kasperson, J.X., Kasperson, R.E., and Turner II, B.L. (Eds.) (1995). Regions at Risk: Comparison of Threatened Environments. Tokyo: United Nations University Press.

Kates, R.W., Clark, W.C., Corell, R., Hall, J.M., Jaeger, C.C., Lowe, I., McCarthy, J.J., Schellenhuber, H.J., Bolin, B., Dickson, N.M., Faucheaux, S., Gallopin, G.C., Grübler, A., Huntley, B., Jäger, J., Jodha, N.S., Kasperson, R.E., Mabogunje, A., Matson, P., Mooney, H., Moore, B., III, O'Riordan, T., and Svedin, U. (2001). Sustainability Science, Science 292: 641-42.

Kaufmann, R. and Seto, K. (2001). Change Detection, Accuracy, and Bias in a Sequential Analysis of Landsat Imagery in the Pearl River Delta, China: Economic Techniques, Agriculture, Ecosystems, and Environment 85: 95-105.

Laird, N.M. and Ware, J.H. (1982). Random-Effects Models for Longitudinal Data, Biometrics 38: 963-74.

Lambin, E.F., Agbola, S.B., Angelsen, A., Bruce, J.W., Coomes, O.T., Dirzo, R., Fischer, G., Folke, C., George, P.S., Homewood, K., Imbernon, J., Leemans, R., Li, X., Moran, E.F., Mortimore, M., Ramakrishnan, P.S., Richards, J.F., Skanes, H., Steffen, W., Stone, G.D., Svedin, U., Veldkamp, T.A., Vogel, C., Xu, J., Turner, B.L., and Geist, H.J. (2001). The Causes of Land-Use and Land-Cover Change: Moving Beyond the Myths, Global Environmental Change 11: 261-69.

Lambin, E.F. (1996). Change Detection at Multiple Temporal Scales: Seasonal and Annual Variations in Landscape Variables, Photogrammetric Engineering and Remote Sensing 62: 931-38.

Lambin, E.F. (1999). Monitoring Forest Degradation in Tropical Regions by Remote Sensing: Some Methodological Issues, Global Ecology and Biogeography 8: 191-98.

Laney, R.M. (2002). Disaggregating Induced Intensification for Land-Change Analysis: A Case Study from Madagascar, Annals of the Association of American Geographers 92: 702-26.

Laris, P. (2002). Burning the Seasonal Mosaic: Preventative Burning Strategies in the Wooded Savanna of Southern Mali, Human Ecology 30: 155-86.

Li, D. and Shao, J. (1994). Wavelet Theory and its Application in Image Edge Detection, International Journal of Photogrammetry and Remote Sensing 49: 4-12.

Li, X. and Yeh, A. G.O. (2002). Neural-Network-Based Cellular Automata for Simulating Multiple Land Use Changes Using GIS, International Journal of Geographical Information Systems 16: 323-43.

Liu, J., Ouyang, Z., Taylor, W., Groop, R., Tan, Y., and Zhang, H. (1999). A Framework for Evaluating the Effects of Human Factors on Wildlife Habitat: the Case of Giant Pandas, Conservation Biology 13: 1360-70.

Luhman, N. (1985). A Sociological Theory of Law. London: Routledge and KeganPaul.

Malanson, G.P. (2003). Habitats, Hierarchical Acales, and Nonlinearities: An Ecological Perspective on Linking Household and Remotely Sensed Data on Land-Cover/Use Change. In Fox, J., Rindfuss, R.R., Walsh, S.J., and Mishra, V. (Eds.) Linking Household and Remotely Sensed Data: Methodological and Practical Problems, 265-83. Boston: Kluwer Academic Publishers.

Malanson, G.P. (1999). Considering Complexity, Annals of the Association of American Geographer 89: 746-53.

Malanson, G.P., Butler, D.R., and Walsh, S.J. (1990). Chaos theory in Physical Geography, Physical Geography 11: 293-304.

Malanson, G.P., Butler, D.R., and Georgakakos, K.P. (1992). Nonequilibrium Geomorphic Processes and Deterministic Chaos, Geomorphology 5: 311-22.

Mallat, S. (1989a). A theory for Multiresolution Signal Decomposition: The Wavelet Representation, IEEE Transactions on Pattern Analysis and Machine Intelligence 11: 674-93.

Mallat, S. (1989b). Multifrequency Channel Decompositions of Images and Wavelet Models, IEEE Transactions on Acoustics, Speech, and Signal Processing 37: 2091-2110.

Manson, S.M. (2001). Simplifying Complexity: A Review of Complexity Theory, Geoforum 32: 405-14.

Marcano-Vega, H., Aide, T.M., and Baez, D. (2002). Forest Regeneration in Abandoned Coffee Plantations and Pastures in the Cordillera Center of Puerto Rico, Plant Ecology 161: 75-87.

Matson, P.A., Parton, W.J., Power, A.G., and Swift, M.J. (1997). Agricultural Intensification and Ecosystem Properties, Science 277: 504-08.

McCracken, S.D., Brondizio, E.S., Nelson, D., Moran, E.F., Siqueira, A.D., and Rodriguez-Pedraza, C. (1999). Remote Sensing and GIS at Farm Property Level: Demography and Deforestation in the Brazilian Amazon, Photogrammetric Engineering and Remote Sensing 65: 1311-20.

McGarigal, K. and Marks, B.J. (1993). Fragstats: Spatial Pattern Analysis Program for Quantifying Landscape Structure. Corvallis, OR: Forest Science Department, Oregon State University.

Mertens, B., Sunderlin, W., Ndoye, O., and Lambin, E.F. (2000). Impact of Macroeconomic Change on Deforestation in South Cameroon: Integration of Household Survey and Remotely-Sensed Data, World Development 28: 983-99.

Messina, J.P. and Walsh, S.J. (2001). 2.5D Morphogenesis: Modeling Landuse and Landcover Dynamics in the Ecuadorian Amazon, Plant Ecology 156: 75-88.

Millington, A.C, Walsh, S.J., and Osborne, P.E. (Eds.) (2001). GIS and Remote Sensing Applications in Biogeography and Ecology. Boston: Kluwer Academic Publishers.

Mohanty, K.K. (1997). The Wavelet Transform for Local Image Enhancement, International Journal of Remote Sensing 18: 213-19.

Myers, N. (1997). Mass Extinction and Evolution, Science 278: 597-98.

National Research Council. (1999). Our Common Journey: A Transition Toward Sustainability. Washington, DC: National Academy Press.

National Research Council (2001). Grand Challenges in Environmental Sciences. Washington, DC: National Academy Press.

National Research Council (2002). Down to Earth: Geographic Information for Sustainable Development in Africa. Washington, DC: National Academy Press.

O'Neill, R.V., Krummel, J.T., Gardner, R.H., Sugihara, G., Jackson, B., DeAngelis, D.L., Milne, B.T., Turner, M.G., Zygmunt, B., Christensen, S.W., Dale, V.H., and Graham, R.L. (1988). Indices of Landscape Pattern, Landscape Ecology 1: 153-62.

Pacala, S.W., Hurtt, G.C., Baker, D., Peylin, P., Houghton, R.A., Birdsey, R.A., Heath, L., Sundquist, E.T., Stallard, R.F., Ciais, P., Moorcroft, P., Caspersen, J.P., Shevliakova, E., Moore, B., Kohlmaier, G., Holland, E., Gloor, M., Harmon, M.E., Fan, S.M., Sarmiento, J.L., Goodale, C.L., Schimel, D., and Field, C.B. (2001). Consistent Land and Atmosphere-Based U.S. Carbon Sink Estimates, Science 292: 2316-20.

Pan, W.K.Y, Walsh, S.J., Bilsborrow, R.E., Frizzelle, B.G., Erlien, C.M., and Baquero, F.D. (in press). Farm-Level Models of Spatial Patterns of Land Use and Land Cover Dynamics in the Ecuadorian Amazon, Agriculture, Ecosystems and Environment.

Parker, D, C., Berger, T., and Manson, S.M. (2002). Agent-Based Models of Land-Use and Land-Cover Change. In McConnell, W.J. (Ed.) Report and Review of an International Workshop, Irvine, CA, October, 2001, LUCC Report Series No. 6.

Parker, D.C., Manson, S.M., Janssen, M.A., Hoffman, M.J., and Deadman, P. (2003). Multi-Agent Systems for the Simulation of Land Use and Land Cover Change: A Review, Annals of the Association of American Geographers 93.

Pastor, J. and Broschart, M. (1990). The Spatial Pattern of a Northern Conifer-Hardwood Landscape, Landscape Ecology 4: 55-68.

Phillips, J.D. (1995a). Time Lags and Emergent Stability in Morphogenic/Pedogenic System Models, Ecological Modelling 78: 267-76.

Phillips, J.D. (1995b). Self-organization and Landscape Evolution, Progress in Physical Geography 19: 309-21.

Phillips, J.D. (1999a). Spatial Analysis in Physical Geography and the Challenge of Deterministic Uncertainty, Geographical Analysis 31: 359-72.

Pitt, D. G. and Nassauer, J.I. (1992). Virtual Reality Systems and Research on the Perception, Simulation and Presentation of Environmental-Change, Landscape and Urban Planning 21: 269-71.

Read, J.M., Clark, D.B., Venticinque, E.M., and Moreira, M.P. (2003). Application of Merged 1-m and 4-m Resolution Satellite Data to Research and Management in Tropical Forests: An Evaluation of IKONOS Imagery, Journal of Applied Ecology 40: 592-600.

Read, J.M., Lam, N.S-N. (2002). Spatial Methods for Characterizing Land Cover and Detecting Land-Cover Change for the Tropics, International Journal of Remote Sensing 23: 2457-74.

Riitters, K.H., O'Neill, R.V., Hunsaker, C.T., Wickman, J.D., Yankee, D.H., Timmins, S.P., Jones, K.B., and Jackson, B.L. (1995). A Factor Analysis of Landscape Pattern and Structure Metrics, Landscape Ecology 10: 23-39.

Rindfuss, R.R., Prasartkul, P., Walsh, S.J., Entwisle, B., Sawangdee, Y., and Vogler, J.B. (2003). Household-Parcel Linkages in Nang Rong, Thailand: Challenges of Large Samples. In Fox, J. Rindfuss, R.R., Walsh, S.J. and Mishra, V. (Eds.) People and the Environment: Approaches for Linking Household and Community Surveys to Remote Sensing and GIS, 131-72. Boston: Kluwer Academic Publishers.

Robbins, P. (1998). Authority and Environment: Institutional Landscapes in Rajasthan, India, Annals of the Association of American Geographers 88: 410-35.

Rundquist, D. (2001). Field Techniques in Remote Sensing: Learning by Doing, GeoCarto International 16: 83-88.

Sala, O.E., Chapin, F.S.I., Armesto, J.J., Berlow, E., Bloomfield, J., Dirzo, R., Huber-Sanwald, E., Huenneke, L.F., Jackson, R.B., Kinzig, A., Leemans, R., Lodge, D.M., Mooney, H.A., Oesterheld, M., Poff, N.L., Sykes, M.T., Walker, B.H., Walker, M., and Wall, D.H. (2000). Global Biodiversity Scenarios for the Year 2100, Science 287: 1770-74.

Sengupta, R.R. and Bennett, D.A. (2002). Agent-Based Modelling Environment for Spatial Decision Support, International Journal of Geographical Information Science.

Seto, K.C., Kaufmann, R.K., and Woodcock, C.E. (2002). Monitoring Land Use Shange in the Pearl River Delta, China. In Walsh, S.J. and Crews-Meyer, K.A. (Eds.) Linking People, Place, and Policy: A GIScience Approach, 69-90. Boston: Kluwer Academic Publishers.

Simpson, D. M. (2001). Virtual Reality and Urban Simulation in Planning: A Literature Review and Topical Bibliography, Journal of Planning Literature 15: 359-72.

Siu, D.Z. and Zeng, H. (2001). Modeling the Dynamics of Landscape Structure in Asia's Emerging Desakota Regions: A Case Study in Shenzhen, Landscape and Urban Planning 53: 37-52.

Soares-Filho, B.S., Cerqueira, G.C., and Pennachin, C.L. (2002). DINAMICA-A Stochastic Cellular Automata Model Designed to Simulate the lLandscape Dynamics in an Amazonian Colonization Frontier, Ecological Modelling 154: 217-35.

Steffen, W., Jäger, J., Carson, D., and Bradshaw, C. (Eds.) (2002). Challenges of a Changing Earth: Proceedings of the Global Change Open Science Conference, Amsterdam. Heidelberg: Springer-Verlag.

Stephenne, N. and Lambin, E.F. (2001). A Dynamic Simulation Model of Land-Use Changes in Sudano-Sahelian Countries of Africa (SALU), Agriculture, Ecosystems, and Environment 85: 145-61.

Tucker, C.J. and Sellers, P.J. (1986). Satellite Remote Sensing of Primary Production, International Journal of Remote Sensing 7: 1395-1416.

Tucker, C.J. and Townshend, J.R.G. (2000). Strategies for MonitoringTropical Deforestation Using Satellite Data, International Journal of Remote Sensing 21: 1461-71.

Turner, M.D. (1999). Merging Local and Regional Analyses of Land-Use Change: The Case of Livestock in the Sahel, Annals of the Association of American Geographers 89: 191-219.

Turner, M.G. (1990a). Landscape Changes in Nine Rural Counties in Georgia, Photogrammetric Engineering and Remote Sensing 56: 379-86.

Turner, M.G. (1990b). Spatial and Temporal Analysis of Landscape Patterns, Landscape Ecology 4: 21-30.

Turner, B.L. II. (1994). Local Faces, Global Flows: The Role of Land-Use/Cover in Global Environmental Change, Land Degradation and Society 5: 71-78.

Turner, B.L. II. (2002). Toward Integrated Land-Change Science: Advances in 1.5 Decades of Sustained International Research on Land-Use and Land-Cover Change. In Steffan, W., Jäger, J., Carson, D., and Bradshaw, C. (Eds.) Challenges of a Changing Earth: Proceedings of the Global Change Open Science Conference, Amsterdam. Heidelberg: Springer-Verlag.

Turner, B.L. II. and Ali, A.M.S. (1996). Induced Intensification: Agricultural Change in Bangladesh with Implications for Malthus and Boserup, Proceedings, National Academy of Sciences 93: 14984-91.

Turner B.L. II, Geoghegan, J. and Foster, D.R. (2003). Integrated Land-Change Science and Tropical Deforestation in the Southern Yucatan: Final Frontiers. Oxford, UK: Clarendon Press.

Turner, B.L. II and Geoghegan, J. (2003). Land-Cover and Land-Use Change (LCLUC) in the Southern Yucatan Peninsular Region (SYPR): An Integrated Approach. In Fox, J., Rindfuss, R.R., Walsh, S.J., and Mishra, V. (Eds.) People and the Environment: Approaches for Linking Household and Community Surveys to Remote Sensing and GIS, 31-60. Boston: Kluwer Academic Publishers.

Ullah, A., Rundquist, D., and Derry, D. (2000). Characterizing Spectral Signatures for Three Selected Emergent Macrophytes: A Controlled Experiment, GeoCarto International 15: 29-39.

Veldkamp, A. and Fresco, L.O. (1996). CLUE: A Conceptual Model to Study the Conversion of Land Use and its Effects, Ecological Modelling 85: 253-70.

Veldkamp, A. and Lambin, E.F. (2001). Predicting Land-Use Change, Agriculture, Ecosystems, and Environment 85: 1-6.

Verburg, P.H., de Koning, G.H.J., Kok, K., Veldkamp, A., and Bouma, J. (1999). A Spatial Explicit Allocation Procedure for Modeling the Pattern of Land Use Change Based Upon Actual Land Use, Ecological Modelling 116: 45-61.

Vitousek, P.M., Matson, P.A., Schindler, D.W., Schlesinger, W.H., Tilman, D.G., Aber, J.D., Howarth, R.W., and Likens, G.E. (1997). Human Alteration of the Global Nitrogen Cycle: Sources and Consequences, Ecological Applications 7: 737-50.

Walsh, S.J., Bian, L., McKnight, S., Brown, D.G., and Hammer, E.S. (2003a). Solifluction Steps and Risers, Lee Ridge, Glacier National Park, Montana, USA: A Scale and Pattern Analysis, Geomorphology 1399: 1-18.

Walsh, S.J., Bilsborrow, R.E., McGregor, S.J., Frizzelle, B.G., Messina, J.P., Pan, W.K.T., Crews-Meyer, K.A., Taff, G.N., and Baquero, F.D. (2003b). Integration of Longitudinal Surveys, Remote Sensing Time-Series, and Spatial Analyses: Approaches for Linking

People and Place. In Fox, J. Rindfuss, R.R., Walsh, S.J., and Mishra, V. (Eds.) Linking Household and Remotely Sensed Data: Methodological and Practical Problems, 91-130. Boston: Kluwer Academic Publishers.

Walsh, S.J., Butler, D.R., and Malanson, G.P. (1998). An Overview of Scale, Pattern, and Process Relationships in Geomorphology: A Remote Sensing and GIS Perspective, Geomorphology 21: 183-205.

Walsh, S.J., Crews-Meyer, K.A., Crawford, T.W., Welsh, W.F., Entwisle, B., and Rindfuss, R.R. (2001). Patterns of Change in Land-Use and Land-Cover and Plant Biomass: Separating Intra- and Inter-Annual Signals in Monsoon-Driver Northeast Thailand. In Millington, A.C., Walsh, S.J., and Osborne, P.E. (Eds.) GIS and Remote Sensing Applications in Biogeography and Ecology, 91-108. Boston: Kluwer Academic Publishers.

Walsh, S.J. and Crews-Meyer, K.A. (Eds.) (2002). Linking People, Place, and Policy: A GIScience Approach. Boston: Kluwer Academic Publishers.

Walsh, S.J., Welsh, W.F., Evans, T.P., Entwisle, B., and Rindfuss, R.R. (1999). Scale Dependent Relationships Between Population and Environment in Northeastern Thailand, Photogrammetric Engineering and Remote Sensing 65: 97-105.

Watson, R. T., Noble, I.R., Bolin, B., Ravindranath, N.H., Verardo, D.J., and Dokken, D.J. (Eds.) (2000). Land Use, Land-Use Change and Forestry. Special Report of the IPCC (Intergovernmental Panel of Climate Change). Cambridge: Cambridge University Press.

White, R. and Engelen, G. (1997). Cellular Automata as the Basis of Integrated Dynamic Regional Modeling, Environment and Planning B: Planning and Design 24: 235-46.

Wolfram, S. (1984). Cellular Automata as Models of Complexity, Nature 311: 419-24.

Wood, C.H. and Skole, D. (1998). Linking Satellite, Census, and Survey Data to Study Deforestation in the Brazilian Amazon. In Liverman, D., Moran, E.F., Rindfuss, R.R., and Stern, P.C. (Eds.) People and Pixels: Linking Remote Sensing and Social Science: 70-93. Washington, DC: National Academy Press.

Wu, F. (1998). Simulating Urban Encroachment on Rural Land with Fuzzy-Logic-Controlled Cellular Automata in a GIS, Journal of Environmental Management 53: 293-308.

Yocky, D.A. (1996). Multiresolution Wavelet Decomposition Image Merger of Landsat Thematic Mapper and SPOT Panchromatic Data, Photogrammetric Engineering and Remote Sensing 62: 1067-74.

Zhan, X., DeFries, Townshend, J.R.G., DiMieeli, C., Hansen, M., Huang, C., and Sohlberg, R. (2000). The 250m Global Land Cover Change Product from the Moderate Resolution Imaging Spectroradiometer of NASA's Earth Observing System, International Journal of Remote Sensing 21: 1433-60.

CHAPTER 22

D.R.F. TAYLOR

CAPACITY BUILDING AND GEOGRAPHIC INFORMATION TECHNOLOGIES IN AFRICAN DEVELOPMENT

Abstract   This chapter argues that the potential of geographic information processing (GIP) technologies for development in Africa is considerable, but that there are many barriers to be overcome before that potential can be fully realized. Building on the National Academy's study *Down to Earth: The Geographical Foundations of Sustainable Development in Africa* (2002), I consider the barriers and argue that developing and expanding geospatial capacity at societal, organizational, and individual levels is indispensable to the increased use of utility of GIP technologies. The development of indigenous capacity is key to making the potential of GIP to African development a reality.

Keywords   African development, geographic information processing, indigenous capacity, human capital, social capital

1. INTRODUCTION

The potential of geographic information processing technologies (GIP) for development in Africa is considerable, but there are many barriers to be overcome before that potential can be fully realized. One of the most comprehensive studies of this topic was published in 2002 by the National Research Council of the National Academy's of the National Academy of Sciences. This study, entitled *Down to Earth: The Geographic Foundations of Sustainable Development in Africa* (NRC 2002) outlined the importance of geographic data for African development and the key technologies involved in utilizing this data, including remote sensing, Geographic Information Systems (GIS), Geographic Global Positioning Systems (GPS) and geographically based decision support systems. The committee which prepared the study observed:

Stanley D. Brunn, Susan L. Cutter, and J.W. Harrington, Jr. (Eds.), Geography and Technology,
521-546. © 2004 Kluwer Academic Publishers. Printed in the Netherlands.

Geographic data are obtained from ground-based (in situ) measurements or from remote-sensing systems. These data are of little practical value in sustainable development decision-making if they cannot be analyzed in conjunction with development data, such as economic or health data, that are geographically referenced...Data that describe environment and development can be linked by geographic location to provide greater understanding of complex issues, and GIS's were developed specifically for this purpose...If a GIS involves the integration of geographically referenced data in a problem-solving situation, it can become important for decision making, or a 'decision support system'... (NRC 2002: 2-3)

The nature of these geographic information technologies was fully described in *Down to Earth* and also in several chapters in this book and will not be reconsidered here. There are four critical needs that lie at the heart of efforts to expand the use of geographic information for African development: (1) creating infrastructure to support the use of geographic datasets; (2) ensuring availability and accessibility of the geographic data used for many applications; (3) developing decision-support systems that effectively use geographic data; and (4) expanding geospatial capacity at individual, organizational, and societal levels (NRC 2002). Regardless of which geographic information technologies are to be used, or whether these technologies are the most modern available (which was the main focus of *Down to Earth*), or more traditional technologies such as the use of transparent map overlays on paper maps, these four needs still apply.

This chapter will argue that developing and expanding geospatial capacity is indispensable if the potential utility of geographic information technologies is to be realized. In particular, the development of indigenous capacity is key. African development is about people, and technology is a means to an end, not an end in itself. People cannot be "developed": they can only develop themselves. Appropriate technology can, perhaps, best be described as the technology most appropriate to the task at hand. That task is socioeconomic development defined and determined by Africa's people themselves (Taylor and Mackenzie 1992). In some instances, this may be the highest technology available, such as the analysis of AVHRR images. In others, it may be the creation of a hand-drawn poverty map by a group of villagers as part of a participatory research process. Regardless of the technology used, it must be in the hands of the people most directly involved in the struggle for development on the ground and who understand the socioeconomic and cultural context in which the technology is to be used. "Empowerment" is an over-used and often ill-defined term in the development literature but, however defined, the building and human and social capital is a central part of the process (Taylor 1991, 1998; Ekong 1998; Juma et. al. 2001; Juma 2001; Labatut 2002; ECA 2003f).

Much of the geographical data of potential utility to African development has been generated outside of Africa. The U.S. alone has spent millions of dollars generating at least 11 major geographic datasets of relevance to African development (NRC 2002, 3) on a wide range of topics from land cover to establishing population distribution to early-warning fire and famine systems. The indigenous capacity and understanding of the technologies required to create, access, and utilize these datasets is of paramount importance. The reality is that many of the technologies required are advanced geospatial technologies, and if they are to be used, additional capacity must be built to do this. Of special interest is how these technologies can be used at local scales in support of decentralization and community empowerment (ECA 2003f) which are seen as key factors in the struggle for development (Taylor and Mackenzie 1992; Chambers 1997).

1.1. The Development Context

The New Partnership for Africa's Development (NEPAD) (OAU 2001a) recognizes the importance of education and capacity building in Africa's development future in several of its clauses. This theme runs through this key document, which grows out of the new African Initiative announced in Pretoria, South Africa in July 2001 (OAU 2001b). Education and capacity building in Africa was a major focus at the G8 meeting held in Kananaskis, Alberta, Canada in July 2002. The chair of the meeting, Prime Minister Chrétien of Canada, identified African development as a major agenda item (Chrétien 2002) and the G8 developed an African Action Plan in response to NEPAD in which education and capacity building are major elements. This plan has not received the funding promised by the G8 nations, with notable exceptions such as Canada, and African leaders met with the G8 in France in June 2003 to discuss effective implementation of this plan.

African leaders intend to "revitalize and extend the provision of education and training" (NEPAD para. 49) and have developed a program of action to achieve sustainable development. "Human resources, including education, skills development and reversing the brain drain" is one of the five key sectors of NEPAD's action plan as is "infrastructure, especially information and communications technology" (NEPAD para. 97).

Geographic information processing is part of Information and Communications Technology (ICT) and para. 108 of the NEPAD document draws specific attention to this.

ICTs can be helpful tools for a wide range of applications, such as remote sensing for environmental, agricultural and infrastructural planning; The existing complementarities can be better utilised to provide training that would allow for the production of a critical mass of professionals and the use of ICTs; In the research sector, we can establish African programmes as well as technological exchange programmes capable of meeting the continent's specific needs... (NEPAD para. 108)

The Economic Commission for Africa has also recently outlined the role of geoinformation in NEPAD (ECA 2003c). It is noteworthy that the emphasis is on the application of technology to developmental problems, not on the technologies themselves.

1.1.2. Education and African Development

The "Education Gap" in Africa is a wide one, and this affects all levels of education. The primary objective of NEPAD in this respect is to achieve universal primary education by 2015 (NEPAD para. 120). This follows from the argument that the greatest individual returns to investment come from basic education (Psacholopoulos 1994) but the plan also talks of the need to "expand access to secondary education and improve its relevance to African development; [and to] Promote networks of specialized research and higher education institutions (NEPAD 2001 para. 120)."

Explicit recognition of the importance of university education is also given:

> The plan supports the immediate strengthening of the university system across Africa, including the creation of specialized universities where needed, building on available African teaching staff. The need to establish and strengthen institutes of technology is especially emphasized. (NEPAD, para 123)

This plan fits in well with the arguments of two major studies on the importance of university education (Bourne 2000; World Bank and UNESCO Task Force 2000) which challenge the assumption that investment in basic education will have the greatest return and argue that although individual returns may be greater for basic education, the social returns to higher education are significant. Akilagapa Sawyerr (2003) argues that recent studies have shown that the methodologies and conclusions that returns to basic education are greater than higher education are seriously flawed. Higher education is particularly important for geographic information processing. "The key problems in education in Africa are the poor facilities and inadequate systems under which the vast majority of Africans receive their training" (NEPAD para. 122).

Higher education in particular has suffered very badly and African universities are in a state of crisis (Swartz 2000; Association of African Universities 1997). Sawyerr (2003) argues that African universities have been "intellectually decapitated." "This has led to an increasing 'brain drain' from Africa especially at the university and postgraduate levels" (Downes 2000;

Carrington and Detragiache 1998; Ekong 1998). A major challenge is "to reverse the brain drain and turn it into a 'brain gain' for Africa" (NEPAD para. 124) and this requires capacity building. It has been argued that "brain circulation" is perhaps a more appropriate concept (Juma et al. 2001, NRC 2002).

1.2. Components of Capacity Building

As Beerens has argued, "capacity building comprises human resources development, organisational strengthening, and institutional strengthening– of which education is part and parcel" (Beerens, 2002, 2). The purpose and focus of each of these components is outlined in Table 22-1.

These three components are closely related, and capacity building needs to consider all three to be effective. In the African context, this is rarely the case and as a result there are sometimes discontinuities and contradictions between the capacity building strategies in operation. The emphasis has been on human resource development and, to a lesser degree, on organizational strengthening. The importance of institutional strengthening for more effective application of geographic information processing has not been fully recognized. If the relationships among these three components are not explicitly considered, then attempts to build capacity can be counterproductive. For example, if staff from an organization receives scholarships to study overseas,

Table 22-1. Components of geospatial capacity building.

Purpose	Focus
Human Resources Development	Supply of technical and professional personnel
Organizational Strengthening	Strengthening the management capacity of organizations; imbedding geo-information communication technology solution (systems and processes) as well as strategic management principles
Institutional Strengthening	Strengthening the capacity of an organization to develop and negotiate appropriate mandates and modus operandi as well as appropriate (new) legal and regulatory frameworks

Source: Groot and Georgiadou 2001, quoted in Beerens (2002, 2).

this can weaken the organization unless replacements are found during their absence. If they become part of the "brain drain," the negative impact can be compounded (Carrington and Detragiache 1998; Downes 2000; Ekong 1998) and only partially offset by "brain circulation."

The situation is further complicated by the fact that organizations responsible for capacity building in GIP are rarely the organizations involved in the application of GIP to key development problems such as health or the environment. Building inter-institutional linkages is part of the institutional strengthening process. In addition, strengthening the capacity of organizations such as ministries of health, environment, transportation, and planning to utilize and apply geographic information processing is equally, if not more, important than strengthening the capacity of organizations to deal with the technical aspects of GIP. The development of human capital is important for GIP, but this is only part of the development of important social capital, which also includes the development of institutional structures and interinstitutional relationships. If geographic information processing is not effectively applied, then its contribution to African development will be marginal. This is not primarily a technical problem, nor is it confined to the African context. The development of spatial data infrastructures (SDI), which is a major focus in the application of geographic information technologies in Africa, requires technical expertise (ECA 2001e; NRC 2002; ECA 2003e). Although this is necessary, it is by no means sufficient to ensure that geographic information from key application areas is attached to the SDI. This requires the development of social capital to create the critical institutional linkages.

1.3. The Current Situation

1.3.1. Approaches and Challenges

Capacity building in GIP in Africa has largely been concentrated on the development of human capital. Bassolé (2002) has estimated that over half of current training in geographic information processing in West Africa is funded by various development projects where it is a part of the project. If these estimates are correct, then this is the major current human resource development process. Bassolé comments:

> The level of GIS and RS expertise though increasing [in numbers], is still very low in West Africa. A considerable amount of money is spent from project budgets to train operational level staff. The number of high-level experts is very low, keeping the few ones available very busy with operational activities. Therefore, the strategic thinking, managerial initiatives, and policy development work is neglected. (Bassolé 2002, 10)

A strength of this form of training is that it is "hands on" and applied to a specific development problem. The major disadvantage is that when the project stops, the training often terminates. Such approaches are rarely sustainable and, in addition, the training tends to be highly technical and restricted to the particular technologies used for the project. These may or may not be transferable to other situations. Projects funded by major national aid agencies often utilize technology from the funding nations. Human resource development is a long-term process which does not always fit well with the results-based approach over a restricted timeframe favored by many aid projects.

Nongovernmental organizations are also players in this field, and Bassolé (2002) estimates that 2 percent of training in West Africa is funded by these sources. NGOs often work at the local level and demonstrate the potential of GIP in "grass roots" applications. They have some of the same weaknesses as the larger aid projects described above, and in addition, such projects are rarely strategic in their intent, thereby weakening their cumulative impact. There are numerous examples of poverty mapping initiatives and small scale GISs but few examples of sustainable strategic initiatives growing out of these. This is a major weakness of "bottom up" capacity building approaches (Taylor and Mackenzie 1992).

Another element of human resource development in geographic information processing is "on-the-job" training. Bassolé (2002) estimates this at 5 to 8 percent, and most of it takes place within national mapping agencies. This includes training by commercial firms which sell their technology to African organizations with some instruction as part of the contract. On-the-job training within a national mapping agency has the advantage of relevance to the agency's immediate needs and helps strengthen the organization in a direct way. Much, however, depends upon the resources available for on-the-job training, and these are rarely adequate. Training in the use of systems by private sector vendors often tends to be a "turnkey" operation. Participants are given short-term intensive instruction on how to operate the system but rarely in the concepts and principles of GIP.

1.3.2. The Role of the Formal Education Sector

Bassolé (2002) estimates that 35 to 45 percent of human resource development in GIP takes place in the formal education system either in Africa or abroad that is funded by both national and international scholarships. The major institutions involved are universities or polytechnics, in which GIP is part of the curriculum, and national and regional training centers which specialize in education in the mapping sciences. The universities are key to human capacity building for the knowledge-based economy of which GIP is a part. But numerous studies have documented the crisis facing many African universities today (Ajayi et al.1996; Bourne 2000; Association of African Universities and the World Bank 1997; Davenport 2000; Downes 2000; Mehta 2000; Swartz 2000; Task Force on Higher Education and Society 2000; Labatut 2002). There have been massive funding cuts to universities throughout the 1980s and 1990s and, although university enrollments have increased in many African countries, the provision of the human and physical resources required to deal with these increases has not kept pace. As a result, the institutional base in 2003 is, in many respects, weaker than it was in 1980. The research function has been particularly hard hit.

Capacity building in GIP in African universities, as in universities in North America, is found in a number of different faculties and departments. Applied research in GIS and decision support systems is often found in departments of geography whereas research in remote sensing, photogrammetry, surveying, geodesy, and other mapping sciences is usually found in faculties of engineering. Although there has been convergence of mapping sciences marked by the adoption of terms such as "geomatics," this convergence has yet to have major impact on the curricula and academic organization of most African universities.

When a university is struggling to survive and lacks sufficient resources in terms of infrastructure, equipment, library, and faculty, introducing new capacity building in geographic information processing is difficult. Improvements in this respect will depend upon strengthening universities as organizations. Higher education has not been a recent priority for African countries or for the major funding agencies such as the World Bank, where considerable emphasis has been placed on basic education. The importance of higher education is recognized in rhetorical terms but adequate resources have not been made available to the higher education sector.

The most effective application of GIP requires improved instruction but also improved research capacity, and this is a major challenge. In many African universities, increased enrollments without a commensurate increase in resources has meant that most African academics face very heavy teaching loads, leaving little time for research:

> research has suffered massive funding cuts through the 1980s and 1990s, with the result that, although there are more and better trained people, the institutional base for research is in many respects weaker than it was previously. In some countries and sectors research is heavily dependent on funding by donors, some of whom impose strict conditions on research questions, methodologies and composition of research teams. Such research is often done under contract and is not publicly available. Private sector funding for research and development is negligible...The

continuing gulf between North and South in terms of research resources; the increasing importance of knowledge and ideas for the solution of stubborn development problems–especially in a time of declining budgets; the necessity for a domestic science and technology capacity in all [are important issues]...It is vital that the peoples in Africa be in a position to control their own "knowledge-based" development. Therefore strengthening the capacity for research, independent policy analysis and accessing knowledge are critical. Analytical capacity must allow Africans to contribute as informed participants in major international debates...They must be able to deal directly with issues of direct domestic concern...where, in the absence of indigenous capacity, the analysis of external actors may be all that is available and will carry undue weight. (Labatut 2002, 2-3)

This is certainly the case with GIP which is dominated by external actors. Little of the impressive research on the application of geographic information technologies as outlined in *Down to Earth* (NRC 2002) is carried out in Africa by Africans. African institutions even lack the capability to effectively access and apply this research, thus seriously limiting its utility. African universities are key institutions in this respect (NRC 2002) and GIP is unlikely to make its full contribution to African development problems unless the capacity of African universities is strengthened. African universities, despite the difficulties they face, are still key generators of new knowledge in Africa and have the capability of creating the institutional linkages so important for the future (Juma 2002). It is in the universities where the applications of geographic information to African development problems should be developed. The work by Akinyemi (2001) is an interesting example in the context of urban poverty. Outside analysis is not without value, but effective application of GIP is best done by individuals who understand both the technology and the development context in which it is to be applied. Developing technology is a necessary but by no means sufficient step. Much human resource development has concentrated on improving technological knowledge but relatively little into understanding the cultural and economic context in which this technology is to be applied. This is, perhaps, the main challenge in capacity building for geographic information processing in Africa. This challenge is not primarily a technical one.

1.3.3. The Role of Regional Training Centres

The Regional Centres established by the U.N. Economic Commission for Africa are making an important contribution to increasing technological capacity in geographic information processing in Africa. Many of these centres have been in existence for over three decades. They include the Regional Centre for Mapping of Resources for Development (RCMRD), the Regional Centre for Training in Aerospace Surveys (RECTAS), the Southern and Eastern Africa Mineral Centre (SEAMIC), and the African Centre for Meteorological

Applications and Development (ACMAD). A case study of RECTAS helps to illustrate the contribution that Regional Centres are making (Kufoniyi 2001). The Centre is involved in training in geoinformatics as well as in carrying out studies and research in this field. It also conducts seminars and courses to introduce government officials to geoinformatics and provides advisory and consulting services to ECA member states. The Centre underwent substantial reorganization and modernization in 2000. The training courses given are bilingual and are at the level of technicians and technologists, and more recently, "the Post Graduate course was modernised with the introduction of the integrated programme in Geoinformation Production and Management" (Kufoniyi 2001, 2). Between 1973 and 2000, RECTAS graduated more than 900 people.

One interesting new initiative is the development of a new capacity building project which has been given the name CABGLEN (Capacity Building for Geoinformation Production and Management for Sustainable Local Environment and Natural Resources Management). This initiative is being developed in cooperation with the Federal School of Surveying, Ojo, Nigeria (FSS), the International Institute for Aerospace Survey and Earth Sciences (ITC), the Netherlands, and the Groupement pour le Développement de la Télédection Aerospatiale (GDTA), France. There are four program components:

1. An M.Sc. in Geoinformation Production and Management as a sandwich program in modular format with ITC or GDTA
2. A Professional Masters (P.M.) in Geographic Production and Management to be run at FSS and RECTAS
3. A three-month certificate course in any of the four specializations of the P.M. course in photogrammetry and remote sensing, cartography, and geoinformation visualization, spatial information systems, and optimization of geoinformatics processes. This course is designed for staff upgrading of the various agencies in the region.
4. Refresher courses and workshops of one week or for a month given at RECTAS and FSS that are designed for the upgrading of executive level and senior managers given at RECTAS and FSS.

This innovative new project is a good example of the type of capacity building required to ensure that African ability to utilize geographic information is improved, and of the partnerships required to achieve this.

Until 2000 RECTAS's main contribution to capacity building was in the development of technical human resources in photogrammetry and remote sensing. The reorganization and reorientation of the Centre's activities together with the new CABGLEN project marks an important new direction. As the

Director of RECTAS observed: "The need for capacity building in Geoinformation Production and Management in Africa cannot be over-emphasised as geospatial information is definitely the sine-qua-non for sustainable national development" (Kufoniyi 2001, 12). RECTAS is "Blazing the Trail in Capacity Building for Geoinformation Production and Management for Sustainable Management of Local Environment and Natural Resources in Africa" (Kufoniyi 2001). But the trail has only just been identified and the resources to establish and develop it have yet to be found. Africa is also a large continent, and many more innovative "trail blazers" are required.

1.3.4. The Role of the ECA Committee on Development Information

The Regional Centres of the ECA are supported by Geoinformation activities at ECA headquarters in Addis Ababa (Ezigbalike 2002). Harnessing Information for Development is one of six core programs of ECA and geoinformation is a part of the Development Information Services Division. In addition to supporting the Regional Centres, activities include advisory services, developing inventories and databases, and organizing conferences and meetings.

Support is also given to the other five core program areas of ECA. From 1963-96 the ECA organized a number of U.N. Cartographic Conferences which, in 1999, became part of the activities of the Committee on Development Information (CODI). This Committee provides policy and technical guidance for the implementation of the core program on informatics for development; these recommendations are passed on to the Conference of African Ministers of Planning and Development. CODI has a number of subgroups in addition to Geoinformation and these include Library and Documentation Services, Information and Communication Technologies, Statistical Development and Database Development (ECA 2001c; ECA 2003b). Geoinformation is thus part of a more general thrust of information provision for development, and the potential synergies and linkages between various parts of the development Information Services Division are important ones. The reorganization of ECA is relatively recent and the organizational structures and processes to ensure effective interaction between the various groups interested in information for development is a work in progress. Drawing what were effectively separate activities together, including the Information Technology Centre (ECA 2001d), in a new organizational structure has the potential to increase their impact. The Geoinformation group has the challenge of demonstrating the importance of their contribution and integrating this with other elements of the Development Information Services Division and with the core ECA programs.

Although progress has been made in this respect, more remains to be done (ECA 2003a, 2003b).

The Regional Centres of the ECA are complemented by a number of other regional national and subnational centres together with a small but growing number of private-sector consulting and training firms. Bassolé (2002) has shown how a network known as EIS-Africa is making an important contribution to the application of GIS and remote sensing to natural resource management and development issues in Africa. EIS-Africa is a network for the cooperative management of environmental information on Africa and draws together a number of private and public sector institutions and experts promoting "access to and use of environmental information in the SD [sustainable development] process" (Bassolé 2002, 3).

Professional organizations of individuals interested in geoinformation are also important in building capacity in Africa. These exist at the Pan-African as well as national levels. The African Organization for Cartography and Remote Sensing is supported by the ECA, and similar societies exist at the national level in various countries such as Nigeria. An increasing number of organizations in the ICT field is emerging in Africa.

1.3.5. International Players and Agencies

Although indigenous African capacity-building efforts are growing, the application of GIP to development problems in Africa is dominated by international players, which influences capacity building both directly and indirectly. Reference has already been made to the fact that over half of the people trained in GIP receive their training as part of an aid project which, despite rhetoric to the contrary, is often controlled and directed by the agency concerned. One version of the "golden rule" can be stated as "whoever controls the gold rules!" In these cases, influences are direct and, as mentioned earlier, the capacity building is rarely sustainable. There are, of course, exceptions but the empirical evidence available suggests that a lack of sustainability is unfortunately too often the case. Aid projects can and do make important contributions to African development, but as outlined earlier, they are not the best way to build capacity in GIP. The need to achieve quantifiable results within a restricted timeframe often sees GIP introduced into projects in ways which are ineffective for the capacity-building process. To effectively use modern GIP a relatively slow learning curve is the norm. Operating the hardware and software in far from optimum conditions with frequent power outages, delays in importing equipment because of cumbersome customs procedures and other constraints which can lead to further delays. In these circumstances, to meet the needs of the project and to satisfy the accountability requirements of the donor agency, international experts sent to build local capacity often resort to doing the work themselves, especially when project time is running out.

1.3.6. The GDTA

Two international agencies have played a key role in building human capacity in geographic information processing in Africa–GTDA of France and ITC of the Netherlands. GTDA operates mainly in French-speaking Africa and ITC mainly in English-speaking Africa, although this linguistic division is by no means absolute.

The GDTA (The Aerospace Remote Sensing Development Group in English) (GTDA website 2002) was established in 1973 and is a French Economic Consortium whose mission is to promote the development of remote sensing, space imaging, and GIS worldwide through training courses and workshops. There are four public member organizations: the Institut Géographique National (IGN), France's national mapping agency, the French Space Agency (CNES), the Bureau des Recherches Géologique et Minières (BRGM) and the French institute for oceanographic research (IFREMER). Since its inception in 1973, GTDA has trained almost 4,000 students but only a relatively small proportion of that total are from Africa. The general mandate of the organization is:

> to promote the general use of satellite-based remote sensing, a technology currently limited on a worldwide scale by the lack of specialists, by training professionals from all disciplines to use space data, digital images and derived products and incorporate them into information systems (such as a GIS) for decision making and management purposes. (GDTA website February 2002)

As an economic consortium, GDTA utilizes and promotes French technology and, unlike ITC, does not confine its activities to developing nations.

1.3.7. The ITC

ITC was founded in the Netherlands in 1950 by the Dutch Government at the request of the United Nations in order to build capacity through educating and training mid-career professionals from developing countries in aerial surveys, hence its initial name, International Training Centre for Aerial Survey (Beerens 2002, 3). The Institute initially concentrated on photogrammetry and cartography, but as technology in the mapping sciences developed, new activities were added including the analysis of satellite imagery and GIS. ITC has been in the forefront of curriculum development for GIP and recognizing:

> that geo-information science had 'come of age', the scope and the name of the Institute were accordingly adjusted to 'International Institute for Geo-Information Science and Earth Observation'...on January 1, 2002. (Beerens 2002, 3)

ITC receives its core funding from the Dutch Ministry of Education, Cultural Affairs and Science and the funding for the scholarship comes from the Dutch Ministry for Development Cooperation. It is exclusively a postgraduate institution teaching in English. In a normal year, between 400 and 500 students from 75 countries are in residence, including 50 Ph.D. researchers. There are six separate educational programs, including Geoinformatics, Geoinformation Management, Urban Planning and Land Administration, and Natural Resources Management. For each of these program areas, there are four different education and training possibilities: an 18 month M.Sc. which also qualifies individuals for entry into a Ph.D. program; a one-year Professional Masters program; short tailor-made courses responding to the needs of specific organizations either in Holland, in the home country, or a combination of both; and six two-week refresher courses to update alumni on new developments held either at ITC or in the home country of the alumni, often in collaboration with local partner institutions. Between 1950 and 2002, ITC graduated 4,433 individuals from 45 African countries (Beerens 2002, Appendix 2, 18). All were mid-career professionals, the primary target market for ITC.

Although ITC has been involved in strengthening organizations since the 1960s, its main achievements have been in the field of education and training. In a frank analysis of ITCs achievements, Beerens (2002) lists a number of impressive achievements: a high success rate of over 95 percent in terms of graduation; relatively little "brain drain"; the provision of cheap user-friendly software developed in-house by ITC, and more recently, a journal to keep alumni informed of recent advances; and the provision of refresher courses. However,

> The initial practice of selecting individual mid-career professionals for our fellowship programmes has resulted in expertise being spaced rather thinly over Africa, not concentrated and resulting in a critical mass required for capacity building within a single organization. (Beerens 2002, 6)

ITC has also worked to strengthen individual organizations in Africa, and RECTAS is one of ITC's five sister institutions. In addition, in 2002, ITC had eight projects to strengthen organizations or professional bodies in Africa. Here ITC's achievements have been more mixed. Beerens (2002) observes that strengthening an organization requires an eight- to ten-year commitment from both sides and even then, sustainability remains a problem. He also comments on the poor results in terms of sustainability of projects that were initially heavily dependent on massive expatriate staff involvement.

ITC has analyzed the changing situation and has concluded that, "In the next five to 10 years, ITC faces changes that will have implications far greater than all those of the first 50 years put together" (Beerens 2002, 8) and is making fundamental changes in its strategic goal and approach. A major element in this process is a shift in emphasis from education and training to capacity building which is defined as "improvement in the ability of public and private sector organizations to perform appropriate tasks either singly or in cooperation with other organizations" (Beerens 2002, 10). In essence, this is an emphasis on the development of social capital rather than on human capital.

> ITC will put particular emphasis on support to those organizations that have been assigned national responsibility for establishing and maintaining GDIs and which should be strengthened in their capacity to lead in the development and maintenance of foundation data within the broad set of principles of a national GDI. (Beerens 2002, 10)

ITC has recognized the need to redefine the meaning and nature of partnerships and to move to a "knowledge development" approach. ITC plans to develop new linkages within the Netherlands with Dutch universities and other "knowledge institutions" including joint research programs, staff exchanges, and dual assignments. Increased linkage between institutions dealing with GIP and other institutions involved in understanding ICTs and their application to development problems is long overdue, and to its credit, ITC has concluded that action is required at home. It is planning steps to increase the institutional linkages so important in developing social capital. The plan is "that ITC will become a centre for the exchange of expertise rather than mainly an institution for education and training" (Beerens 2002, 11). The plan does, however, face problems of implementation and requires the cooperation of other agencies.

The nature of partnerships will be changed both domestically and internationally.

> We intend to transfer [part of] our education programmes to the home countries of our clients under the title 'decentralisation'. We intend to do that through establishing joint educational programmes with scientific institutions, implemented in collaboration with and by staff of partner institutions in the home countries or regions of our clients. (Beerens 2002, 12)

This decision will involve flexible modular programs, "sandwich" programs, and courses and support on the Internet. The principle is one of equal partnership with degrees offered jointly.

This strategy is an important change from what was predominantly an "aid" relationship. But one difficulty in implementing such a strategy is that ITC expects partner institutions to have the capability and funding required to deliver their part of the bargain. Partnership will be primarily with universities in Africa or with institutions associated with universities, such as the ECA Centres, to facilitate the awarding of degrees. There are a number of challenges to be met in the implementation of this innovative decentralization strategy. The most obvious one is funding and the provision of adequate resources both to the ECA Centres and to the universities. As outlined earlier, African universities are in crisis in this respect.

A second challenge is for the African organizations involved to create the institutional linkages required within their countries and regions to make their efforts more relevant to the application of GIP to the development process. They have been, and largely are, technical training institutions. The kind of new partnerships envisaged by ITC within the Netherlands will have to be mirrored inside Africa if the partnerships are to be truly effective in building capacity. This is not a technological challenge. It is a challenge to the organizational and institutional vision and management of the institutions involved.

A third challenge relates to the relative strength of ITC vis-à-vis the partner institutions. ITC was founded almost a decade before most African countries achieved independence from their colonial masters. It has been well funded and has supplemented that funding with additional resources from various development projects over the years. Indeed, ITC policy requires staff members to raise a portion of their own salaries from project and consulting activities. ITC has a staff of highly qualified individuals and is extremely well equipped with state-of-the-art software and hardware. It has developed an innovative curriculum and is adapting to the fast pace of technological change extremely well. It has successfully trained thousands of mid-career professionals. It is the leading training institute for geographic information processing for developing nations and has made a major contribution to the development of human capital in this field. Yet this success is paradoxical in terms of capacity building in geographic information processing in Africa. The very existence of ITC has made organizational and institutional development in Africa difficult and in some instances perhaps unnecessary. Given the scarcity of resources and the conflicting demands on these resources, the need and incentives to develop indigenous training capacity was reduced. Given the generous scholarships available to Africans wishing to study at ITC, it was often easier for Africans to attend ITC than most institutions in Africa and the quality of education available in the Netherlands was superior to most African equivalents. Adelemo, writing in the 1980s, comments "it is easier for some Nigerians to study cartography in the Netherlands than at Kaduna Polytechnic" (Adelemo et al. 1985, 236). The situation is further compounded by the fact that supply and demand in GIP were often out of balance. A comprehensive study of the situation in Nigeria in the mid-1980s (Adeniyi 1985; Adalemo et al. 1985; Duru 1985), for example, showed that the problem was not only the shortage of trained human resources but also the institutional capacity to effectively utilize that manpower. Adeniyi reports that of 93 Nigerians trained externally in remote sensing, many at ITC, only 22 were in a position to apply remote sensing techniques and only five of these had equipment to work with. (Adeniyi 1985).

ITC has recognized this dilemma.

> Perhaps donors themselves should not set standards too high, standards that require African countries to look for help from the outside, either in the form of expatriate technical assistance or overseas education and training. The problem then is that this type of assistance, although at first temporary and targeted, becomes structural. We have to accept that development takes place not by throwing money, projects and expatriate technical assistance at problems but by recognising the need to start from local conditions and capacities. (Beerens 2002, 14)

This situation reflects the general view of the role of Official Development Assistance by the Government of the Netherlands. The Netherlands Minister of Development Cooperation recently commented:

> Despite massive technical assistance, aid programmes have probably weakened capacity in Africa. Technical assistance has displaced local expertise and drawn away civil servants to administer aid-funded programmes in precisely the opposite of the capacity building intentions of donors and recipients. (quoted by Beerens 2002, 14)

The Netherlands is setting an important example of a change in partnership relationships, reflected in ITC's plans, which is much closer to the new vision of an Africa-centered development envisaged by NEPAD. Whether other industrialised nations will follow suit remains to be seen.

1.3.8. Professional Associations

In recent years, international professional associations in the mapping sciences such as FIG, ISPRS, ICA, IAG, and IHO, have played a small but increasing role in capacity building in Africa in a variety of ways. This role has included holding commission and working group meetings on the continent which allow African professionals access to the latest developments in the field. The International Cartographic Association (ICA) is a good example. It held its first ever General Assembly on the African continent in South Africa in 2003 and for the last eight years, has provided young African cartographers with scholarships to attend ICA general assemblies to present their work. Felicia Akinyemi's work on GIS and Urban Poverty mentioned earlier is an example (Akinyemi 2001).

1.3.9. Foreign Universities

Universities and colleges in the industrialised nations have also been important players by providing scholarships to Africans to study GIP. The

Fulbright scholarship program is an example of a number of programs which help to fund these individuals. In Canada, the Canadian International Development Agency has funded a substantial program of university linkages for capacity building in Africa and other developing nations, although GIP is only a small part of such programs.

In terms of research, the International Development Research Centre (IDRC) based in Canada, with its philosophy of supporting in situ development research, is making a major contribution to building research capacity in African universities in ICTs for development. GIP is an integral part, through its Acacia program (IDRC 1999, Labatut 2002)

1.3.10. The Private Sector

The private sector is playing an important and growing role in the GIP field in Africa, and the GIS Africa meetings reflect this role. American firms compete with French, German, British, Canadian, Swedish companies, among others, for a growing market. Hardware and software sales are growing, but the most rapidly growing sector is in the provision of consulting and support services, often funded by the World Bank and other national and international agencies through projects. Although the project approach to capacity building in Africa has drawbacks as outlined earlier, it is still the major source of training in GIP in the continent. Several American firms, such as ESRI, are making an important contribution by donating software free to African universities and educational institutions.

1.4. Future Directions in Capacity Building

A key document in this respect is the plan developed by the Committee on Development Information (CODI-2) of the ECA in Addis Ababa in September 2001 entitled "The Future Orientation of Geoinformation Activities in Africa" (ECA 2001a). This comprehensive document covers all aspects of geoinformation in Africa and sees what is described as "The Personnel Problem" as key. Not only are professionals in short supply,

> they are thinly spread and there are still few with enough cross-disciplinary mix required for the maintenance and application of spatial data infrastructures. Most of the training programmes tended to emphasise the technical aspects of using GIS, and not enough in the holistic infrastructure concepts.

Under "Capacity Building," 13 specific recommendations are made:

> 1. Establish an African geoinformatics curriculum of a modular system of short courses. These courses are aimed at users as well as professionals/developers. African institutes of higher learning should modernize their geomatics curricula in modular form to be responsive to long-term education at various levels (Ph.D,

MSc, PGD, Technologist, Technician) as well as short-term training and retraining in geoinformatics. To enhance postgraduate education we should explore and set up articulated MSc and Ph.D. degree programs consisting of African universities and foreign universities.

2.   Conduct a baseline survey of universities and institutions that have the capability to deliver the modular geoinformatics curriculum (MGC). A survey of the ad hoc, user-oriented short courses being offered outside the traditional geoinformatics community should also be conducted.

3.   The MGC will be run on a cost recovery basis to allow participating institutions to sustain the facility.

4.   Establish a network of African universities and training facilities, the Geoinformatics Education Network (GeoEdNet), that will deliver the MGC.

5.   Ensure that MGC is marketed effectively to help ensure access and avoid duplication.

6.   Include foreign affiliates in the network to share curriculum ideas and logistical experiences.

In addition to formal education and training programmes, there is a further need for professional forums on geoinformatics and for coordination of research and development activities in related areas. The areas of research should include implementation and management of geoinformation projects, and the assessment of outcomes of geoinformation technologies (GIT). Specific recommendations relating to research and coordination activities are:

7.   Establish a Network of African Associations of Geoinformatics (NAAG) to facilitate the exchange of ideas within and between the land surveying, remote sensing and other geoinformation production and application communities

a. Explore the possibility of establishing a Geoinformatics journal (*African Journal of Geoinformatics*) to serve as a tool for distributing information and to build a larger sense of community between the many active national, regional, and specialist communities.

b. Explore the possibility of expanding the existing biannual Africa GIS conference. Include selected training modules.

8.   The regional centers of the ECA (RECTAS, Ile-Ife, RCMRD, Nairobi); AOCRS, Algiers, and other similar organizations may coordinate the GeoEdNet and the NAAG, and their associated activities.

9.   The GeoEdNet will establish an outreach program to raise the awareness of Geoinformatics in potential user communities. A kiosk that includes posters, leaflets, and computer demonstrations of geoinformatics applications should be created and taken to symposia and meetings of potential user groups, such as medicine, agriculture, and transportation.

10. Establish a set of pilot applications of GIT of particular relevance. Nodes in the GeoEdNet will carry out these activities. The applications will be used in a

GIT applications module. The applications will be documented in an African Geoinformatics casebook. This can be used in the Geo-Kiosk and distributed to selected decision-makers.

11. Establish research projects that:
    a. Benchmark GIT vs. traditional methods for applications,

    b. Assess the economics of geoinformation.

12. Introduce a Geoinformatics Project Management module to GeoEdNet.

13. Ensure that training modules and demo applications are matched to available technologies, e.g., Internet and computer hardware. (ECA 2001a, 13-14)

These are important steps that deal primarily with the development of human capital and, to a lesser extent, organizational capacity, but they do not address the need to develop and strengthen linkages that could help to make geographic information processing central to African development problems. Progress is outlined in the documentation of the CODI-3 meetings held in May 2003 (ECA 2003a, 2003d). Resource constraints continue to affect implementation of the plans outlined by CODI-2.

1.4.1. ITC Strategies

As argued earlier, GIP must avoid being seen as an isolated technical specialization and become part of the mainstream ICT strategies for development. The importance of information and telecommunications for development was clearly articulated at the African Development Forum in 1999 (ECA 1999) and recognized in the Okinawa Charter by the G8 leaders (Government of Japan 2000) which established the Digital Opportunities Taskforce (DOT) as an implementation medium. The ECA has continued its work in this area, a major manifestation being the creation of the African Learning Network (ECA August 2001b). Further progress is outlined in the common position for Africa's digital inclusion (ECA 2003g) and Africa's position on the world summit on the information society (ECA 2003h).

There are three "pillars" in this program:
1. ICTs in schools and the creation of a regional SCHOOLNET AFRICA structure.
2. Creation of VARSITY NET linking universities and research institutions.
3. A national network which deals with youth who are not in school (Out of School Youth) entitled OSSYNET.

This is a comprehensive initiative which reaches into the education system, both formal and informal, at all levels. GIP and the importance of the geographic foundations for sustainable development, both conceptual and technical, should be a larger part of this initiative. Geography should be seen as a central discipline for the new African Learning Network (ECA 2001b) and the technologies of GIP an integral part of the development and use of ICTs in African development (ECA 2003b).

This is currently not the case, and GIP is being developed primarily as an independent higher-level technical specialization. The proposed change in ITC's curriculum to emphasize the provision of geoinformation for strengthening civil society is both innovative and necessary:

> Programme formulation was embarked on by the end of 2000 and in the course of 2001 resulted in a research programme that...comprises five overlapping spearheads with the shared overall aim of strengthening civil society.

1. geo-information provision for strengthening civil society.
2. geo-information for the multifunctional use of space.
3. geo-information provision for natural disasters and environment.
4. geo-information provision for food and water security.
5. geo-information for global change monitoring. (Beerens 2002, 11)

1.5. African Universities and GIP

Research, teaching, and social service are three key functions of African universities. Of the three, teaching has been predominant in recent years and, as outlined earlier, research has suffered badly. It may be true, as Downes (2000) suggests, that the perception of African universities as isolated "ebony towers" is outdated, but there is still a need for substantial change to make universities more locally relevant, as Pearson (2000) argues.

For the application of GIP, the geography departments of Africa's universities are key and their teaching and research capacities should be strengthened (NRC 2002). The New Partnership for African Development (NEPAD) provides a vision for an African-led development. Unlike many previous plans, it was drafted by African leaders and thinkers. To succeed, it will require new forms of partnership. It has been endorsed by some key players in the G8 including Prime Minister Chrétien of Canada and Prime Minister Blair of the U.K. (Chrétien 2002; Blair 2002). Education, capacity building, and the role of Information and Communications Technology, of which GIP is an integral part, will be central to the success of NEPAD (ECA 2003c). A new opportunity exists to demonstrate the centrality of geographic information for sustainable development in Africa.

Human capacity and capacity building demand new approaches including the development of South-South cooperation and networking (Mehta 2000; Ching 2000). Nations such as China have developed considerable capacity in GIP and their experience could be of utility to African institutions. Mexico, through SEP-CONACYT (the Ministry of Education and the National Research Council) has established CentroGEO, a center with growing research and teaching potential in geoinformatics which introduced its own Master's and Ph.D. programs in 2002 and could be a partner for African institutions with similar interests. Institutional role models in developing nations are important and share common problems and understanding with emerging African institutions in ways which institutions in the north cannot emulate. South-South cooperation has considerable potential, as does more effective networking among African institutions.

1.5.1. Private Sector Partnerships

New opportunities also exist for private sector partnerships and joint initiatives in capacity building in Africa. African players which offer their own training and certificate programs, such as CAROGRAPHX in Nigeria, are emerging (Balogun and Soneye 1999). If GIP is to grow in significance in Africa, there must be private sector growth from within Africa itself. Africa is a growing market of 750 million people and effective applications and capacity building in GIP will lead to increasing demand from the marketplace. GIP is a growth industry, and so is the African private sector at present. Increased capacity is also required in the African private sector, and capacity building, business partnerships, and joint ventures in this sector are just as important as academic and research linkages, although rarely if ever considered in the literature on capacity building in GIP for Africa.

For nations such as the U.S., Canada, the U.K., and France, this poses a dilemma. They are competing with each other for the market in GIP. As the international community increasingly turns its attention to Africa, the market will grow. Firms from industrialized nations already compete vigorously for the growing number of contracts in this field, led by international agencies and national governments alike. African firms and agencies have difficulty competing in this growing marketplace and often lack the human and organizational capacity to do so. One possible approach might be joint ventures or partnerships with small and medium-sized enterprises from developed countries to build up the knowledge and resource base required to grow. Preference might also be given to African firms by both African and international agencies funding projects. Such preferences can be developed within the rules of the World Trade Organization with sufficient political will. Infant African industries must be nurtured if they are to grow and to contribute to the economic growth so vital for Africa's future.

1.5.2. Brain Drain, Brain Gain, Brain Circulation

Here, Africans who have been part of the "brain drain" may have an important role to play. One of NEPAD's aims is to turn the current "brain drain" into a "brain gain," and utilizing the resources of what has come to be known as the "African diaspora" through "brain circulation" can play a part in this process (Juma et al. 2002). The success of human capacity building has seen many outstanding individuals in the GIP field take up permanent residence in the U.S. and other countries. This includes individuals who are playing a key role in the private sector. Some of these individuals may be persuaded to return to Africa on a permanent basis, but many more may be prepared to give their expertise on a shorter term, although continuing basis, in the form of consultancies. Hawley (2001) has argued for the creation of a "Technology Corps" to mirror the creation of the "Peace Corps" in the 1960s. Such a corps could include Africans currently resident in the U.S., and the concept could be expanded to include individuals with expertise to help the African private sector in GIP to develop. An organizational structure such as a "technology corps" is required in order to ensure that the cumulative effect of such efforts is maximized. The U.N. system already has a structure in place to involve expatriate nationals in the development of such nations. Informal institutional structures already exist such as the "town associations" of Ghana which have branches in North America dedicated, as their name suggests, to improving the welfare of their home towns in Africa.

Indigenous capacity building in GIP is key to making the potential for the application of GIP to African development a reality. At present the rhetoric exceeds the reality (Taylor 1998). A full-page advertisement taken out in major Canadian newspapers by the International Development Research Centre entitled "And What About Africa?" (*Globe and Mail*, February 8, 2002, B8) captures the essence of the situation well: "Ultimately, development advances when people have the knowledge–and the power–to identify their own problems and agree on their own solutions."

REFERENCES

Adeniyi, Peter, O. (1985). Remote Sensing, Resource Development and Education in Africa. In Taylor, D.R.F. (Ed.) Education and Training in Contemporary Cartography. Chichester, U.K.: Wiley.

Ademlemo, I.A., Ayeni, O.O. and Balogun, D.Y. (1985). Manpower and Curriculum Development in Cartography in Developing Nations. In Taylor, D.R.F. (Ed.) Education and Training in Contemporary Cartography. Chichester: Wiley.

Ajayi, J.F.A., Goma, L.K.H., and Johnson, G.A. (1996). The African Experience with Higher Education. Ghana: Association of African Universities, Ghana.

Akinyemi, F.O. (2001). Geographic Targeting for Poverty Alleviation in Nigeria: A Geographical System (GIS). Approach. In Proceedings of the 20th International Cartographic Conference, vol. 2: 1259-70, Beijing, China.

Association of African Universities and the World Bank (1997). Revitalizing Universities in Africa: Strategies and Guidelines. Washington, DC: World Bank.

Balogun, O.Y. and Soneye, A.S.O. (1999). Cartography in the Service of Government, NCA Special Publication, Cartografx Ltd., Lagos.

Bassolé (2002a). West Africa: Workforce Issues in the Context of Capacity Building and Decision Making, paper presented to the National Academies Committee, Washington, January 6, p. 6.

Bassolé (2002b). Upper Niger Region: Contribution of GIS and RS to Addressing RRM and Development Issues. Accessed May 2002 at www.opengis.org/gisd/cgi/listgisdtxt.pl.

Beerens, S.J.J. (2002). Capacity Building for Geospatial Information Handling in Africa: The ITC Perspective, Presentation to Committee on the Geographic Foundation for Agenda 21, p. 18.

Blair, Tony (2002). Partnerships for African Development, Speech by U.K. Prime Minister to the Nigerian National Assembly, Lagos, February 7.

Bourne, R. (Ed.) (2000). Universities and Development, London: Association of Commonwealth Universities.

Carrington, W.J. and Detragiache, E. (1998). How Big is the Brain Drain? IMF Working Paper 98/l02, Washington.

Chambers, R. (1997). Whos Reality Counts? Putting the First Last. London: Intermediate Technology.

Ching, L.Y. (2000). Networks and Networking. In Bourne, E. (Ed.), Universities and Development, Ch. 6. London: Association of Commonwealth Universities.

Chrétien, J. (2002). Address by Prime Minister Jean Chrétien to the World Economic Forum Plenary Session New York February 1st, Prime Ministers Office, Ottawa.

Davenport, P. (2000). Development and the Knowledge Economy. In Bourne, E. (Ed.), Universities and Development, Ch. 1. London: Association of Commonwealth Universities.

Downes, A. (2000). University Graduates and Development. In Bourne, E. (Ed.) Universities and Development, Ch. 5. London: Association of Commonwealth Universities.

Duru, R.C.B. (1985). Status and Constraints of Automated Cartography Training in Africa: The Nigerian Example. In Taylor, D.R.F. (Ed.) Education and Training in Contemporary Cartography, 243-58. Chichester: Wiley.

Economic Commission for Africa (1999). Strengthening Africas Information Infrastructure, African Development Forum 1999, Theme 3. www.un.org/depts/eca/adf/infrastructure.htm

Economic Commission for Africa (2001a). The Future of Geoinformation Activities in Africa: A Position Paper. Development Information Services Division. ECA/DISD/GEOINFO/DOC/01, p. 30. Addis Ababa.

Economic Commission for Africa (2001b). The African Learning Network: Emerging from Behind the Knowledge Curtain, E/ECA/DISD/CODI.2/21, August. Addis Ababa.

Economic Commission for Africa (2001c). Report on ECA Activities in the Area of Information and Communication Technologies, Information Systems and Libraries (1999-2001). and Programme of Work for the Biennium (2002-2003), E/ECA/DISD/CODI.2/26, August. Addis Ababa.

Economic Commission for Africa (2001d). The Information Technology Centre for Africa (ITCA), Business Plan for 2001-2002, E/ECA/DISD/CODI.2/28, March. Addis Ababa.

Economic Commission for Africa (2001e). Report on the Sub-Committee on Geoinformation, Second Meeting of the Committee on Development Information (CODI-II), September. Addis Ababa. www.uneca.020/codi/codi2.htm

Economic Commission for Africa (2001f). Report of the Workshop on Spatial Data Infrastructures (SDIS): Technical and Institutional Components, DISD/GEOINF/WS_SDI/01, October. Addis Ababa.

Economic Commission for Africa (2003a). Report of the Follow Up Activities of CODI 2, E/ECA/DISD/CODI.3/2, May. Addis Ababa.

Economic Commission for Africa (2003b). Report of the African Technical Advisory Committee (ATAC). on the Implementation of the IASI, E/ECA/DISD/CODI.3/6, May. Addis Ababa.

Economic Commission for Africa (2003c). Sub-committee on Geo-information Policy Issues–Geoinformation and NEPAD: Executive Summary, E/ECA/DISD/CODI.3/9, May. Addis Ababa.

Economic Commission for Africa (2003d). National and Regional Capacity Building for Geoinformation Technology, E/ECA/DISD/CODI.3/28, May. Addis Ababa.

Economic Commission for Africa (2003e). Sub-committee on Geoinformation Technical Issues, Spatial Data Infrastructures: Getting Started, E/ECA/DISD/CODI.3/30, May. Addis Ababa.

Economic Commission for Africa (2003f). GI in support of decentralisation and community empowerment, E/ECA/DISD/CODI.3/27, May. Addis Ababa.

Economic Commission for Africa (2003g). The Common Position for Africas Digital Inclusion: Recommendations of the meeting on Africas Contribution to the GI DOT Force and the UN ECOSOC Panel on Digital Divide. www.uneca.org/aisi/docs/commonpositiondoc.pdf.

Economic Commission for Africa (2003h). Bamako 2002 and Africas Position on the World Summit on the Information Society (WSIS). www.uneca.org/aisi/bamako2003/index.htm.

Ekong, D. (1998). Sustainable Development and Graduate Employment: The African Context. In Holden Ronning, A. and Kearney, M. (Eds.), Graduate Prospects in a Changing Society. Paris: UNESCO Publishing.

Ezigbalike, D. (2002). Selected Geoinformation Activities at ECA. Presentation to the Committee on the Geographic Foundations for agenda 21, Washington, January, p. 25.

Government of Japan (2000). The Okinawa Charter on the Global Information Society, Kyushu-Okinawa G8 meetings, p. 8. Tokyo.

GTDA website (2002). http://ceos.cnes.fr: 8100/CDROM-00/GESI/infosrc/gdtc.htm.

Hawley, M. (2001). Things That Matter: A Technology Corps. MIT Technology Review, November: 1-4.

International Development Research Centre (1999). Internet for All: The Promise of Telecentres in Africa, IDRC Briefing Paper 3, Ottawa.

International Development Research Centre (2002). And What About Africa? Globe and Mail, Toronto, February 8, p. B8.

Juma, C. (2001). The Global Sustainability Challenge: From Agreement to Action, International Journal of Global Environmental Issues 2: 1-14.

Juma, C., Fang, K., Honda, D., Huete-Perez, J., Konde, V., and Lee, S.H. (2001). Global Governance of Technology: Meeting the Needs of Developing Countries, International Journal of Technology Management 22: 629-55.

Kufoniyi, O. (2001). RECTAS: Blazing the Trail in Capacity Building for Geoinformation Production and Management of Land, Environment and Natural Resources in Africa, Presentation to the Second Meeting of the Committee on Development Information (CODI), October, p. 16. Addis Ababa.

Labatut, J.-P. (2002). IDRC and the African Development Challenges, manuscript, Carleton University, Ottawa, Canada, p. 18.

Mehta. G. (2000). Science and Technology Issues.In Bourne, E. (Ed.) Universities and Development, Ch.5. London: Association of Commonwealth Universities.

National Research Council (NRC) (2002). Down to Earth: Geographic Information for Sustainable Development in Africa. Washington, DC: National Academy Press.

Organization for African Unity (2001a). The New Partnership for Africas Development (NEPAD), October. Abuja, Nigeria.

Organization for African Unity (2001b). A New African Initiative: Merger of the Millennium Partnership for the African Recovery (MAP). and Omega Plan, July. Pretoria.

Pearson, R. (2000). The Challenge of Local Development. In Bourne, E. (Ed.) Universities and Development, Ch. 4. London: Association of Commonwealth Universities.

Psacholopoulos, G. (1994). Returns to Investment in Education: A Global Update. World Development 22: 1325-43.

Sawyerr, Akilagapa (2003). Address of the Secretary General of the Association of African Universities to the Association of Universities and Colleges of Canada, May 20, Ottawa.

Swartz, D. (2000). The State of Universities in the Commonwealth. In Bourne, E. (Ed.) Universities and Development, Ch. 2. London: Association of Commonwealth Universities.

Task Force on Higher Education and Society (2000). Peril and Promise: Higher Education in Developing Countries. Washington, DC: World Bank.

Taylor, D.R.F. (1998). Modern Cartography, Policy Issues and the Developing Nations: Rhetoric and Reality. In Taylor, D.R.F (Ed.) Policy Issues in Modern Cartography, 185-214. Oxford: Pergamon.

Taylor, D.R.F. (1991). GIS and Developing Nations. In Maguire, D.B., Goodchild, M.F., and Rhind, D.W. (Eds.), GIS Principles and Applications, vols. I-II: 71-84.

Taylor, D.R.F. and Mackenzie, F.M. (1992). Development from Within: Survival in Rural Africa. London: Routledge.

World Bank/UNESCO (2000). Peril and Promise: Higher Education in Developing Countries. Washington, DC: World Bank.

CHAPTER 23

GRAHAM A. TOBIN
BURRELL E. MONTZ

NATURAL HAZARDS AND TECHNOLOGY: VULNERABILITY, RISK, AND COMMUNITY RESPONSE IN HAZARDOUS ENVIRONMENTS

Abstract   Understanding the causes and effects of natural hazards has been facilitated greatly by technologies such as geographic information systems (GIS), remote sensing, and satellite imagery as well as by advances in data collection and dissemination. In addition, various technologies have been applied to protect areas from damage or to minimize damage when events occur. Too often, however, the available technologies have been directed to quick-fix responses and remedies, thus hindering rather than increasing our understanding of the factors at work. This trend has been changing, in large part due to advances in analytical tools that allow for consideration of complex systems. Yet despite these advances, global losses continue to increase, reflecting the need to incorporate technologies within the context of the political, social, and economic systems that may increase or mitigate vulnerability.

Keywords   disasters, mitigation, resilience, GIS, mapping, risk communication, planning

1. INTRODUCTION

This chapter addresses evolving concepts of human vulnerability in relation to natural hazards, and shows how technological advances have influenced both research on hazards and policies for mitigating disasters. Hazards research has made tremendous progress, moving the rhetoric from one predominantly concerned with natural phenomena and the "technological fix," to one that includes processes of the human-use system, specifically the complex web of social, political, and economic forces that together comprise individual and community vulnerability. It is the combination of risk and vulnerability that reflects the degree to which society is threatened by, or alternatively protected from, the impacts of events. In this regard, technological developments have significantly altered the hazard landscape, not only providing new tools to facilitate research and understanding, but also

Stanley D. Brunn, Susan L. Cutter, and J.W. Harrington, Jr. (Eds.), Geography and Technology,
547-570. © 2004 Kluwer Academic Publishers. Printed in the Netherlands.

broadening the inventory of potential mitigation strategies. Geographic information systems (GIS), remote sensing, and satellite imagery are now fundamental components of hazard research that have created new opportunities and enhanced our scientific approaches. At the same time, other technological advances, including those in communication systems, have greatly improved the range of detection, warning, and mitigation techniques available to hazard managers. Still, losses continue to increase. This chapter develops these ideas, demonstrating how such advances have improved, but at other times hindered, our understanding of and responses to hazards.

2. THEORETICAL AND CONTEXTUAL FRAMEWORKS

A conceptual framework that pulls together the salient contextual features and provides a relational schema of natural hazard mitigation is outlined in Figure 23-1 (Tobin and Montz 1997). An amalgamation of physical characteristics and political factors works to define risk, whereas vulnerability is determined by all the elements in various combinations, thus illustrating the dynamic nature of the hazard environment. Similarly, loss reduction options offer various opportunities and constraints that are interdependent with changing levels of vulnerability. This suggests that if we alter any of the elements, in the top or the bottom of the diagram, we have also changed vulnerability. Measuring the extent to which vulnerability has been altered is difficult, however, because the elements are not equal in their contributions. Indeed, not all may be relevant in a given situation, and some are certainly more significant than others. For instance, political and economic factors, over which an individual may have little control, will dominate in some hazardous situations and in post-event activities and priorities. In other cases, relationships between the physical environment and social characteristics define what happens, either pre- or post-event. Context, therefore, is critical to this conceptualization (Mitchell et al. 1989).

To comprehend the vicissitudes of natural hazards, many researchers have incorporated new paradigms into their studies that focus attention on community sustainability and societal resilience to disasters. Sustainable and resilient communities are defined as societies that are structurally organized to minimize the effects of disasters, and, at the same time, have the ability to recover quickly by restoring the socioeconomic vitality of the community. Of necessity, we have developed refined concepts of vulnerability and risk that include all sectors of society, including those traditionally marginalized. Furthermore, we have seen this interest transferred through decisionmakers and planners into innovative policies that encompass broader planning goals and more sophisticated hazard mitigation programs than previously implemented.

The relationship between community sustainability and hazards is complex and cannot be addressed adequately without consideration of the range of factors illustrated in Figure 23-1. Clearly, technology influences a number of the components of this model, including relationships between technological change and levels of vulnerability, new opportunities for loss reduction, and enhanced understanding of complex interactions through

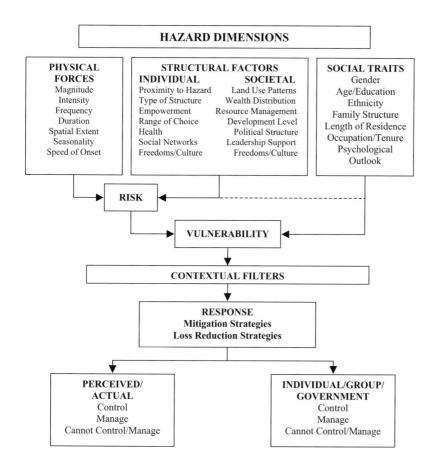

Figure 23-1. Natural hazards and human response in context. Technological advances play a role in each of these elements (Source: adapted from Tobin and Montz 1997).

improved analytical tools. As might be expected, not all are positive changes. Technological advances can exacerbate human vulnerability by facilitating the movement of people into hazardous zones. Take Florida, for example, where the development of air conditioning and pesticides such as DDT are often credited with opening the south of the state to high levels of immigration. There are now millions of people exposed to storm surge and flooding associated with tropical storms, with little or no experience of such events. At the same time, of course, improvements in data collection, such as through satellite imagery, and more sophisticated, computer-based, analytical techniques mean that understanding of such storms is far superior to what it was twenty or thirty years ago. Thus, in Figure 23-1, technological developments can play different roles in virtually every box, shaping not only our scientific understanding of the geophysical characteristics of the hazard, but also influencing the political, social, and economic traits that constitute society. As far as hazard mitigation is concerned, it is how such technological developments are used that is most important.

Because of the numerous and varied roles of technology, it is important to consider the range of elements that comprise natural hazards research and mitigation. Thus, we start with physical characteristics and move through structural, temporal, and spatial issues to consideration of integration and unifying themes.

3. PHYSICAL CHARACTERISTICS OF NATURAL HAZARDS

Natural hazards researchers have learned much from examining the physical forces that comprise disasters, and such work has leaned heavily on the natural sciences. There are many common features among natural hazards, even those as apparently divergent as floods and droughts, or tornadoes and earthquakes. For example, it is possible to classify natural hazards according to measures of hazard frequency, intensity, duration, and speed of onset, each of which can have a direct impact on response patterns. It is in this capacity that technology has played an important role in hazards research. Development of Doppler radar, for instance, has greatly improved weather monitoring and forecasting, while satellite-based remote sensing provides enhanced images of such phenomena as hurricanes and wildfires as they develop and spread. Indeed, much work has been undertaken into the physical dimensions of natural hazards, and virtually all now incorporates some of the latest technology, from computer databases to remote sensing and GIS. To take one case, technological advances have vastly increased our knowledge of volcanism and its potential impacts, which ultimately may permit more accurate warnings of volcanic activity over time and place. Specifically, Connor et al. (1993, 2001) developed an ash dispersion model to map the deposition of tephra from volcanic eruptions through refined remote sensing techniques, while Hepner and Finco (1995) modeled gaseous contaminant pathways over complex terrains using GIS. Some work is not hazard specific, but rather examines the uses of GIS in different hazard contexts (Carrara and Guzzetti 1995). Thus, technological advances, based on sound scientific theory, can facilitate data collection and analysis, and significantly enhance mitigation planning.

Specific physical criteria often form the basis for mitigation strategies. In the U.S., for example, the 100-year floodplain has become the standard for flood alleviation, such that the National Flood Insurance Program utilizes it with respect to regulation and zoning ordinances. Defining the precise location of the 100-year floodplain, however, is not easy and was originally based on little more than an approximation of certain contour lines. More recently, though, technology has allowed for incorporation of a greater number of variables in hydrologic and elevation models, resulting in more realistic floodplain boundaries. In this way, GIS technology is helping to redraw such boundaries and has the potential of frequent updates as hydrological conditions change. Furthermore, the application of GIS also allows for a focus on local issues.

GIS technology has also been applied to interactive mapping on the Internet, addressing both natural and technological hazards (Hodgson and Cutter 2001). Users can input locations, such as zip codes, communities, or states, and maps are provided that show the historical occurrences of events, sites of known lead contamination, and toxic release inventory data. These websites are the products of the federal government, for example, the Environmental Protection Agency (www.epa.gov/enviro/html/em) and the USGS (www.usgs.gov), nonprofit organizations, including the Environmental Defense Fund (www.scorecard.org), and public/private collaborations, such as the one between FEMA and the Environmental Systems Research Institute (www.esri.com/hazards/index.html). These and other sites offer information at various scales and with varying abilities for manipulation. Some are more useful for research than others, because of the levels of accuracy available; others are more suited to increasing public awareness.

Remote sensing techniques provide more accurate data for the mapping of events and hazard zones. For instance, Lougeay et al. (1994) use two different approaches, both based on remotely sensed imagery, to calculate the extent of flooding in the 1993 U.S. Midwestern floods. Similarly, Helfert and Lulla (1990) demonstrate how photographs taken from the space shuttle could be used to map continental scale biomass burning in the Amazon Basin. In

Florida, new storm surge zones are being established from LIDAR data, which will significantly change the spatial extent of these hazard areas. If incorporated into hurricane policy, these new zones are sufficiently refined that fewer people will need to evacuate the area, reducing costs and confusion on roads during emergencies.

Even with these important uses of technology in understanding the physical characteristics of natural hazards, decisions such as determining design levels for mitigation are part of the human realm. Managers must ultimately decide on appropriate courses of action, and in this regard, physical criteria may not be as clearly defined as managers would wish. Harwell (2000) expounds on this in her discourse on remote sensing technology and forest fires in Indonesia, pointing out that technology has become the lingua franca for discussions on the disaster. She demonstrates that even identical, remotely sensed images produced significantly different interpretations as to the crux of the fire problem. Indeed, competing analyses of the images actually served to increase uncertainty about the extent of damages, and several questions concerning the hazard remained unanswered. For instance, users of the remotely sensed imagery could not identify the local impacts of the forest fires on traditional food crops. On the other hand, the technology did help provide valuable insights into other aspects of the fire hazard. To illustrate this, Non-Governmental Organizations (NGOs) in Indonesia used satellite images from websites and GIS technology to document and publicize that it was not small farmers setting fire to their fields who were to blame for the smoke, but rather it was large plantation owners clearing forests for new land.

The example above suggests that, while technological developments have certainly increased the amount and accuracy of information available on natural hazards, there are significant constraints and limitations associated with them as well (McMaster et al. 1997). Carrara et al. (1999) are quite critical of the way GIS has become a buzzword that generates "great expectations and promises much" while not always delivering. They attribute this to misconceptions about the technology and/or knowledge of hazard theory. They argue that: (1) computer-generated maps are considered more accurate and credible than hand-drafted ones even though conversion of data to digital form may incorporate random or systematic errors; (2) a map portraying natural resources or hazards obtained by data manipulation within a GIS is assumed to be more objective and unbiased than other maps; and (3) handling of geographical data in a GIS is simplified for users who are not experts in technology and not knowledgeable of the theory behind it. Thus technology can lend an air of infallibility and scientific validity that goes far beyond the reliability and accuracy of the data.

An additional problem stems from a disjuncture between the GIS/ technology community and hazards researchers. A mutual understanding between the use of the latest technological tools and academic theory is essential if progress is to be made in hazard mitigation. Carrara et al. (1999) suggest that hazard models are often developed by technical experts without sound theoretical underpinnings, while hazards researchers do not acquire the necessary skills to apply the technologies effectively. In other words, technology is not used efficiently and important contextual information is omitted from analyses. In addition, the temporal and spatial scales at which data are available may compromise the utility of many GIS-based products for hazard management applications (Hodgson and Cutter 2001). While these criticisms have some validity, there are now many examples of GIS mapping and remotely sensed data being incorporated into sound hazard policy. For instance, mapping of hazard areas has assumed much greater importance in policy as technology has provided the means to update and refine data (Cutter et al. 1999; Monmonier 1997). Thus, improved understandings of physical criteria are important, but more is needed for hazard assessment and mitigation practices.

4. TECHNOLOGICAL SOLUTIONS AND BIASES

Technology has always been an important consideration in hazards research, in part because hazard mitigation was originally based on the belief that disaster reduction was possible only through technology and engineering. The construction of dams and levees, tornado shelters, and seawalls, for example, was seen as the most effective means to protect the public from specific vagaries of nature. Up to design levels, some engineering works have been very effective in preventing losses (Interagency Floodplain Management Review Committee 1994; Tobin 1995). One would hardly argue that being able to build roads and structures to withstand the shaking from earthquakes or high wind speeds is not an appropriate application of technology. Thus technological advances, at different points in time, have been effective mitigation strategies.

This trend was reflected in the early development of the U.N. International Decade for Natural Disaster Reduction (IDNDR) in the 1990s, when the emphasis was placed on technology and the dissemination of scientific knowledge, based on "the promise scientific and technical progress holds for understanding these hazards and mitigating their effects" (National Research Council 1989, 2). Indeed, the initial impetus for the IDNDR came from the earthquake community with technological issues placed at the forefront. As noted by Housner and Chung (2000, 64), a key task of the decade was to "develop cost effective methods, using locally available materials, techniques, and human resources to retrofit existing buildings and structures against seismic forces."

This bias towards technology and technological solutions generated considerable criticism. Alexander (1993, 617) in reference to this technological emphasis stated, "An extreme interpretation would be that the Decade represents an attempt by engineers and physical and natural scientists to concentrate their academic power and funding opportunities into their own hands in the name of applying their sciences." This is a valid criticism if one considers where funding has been applied. Of course, this is not to overlook the significance of technology, and Alexander further states (p. 617) that, "This, however, should not negate the fact that science and technology have very important roles to play in monitoring and mitigation efforts." Indeed, as the decade progressed, more attention was devoted to mitigation strategies as well as global telecommunications (Hamilton 2000). Follow-up activity by the U.N., the International Strategy for Disaster Reduction, continues this trend.

Inherent to the U.N. Decade, which has been oriented towards disaster and disaster mitigation, is a particular bias towards the technological fix as an overriding solution. The control of nature rather than human activities is the fundamental premise here. White et al. (2001) point out that the emphasis on disasters has undoubtedly contributed to the reduction in fatalities, but it has also probably limited the advancement of comprehensive mitigation strategies. They argue that a broader hazards approach, as opposed to one focused on disasters, would be more productive by fostering scientific investigations of vulnerability. This argument is also taken up by Mileti (1999) in the second assessment of hazard research in the U.S. The common theme is that both technological advances and understanding of social forces are important in mitigating natural hazards. Indeed, it is a sound comprehension of vulnerability, as shown in the theoretical framework (see Figure 23-1), which may prove the key to implementing effective mitigation strategies.

An area that reflects the issues noted above, and in which improvements in technology have been seen as critically important, is forecasting and warning systems. Warnings depend on the collection of geophysical data, the careful analysis of those data, the broad dissemination of the warning message, and, ultimately, effective response and remedial action. The accuracy of such warnings depends, in turn, on the speed and reliability of the scientific models. All elements of the forecasting and warning system must function if such mitigation measures are to work. For example, for flash flood warnings, both meteorological monitoring and communication systems must be in place and ready at all times. In the U.S., real-time meteorological and hydrological data are available, which can be combined with soil moisture data and other factors in sophisticated models, to produce information on which accurate forecasts can be made and warnings issued. Yet it is the flow of information through the communication system that facilitates or complicates the dissemination of the warning message (Gruntfest and Waterincks 1998).

It is clear that forecasting and warning systems have significantly reduced fatalities and injuries from many natural hazards, especially in the U.S. Unfortunately, as pointed out by Sorenson (2000), this has not been true for all countries, especially, the less wealthy nations. The problem remains that the technology required for sophisticated forecasting and warning systems is prohibitively expensive to develop and maintain. To address this, Sorenson (2000) advocates improving local warning systems by providing low-cost or no-cost warning dissemination techniques, and by providing technical assistance and/or cost-sharing strategies for additional resources for better warning and related communications equipment. It has also been suggested that NGOs should play a role in this venture (Benson et al. 2001), although questions of accountability and quality of NGOs in many arenas are frequently raised (see Hilhorst 2002 for a review).

The examples given above provide brief illustrations of the critical importance of technology and technological advances in understanding events and in providing mitigation alternatives. Failures do occur, however, and advances in technology do not necessarily equate with less risk, so losses continue to increase. Thus there is more that needs to be considered than simple technological solutions to mitigation.

4.1. Social Science Perspectives

Missing from the equation has been the application of social science, as so aptly demonstrated by White (1945) when looking at structural solutions to flooding, Mitchell (1989) with respect to IDNDR, and Gruntfest and Carsell (2000) on warnings, to cite some examples. We know from studies of perception and adjustments that technological remedies for natural events influence perceptions of the problem and tend to distort views of potential victims and decisionmakers about the possible severity of events, who then may see the problem as solved (Burton et al. 1993). Various factors, including applications of technology to mitigation and warning, may lead to

> the resignation that there is no other place to relocate, the misunderstanding that the area is safe, the attitude that the hazard can be humanly conquered, and the reliance on insurance and institutional relief on the remote chance that something doesn't go as designed. (Nelsen 1994, 6)

Forecasting and warning technology, for instance, does not solve the problem nor does it necessarily get people to relocate. Indeed, research has established that individuals will only respond to evacuation warnings if they believe the threat to be real, that the impacts will be experienced by them and their families, and that responding to the warning will result in protection (Quarantelli 1980; Smith 1996). The literature also suggests that while some people may voluntarily leave when a hazard threatens, many await an official warning or order to do so (Fischer et al. 1995; Perry 1985) and that compliance is more likely if the source is considered reliable (Lindell and Perry 1992). It is the human dimension, therefore, that ultimately determines the relative success of a forecasting and warning system.

The levee effect is another classic consequence of the technological fix that has contributed to increased disaster losses and certainly exacerbated hazard problems (Segoe 1937; White 1961). The construction of structural measures to mitigate hazards or even the implementation of nonstructural adjustments, such as zoning ordinances, can lead to increased development on the "safe" side of the project (Tobin 1995). Consequently, losses can be catastrophic and, over the long term, possibly greater than if the project had never been implemented because of the increased development (Tobin and Montz 1994a).

Just as flood control structures illustrate a misunderstanding about safety, so too do lay understandings of technical terms. It is not unusual, for instance, to hear flood survivors state that they have experienced the "one in one-hundred year" event, and consequently will not experience a similar flood in their lifetimes. As many survivors of the 1993 upper Mississippi River floods in the U.S. found out, this is a fallacious argument, since flooding occurred again that summer in some communities and just two years later in others. This attitude can also be reflected in behavior. During the postdisaster period in 1993, some of these people took out flood insurance, only to let it lapse one year later. As the period between disasters increases, this problem gets even worse since the hazard is replaced by more day-to-day concerns. Thus socioeconomic forces can work against effective mitigation. A similar influence is seen with hazard warnings with a general mistrust of official messages. As Gruntfest and Carsell (2000) point out, despite the information available through the World Wide Web and other sources, some individuals prefer to rely on their own observations. Even local officials differ in the extent to which they rely on different sources and different types of information in issuing warnings (Baker 1995).

A prevailing feature of disaster response is the ad hoc or piecemeal approach usually given to hazard mitigation. To a certain extent, focusing on technological and engineering options, as has been the pattern, serves to avoid the comprehensive view that is required. Again, there are many good reasons for this, the most significant, perhaps, being the humanitarian one of trying to help those in need as quickly as possible. This approach only serves to fuel the disaster-damage-repair-disaster cycle, however, and of course raises the question of why people continue to live in hazardous areas, given all of the information that is available.

There are various explanations to account for continued occupancy of hazardous environments, including structural constraints, such as lack of choice of alternatives, perceptions that benefits of location outweigh costs, and lack of awareness of the hazard (Hewitt 1997; Kasperson et al. 1995). It may well be that the perceived economic advantages and other benefits to remaining in hazard-prone areas outweigh perceived costs. Applications of technology have enhanced our studies in this regard. Hodgson and Palm (1992), for example, examined attitudes towards disaster through a spatial analysis of human responses to the earthquake hazard in California. Using a GIS design that included geographically coded data, combined with survey questionnaires, they looked at location, attitudes toward the hazard, and purchase of earthquake insurance. Such studies have provided new insights into hazard-zone behavior that may eventually lead to improved mitigation.

Researchers have evaluated the relative benefits and costs of hazardous locations. Through applications of various GIS methods and statistical software, the impacts of hazards on property values and the extent to which risk may be capitalized into the value of land has been analyzed. Chao et al. (1998) reviewed 13 such studies and found that all used spreadsheets and sophisticated statistical techniques. Several found that negative impacts from the floods could last for many years, as witnessed by the studies of floodplains and land values by Montz and Tobin (1996, 1999) and Tobin and Montz (1994b). The consequences of this capitalization may mean that hazard-prone areas develop different socioeconomic traits compared to other areas. For example, many floodplains are inhabited by lower-income households, elderly, young families, people on assistance, and a greater proportion of mobile homes (Interagency Floodplain Management Review Committee 1994), characteristics that are known to exacerbate hazard vulnerability. In this case, therefore, georeferenced data, GIS technology, and statistical software have greatly facilitated detailed spatial analyses of socioeconomic factors that affect losses to events (see Figure 23-1).

These consequences do not occur separate from other forces. It is worth noting the role insurance and insurance companies play in hazard mitigation and policy development. Because they have much to lose when disaster strikes as well as much to gain from mitigation, and thus avoided losses, major insurance companies regularly estimate total insurance losses from past disasters and seek to document potential future losses. Such estimates, unfortunately, provide an incomplete assessment of global losses for a number of reasons. Specifically, there are many hidden costs, no real standards for data collection, and no criteria for estimating indirect losses and depreciated values. In addition, the costs of emergency measures are generally not included in such estimates (White et al. 2001). Munich Re regularly publishes an annual review of natural catastrophes, with such estimates being part of a larger effort aimed at addressing risk management (Munich Re 2003). Among other approaches, Munich Re addresses the topic of geocoding of risks and asks the question: "Does geographic underwriting improve risk management?" The implications of this question for hazard mitigation are important because insurance can be an important part of the mitigation arsenal. The participation of more than 18,000 communities in the National Flood Insurance Program in the U.S., overseen by the Federal Emergency Management Agency (FEMA) and the activities of the Institute for Business and Home Safety (IBHS) are examples of this issue. FEMA utilizes the latest GIS technology to help define the 100-year floodplains for zoning and insurance purposes. Among other activities and programs, IBHS uses a geocoded database on paid losses to events in an effort to encourage loss reduction. Of course, such strategies do not eliminate the problem, but they do contribute to awareness and mitigation efforts.

This section illustrates the importance of technological applications to natural hazards research. Advances in technology have contributed to the mitigation of the adverse effects of events by providing protective structures and building standards and facilitating advances in knowledge about the dynamics of hazards. These are not solutions, however, but rather steps in a process that must include socioeconomic and political factors. Geographic research in these areas has incorporated various technologies, such as GIS, remote sensing, and spatial statistics in an attempt to further our understanding of the relationships at work. The rest of this chapter focuses on the ways in which technology has facilitated research rather than mitigation, though it is sometimes difficult to separate the two.

5. DATA NEEDS AND DEFICIENCIES

Successful applications of technology and science to solving hazard-related problems are based on organized strategies that include definitions of severity of the problem and need, which in turn are dependent on adequate baseline data. To this end, Sapir and Misson (1992) advocate a systematic data collection strategy, consistent among places, to improve preparedness and response programs at the global level. As an example, GIS software known as Emergency Management Information Systems (EMIS) was developed and has been employed at emergency management offices in New York. Although problems arose in applications of this software (Monmonier and Giordano 1998), systematic data collection strategies undoubtedly enhance understanding of needs and, ultimately, emergency relief distribution.

Data required for the EMIS models unfortunately are usually incomplete, and data that do exist frequently present conflicting findings (Mitchell et al. 1989). Part of this is due to faulty or inconsistent methodologies utilized to collect data and/or to derive information. For example, it is impossible to determine the extent to which losses have been reduced if we do not know what losses have been historically. Emphasis, then, must be placed on developing methodologies for collecting data that are accurate, consistent, and able to be carried out relatively simply; otherwise, we will continue to make poor decisions because we are basing those decisions on inadequate data.

In spite of these shortcomings, the uses of computer technology have clearly benefited data collection. King (2001) for example, shows how large databases on land use and demographics aid analyses of vulnerability, even though problems associated with such work are recognized, especially aging of data, the arbitrary nature of boundaries, and the problem of weighting indices and categorization of vulnerability. Montz and Evans (2001) have looked at the use of GIS and vulnerability issues with the flood hazard and discussed the difficulties of assigning weights to different components of the model.

When considering data collection needs, it must also be kept in mind that individual decisionmaking is made through the perceived world. This perceptual context is dynamic and variable over time and space, and is perhaps most difficult to model from a scientific perspective. Questionnaire surveys of disaster victims have attempted to elicit perceptual worldviews, but these have been merely snapshots of changing and evolving scenarios. What people say they will do under certain circumstances does not always correspond with what actually happens later (Deutscher 1973), and, in many cases, the influence of broader political, economic, cultural, and social factors on perceptions has sometimes been missed. Despite the many studies that have been undertaken to explore these issues, no clear model has been forthcoming to explain or predict individual behavior, although computer applications, improved data collection, and large databases have all added to our understanding of perception, and offer promise for the future (Cutter et al. 1999, 2000). This is especially true for developing models of vulnerability and community resilience which may hold the keys to effective mitigation policies, particularly at the local level.

Finally, in the quest to facilitate data collection and to improve databases, it must be recognized that access to this information, and thus application of it, will not be equitable. Maps and diagrams do not necessarily provide the answer either, even if they are constructed with the latest GIS technology. For example, in reference to earthquakes, Monmonier (1997, 16) states,

> What most people know about earthquakes is a naïve mix of fact and myth. Although perpetuated by maps, these misconceptions result mostly from misinterpretation of maps, not from factual errors.

The issue, therefore, is not a problem with technology, which can help provide valuable insights, but with understanding and communicating information. Put bluntly, many people have difficulty reading maps.

Another issue that is becoming apparent is the digital divide, and the fact that many people do not have ready access to this technology (Digital Divide Network 2002). In addition, even access does not guarantee use, as shown by Bird and Jorgenson (2002). This divide will likely increase differences in data collection efforts and in applications of technologies to understanding hazards. As a result, the gap between the "haves" and the "have-nots" will widen, this time based on use and availability of appropriate technologies. Thus increasing access to data and facilitating use must be high on the list of priorities as data needs and deficiencies are addressed.

6. TEMPORAL AND SPATIAL ANALYSES OF HAZARDOUSNESS

6.1. Temporal Issues

Temporal considerations influence both the type of hazard mitigation that is adopted and the types of hazards that are the subject of research. In both cases, there is a bias to more intensive events and to quick-fix kinds of solutions. With respect to the adoption of hazard mitigation, the emphasis is invariably on short-term benefits, rather than longer-term goals, such as those associated with a unified program of hazards management. For the same reasons, local interests, such as those of the local taxpayer and voter, win out over the welfare of the general public. An example of this variation is the response to the 1993 flooding in the Upper Mississippi River basin in the U.S. While some residents were evaluating long-term land management options (Interagency Floodplain Management Review Committee 1994), levees were being rebuilt, many with increased design levels (Myers and White 1994). Thus the quick, technological response prevailed, despite our knowledge of the false sense of security that such responses generate.

Similar decisions can be seen worldwide where short-term economic pressures place priorities on issues other than the possible impacts of an event with a low probability of occurrence. Development projects, funded if only in part by international agencies, bring in financial and new technological resources to countries in an attempt to further the economic base. Of course, there is great political currency associated with such projects because of the promise they hold for economic development. At times, the fact that they may increase vulnerability, as the 1984 chemical disaster in Bhopal, India, illustrated too clearly, or otherwise alter relationships between people and the environment, takes a backseat to development goals. In addition, it is not unusual to see politicians exploiting disasters for political gain by posing for photographs at disaster sites. This image is then broadcast across the nation by media outlets and on the World Wide Web, demonstrating the politician's "commitment" to compassion and concern for local survivors. Technology in this way may increase hazard awareness, but it also carries another burden, namely the manipulation of those most vulnerable.

While increasing the awareness of hazards, IDNDR was similarly biased toward quick-onset events and the collection of data about them. Less attention was given to pervasive problems such as famine and drought, and initially there was little concern for global climate change (Mitchell 1989). Greater consideration is now evident for slow onset, more chronic hazards as they contribute to hazardousness and vulnerability, and efforts are being directed to collecting data and developing models associated with them. For example, Whiteford et al. (2002) and Tobin and Whiteford (2002) looked at the ongoing impacts of the chronic volcano hazard of Mount Tungurahua in Ecuador. Climate change models are other important steps in this direction (Stern and Easterling 1999), but application to impacts on the ground remains a critical need. The lack of work in this area represents another example of political factors influencing disaster management, because pervasive hazards do not usually have the same visual impact as, for example, an area devastated by an earthquake. This bias also reflects an emphasis on technological and structural fixes, which may be more relevant for quick-onset events. Nevertheless, technological advances, such as GIS and remote sensing, will significantly enhance our understanding of these chronic events.

6.2. Spatial Issues

Many natural hazards are location-specific. Floods occur mostly in floodplains, those near the coast are subject to storm surge from hurricanes, and seismically active areas can be defined and mapped. Application of technological advances to the mapping of these factors was addressed earlier, and their contributions to advancing our knowledge and understanding of hazards are significant. Those evaluating hazard mitigation strategies must consider not only the spatial extent of hazards or events, but also the broader impacts of events and the effects of implementing mitigation projects on those at risk. In fact, the mapping of risks, hazards, and other technological threats has been undertaken in many countries, including the volcano hazards in Indonesia, tsunami and earthquake zones in Japan, and nuclear threats in Russia. GIS has been particularly helpful to this end (Cutter et al. 1999). For example, census data have been mapped and combined with hazard zones to allow for evaluations of vulnerability and hazardousness (Cutter et al. 2000). From this, we have learned that the spatial pattern of social costs can be distributed differently to actual risk, suggesting that resources (human, financial, and technological) are needed to ensure that correct measures are being taken and implemented in appropriate locations.

In a project to evaluate the hazardousness of place, Hewitt and Burton (1971) pointed out that any given location is subject to multiple hazards, but acknowledged that analytical tools to evaluate them did not exist at that time. GIS, however, now provides the necessary tools through which such analyses can occur, and several studies have addressed issues associated with multiple hazards. Cutter et al. (2000), Montz (1994, 2000), and Emrich (2000) provide examples of various ways of defining hazardousness. These sources and others show that there are different ways to define hazard areas, sometimes based on the spatial extent of a combination of events, other times based on the probabilities of occurrence. While great strides have been made in large part based on the capabilities of GIS software, theoretical difficulties remain.

We have, undoubtedly, come a long way in evaluating spatial dimensions of hazards. The insights garnered must feed into development plans, however, which then must coordinate with hazard planning to prevent transferring frequent local problems into large-scale catastrophic losses. For instance, some protection might be required of property located in "safe" areas but still at risk should a structure, or project, fail. GIS will define these areas, within bounds. Similar decisions could be made for other projects and different hazards—the mitigation measure does not eliminate the risk completely, and people are still vulnerable to disaster.

Technology transfer is important, but so too is the redistribution of resources. Blaikie et al. (1994) suggest that mitigation be prioritized towards specific sectors of society so that the maximum number of people are protected for given resources, and that mitigation measures be "active," at least in developing countries, to encourage participation in hazard reduction practices. Passive actions include planning controls and structural measures, whereas active measures include resource transfers as incentives to undertaking remedial action. For example, in Bangladesh, the control of flooding has been improved in some locations by compartmentalization projects that allow measured flooding of fields, so benefits to agriculture could still be enjoyed with projects, regulated at the local level, reflecting the implementation of low-level technology. The Association of State Floodplain Managers (1992) in the U.S. also urges state and local-level controls for flood alleviation plans, rather than federal involvement. Once again, technology and expertise must be available at the local level if mitigation is to be successful, and for the transfer to be economically viable (Sorenson 2000).

7. VULNERABILITY, SUSTAINABILITY, AND RESILIENT COMMUNITIES

The discussion above begs the question of vulnerability. Technological advances have clearly facilitated our abilities to understand the nature of events and hazardous locations, but there remains a question of losses. One could argue that particularly vulnerable groups should be identified in hazard mitigation planning (Mustafa 1998). Where are the elderly, the young families, the poor, those needing special physical and mental attention located? In addition, a primary priority in many places should be saving lives, not property, and, as Blaikie et al. (1994) point out, the property of the poor is frequently more critical than that of the wealthy. GIS, remote sensing, and other technologies are available to assist in this endeavor, but as noted above, local governments are best placed to answer these questions and set priorities, within their planning activities.

In theory, sustainable and resilient communities should be able to withstand extreme geophysical processes and recover rapidly from disasters whenever they occur. Sustainability and resilience, then, are contingent upon careful planning and organization of society, to ameliorate the impacts of disasters and to facilitate recovery processes. Such comprehensive planning must encompass mitigation strategies to reduce risk and exposure, post-disaster plans to promote short- and long-term recovery, and careful consideration of structural and cognitive factors that will influence program effectiveness. The technological advances discussed in this chapter can facilitate this process, but it must be recognized that they come with a cost. Those least able to underwrite that cost are, too often, those at the greatest risk.

8. PROGRESS IN UNIFYING CONCERNS

Losses have continued to rise, even in the face of considerable investment in alleviation adjustments and improved detection and analytical capabilities. A new focus is required to limit reconstruction and development in known hazardous areas. To this end, Figure 23-2 takes into account a more comprehensive view of hazard mitigation necessitating a change in the accepted norms of society (Tobin and Montz 1997). Post-disaster relief and rehabilitation cannot be allowed to perpetuate the problems of the past by merely rebuilding what has been damaged. To accomplish this, a change in perception and attitude is required at all administrative levels, and by all parts of society. Politicians and hazard managers must address the needs and interests of marginalized groups directly and seek to minimize the vulnerability of those in hazard-prone areas. At the same time, individuals must be made aware of the hazardousness of particular locations and be willing to change behavior, and if necessary, location, to reduce hazard losses. In effect, misperceptions of reality must be changed. Clearly, this is no easy task, since it requires more than simply implementing a structural or nonstructural alleviation project. Instead, mitigation goals, plans, and projects must encompass sociological and economic realities, which define the context of a given place, and thus

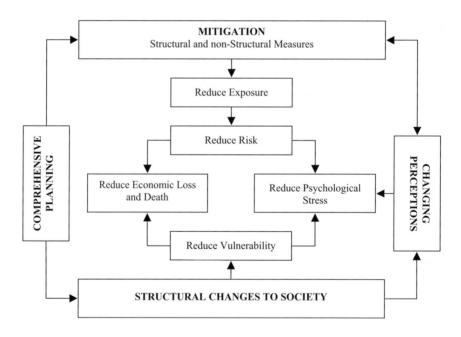

Figure 23-2. A revised model of hazard management. Loss reduction can occur only through structural change in society so that people are not forced into hazardous areas, and short-term benefits do not turn into significant losses over the long-term. Technology is an important component in hazard mitigation and response, but it is only one of many important factors.

integrate new concepts of hazard loss reduction. To achieve all of this, the technological advances that have been discussed throughout this chapter must be applied to various parts of the diagram. Figure 23-2, then, broadens our definitions of hazard mitigation.

Coordinated planning at all levels of government to reduce hazard losses is certainly not a new concept (UNISDR 2001). The innovation of Figure 23-2 is the focus on reduction of vulnerability and the need for comprehensive planning. In a wider sense, hazard reduction plans must be reoriented to address long-term societal needs, rather than concentrating on immediate short-term fixes. Integrated approaches to enhance the decisionmaking process have been advocated by Pareschi et al. (2000), who use a GIS combined with remote sensing technology and telecommunications to examine a warning system in a volcanic risk area. The importance of the GIS was to provide (1) risk mitigation information, (2) suitable tools during the impending crisis, and (3) a basis for emergency plans. All of these are important components in forecasting and warning systems. Similarly, the availability of disaster information on the Internet (Cross 1997) and its growing use in emergency management (Gruntfest and Weber 1998) illustrate technological advances that can enhance mitigation planning and response.

In the best scenario, mitigation strategies should be assimilated into standard planning practices and other development projects such that hazards will become part of the normal functioning of society and will be treated in context. This requires, however, a change in political will and a redirection of goals in which hazards are truly normalized. Hazards cannot be viewed as exceptional circumstances or unique events, but must become part of the overall planning process; they must become a component of "doing business." This requires a long-term view which is not the norm in most political processes. Furthermore, mitigation strategies must also be aimed at multiple hazards, because so many disasters are the consequence of multiple events, such as the flood that initiates a gasoline fire, the earthquake that ignites large urban fires, the thunderstorm that also brings tornadoes and large hail damage (Alexander 1993). Quick-fix, technological approaches solve immediate problems and carry a great deal of political currency. Yet planning for multiple hazards at a place is an important principle to follow and fits with the vulnerability arguments made by Blaikie et al. (1994). Unless multiple hazards are incorporated into planning and hazard mitigation activities, efforts to reduce vulnerability of the population may well be wasted. The challenges presented here have been addressed through GIS and other technological advances; that is, the techniques to depict and analyze the complex interactions that are at work in defining hazardousness now exist. How political and economic factors play out is another story.

9. CONCLUSIONS

Hazard problems are not easily defined because they interact with all aspects of society, suggesting a complexity that makes understanding, modeling, and mitigation difficult. All of these are complicated by scale and the spatial organization of society. The different ways in which places are organized lead to communities divided by religion, language, class, ethnicity, historical context, distribution of wealth, conflicting cultural norms, and political structures. In essence, behavior and response to natural hazards are contingent upon the social, economic, and political realities of place, as well as on variations in the physical environment. Hazards, therefore, must be examined through filters that account for these spatial differences in physical, political, and economic contexts, as well as broader social needs. The advances in technology, especially GIS, remote sensing, and computerized databases, therefore, have expanded our research endeavors.

It can be argued that natural hazards scientists have a moral obligation to pursue applied research, seeking out real solutions to societies' problems, rather than just publishing research findings in obscure academic journals. A proactive, participatory agenda is required that provides venues for scientists to work with politicians, planners, emergency managers, as well as those at risk to develop understandings of the nature of the problem and of potential solutions. This effort will necessarily involve combining technological analyses and approaches with local knowledge and goals. Rising populations, finite resources, spatial inequalities in resource distribution, the marginalization of specific groups, along with greater global interdependence, all contribute to a hazardous world beset with continuing risk and increasing vulnerability. There is an ongoing conflict between the extremes of the geophysical world, which do not recognize political and administrative boundaries, and the human use system. Technological advances have allowed us to begin to take the broader view that is required, in terms of understanding the complex mix of factors that define risk and vulnerability at a place and of developing mitigation strategies that can be sustained over the long-term.

Given past history, however, a cautionary warning concerning possible overdependence on the latest technological developments is warranted. It is possible that the "technological fix" of the past, so eloquently discussed by White (1945) and others over the years, is now being replaced with a "cybernetic simplification" of reality. Indeed, the appearance of truth is beguiling, and the numbers spit out from a machine can provide a new false sense of security. There is a constant danger that electronically based, technological explanations of hazards may be the new equivalent of Orpheus hailing the decisionmaker. Consequently, it is essential that hazard assessment be based squarely on hazard theory and sound scientific endeavors, incorporating natural and social science outlooks, as well as an understanding of the humanities. The latest technology should be used for illumination and not to support preconstructed ideas. To misquote Andrew Lang (1844-1912): "We should use technology as a drunken man uses lamp posts–for support rather than illumination."

REFERENCES

Alexander, D. (1993). Natural Hazards. New York: Chapman Hall.

Association of State Floodplain Managers (1992). Floodplain Management, 1992, State and Local Programs. Madison, WI: Association of State Floodplain Managers.

Baker, E.J. (1995). Public Response to Hurricane Probability Forecasts, The Professional Geographer 47: 137-47.

Benson, C., Twigg, J. and Myers, M. (2001). NGO Initiatives in Risk Reduction: An Overview, Disasters, The Journal of Disaster Studies, Policy and Management 25: 199-215.

Bird, S.E. and Jorgenson, J. (2002). Extending the School Day: Gender, Class and the Incorporation of Technology in Everyday Life. In Consalvo, M. and S. Paasonen (Eds.), Women and Everyday Uses of the Internet: Agency and Identity, 255-74. New York: Peter Lang.

Blaikie, P., Cannon, T., Davis, I., and Wisner, B. (1994). At Risk: Natural Hazards, Peoples Vulnerability, and Disasters. New York: Routledge.

Burton, I, Kates, R.W., and White, G.F. (1993). The Environment as Hazard. New York: Guilford Press.

Carrara, A. and Guzzetti, F. (Eds.) (1995). Geographical Information Systems in Assessing Natural Hazards. Dordrecht: Kluwer Academic Publishers.

Carrara, A., Guzzetti, F., Cardinal, M., and Reichenbach, P. (1999). Use of GIS Technology in the Prediction and Monitoring of Landslide Hazard, Natural Hazards 20: 117-135.

Chao, P.T., Floyd, J.L., and Holiday, W. (1998). Empirical Studies of the Effect of Flood Risk on Housing Prices. Institute for Water Resources Report 98-PS-2. Alexandria, VA: U.S. Army Corps of Engineers Institute of Water Resources.

Connor, C.B., Hill, B.E., Winfrey, B., Franklin, N.M., and La Femina, P.C. (2001). Estimation of Volcanic Hazards from Tephra Fallout, Natural Hazards Review February: 33-42.

Connor, C.B., Powell, L., Strauch, W., Navarro, M., Urbina, O., and Rose, W.I. (1993). The 1992 Eruption of Cerro Negro, Nicaragua: An Example of Plinian-Style Activity at a Small Basaltic Cinder Cone, EOS: Transactions of the American Geophysical Union 74: 640.

Cross, J. A. (1997). Natural Hazards and Disaster Information on the Internet, Journal of Geography 96: 307-14.

Cutter, S.L., Thomas, D.S.K., Cutler, M.E., Mitchell, J.T., and Scott, M.S. (1999). South Carolina: Atlas of Environmental Risks and Hazards. Department of Geography Hazards Research Lab, University of South Carolina. Columbia: University of South Carolina Press.

Cutter, S.L., Mitchell, J. T., and M.S. Scott. (2000). Revealing the Vulnerability of People and Places: A Case Study of Georgetown County, South Carolina, Annals of the Association of American Geographers 90: 713-37.

Deutscher, I. (1973). What We Say/What We Do: Sentiments and Acts. Glenview, IL: Scott, Foresman.

Digital Divide Network. (2002). Bringing a Nation Online: The Importance of Federal Leadership, Digital Divide Network Homepage. www.digitaldividenetwork.org/content/stories/index.cfm?key=248.

Emrich, C.T. (2000). Modeling Community Risk and Vulnerability to Multiple Natural Hazards: Hillsborough County, Florida. MA Thesis, University of South Florida, Tampa.

Fischer, H.W., Stine, G.F., Stoker, B.F., Trowbridge, M.L., and Drain, E.M. (1995). Evacuation Behaviour: Why Do Some Evacuate, While Others Do Not? A Case Study of the Ephrata, Pennsylvania (USA), Evacuation, Disaster Prevention and Management 4: 30-6.

Gruntfest, E. and Carsell, K. (2000). The Warning Process: Toward an Understanding of False Alarms. Washington, DC: U.S. Bureau of Reclamation.

Gruntfest, E. and. Waterincks, P. (1998). Beyond Flood Detection: Alternative Applications of Real-Time Data. Washington, DC: U.S. Bureau of Reclamation.

Gruntfest, E. and Weber, M. (1998). Internet and Emergency Management: Prospects for the Future, International Journal of Mass Emergencies and Disasters 16: 55-72.

Hamilton, R.M. (2000). Science and Technology for Natural Hazard Reduction, Natural Hazards Review 1: 56-60.

Harwell, E.E. (2000). Remote Sensibilities: Discourses of Technology and the Making of Indonesias Natural Disaster, Development and Change 31: 55-72.

Helfert, M.R. and Lulla, K.P. (1990). Mapping Continental-Scale Biomass Burning and Smoke Pails Over the Amazon Basin as Observed from the Space Shuttle, Photogrammetric Engineering and Remote Sensing 56: 1367-73.

Hepner, G.F. and Finco, M.V. (1995). Modeling Dense Gaseous Contaminant Pathways Over Complex Terrain Using a Geographic Information System, Journal of Hazardous Materials 42: 187-99.

Hewitt, K. (1997). Regions of Risk: A Geographical Introduction to Disasters. Singapore: Longman.

Hewitt, K. and Burton, I. (1971). The Hazardousness of Place: A Regional Ecology of Damaging Events. Toronto: Department of Geography, University of Toronto.

Hilhorst, D. (2002). Being Good at Doing Good? Quality and Accountability of Humanitarian NGOs, Disasters: The Journal of Disaster Studies, Policy and Management 26: 193-212.

Hodgson, M.E. and Cutter, S.L. (2001). Mapping and the Spatial Analysis of Hazardscapes. In Cutter, S.L. (Ed.) American Hazardscapes: The Regionalization of Hazards and Disasters, 37-60. Washington: Joseph Henry Press.

Hodgson, M.E. and Palm, R. (1992). Attitude and Response to Earthquake Hazards: A GIS Design for Analyzing Risk Assessment, Geo Info Systems 2: 40-51.

Housner, G.W. and Chung, R.M. (2000). What the IDNDR Has Meant to the Earthquake Community, Natural Hazards Review 1: 61-4.

Interagency Floodplain Management Review Committee. (1994). Sharing the Challenge: Floodplain Management into the 21st Century. Report to the Administration Floodplain Management Task Force. Washington, DC: U.S. GPO.

Kasperson, J.X., Kasperson, R.E., and Turner B.L. (Eds.) (1995). Regions at Risk: Comparisons of Threatened Environments. Tokyo: United Nations University Press.

King, D. (2001). Uses and Limitations of Socioeconomic Indicators of Community Vulnerability to Natural Hazards: Data and Disasters in Northern Australia, Natural Hazards 24: 147-56.

Lindell, M.K. and Perry, R.W. (1992). Behavioral Foundations of Community Emergency Planning. Philadelphia: Hemisphere Publishing.

Lougeay, R., Baumann, P., and Nellis, M.D. (1994). Two Digital Approaches for Calculating the Area of Regions Affected by the Great American Flood of 1993, Geocarto International 9: 53-9.

McMaster, R.B., Leitner, H., and Sheppard, E. (1997). GIS-Based Environmental Equity and
    Risk Assessment: Methodological Problems and Prospects, Cartography and Geographic
    Information Systems 24: 172-89.
Mileti, D.S. (1999). Disasters by Design: A Reassessment of Natural Hazards in the United
    States. Washington, DC: Joseph Henry Press.
Mitchell, J.K. (1989). Where Might the International Decade for Natural Disaster Reduction
    Concentrate its Activities? Decade for Natural Disaster Research, DNDR 2. Boulder,
    CO: Institute of Behavioral Science, University of Colorado.
Mitchell, J.K., Devine, N., and Jagger, K. (1989). A Contextual Model of Natural Hazard, The
    Geographical Review 79: 391-409.
Monmonier, M. (1997). Cartographies of Danger: Mapping Hazards in America. Chicago:
    University of Chicago Press.
Monmonier, M. and Giordano, A. (1998). GIS in New York State County Emergency
    Management Offices: User Assessment, Applied Geographic Studies 2: 95-109.
Montz, B.E. (1994). Methodologies for Analysis of Multiple Hazard Probabilities: An
    Application in Rotorua, New Zealand. Fulbright Research Scholar Report. Prepared for
    the Centre for Environmental and Resource Studies, University of Waikato, Hamilton,
    New Zealand.
Montz, B.E. (2000). The Hazardousness of Place: Risk from Multiple Natural Hazards. Papers
    and Proceedings of Applied Geography Conferences 23: 331-39.
Montz, B.E. and Evans, T.A. (2001). GIS and Vulnerability Analysis. In Gruntfest, E. and J.
    Handmer (Eds.) Coping with Flash Floods: 37-48. Dordrecht: Kluwer Academic
    Publishers.
Montz, B.E. and Tobin, G.A. (1996). The Environmental Impacts of Flooding: A Case Study of
    St. Maries, Idaho. Papers and Proceedings of Applied Geography Conferences 19: 15-
    23.
Montz, B.E. and Tobin, G.A. (1999). Ten Years After: A Longitudinal Study of the Effects of
    Flooding on Residential Property Values. In Bell, R. (Ed.) Real Estate \Damages: An
    Analysis of Detrimental Conditions, 245-56. Chicago: Appraisal Institute.
Munich Re. (2003). Annual Review of Natural Catastrophes, 2002. Munich, Germany: Munich
    Re. www.munichre.com
Mustafa, D. (1998). Structural Causes of Vulnerability to Flood Hazard in Pakistan, Economic
    Geography 74: 289-306.
Myers, M.F and White, G.F. (1994). The Challenge of the Mississippi flood, Environment 35:
    7-35.
National Research Council. (1989). Reducing Disasters Toll: The United States Decade for
    Natural Disaster Reduction. Washington, DC: National Academy Press.
Nelsen, E.W. (1994). The International Decade for Natural Disaster Reduction (IDNDR): A
    Global Concept Yet to Reach its Potential. Unpublished Independent Research Paper,
    Binghamton, NY: Environmental Studies Program, Binghamton University.
Pareschi, M.T., Cavarra, L., Favalli, M., Giannini, F., and Meriggi, A. (2000). GIS and Volcanic
    Risk Management, Natural Hazards 21: 361-79.
Perry, R.W. (1985). Comprehensive Emergency Management: Evacuating Threatened
    Populations. London: JAI Press.
Quarantelli, E.L. (1980). Some Research Emphases for Studies on Mass Communication
    Systems and Disasters. In Disasters and Mass Media, 239-299. Washington, DC: National
    Academy of Sciences.
Sapir, D.J. and Misson, C. (1992). The Development of a Database on Disasters, Disasters 16:
    75-80.
Segoe, L. (1937). Flood Control and the Cities, The American City 52: 55-6.

Smith, K. (1996). Environmental Hazards: Assessing Risk and Reducing Disaster. New York: Routledge.

Sorenson, J.H. (2000). Hazard Warning Systems: A Review of 20 Years of Progress, Natural Hazards Review 1: 119-25.

Stern, P. and Easterling, W.E. (Eds.) (1999). Making Climate Forecasts Matter, Panel on the Human Dimensions of Seasonal-to-Interannual Climate Variability. Washington, DC: National Academy Press.

Tobin, G.A. (1995). The Levee Love Affair: A Stormy Relationship, Water Resources Bulletin 31: 359-67.

Tobin, G.A and Montz, B.E. (1994a). The Great Midwestern Floods of 1993. Fort Worth, TX: Saunders College Publishing.

Tobin, G.A. and Montz, B.E. (1994b). The Flood Hazard and Dynamics of the Residential Land Market, Water Resources Bulletin 30: 673-85.

Tobin, G.A. and Montz, B.E. (1997). Natural Hazards: Explanation and Integration. New York: Guilford Press.

Tobin, G.A. and Whiteford, L.M. (2002). Community Resilience and Volcano Hazard: The Eruption of Tungurahua and Evacuation of the FalDas in Ecuador, Disasters: The Journal of Disaster Studies, Policy and Management 26: 28-48.

UNISDR. (2001). Targeting Vulnerability: Guidelines for Local Activities and Events. United Nations International Strategy for Disaster Reduction.

White, G.F. (1945). Human Adjustments to Floods. Research Paper 29. Department of Geography, University of Chicago. Chicago: University of Chicago Press.

White, G.F. (Ed.) (1961). Papers on Flood Problems. Research Paper 70. Department of Geography, University of Chicago. Chicago: University of Chicago Press.

White, G.F., Kates, R.W., and Burton, I. (2001). Knowing Better and Losing More: The Use of Knowledge in Hazards Management, Environmental Hazards 3: 81-92.

Whiteford, L.M., Tobin, G.A., Laspina, C., and Yepes, H. (2002). In the Shadow of the Volcano: Human Health and Community Resilience Following Forced Evacuation. Technical Report prepared for The Center for Disaster Management and Humanitarian Assistance. Tampa: The University of South Florida.

PART V

THE WORLDS BEFORE US

CHAPTER 24

JEROME E. DOBSON

THE GIS REVOLUTION IN SCIENCE AND SOCIETY

Abstract   Geography, through geographic information systems (GIS), is changing science and society in fundamental, pervasive, and lasting ways. The GIS-based precision bombing of Baghdad changes the nature of warfare, foreign policy, and international relations as profoundly as did the atomic bombing of Hiroshima in 1945. In the U.S. proven applications already affect just about everything that involves location, movement or flow. Most GIS applications are beneficial, but some possess enormous power for good or evil, depending on how they are used. GIS-based human-tracking technologies, for instance, threaten to alter age-old social relationships- parent/child, husband/wife, employer/employee, and master/slave. Society cannot afford to continue "business as usual" with regard to geography and GIS, but remedies will require corrective actions as dramatic as those accorded to physics and nuclear engineering after World War II. Yet popular misconceptions about geography and simplistic views of GIS hamper public debate.

Keywords   geography, geographic information system (GIS), GIS and society, privacy, slavery, warfare, science, macroscope.

The ancient discipline of geography, manifested through modern geographic information systems (GIS), is impacting science and society so profoundly that the word "revolution" seems inadequate to describe what is happening. Compared to its true potential, however, GIS may be the most underutilized technology available today; rather "evolution" better describes the creeping acceptance and grudging recognition of GIS that are occurring in many sectors.

For proven GIS success, nothing compares to the precision-guided weapons, demonstrated so sensationally in the March 2003 barrage of Baghdad. For its latent potential, nothing compares to the specter of "geoslavery" based on human-tracking devices currently for sale (Dobson and Fisher 2003). In the U.S., proven applications already affect just about everything that involves location, movement, or flow. Other technologically advanced nations are advancing towards that same end. In the rest of the world, adoption is slow and selective. The potential for Public Participation

573

Stanley D. Brunn, Susan L. Cutter, and J.W. Harrington, Jr. (Eds.), Geography and Technology,
573-587. © 2004 Kluwer Academic Publishers. Printed in the Netherlands.

GIS (PPGIS) is great (Pickles 1999), even in the least developed nations, but so is the potential for corporate or government dominance of GIS adoption. Even in the U.S., many beneficial applications are delayed or halfheartedly undertaken due to a poor understanding of geography and GIS, ambivalence, or outright hostility.

Most GIS applications are overwhelmingly beneficial, but some-notably precision-guided weapons and human-tracking devices-possess enormous power for good or evil, depending on how they are used. Thus a plethora of new GIS-related ethical issues calls for urgent public debate, even while popular misconceptions about geography and simplistic views of GIS hamper intelligent discourse among scholars, practitioners, and individuals throughout society.

This chapter demonstrates the revolutionary impact of GIS through examples in warfare, science, and society. Also, it questions the nature of the geographic revolution, raises ethical issues, and urges public debate. Some of the issues raised are obvious to practically everyone, others only to specialists in particular aspects of the ongoing revolution. Some may come as sudden revelations glaring from dramatic events, but most have materialized slowly over the decades. I conclude with a summary statement ambitiously intended to characterize the status of GIS in science and society today.

1. GIS IN WARFARE

In 1867 Alfred Nobel invented dynamite, and the nature of warfare changed forever. In 1945 the first atomic bomb detonated over Hiroshima, and the nature of warfare changed forever. On the night of March 21, 2003 and for two weeks thereafter, the largest barrage of conventional weapons in history rained down on Baghdad, Iraq, with such precision that the lights stayed on. This event too has changed the nature of warfare forever. Each of these historic advances-one in chemistry, one in physics, and one in geography-caused deep, fundamental changes not only in warfare but throughout science and society. Hence, geography now joins chemistry and physics among the most beneficial and yet most dangerous of sciences.

Throughout history, geography and cartography have been integrally linked with warfare. Herodotus (ca. 490-420 B.C.), for instance, reported that Athens fought Persia in the Peloponnesian Wars while Sparta avoided Persia's wrath because Sparta's king understood cartographic scale while Athens's leaders did not. That particular "map of the world engraved in bronze" has never been found, but one of the earliest maps ever found depicts the consequences of war. Drawn in the second century B.C. after 13 years of fighting, this silk map from ancient China depicts military encampments and defensive features with poignant notes next to some towns saying, "35 families, all moved away...108 families, none back...now nobody." (Wilford 2000, 8) More recently, Ambrose Bierce (1842-1914) stated, "War is God's way of teaching Americans geography." Bierce, no doubt, was thinking of place-name geography, but the same principle applies to scientific geography, geographic information science, and GIS.

An objective appraisal of what's behind the revolutionary precision of warfare today would regard GPS coordinates only as one of many essential databases. Indeed, the most impressive progress has been in the rapid expansion of global geographic data coverage, viz., the cartographic base plus thematic data characterizing population, elevation, transportation networks, hydrographic features, and land cover. Without these databases, a cruise missile would not know what to hit and what to miss. It would, in fact, not be able to distinguish Baghdad from Boston.

The recent prominence of military applications leads many people to assume that GIS arose primarily from the military, but that is not the case. The military establishment excelled from the beginning in many aspects of remote sensing and cartography, and those new capabilities contributed mightily to the advancement of GIS. The same agencies, however, were slow to shift from automated mapping to analytical GIS. Military products consisted overwhelmingly of maps intended for visual inspection. Only since the late 1990s has digital GIS functionality been substantially integrated into military operations.

One impetus may have been the infamous May 1999 NATO bombing of the Chinese Embassy in Belgrade. Many agreed that this tragic incident was a case of "wrong address," and there has been much conjecture as to how the mistake could have been made. In fact, such errors are to be expected when analysts systemically choose automated mapping—which enables one to make the same mistake very efficiently a thousand times in a row—over GIS— which allows one to compare many geographic databases one against another. Circumstantially, the incident appears to fit these general principles since some databases contained the correct address, while the one that mattered did not. Even conspiracy theorists who suspect the attack was intentional should recognize that such a rogue act likely would be thwarted if multiple analysts shared the ability to compare databases through GIS. Whether this particular incident was influential or not, it is clear that U.S. military planners learned much from their experience in Bosnia and Kosovo, and GIS has played an integral, analytical role ever since. A prominent example is the increased role of GIS in protecting civilian populations.

2. PROTECTING CIVILIAN POPULATIONS

The number of civilian casualties appears to have been historically low in the recent War in Iraq. While no one knows the actual number, many estimates place the death toll among noncombatants far below what would have been expected from the experience in previous conflicts. That is no comfort to those who died or those who mourn them, but it is a matter of great importance to survivors and those who rejoice with them. This section explains how long-term geographic and cartographic traditions contributed to the low losses in human life.

It should be self-evident that, to avoid targeting people, one first must know where they are, which means having more reliable and accurate population data. Worldwide coverage comes from the LandScan Global Population Database (Dobson et al. 2000, 2003), enhanced locally through fieldwork (Dobson 2003). LandScan is a prime example of how valuable and enduring the legacies of geographic research can be—or, conversely, how deep current information technologies reach into geography's collective kitbag. The global LandScan database heavily relies on dasymetric interpolation invented by John K. Wright of the American Geographical Society more than 65 years ago (Wright 1936). Local enhancements rely on methods employed in settlement geography by French geographer Albert Demangeon (1872-1940) and others more than 70 years ago (Demangeon 1930; Demangeon and Weiler 1937; Nopsca 1925) and more recently by Rapoport (1969).

The primary determinants of accuracy and precision in the LandScan model are (a) size of census units, (b) age and quality of census counts, and (c) precision of cartographic and other ancillary databases. Thus, good census data are essential, but so are global databases providing roads, land cover, slope, and nighttime lights (Dobson 2001).

The global cartographic database supporting precision warfare, in general, and protection of civilian populations in particular, derives in a very real sense from an effort begun by geographers more than a century ago. At three International Geographical Congresses from 1891 through 1913, Albrecht Penck proposed a concerted international effort to map the world at 1:1,000,000 scale. Cartographic standards were established, but the work was never completed. The greatest success was the "Millionth" *Map of Hispanic America*, consisting of 107 sheets, completed by the American Geographical Society from 1920-45. After World War II, Penck's dream was superceded by the Defense Mapping Agency's 1:250,000 topographic map series, which improved on precision but failed to meet Penck's thematic goals.

For global applications, even today, most civilian users remain at Penck's 1:1,000,000 scale, based on cartographic data produced by the National

Imagery and Mapping Agency (formerly Defense Mapping Agency). It is readily available for the entire earth and it is a tremendous aid on which many global studies (military and civilian) rely. For military applications, however, digital data are available at 1:250,000 scale, and the difference is astounding. The coarser database, for instance, shows only major highways, while the new database shows secondary and tertiary roads and even many trails. Also the number of "point" locations for populated places is at least an order of magnitude greater. Populated polygons in the old database are out-of-date and vastly underrepresent city size, yet the new database is precise enough to distinguish built-up areas within cities. For a while, the new database was available to the public via the Internet. Since 9/11, however, it is restricted to official use only. Coverage is complete for most of the world, but substantial areas remain to be done in scattered regions. LandScan's quality depends so heavily on NIMA's global database for roads, waterbodies, and administrative boundaries that LandScan's developers routinely distinguish its accuracy between tiles with the finer cartographic data and those without it.

The best land-cover database available to LandScan's developers worldwide is the U.S. Geological Survey's Global Land Cover Characteristics database derived from relatively coarse (1 km-resolution) satellite imagery. It is reasonably reliable for all land-cover types except wetlands and developed lands, but there is considerable variation in accuracy from cell to cell. Initially, LandScan developers modified this land cover database extensively. They geo-registered the data to a common grid for the entire globe and devoted considerable effort to reconciling positional accuracy against other global databases, especially shoreline data at 1:250,000 scale. They replaced the "urban" class with two new classes—"developed" and "partly developed." The "partly developed" class typically includes suburban areas, small towns, and scattered industries, airports, etc. Global land cover databases are expected to improve as new satellite data become available.

Slope calculations depend on digital elevation data. Digital elevation data at 30 arc second resolution (1 km or finer) are complete for the entire world and readily available over the Internet. Far more precise data are available for official use only. That coverage is complete for most of the world, but substantial areas remain to be done in scattered regions. There is a great deal of excitement about new shuttle radar elevation data, but radar processing is complex and products have come slowly. Ultimately, the shuttle data will cover most of the world, but fine resolution data are unlikely to be released for quite some time.

Nighttime lights are the best available global indicator of where people live, work, and play, and the amount of light emitted is roughly proportional to the number of people. A military satellite records the light emitted from each square kilometer on Earth, and the results have been made available to military and civilian users alike. Global modelers owe a great debt to Chris Elvidge at the National Geophysical Data Center for producing the original frequency data and his new radiance-calibrated version that separately distinguishes low-intensity lights in the countryside, saturated lights in cities, and everything in between. Nighttime lights are an invaluable resource to LandScan and potentially to many other global applications.

3. RECOGNITION

From these explicit examples focusing on protection of civilian populations, it should be evident that precision warfare involves far more than GPS or even GIS. It is, quintessentially, a continuation of geographic and cartographic legacies that go back thousands of years. As Lanham Lister, a Ph.D. student in geography at the University of Kansas, points out, geographic information technology truly has come full circle. The oldest known map is a cadastral plot of the city of Babylon dated about 2200 B.C. Four thousand years later, that same technology has changed the world again, and it happened in Mesopotamia less than 90 km (55 mi) north of Babylon's ancient ruins.

Yet geography is the stealth discipline, always flying beneath politicians' and journalists' "radar." National leaders marvel at their newfound precision without knowing what lies behind it. It is as if in 1945, their counterparts— President Truman, Secretary of War Stimson, and General MacArthur—had praised the A-Bomb without knowing the role that physics or nuclear engineering had played in its invention.

Officials, reporters, and the public tout the Global Positioning System (GPS) alone as if nothing matters but a sterile string of lat/lon coordinates prescribing each weapon's trajectory. They ignore the GIS required to make sense of those coordinates and to represent and analyze geographic features, both physical and cultural, for vast regions and the globe. They cannot bring themselves to say "GIS," even though GPS coordinates would be useless without it. They are oblivious to geography, the intellectual home of GIS.

Reporters avoided the proper acronym "GIS," in favor of the simpler, more commonly held "GPS." They parroted with surprising alacrity a plethora of new acronyms, RPG, BDA, AAA, EPW, arising daily from the battlefield, while ignoring GIS, which has existed for decades. They do not recognize geography as the venerable science engaged in this technological pursuit for thousands of years starting with those ancient Babylonians and progressing through ancient Greeks, Romans, and Chinese; medieval Arabs; and thousands of geographers around the world today.

Does proper recognition matter? Yes, it does. Imagine how the world might have suffered if political leaders and the public really had not known chemistry was behind dynamite or physics behind the A-Bomb. Public funds would not have poured into universities and government laboratories to advance "the sciences," especially chemistry and physics. Little public support would have gone to the development of beneficial, peaceful uses. Worst of all, society would not have understood how to protect itself against the terrible power unleashed by such frightening technologies. A knowledgeable coven of engineers would have controlled the technology, and society would have been at their mercy. This is precisely where society finds itself today in regard to precision weaponry, geography, and GIS.

4. SOCIETY

Society has been a central theme since the inception of GIS and a driving force behind continued development. Foresman's (1998) "geomander" shows almost 40 distinct origins that converged to create GIS from the 1960s through the 1990s, with precursors stretching back even to the 1890s. Many of these origins directly served social functions. The U.S. Census, for instance, was a primary motivation for automation, per se, starting with machine tabulation of the 1890 census and peaking with the invention of topological GIS structures in the 1960s. Elsewhere, prominent themes included environmental protection, agriculture, natural resource management, and state planning, all of which directly relate to the interests of society.

The earliest example of distributed mapping based on the regular telecommunication of geographic information seems to have been the transmission of tri-daily weather maps by the Weather Bureau. Larson (2000) notes that the service could not have existed before the telegraph was in place to collect simultaneous readings and to distribute predictions in advance of predicted events. A special cartographic code was devised. By the 1890s, in railway stations across the nation, hundreds of draftsmens simultaneously received the code by telegraph, and each one separately transcribed that information to a map. Information technology was as important then as now for protecting communities, planning daily activities, and growing food.

Today, deliberations on GIS and society often emphasize decision support, public participation (PPGIS), privacy assurance, fairness, and equity. All of these noble objectives require equitable public access to GIS data and software. Such access was limited until at least the early 1990s due to the high cost of equipment and labor, primitive state of software, dearth of data, and general lack of expertise. Some current rhetoric speaks of dominance by governments and corporations, but even the best endowed in those days were inhibited, if not by funding, then by all the other factors. A telling example was the failed attempt by the Carter Administration to establish a GIS-based decision support system in the White House (Cowen 1983).

By 2000, the world GIS market had reached an estimated 500,000 installed units and about 2 million users. Simultaneously, distributed mapping expanded rapidly in advanced nations due to the explosion of information disseminated in graphic form or downloaded in digital form via the Internet (Crampton 2000). Clearly, the nature of the user base had changed. GIS, distributed mapping, and the Internet were seen increasingly as a democratizing force.

Since 1993, GIS and society has marched prominently, though not exclusively, under the banner of "critical GIS" (Pickles 1995). As the name implies, critical GIS arose from concerns held initially by GIS opponents. Subsequently, many of those same individuals came to see certain benefits, but also GIS proponents increasingly acknowledged limitations and risks. A not-quite-complete meeting of the minds eventually settled into productive collaboration, or at least deliberation, as to how benefits might be promoted and risks minimized to the lasting good of society.

To illustrate, consider the issues of privacy and surveillance that have troubled many GIS developers, practitioners, and opponents. Pickles (1991) was the first geographer to warn about GIS as an instrument of surveillance, sounding an alarm that reverberates to the present (Crampton 2003, Goss 1995, Monmonier 2002). On March 14, 1998, Charlie Zdravesky of radio station KUNM 89.9 FM in Albuquerque, New Mexico, hosted an hour-long discussion of GIS, mostly focusing on privacy. He interviewed twelve GIS practitioners in his own state plus eight outsiders prominent in national and international GIS affairs. Without exception, they expressed serious concerns about privacy. Remarkably, the statements made by eminent GIS proponents were virtually identical to those made by outspoken GIS critics. The two camps differed primarily in how they viewed tradeoffs. Proponents generally perceived that benefits heavily outweighed risks. Conversely, opponents discounted many of the benefits or perceived fewer of them.

Recently, two long-term GIS insiders (Dobson and Fisher 2003) went even further, suggesting that GIS technology has progressed to the point that society must contemplate a new form of slavery characterized by location control. Geoslavery is defined as a practice in which one entity, the master, coercively or surreptitiously monitors and exerts control over the physical location of another individual, the slave. Inherent in this concept is the potential for a master to routinely control time, location, speed, and direction for each and every movement of the slave or, indeed, of many slaves simultaneously. Enhanced surveillance and control may be attained through complementary monitoring of functional indicators such as body temperature, heart rate, and perspiration.

Such concerns were prompted, not by some vague Orwellian dream, but by human-tracking products currently sold without restriction for $400 or less. These products go far beyond privacy to issues of dominance, subjugation, and control. The authors suggest that such devices are, in fact, the greatest threat to personal freedom ever faced by humankind. They raise disturbing issues of human rights, especially women's rights.

Yet this same family of products offers enormous benefits to society, for example, in terms of protecting property and improving the efficiency of goods in transport. As with many other surveillance technologies, the trade-off is safety and security on one hand versus privacy and freedom on the other.

Now both camps, GIS developers and critical theorists, are collaborating to raise awareness and foster debate. They differ in language, audience, and approach, while they pursue the same goal.

5. SCIENCE

Decades from now, what will future scientists consider to be GIS' greatest contribution to science? They probably will not say much about data management or better maps, though both are important. They will recognize the heightened role of spatial thinking, but even that may not be at the top of the list. Instead, they will acknowledge that GIS revolutionized science by forcing the disciplines to talk to one another (Dobson 1999b).

Integration remains the paramount need of science today, but solutions are difficult. An observation by historian Thomas Cahill (1995, 5) provides a striking illustration of how completely the dearth of integration pervades science. Writing on the "precarious transition" from classical Rome to medieval Europe, he noted that most experts specialize in one period or the other so that "analysis of the transition itself falls outside their-and everyone's?-competence...I know of no single book now in print," he wrote, "that is devoted to the subject...nor even one in which this subject plays a central part." Thus, it is not only geographical integration or interdisciplinary studies but *all science on the cusp* that is missing.

Recognizing this universal deficiency, Joël de Rosnay (1975, xiii) said "We need, then, a new instrument. The microscope and the telescope have been valuable in gathering the scientific knowledge of the universe. Now a new tool is needed by all those who would try to understand and direct effectively their action in this world, whether they are responsible for major decisions in politics, in science, and in industry or are ordinary people as we are." He proposed "the macroscope: a new world scientific system."

The "system" de Rosnay describes sounds a lot like geography and the instrument like GIS. "For more than a century, science has been dominated by the concept of the microscope. We have sought more and more detail about everything, and integration has suffered through neglect. Now with remote sensing and GIS, we have, for the first time in history, a rudimentary macroscope." (Dobson 1993, 1494) It is still rudimentary, but progress is being made as researchers and developers extend GIS to the third dimension, attempt to better represent time, and improve linkages to process and transport models.

Even if perfected, the macroscope alone will not ensure integration. A greater challenge lies with the culture of science. What major scientific discoveries can geographers make, what new theories can they develop that do not fall within turf already claimed by one or more highly specialized disciplines? Thus, integrative geographers inevitably find themselves cast as outsiders, sometimes as provocateurs, challenging authoritative experts who may, in fact, be profoundly oblivious of vital factors outside their own narrow specialties. Geographers have long faced this problem, and it will surely vex any geographic information scientist who manages to discover new scientific theory using GIS as the macroscope.

If the macroscope is as powerful as the microscope and telescope, it can be expected to generate revolutionary new theories in rapid succession just as those earlier inventions did. Many conventional theories—developed in isolation by specialized disciplines with little thought for geographic relationships, spatial logic, or integration—have stood unchallenged for decades. The time is right for geographers and geographic information scientists to enter the fray, and they have much to offer.

In the Tom Robbins (1971) novel, *Another Roadside Attraction*, a band of hippies stumbles across an artifact that contradicts the core beliefs of a major worldwide religion. The potential impact is so great that they ultimately decide not to reveal what they found. That discovery was fictional, but it is not much different from what happened earlier when the telescope revealed the motion of planets or the microscope revealed the secrets of blood circulation or pathogenic disease. At that exalted level, science is society.

Will scientists using GIS as their macroscope discover another roadside attraction, some new piece of knowledge that shakes the very foundations of what we believe, something so profound that science cannot advance without changing society? Probably so. And when it does, the impact of GIS on science and society will eclipse anything that has been discussed to date in GIS literature.

In our past, there is ample proof that geographers can spark major scientific revolutions, sometimes long before society is ready to accept them.

Geographer Abraham Ortellius, for instance, proposed the theory of continental drift in no uncertain terms as early as 1596 (Romm 1994) based on quintessentially geographical reasoning. Geographer Antonio Snider-Pellegrini took the case further in 1858, mapping South America and Africa with no ocean between. Climatologist Alfred Wegener stunned the world in 1912 with his compelling case based on new lines of evidence (all geographical) within a paradigm that was undeniably geographical. In the early 1960s, some 365 years after Ortellius's initial proposal, the putative authorities finally conceded, but the battle is not over. Recently, a geographer discovered at least five additional continental fits of magnitude and quality equal to those that sparked the original debate. The discoveries have been welcomed in geographical journals (Dobson 1992, 1996), but squelched elsewhere. No one denies the geometric fact of their existence, but still no geologist, geophysicist, or paleontologist has addressed their implications.

Hellman (1999) lists continental drift as one of science's ten greatest feuds of all time. Evolution appears as well, of course, and that too is a case in which a geographer proposed a theory that society was not ready to accept. Geographer Alfred Russell Wallace long has been recognized for independently discovering evolution. Hellman's account and a more recent one by Winchester (2003) vividly demonstrate why some scholars now believe Wallace should be accorded equal credit as codiscover with Darwin.

Of the remaining eight items on Hellman's list, four are fundamentally geographical, two archaeological, one biological, and one mathematical. Galileo's heliocentric Earth and Kelvin's estimate of the age of the Earth surely fall within the traditional purview of physical geography. Mead's nature versus nurture readily falls within the purview of cultural geography. Wallis and Hobbes's squaring the circle would be grist for the mill of modern GIS. Thus, more than half are geographical in one way or another. If that is any indication of revolutions to come, much mischief is yet to be done by geographers, armed now with better tools than ever.

It is foolhardy to speculate, of course, on discoveries not yet made, but surely geographers will find new evidence and propose new theories as radical and soul-stirring as those of the past. A safe bet would be to look for those places and topics where least is known today. For sheer geographic ignorance, no place on Earth compares to the world ocean, and no other part of the ocean compares to Aquaterra, the portion that was exposed repeatedly during the ice ages (Dobson 1999a). This global feature-125 meters deep, as large as North America in size, all flat, all coastal, and mostly tropical, is as distinctive as, for example, the Continental Shelf, the Russian Steppe, or any individual continent. Yet it was named only four years ago and has never been explored. Over the past 120,000 years, when modern humans are known to have existed,

Aquaterra was at times a vast coastal lowland; at times a vast shallow sea. During each of the four ice ages, it would have been the most hospitable place on Earth and was, undoubtedly, the ancestral home of most, perhaps all, of the world's population. There, perhaps, lie the secrets of human origins, the causes of many ancient migrations and battles, the origins of ancient religions, the precursors of many physical and cultural inventions, and the evolutionary origins of human traits as fundamental as our well-known penchants for violence, warfare, and prejudice.

6. A GEOGRAPHIC REVOLUTION?

Since 1992, I have written about the "geographic revolution." In 1999, the National Council for Geographic Education picked "Advancing the Geographic Revolution" as the theme for its annual meeting. Later, David Lanegran of Macalester College used the term and was quoted in a newspaper article entitled "A Welcome Renaissance in Geography" (Kersten 2001). Otherwise, it is difficult to find any published use of the phrase, much less proclamations ascribing any significance to it. A recent report in The Economist, for instance, proclaimed "The Revenge of Geography," stated that "geography [i.e., distance as an economic constraint] is far from dead," and cited work by two of our students (www.ittc.ku.edu/wlan)—without mentioning geography as a discipline or field (*The Economist* 2003).

Are we in a geographic revolution or not? Surely, there can be little doubt after what geographic technology has done to warfare. To any thoughtful person, the revolution should already have been evident from the penetration of that same technology into practically every aspect of ordinary life. Yet few people seem to have realized it.

Perhaps it is always difficult for society to recognize a revolution while it is underway. How many people in the late 1400s, for instance, suspected that the printing press, invented at mid-century, would spark an intellectual revolution and lead eventually to widespread literacy?

In the 1600s, it became fashionable to hold parties at which the center of attention was a microscope, and guests would take turns viewing some tiny object. One of the most popular objects was a small transparent fish, and party-goers were fascinated with the spectacle of blood coursing though it's veins. Previously, there had been heated debate over whether blood circulated cyclically through the body or, alternatively, was generated in one organ and disposed in others. Suddenly, every socialite knew the truth, but how many recognized that the microscope would foster wholesale revolutions in medicine, biology, and other sciences?

It is disappointing and dangerous that current leaders of science and society fail to recognize what is happening in geography. It is our task as geographers and geographic information scientists to educate and inform, and also to guide society toward responsible stewardship of awesome new powers emerging from our field. At present, there is great enthusiasm for instituting new GIS technologies, but little understanding of how to use them in intelligent ways and virtually no critical thinking about their societal implications.

7. GEOGRAPHY AS DISCIPLINE AND FIELD

How has GIS changed geography as an academic discipline and as a professional field? Clearly, GIS has made explicit, repeatable, and saleable many of the methods and techniques that geographers have employed manually in cartography and intuitively in all sorts of geographic analyses for thousands of years. In many quarters, GIS is generating new peer respect for our craft and sometimes for our theory. GIS has opened new doors for collaboration with other disciplines, often placing geographers in the position of integrating science in ways that could hardly be imagined before.

8. CONCLUSIONS

Geography is the intellectual home of geographic information science and geographic information systems (GIS). GIS is the medium through which most people know and practice the methods and techniques traditionally claimed by geographers. It is profoundly changing society, yet it may well be the most underutilized, relative to its proven capability, of any technology in existence today. It is one of the most promising and, simultaneously, one of the most dangerous technologies emerging today.

GIS constitutes a new scientific tool, a macroscope that ultimately may prove as powerful as the microscope and telescope. Like them, the macroscope can be expected to spark cascading revolutions throughout science and society. GIS has barely tapped its potential in terms of impact on scientific theory. Still, GIS diffusion is constrained by the inability of current products to represent and analyze the third and fourth dimensions, to link process paradigms with spatial paradigms, and to represent and analyze fuzziness and uncertainty.

REFERENCES

Cahill, T. (1995). How the Irish Saved Civilization: The Untold Story of Irelands Heroic Role from the Fall of Rome to the Rise of Medieval Europe. New York: Nan A. Talese.

Cowen, D.J. (1983). Automated Geography and the DIDS Experiment, The Professional Geographer 35: 339-40.

Crampton, J.W. (2003). Cartographic Rationality and the Politics of Geosurveillance and Security, Cartography and Geographic Information Science 30: 131-44.

Crampton, J.W. (2000). A History of Distributed Mapping, Cartographic Perspectives 35: 48-65.

de Rosnay, J. (1975). The Macroscope: A New World Scientific System. (R. Edwards, trans.). New York: Harper and Row.

Demangeon, M.A. and Weiler, A. (1937). Les Daisons des Homes: de la Hutte au Gratte-ciel. (Éditions Bourrelier and Cie: Paris).

Demangeon, M.A. (1930). LHabitat Rural en Égypte. Imprimé par lImprimerie de lInstitut Français dArchéologie Orientale du Caire pour La Société Royale de Géographie dÉgypte: Paris.

Dobson, J.E. (2003). Estimating Populations at Risk. In Cutter, S.C., Richardson, D.B., and Wilbanks, T.J. (Eds.) Geographical Dimensions of Terrorism, 161-67. New York and London: Routledge.

Dobson, J.E. (2001). Global Data Coverage Makes Progress, GeoWorld 14: 26-27.

Dobson, J.E. (1999a). Explore Aquaterra-Lost Land Beneath the Sea, GeoWorld 12: 30.

Dobson, J.E. (1999b). Science Needs a Better Macroscope, GeoWorld 12: 26.

Dobson, J.E. (1996). A Paleogeographic Link Between Australia and Eastern North America: A New England Connection? Journal of Biogeography 23: 609-17.

Dobson, J.E. (1993). A Conceptual Framework for Integrating Remote Sensing, GIS, and Geography, Photogrammetric Engineering and Remote Sensing 59: 1491-96.

Dobson, J.E. (1992). Spatial Logic in Paleogeography and the Explanation of Continental Drift, Annals of the Association of American Geographers 82: 187-206.

Dobson, J.E., Bright, E.A., Coleman, P.R., and Bhaduri, B.L. (2003). LandScan2000: A New Global Population Geography. In V. Mesev. V. (Ed.) Remotely-Sensed Cities, 267-79. London: Taylor and Francis, Ltd.

Dobson, J.E., Bright, E.A., Coleman, P.R., Durfee, R.C., Worley, and B.A. (2000). LandScan: A Global Population Database for Estimating Populations at Risk, Photogrammetric Engineering and Remote Sensing 66: 849-57 (plus front cover).

Dobson, J.E. and Fisher, Peer F. (2003). Geoslavery, IEEE Technology and Society Magazine 22: 47-52.

The Economist. March 15, 2003. The Revenge of Geography, 366, 8315: 19-22. http://www.economist.com/science/tq/displayStory.cfm?story_id=1620794

Foresman, T. (Ed.) (1998). The History of GIS: Perspectives from the Pioneers. Upper Saddle River, NJ: Prentice-Hall.

Hellman, H. (1999). Great Feuds in Science: Ten of the Liveliest Disputes Ever. New York: John Wiley and Sons.

Goss, J. (1995). We Know Who You Are and We Know Where You Live: The Instrumental Rationality of Geodemographic Systems, Economic Geography 71: 171-98.

Monmonier, M. (2002). Spying with Maps: Surveillance Technologies and the Future of Privacy. Chicago: University of Chicago Press.

Nopcsa, F.B. (1925). Albanien Bauten, Trachten und Geräte Nordalbaniens. Berlin: Verlag von Walter de Gruyter and Co.

Kersten, K. (2001). A Welcome Renaissance in Geography, Star Tribune, May 23, 2001. http://www.amexp.org/Publications/Archives/Kersten/kersten052301.htm

Larson, E. (2000). Isaacs Storm. New York: Vintage Books.

Pickles, J. (1991). Geography, GIS, and the Surveillant Society. In Frazier, J.W., Epstein, B.J., Schoolmaster, F.A., and Moon, H. (Eds.) Papers and Proceedings of Applied Geography Conferences 14: 80-91.

Pickles, J. (Ed.) (1995). Ground Truth: The Social Implications of Geographic Information Systems. New York: Guilford Press.

Pickles, J. (1999). Arguments, Debates, and Dialogues: The GIS-Social Theory Debate and the Concern for Alternatives. In Longley, P.A., Goodchild, M.F., Maguire, D.J., and Rhind, D.W. (Eds.) Geographical Information Systems. 2nd ed. 49-60. New York: John Wiley.

Rapoport, A. (1969). House Form and Culture. Englewood Cliffs, NJ: Prentice-Hall.

Robbins, T. (1971). Another Roadside Attraction New York: Bantam Books.

Romm, J. (1994). A New Forerunner for Continental Drift, Nature 367: 407-08.

Wilford, J.N. (2000). The Mapmakers. New York: Alfred A. Knopf.

Winchester, S. (2003). Krakatoa: The Day the World Exploded: August 27, 1888. New York: HarperCollins.

Wright, J.K. (1936). A Method of Mapping Densities of Population with Cape Cod as an Example. Geographical Review 26: 103-10.

CHAPTER 25

MICHAEL CURRY

WHY TECHNOLOGY? NARRATIVES OF SCIENCE AND THE BEWITCHMENT OF AN IMAGE

Abstract   Is technology a necessary element of geography? Although many would answer in the affirmative, pointing to the map, and now geographic information systems, geography in currently identifiable forms, predated the availability of permanent portable maps. In fact, the idea of the map or its successors as forming essential elements of geography has a distinctively negative consequence. Where geography focuses on these technologies, the future of the discipline comes to be defined in terms not of ongoing and emerging research questions, but rather of a purified set of images, of data structures, and forms of representation. The technologies are displaced from the everyday practice of geography. The question for geographers is not "should we develop new technologies," but rather "how will we prevent the technologies from obscuring the ways in which important questions are framed?"

Keywords   technology, geographic information systems (GIS), cartography, narrative

For many, geography is about the mapped world. And so, it is a discipline that is in some basic way grounded in a technology; geographers create representations that are in the first instance mediated by a set of technologies, by measuring devices and compasses, and pens, and then printing presses, cameras, and finally computers.

From this perspective the question, why technology, has a simple answer: without it there would be no geography. There might be something else, perhaps a bunch of whiners, sighing deeply about the loss of a "sense of place." But that would be it, the sort of stuff about which David Hume declared, "Commit it then to the flames: for it can contain nothing but sophistry and illusion" (Hume 1975 [1777], Sec. 12, Part 2). Not the stuff of the National Science Foundation, of the twenty-first century.

Yet we might, here, do well to reconsider this idea, of the primacy for geography of cartography, and indeed of any technology. Here a place to start

Stanley D. Brunn, Susan L. Cutter, and J.W. Harrington, Jr. (Eds.), Geography and Technology,
589-601. © 2004 Kluwer Academic Publishers. Printed in the Netherlands.

is Strabo. Said by many to be the "father of geography," he denied his paternity, and declared that one needed to look back further, to Homer, to find the roots of geography. And in fact, in the *Iliad,* we find what is taken by some to be the first literary representation of a map, the shield of Achilles:

> He made the earth upon it, and the sky, and the sea's water, and the tireless sun, and the moon waxing into her fulness, and on it all the constellations that festoon the heavens, the Pleiades and the Hyades and the strength of Orion and the Bear, whom men give also the name of the Wagon, who turns about in a fixed place and looks at Orion and she alone is never plunged in the wash of the Ocean. (Homer 1951, 18 483-89)

Yet here we have reached a moment at which cartography was surely a thin reed on which to build a discipline.

Here an object lesson is a recent, and otherwise admirable—translation of the *Odyssey* (Homer 1996), by Robert Fagles. It begins with what on reflection is a jarring note: In the front matter is a map, of "Homer's world." Why is it jarring? Granted that Achilles's shield appears to be decorated with a map. But the very fact that it is notable, a first, points to the answer. For if one follows the line of interpretation offered in the 1920s by Milman Parry (1928b, 1928a, 1971) and continued by Albert Lord (1960; Lord et al. 1981) and others, the Homeric epics were, in the first instance, oral works. Their structure, one of rhyme and of the appeal to stock phrases or formulae, was well suited to their presentation, in the form of stories to be recited by speakers who were themselves illiterate. On Parry's view, which in its main lineaments has held up seventy-five years, (see the large body of literature on orality, and especially Ong (1982) and Havelock (1963, 1971)), the works that have been passed to us were works that were put into written form only after a long period in which they existed only in the minds of those who recited them and those who were within earshot, and heard. For Parry, Homer's works, originally oral in form, could not have contained maps; but this made no difference to those who heard them, for they lived in a world without maps—yet in a geographical world, one that could be described whom Strabo termed the first geographer.

It is this matter, the possibility of a geography without maps, that I shall be considering in what follows. I shall suggest that in the ways that geographers have given special status to the map, they have often abandoned the normal structure of scientific inquiry, while at the same time failing to see the importance, even centrality, of other technologies to their inquiries. As we begin to think about how to make people more geographically literate, this lesson ought to be at the forefront of our thoughts.

1. NARRATIVE, SYMBOL, AND GEOGRAPHIC KNOWLEDGE

1.1. Narrative and a World of Places

In Homer's *Odyssey*, we find a tale of a great sailor. And this genre persists, as in Sebastian Junger's nonfiction *The Perfect Storm* (1997), or perhaps Robert Stone's novel, *Outerbridge Reach* (1992). The reader will quickly note that each is in part a tale of technology, as is Edwin Hutchins's account in *Cognition in the Wild* (1995) of the (pre-GPS) landing of a modern-day naval vessel. But where is the technology in the *Odyssey*? There is precious little. This is a story of a world in which technology was not central, itself told in a form in which technology was not central. This, of course, is why the map in the Fagles translation is jarring; the *Odyssey* is a story of a world without maps.

Rather, it is in part a story of a world through which people made their ways by the use of stories, which were fundamental repositories of knowledge. I have discussed this matter elsewhere at greater length (Curry 1996), and will not here repeat what was said there, but would point to two central features of this discourse as it pertains to geography. First, just as this is a nontechnological account of a world represented as largely lacking in technology, it is an account in the form of a narrative of a place in which narratives were a primary means of storing knowledge.

In both cases, the knowledge is fundamentally emplaced. This is true in one sense if we look at the story told *in* the *Odyssey*. Against the background of a world inhabited by "the immortal gods who rule the vaulting skies" (Homer 1996, I, 80), gods who were likely to appear at any time, and in a variety of guises, who were sometimes irritable and even petulant, and who were able to unleash physical forces to meet their own ends, Odysseus and others act in ways that are not all that surprising. Arriving at Telepylus, "We entered a fine harbor there, all walled around by a great unbroken sweep of sky-scraping cliff and two steep headlands" (X, 80-83); he heads for high ground, "I scaled its rockface to a lookout on its crest" (X, 95-100)," to find a vantage point from which he might see whether the place is inhabited. The work is driven, book by book, through a logic of exile, of heading toward home, a new stopping point, hope and anguish, the homeward push thwarted, new strategies, and then another try.

So in the *Odyssey*, narrative functions in two ways; what we have seen above is that one can read the work not as a biography, but rather as a topography, a geographical account of the nature of a set of places, structured in terms of the simple question, "What did they see next?"

On the other hand, that narrative functions as a mnemonic device, one that renders the work more easily retained in and recalled from memory. As a number of writers have pointed out, users of Western languages are by and large notoriously bad at holding lists of unrelated things in memory, but when those things are embedded in a narrative or associated with symbols, they become far easier to remember. This fact was, of course, the basis for the codified "art of memory," described by Cicero (1964), and analyzed by a number of recent historians, including Frances Yates (1966) and Mary Carruthers (1992).

Briefly put, the art of memory is a system by virtue of which one is able to commit to memory a list of items, either words (such as a poem or speech) or the names of objects. The user of the system first constructs a set of places, which will be a permanent part of that individual's system; the important thing is for the user to have a clear sense of what is next to what, and of how to get from one to another. And though the places are typically rooms within a mansion, they may also be other kinds of places, or series of places. Indeed, in *The Mind of a Mnemonist,* psychologist A.R. Luria described twentieth-century Russian journalist S.V. Shereshevski's use of a system wherein the mansion is replaced by a series of places:

> Frequently he would take a mental walk along that street—Gorky Street in Moscow—beginning at Mayakovsky Square, and slowly make his way down, "distributing" his images at houses, gates, and store windows. At times, without realizing how it had happened, he would suddenly find himself back in his home town (Torzhok) where he would wind up his trip in the house he had lived in as a child. (Luria 1987 [1968], 32)

The "mental walk" that Shereshevski took down Gorky street was very much like the "mental walk" that one reciting the Odyssey would have taken, a walk the recreation of which recalls to memory the story itself.

1.2. Symbol and Order

But there is a second aspect to Homer's world. In the classical art of memory, when one takes a mental walk through a place or set of places, the goal is not simply to recollect the walk, but to use it as a means to recall a set of often disparate objects. When trying to memorize a list of such objects, one mentally places them along one's route, and by tracing the route, one can call them to mind.

And in fact—and as Parry noted—we see just such objects throughout the Odyssey; one routinely comes across objects or people, described in terms of epithets or formulae, in ways that help anchor the story, for author and listener.

For the person using the memory system, these objects and people become memorable just to the extent that they have been consciously placed there, and are therefore in one sense out of place; the memory palace has, for example, an empty table in the hall, and having placed there an object that he wants to remember, the mnemonist notices it because it is in a sense unexpected, like a hunter encountered in an otherwise empty wood, or a wisp of smoke from beyond the crest of a hill.

Indeed, this is a world that Foucault (1973) described as one of similitudes, where objects are pregnant with meaning, and where they always refer to their surroundings, as their surroundings refer to them. This, then points to a second form of geography, one where we understand the nature of a place through the interconnections among the objects, people, and events there; it is the sort of geography that is not so much topographic as chorographic, one that moves from the particular to grasp the nature of a place through the objects in it, as the objects in a place through the nature of that place.

2. ON TECHNOLOGY

So if we look to a work from an oral culture, like that from which the Odyssey emerged, we see two ways of doing geography, one that looks very much like a traveler's account, that describes places in a way that uses narrative, and that can look distinctly ethnographic; and a second that is more concerned with the configuration of a place, with what is where and with what belongs and what does not. Neither concerns a society in which technology was a central feature, and neither uses technology as a means for the construction of its accounts of those societies. Yet the products of both forms of inquiry are not that different from some forms of currently accepted geographical knowledge.

One question that this fact raises is obvious: What do we gain in analytical and representational power by the adoption of new technologies? Here, in the context of geography, one immediately thinks of the move to the map (in forms more useful than the front of a shield), and then to GIS, GPS, and remote sensing. And here, too, there is an obvious answer: The invention of the map allows for the invention of a new form of geography, one that concerns itself not with places and regions, not with ethnographic accounts and descriptions of cultural patterns, but with a measured space on the face of the earth. In fact, it makes sense to see the invention of the map as the invention that allowed the invention of the concept of space.

So on this view we start with place and end with space; we start with the particular and end with the universal; we start with facts and end with theories. We start without technologies, and gradually develop them, as means to enhance our bodies—our eyes and ears, arms and legs, noses, and our sense of touch. And through these technologies, we gradually develop the idea of an integrated understanding of the whole earth, and of the possibility not just of explanation, but also of control.

This is a view of the history of geography that has been enshrined in commonsense. But there are surely problems with it. If we grant that a central motivation of geographers themselves, the people who are written about in the standard histories, and whose names become memorable, is not merely the desire to have "more technology," then how does technology fit into that motivation? Do geographers adopt technologies as means to an end, because they believe that those technologies will help them acquire more knowledge? This too has become a sort of commonsense, the idea that there is a difference between "using tools" and "creating knowledge," and the last decades have seen the faltering of the career of many a geographer, who has been judged to have been too interested in using tools and not interested enough in creating new knowledge. And it suggests that we might apply a model of natural selection to technologies; many are tried, and some succeed. Those that succeed are the ones that have been most successful at increasing the size of the stock of knowledge. Geographers who use faster and more efficient tools become more productive, and they succeed.

2.1. Writing the Future

Comforting as this may be, it is, alas, a fairy tale. And that is because it fails to see technologies as anything more than devices for analysis and representation. This at once fails to see other ways in which technologies are embedded in the lives of geographers, just as it fails to attend to the ways in which "nongeographical" technologies are important to the practice of geography.

Here it will be useful to turn again to Homer, but now in the form of Christopher Logue's luminous *All Day Permanent Red* (2003), a rereading of parts of the *Iliad*:

> The son of Tydeus murderous Diomed aka the Child.
> ......
> Sees Hector far down front. Sees Palt
> His Porsche-fine chariot with Meep on reins
> Arriving with the comet's tail (Logue 2003, 37)

On seeing the juxtaposition of the classic tale of the Trojan War and a modern-day sports car, the skeptic might want to say that Logue is guilty of anachronism. Yet shocking as it is, it is also compelling; it is a metaphor that tells us something, that gives us a sense of what kind of person Palt was.

Or:

> To welcome Hector to his death
> God sent a rolling thunderclap across the sky
> The city and the sea
>   And momentarily—
> The breezes playing with the sunlit dust—
> On other slopes a silence fell
> Think of a raked sky-wide Venetian blind.
> And the receding traction of its slats
> Of its slats of its slats as a hand draws it up.
> Hear the Greek army getting to its feet. (Logue 2003, 11)

Here another juxtaposition, of an everyday household technology and a battle scene, provides a striking aural and visual image. Indeed, after this image, it seems hard to imagine it possible, in Homer's own idiom, to get across to today's reader just what it might have been like to be in battle, facing a rising Greek army.

And finally, one might compare the following with a contemporary newspaper account of almost any natural disaster:

> Such is the fury of the Greeks
> That as the armies joined
> No Trojan lord or less can hold his ground, and
>   Hapless as plane-crash bodies tossed ashore
>   Still belted in their seats
> Are thrust down-slope. (Logue 2003, 37)

One immediate reaction to these passages is that they are anachronistic, that the technologies referred to did not exist in Homer's time. Logue, it seems, is reading the past in terms of the present. But this might as easily be seen as a matter of reading the present in terms of the future. And it is just this that geographers do in their own use of technology.

Joseph Rouse (1990) notes that while it is a commonplace to write the history of science as a narrative, beginning with the past and moving to the present, there is another way in which narrative is essential to science, and indeed, to the very operation of science itself. For the writing of a scholarly article is not simply a matter of the reporting of a completed event, the collection of data, hypothesis testing, and so on. A successful scholarly article looks both to the past, to the research done—and to the future. And it looks to the future in a way that is intrinsically complex. For most authors desire that their work be read, but also desire that that work be appreciated, cited by the right people and seen as important. This is in part a matter of writing works that fit into current scientific concerns. But that is not really enough in science, where what is viewed as important constantly changes. So it is also a matter of writing works that will fit within future concerns.

Sometimes this is a matter of luck; a person writes an article, and the world changes in ways that result in that article's being seen as germane. But as Rouse notes, scholars who are consistently successful manage not only to stay ahead of the curve, but in fact to define the curve. And they do that by a process of narrativization. That is, they write articles that at once describe research results, and describe those results in ways that connect their own work with that of other authors. They define a research community. The author narrativizes her/his work in a way that enlists other authors, that takes their works and renders them part of an ongoing research project. In describing a set of research results, the author tells a story about a research future, one in which the author is involved as an active, even leading, participant.

2.2. Writing Geography's Future

In geography, one very important way in which people have written the future of the discipline has been in terms of technology. Granted, that in other disciplines, we also see an interest, even a strong interest, in technology; and one feature of what is termed "big science" is just an interest in technology. But what is interesting in geography is the way in which a set of technologies, cartography, and more recently GIS, have driven a very visible, even central, narrative of the discipline's future.

Consider the following (rather lengthy) account of such a future:

> Members of the general public, including school children, are obtaining detailed information about any place on Earth through an intuitive interface that looks like a large manipulable globe. They rotate the globe to put any region in the forefront, or simply speak to the system to ask it to show a particular place or region. As they zoom in, they see ever increasing detail. The default view shows what the planet looks like at the current moment from the chosen perspective, but the user can ask for clouds to be removed, for the entire planet to be illuminated, or for thematic information such as political boundaries, population densities, endangered species, or land values to be shown. One person uses the system to travel back in time to look at agricultural patterns in southern Mexico in 1450. Another turns the time back half a billion years, and then watches continents form and move into their present positions. Yet another travels into a possible future world in a global warming scenario; to produce the images, the system invokes a Global Climate Model developed several years earlier in a research center that has been made available to the public through this digital earth. Although people without technical training easily use the digital earth, scientists and policy makers also use data from digital earth as input to their models (Mark 1999, 4).

This is a compelling image, one that, as Denis Cosgrove (2001) has shown, has a long history. And it is one that we see enunciated by nongeographers as well as geographers (Gelernter 1992; Gore 1998). But has this narrative functioned in geography as do the narratives that Rouse describes in other sciences? And if not, what is its function?

As Rouse describes the matter, a narrative in science is typically of the form, "We thought that the world worked in this way, but now we realize that it is more accurate to say that it works in this other way." This new understanding may be a matter of the development of new concepts, or the application of concepts in new ways, or the development of new technologies; it very often is a matter of a reclassification of the objects that are believed to make up the world. So, to take a current case, there came a time when scientists began to say that annual variations in temperature were not "just" annual variations, but were in fact evidence of long-term patterns or cycles. And there came a further point at which they began to say that these long-term cycles were not simply natural phenomena, but were instead connected to human activities. Each version of this move, from seeing the natural world as a stage on which people act, to seeing climate patterns as emerging from the interaction of humans with biological, physical, and chemical processes, was one of developing a new narrative, one that in this case ambled between nuclear winter (Turco et al.1983; Ehrlich 1984; Curry 1985, 1986) and global warming (Miller and Edwards 2001; Collier and Webb 2002; Schneider 2002). Indeed, this is a particularly good example of the phenomenon about which Rouse was writing, because in this case, the narrative has expanded to include a large number of scientists. At the same time, it has become well entrenched in popular culture. And it has become institutionalized, as scientific research centers have been created to focus on the issue, funding agencies have supported that work, and governmental agencies, up to the level of the U.N., have joined a movement for regulatory action.

This is not the only such story to be found in recent geographical practice, though it is surely the most prominent; one sees similar processes at work across the range of subdisciplines. But if we look at the lineaments of this story, it seems rather different from the one surrounding GIS of the mapping of the world. For if there are similarities, viz., a bandwagon effect, the development of a subdisciplinary specialty, the rapid expansion of job opportunities, and the support from government and funding agencies, the narrative structure embodied in the scientific work itself is not the same.

For if a standard narrative in science is of the form, "we thought that the world worked in this way, but now we realize that it is more accurate to say that it works in this other way," this is not at all what we see in the discourse surrounding geographic information systems.

A standard explanation for this difference has been that GIS are simply tools or pieces of technology. Yet there have been recent attempts among those who study and use the systems to re-render GIS as a science, a "geographic information science." But there are in fact more important differences, ones that speak to the place of technology in geography, and in science more generally.

Here it is useful to turn to another recent foray into big science, viz., the human genome project. (Indeed, there would be much to learn from an extended comparison of the genome project and GIS.) There were several striking things about that project. One was that it relied heavily on the idea that the genome could be "mapped." There, to map the genome meant to make its elements visible. But it did not mean to make all of its elements visible, nor to establish the use of those elements. Some, and as it happens, the ones the functions of which were not understood, were deemed "junk DNA," and set aside. And even in the case of those portions that were fully mapped, the process of mapping was not the end of the story. As many scientists put the matter, once it had been announced that the mapping was complete, "Now to find out what we have." What they had done, perhaps paradoxically, was to create a detailed map of terra incognitae.

This discussion is in some respects what has happened in the substantially more difficult task set forth, within the GIS narrative, of mapping the world to the extent that one might within a single technological system, wherein one might obtain

> 'detailed information about any place on Earth through an intuitive interface', 'zoom in [and] ... see ever increasing detail', 'travel back in time', 'ask for clouds to be removed ... or for thematic information such as political boundaries, population densities, endangered species, or land values to be shown', travel 'into a possible future world in a global warming scenario', while 'scientists and policy makers also use data from digital earth as input to their models' (Mark 1999, 4).

3. BEWITCHED BY AN IMAGE

One useful way to describe what is at work in both the genome project and the GIS narrative is by noting that while Rouse sees an essential element of science as being the way in which scientists narrativize their work, in these two cases, the narrative has been replaced by an image, one by which, as Wittgenstein put the matter, the user of the image has been "bewitched" (Wittgenstein 2001). The image of an utterly visible world seems so plausible— just take a globe and attach everything we know to it—that the users fail to stop and ask what the image really might mean, if it means anything at all.

In geography at least, one side effect of this bewitchment has been to fail to attend to other technologies in which current scientific practice is deeply embedded. Those who use them in their everyday work are all aware of the technologies used for chemical and biological analysis in physical geography and biogeography, technologies that are not specific to geography itself, but also used in cognate disciplines such as geology, earth sciences, and ecology. But more generally important and less visible have been the technologies of office production, the typewriter, the photocopier, the fax machine (Beniger 1986). And there are of course military technologies, such as the ballistic and cruise missile (Mackenzie 1990) and the earth-orbiting satellite (Mack 1990; Rip and Hasik 2002). Looming in the background are the airplane, the radio, and the automobile. And finally, looming perhaps largest of all, the computer (Edwards 1997). In each case, the importance of that technology to the practice of geography has been at least in part eclipsed by the image of the map.

There is, though, a second feature of the discourse of GIS that renders it different from Rouse's narrative of science. And that is the following. Scientists typically claim that science is detached from and involved in studying the world. Karl Marx's famous dictum that "The philosophers have only interpreted the world, in various ways; the point is to change it" (Marx 1978) is anathema to the rhetoric of contemporary science.

Yet central to the narrative of GIS has been the introduction of a wide range of means for collecting, storing, analyzing, representing, and communicating information, at scales from the most local to the global. The fulfillment of the ideal proposed by David Mark, and more so by authors like Al Gore and David Gelernter, will involve, or in many cases has already involved, the establishment of an integrated, large-scale information network. In part, this network moves beyond the process of collection and operates automatically to classify the data collected. And these systems in effect render true the classification systems that they themselves have created.

Here, though, the lie is given to this narrative. If the systems are represented as means for the promotion of science and for the development among their users of a sense of appreciation of the order and complexity of the world, they in practice are used to very different ends. Here, the complaints by the developers of the systems that they did not mean for them to be used as tools of surveillance and repression sound eerily like the complaints by those involved in the human genome project that they did not mean to promote eugenics. In fact, having been bewitched by the image of the technology, of the fully mapped earth, or genome, the promoters of those projects have left far behind the narratives of science and their democratic and moral underpinnings.

This is not to say that elsewhere in their work, these individuals do not in fact operate in terms of those narratives, working through their beliefs that "we thought that the world worked in this way, but now we realize that it is more accurate to say that it works in this other way." Nor, of course, is it to say that all technologies lead their users down the path of GIS and of the genome project. But whether they do or not is not a matter of opinion, nor the appropriate moment for unleashing slogans; rather, it is an opportunity for empirical investigation.

REFERENCES

Beniger, J.R. (1986). The Control Revolution: Technological and Economic Origins of the Information Society. Cambridge: Harvard University Press.

Carruthers, M. (1992). Book of Memory: A Study of Memory in Medieval Culture. 2nd ed. Cambridge: Cambridge University Press.

Cicero, M. T. (1964). (Cicero). ad C. Herrennium: de ratione dicendi (Rhetorica ad Herennium): with an English translation by Harry Caplan. Cambridge, MA: Harvard University Press.

Collier, M. and Webb, R.H. (2002). Floods, Droughts, and Climate Change. Tucson: University of Arizona Press.

Cosgrove, D.E. (2001). Apollo's Eye: A Cartographic Genealogy of the Earth in the Western Imagination. Baltimore: Johns Hopkins University Press.

Curry, M.R. (1985). In the Wake of Nuclear War—Possible Worlds in an Age of Scientific Expertise. Environment and Planning D: Society and Space 3: 309-21.

Curry, M.R. (1986). Beyond Nuclear Winter: On the Limitations of Science in Political Debate. Antipode 18: 244-67.

Curry, M.R. (1996). The Work in the World: Geographical Practice and the Written Word. Minneapolis: University of Minnesota Press.

Edwards, P. (1997). The Closed World: Computers and the Politics of Discourse in Cold War America. Cambridge: MIT Press.

Ehrlich, P.R. (1984). The Cold and the Dark: The World After Nuclear War: The Conference on the Long-Term Worldwide Biological Consequences of Nuclear War.1st ed. New York: Norton.

Foucault, M. (1973). The Order of Things: An Archaeology of the Human Sciences. New York: Vintage Books.

Gelernter, D. (1992). Mirror Worlds: Or the Day Software Puts the Universe in a Shoebox: How It Will Happen and What It Will Mean. New York: Oxford University Press.

Gore, A. (1998). The Digital Earth: Understanding Our Planet in the 21st Century. Los Angeles: California Science Center.

Havelock, E.A. (1963). Preface to Plato. Cambridge: Harvard University Press.

Havelock, E.A. (1971). Prologue to Greek literacy. Cincinnati: University of Cincinnati.

Homer. (1951). The Iliad (R.A. Lattimore, trans.). Chicago: University of Chicago Press.

Homer. (1996). The Odyssey (R. Fagles, trans.). New York: Penguin.

Hume, D. (1975 [1777]). Enquiries Concerning Human Understanding and Concerning the Principles of Morals. 3rd ed. Oxford: Oxford University Press.

Hutchins, E. (1995). Cognition in the Wild. Cambridge, MA: MIT Press.

Junger, S. (1997). The Perfect Storm: A True Story of Men Against the Sea. 1st ed. New York: Norton.

Logue, C. (2003). All Day Permanent Red: The First Battle Scenes of Homers Iliad Rewritten. New York: Farrar, Straus, and Giroux.

Lord, A.B. (1960). The Singer of Tales. Cambridge: Harvard University Press.

Lord, A.B. and Foley, J.M. (1981). Oral Traditional Literature: A Festschrift for Albert Bates Lord. Columbus, OH: Slavica Publishers.

Luria, A.R. (1987 [1968]). The Mind of a Mnemonist; A Little Book About a Vast Memory (L. Solotaroff, trans.). Cambridge: Harvard University Press.

Mack, P.E. (1990). Viewing the Earth: The Social Construction of the Landsat Satellite System. Cambridge, MA: MIT Press.

Mackenzie, D. (1990). Inventing Accuracy: An Historical Sociology of Nuclear Missile Guidance. Cambridge, MA: MIT Press.

Mark, D. (1999). Geographic Information Science: Critical Issues in an Emerging Cross-Disciplinary Research Domain: Report from Workshop on Geographic Information Science and Geospatial Activities at the National Science Foundation. Buffalo, NY: National Center for Geographical Information and Analysis.

Marx, K. (1978). Theses on Feuerbach. In R.C. Tucker (Ed.) The Marx-Engels Reader. 2nd ed. 143-45. New York: W.W. Norton.

Miller, C.A. and Edwards, P.N. (2001). Changing the Atmosphere: Expert Knowledge and Environmental Governance. Cambridge, MA: MIT Press.

Ong, W.J. (1982). Orality and Literacy: The Technologizing of the Word. London: Routledge.

Parry, M. (1928a). Lâepitháete Traditionnelle dans Homáere; Essai sur un Probláeme de Style Homâerique. Paris,: Sociâetâe dEditions Les Belles lettres.

Parry, M. (1928b). Les Formules et la Mâetriquâe dHomáere. Paris: Sociâete dâeditîons Les Belleslettres.

Parry, M. (1971). The Making of Homeric Verse: The Collected Papers of Milman Parry (A. Parry, trans.). Oxford: Clarendon Press.

Rip, M.R.and Hasik, J.M. (2002). The Precision Revolution: GPS and the Future of Aerial Warfare. Annapolis: Naval Institute Press.

Rouse, J. (1990). The Narrative Reconstruction of Science. Inquiry 33: 179-90.

Schneider, S.H., Armin Rosencranz, and John O. Niles (Ed.) (2002). Climate Change Policy: A Survey. Washington, DC: Island Press.

Stone, R. (1992). Outerbridge Reach. New York: Ticknor and Fields.

Turco, R.P., Toon, O.B., Ackerman, T.P., Pollack, J.B., and Sagan, C. (1983). Nuclear Winter: Global Consequences of Multiple Nuclear Explosions. Science, 222: 4630.

Wittgenstein, L. (2001). Philosophical Investigations, The German Text with a Revised English Translation (G.E.M. Anscombe, trans. 3rd ed.). Oxford: Blackwell.

Yates, F.A. (1966). The Art of Memory. Chicago: University of Chicago Press.

X

FOR MORE INFORMATION: www.wkap.n/prod/s/GEJL

GEOJOURNAL LIBRARY
Book Series

International Handbook on Geographical Education

Edited by
Rod Gerber

The *International Handbook on Geographical Education* is the first truly international publication in the field of geographical education for several decades ... It is a publication for the thinking geographer and educator who appreciate where international education is traveling to and how its challenges can be met.

BOOK SERIES:
GEOJOURNAL LIBRARY 73
Hardbound, ISBN 1-4020-1019-2
November 2002, 360 pp.
€ 135.00 / $ 149.00 / £ 93.00

FREE online preview available

Modelling Geographical Systems
Statistical and Computational Applications

Edited by
Barry Boots, Atsuyuki Okabe and Richard Thomas

This book presents a representative selection of innovative ideas currently shaping the development and testing of geographical systems models by means of statistical and computational approaches ... this volume would provide a useful supplementary text for courses on quantitative geography and geographical systems modelling in both human and physical geography, and GIS and geocomputation.

BOOK SERIES:
GEOJOURNAL LIBRARY 70
Hardbound, ISBN 1-4020-0821-X
August 2003, 368 pp.
€ 130.00 /$ 130.00 / £ 82.00

Managing Intermediate Size Cities
Sustainable Development in a Growth Region of Thailand

Edited by
Michael Romanos and Christopher Auffrey

This book applies a sustainable development framework to the planning and managing of an intermediate size city in a developing region of a developing nation, and assesses the potential of such a framework to effectively guide the city's development.

BOOK SERIES:
GEOJOURNAL LIBRARY 69
Hardbound, ISBN 1-4020-0818-X
October 2002, 356 pp.
€ 130.00 / $ 143.00 / £ 90.00

For more information visit: www.wkap.nl/prod/s/GEJL

Contact Information:
Customers in Europe, Middle East, Africa, Asia and Australasia:
Kluwer Academic Publishers, Order Dept, P.O. Box 322, 3300 AH Dordrecht, The Netherlands
F +31-78-6576476 T +31-78-6576050 E orderdept@wkap.nl W www.wkap.nl
Customers in the Americas:
Kluwer Academic Publishers, Order Dept, P.O. Box 358, Accord Station, Hingham MA 02018-0358, USA
F +1-781-681-9045 T +1-781-871-6600 or (toll free within US) +-1-866-269-wkap E kluwer@wkap.com W www.wkap.com

kluwer
the language of science